D1187306

TIME SERIES ANALYSIS:

Forecasting and Control

by George E. P. Box and
Gwilym M. Jenkins

Unique problem sets based on real data and applied to a broad range of fields have been incorporated into the revised edition of this classic book in time series. The result is a practical text which shows how to analyze non-stationary as well as stationary series and offers active experience in applying the theory of time series to make forecasts.

The material covered includes the building of models for discrete time series and dynamic systems and their use in forecasting and control. All the techniques are illustrated with examples using economic, scientific and industrial data. In Part I, models for stationary and non-stationary time series are introduced and their use in forecasting is discussed and exemplified. Part II is devoted to model building, and presents procedures for model identification, estimation, and checking which are then applied to the forecasting of seasonal time series. Part III is concerned with the building of transfer function models relating the input and output of a dynamic system corrupted by noise. In Part IV it is shown how transfer function and time series models may be used to design optimal feedback and feedforward control schemes. Part V contains an outline of computer programs useful in making the needed calculations and also includes charts and tables of value in identifying the models.

The book is largely self-contained and requires an elementary knowledge of statistics, calculus and matrix algebra. It is suitable for one- or two-semester courses for graduate students in statistics, engineering, business, management science, and operations research, and has been used successfully in training programs for business and industry.

Holden-Day Series in Time Series Analysis

TIME SERIES
ANALYSIS
forecasting
and
control

HOLDEN-DAY SERIES IN TIME SERIES ANALYSIS

TIME SERIES ANALYSIS
forecasting and control

Revised Edition

GEORGE E. P. BOX
University of Wisconsin, U.S.A.

and

GWILYM M. JENKINS
University of Lancaster, U.K.

HOLDEN-DAY

San Francisco, Düsseldorf, Johannesburg, London,
Panama, Singapore, Sydney, Toronto

2 3 4 5 6 7 8 9 0   HA   0 9 8 7 6
ISBN 0-8162-1104-3

To Joan and Meg

THE AUTHORS

George E. P. Box is Professor of Statistics at the University of Wisconsin. He received a Ph.D. in Statistics and the D.Sc. degree from the University of London. Dr. Box was Head of the Statistical Research Section of Imperial Chemical Industries Dyestuffs Division till 1956 and Director of Statistical Techniques Research Group at Princeton University from 1956-59.

He was visiting Research Professor at North Carolina State College in 1953, Ford Foundation Visiting Professor at the Harvard Business School in 1965-66 and Visiting Research Professor at the University of Essex in 1970-71. He is the recipient of the Guy medal in Silver of the Royal Statistical Society, Professional Progress Award of the American Institute of Chemical Engineers and the Shewhart Medal of the American Society for Quality Control.

Professor Box is a coauthor of the following titles: Design and Analysis of Industrial Experiments; Statistical Methods in Research and Production; Advanced Seminar in Spectral Analysis; The Future of Statistics; and Evolutionary Operation.

Gwilym M. Jenkins is Professor and Head of the Department of Systems Engineering at the University of Lancaster. He received a B.Sc. degree in mathematics and a Ph.D. in statistics from the University College, London in 1953 and 1955 respectively.

From 1955-57 Dr. Jenkins was Junior Research Fellow at the Royal Aircraft Establishment, Farnborough, and from 1957-64, Lecturer and then Reader in Statistics at the Imperial College, University of London. He was Visiting Professor at Princeton and Stanford University during 1959-1960, and at the University of Wisconsin during 1964-65. Professor Jenkins has consulted for a large number of companies in the U.K. and the U.S.A. and is retained by the I.C.I., Ltd. He is currently Chairman and Managing Director of ISCOL Ltd., a consultancy company set up by the University of Lancaster.

He is coauthor with Donald G. Watts of another Holden-Day book entitled Spectral Analysis and Its Applications.

PREFACE TO THE REVISED EDITION

The subject is advancing rapidly and in this revised edition the opportunity has been taken to update some material, to correct some earlier mistakes and to clarify and amplify certain sections. A completely new section has been added at the end of the book, containing exercises and problems for the separate chapters. We hope that this new section will add to the value of the book as a course text. We are indebted to Bovas Abraham, Gina Chen, Johannes Ledolter and Greta Ljung for careful checking and proof-reading.

G. E. P. Box, Madison, U.S.A.
G. M. Jenkins, Lancaster, U.K.
January, 1976

Preface

Much of statistical methodology is concerned with models in which the observations are assumed to vary independently. In many applications dependence between the observations is regarded as a nuisance, and in planned experiments, *randomization* of the experimental design is introduced to validate analysis conducted as if the observations were independent. However, a great deal of data in business, economics, engineering and the natural sciences occur in the form of *time series* where observations are *dependent* and where the nature of this dependence is of interest in itself. The body of techniques available for the analysis of such series of dependent observations is called *time series analysis*.

Spectral analysis, in the frequency-domain, comprises one class of techniques for time series analysis, but we shall say very little here about that important subject. This book is concerned with the building of stochastic (statistical) models for discrete time series in the time-domain and the use of such models in important areas of application. Our objective will be to derive models possessing maximum simplicity and the minimum number of parameters consonant with representational adequacy. The obtaining of such models is important because:

(1) They may tell us something about the nature of the system generating the time series;

(2) They can be used for obtaining *optimal forecasts* of future values of the series;

(3) When two or more related time series are under study, the models can be extended to represent dynamic relationships between the series and hence *to estimate transfer functions*;

(4) They can be used to derive *optimal control policies* showing how a variable under one's control should be manipulated so as to minimize disturbances in some dependent variable.

The ability to forecast optimally, to understand dynamic relationships between variables and to control optimally is of great practical importance. For example, optimal sales forecasts are needed for business planning, transfer function models are needed for improving the design and control of process plant and optimal control policies are needed to regulate important process variables, both manually and by the use of on-line computers. Over the last ten years the authors have worked with real data arising in economics and industry and, by trial and error, and by a long sequence of interactions between theory and practice, have attempted to select, adapt, and develop practical techniques to fulfill such needs. This book is the fruit of these labors.

The approach adopted is, first, to discuss a class of models which are sufficiently flexible to describe practical situations. In particular, time series are often best represented by *nonstationary* models in which trends and other psuedo-systematic characteristics which can change with time are treated as statistical rather than as deterministic phenomena. Furthermore, economic and business time series often possess marked *seasonal* or periodic components themselves capable of change and needing (possibly non-stationary) seasonal statistical models for their description.

The process of model *building*, which is next discussed, is concerned with relating such a class of statistical models to the data at hand and involves much more than model fitting. Thus, *identification* techniques, designed to suggest what particular kind of model might be worth considering, are developed first and make use of the autocorrelation and partial autocorrelation functions. The *fitting* of the identified model to a time series using the likelihood function can then supply maximum likelihood estimates of the parameters or, if one prefers, Bayesian posterior distributions. The initially fitted model will not, necessarily, provide adequate representation. Hence *diagnostic checks* are developed to detect model inadequacy, to suggest appropriate modifications and thus, where necessary, to initiate a further iterative cycle of identification, fitting and diagnostic checking.

When forecasts are the objective, the fitted statistical model is used directly to generate optimal forecasts by simple recursive calculation. In particular, this model completely determines whether the forecast projections should follow a straight line, an exponential curve, and so on. In addition, the fitted model allows one to see exactly how the forecasts utilize past data, to determine the variance of the forecast errors, and to calculate limits within which a future value of the series will lie with a given probability.

When the models are extended to represent dynamic relationships, a corresponding iterative cycle of identification, fitting and diagnostic checking is developed to arrive at the appropriate transfer function-stochastic model. In the final section of the book, the stochastic and transfer function models developed earlier are employed in the construction of feedforward and feedback control schemes.

The applications given in this book are by no means exhaustive and it is hoped that the examples presented will enable the reader to adapt the techniques to his own problem. In particular the difference equations used to represent transfer functions and stochastic phenomena may be employed as building blocks which when appropriately fitted together can simulate a wide variety of the systems occurring in engineering, business and economics. Furthermore the principles of model building which are discussed and illustrated have very general application.

AN OUTLINE OF THE BOOK

This book is set out in the following parts (from time to time, a vertical line has been inserted in the left margin to indicate material which may be omitted in the first reading):

Introduction and Summary (Chapter 1)

This chapter is an informal and highly condensed outline of topics discussed, defined and more fully explained in the main body of the text. It is intended as a broad mapping of areas to be subsequently explored, and the student may wish to refer back to it as later chapters are read.

Part I   Stochastic models and their forecasting (Chapters 2, 3, 4 and 5)

After some basic tools of time series analysis have been discussed in Chapter 2, an important class of linear stochastic models is introduced in Chapters 3 and 4 and their properties discussed. The immediate introduction of forecasting in Chapter 5 takes advantage of the fact that the form of the optimal forecasts follows at once from the structure of the stochastic models discussed in Chapter 4.

Part II   Stochastic model building (Chapters 6, 7, 8 and 9)

Part II of the book describes an iterative model-building methodology whereby the stochastic models introduced in Part I, are related to actual time series data. Chapters 6, 7 and 8 describe, in turn, the processes of model identification, model estimation, and model diagnostic checking. Chapter 9 illustrates the whole model building process by showing how all these ideas may be brought together to build seasonal models and how these models may be used to forecast seasonal time series.

Part III   Transfer function model building (Chapters 10 and 11)

In Chapter 10 transfer function models are introduced for relating a system output to one or more system inputs. Chapter 11 discusses methods for transfer function-noise model identification, estimation and diagnostic checking. The chapter ends with a description of how such models may be used in forecasting.

Part IV   Design of discrete control schemes (Chapters 12 and 13)

In these two chapters we show how the stochastic models and transfer function models previously introduced may be brought together in the design of simple feedforward and feedback control schemes.

A first draft of the book was produced in 1965 and subsequently was issued in 1966 and 1967 as Technical Reports Nos. 72, 77, 79, 94, 95, 99, 103, 104, 116, 121 and 122 of the Department of Statistics, University of Wisconsin, and Nos. 1, 2, 3, 4, 6, 7, 8, 9, 10, 11, 13 of the Department of Systems Engineering, University of Lancaster. The work has involved a great deal of research, which has been partially supported by the Air Force Office of Scientific Research, United States Air Force, under AFOSR Grants AF-AFOSR-1158-66 and AF-49 (638) 1608 and also by the British Science Research Council. We are grateful to Professor E. S. Pearson and the Biometrika Trustees for permission to reprint condensed and adapted forms of Tables 1, 8 and 12 of *Biometrika Tables for Statisticians*, Vol. 1, edited by E. S. Pearson and H. O. Hartley, and to Dr. Casimer Stralkowski for permission to reproduce and adapt three figures from his doctoral thesis, University of Wisconsin, 1968. The authors are indebted to George Tiao, David Mayne, Emanuel Parzen, David Pierce, Granville Wilson and Donald Watts for suggestions for improving the manuscript, to John Hampton, Granville Wilson, Elaine Hodkinson and Patricia Blant for writing the computer programs described at the end of the book and also to them and to Dean Wichern, David Bacon and Paul Newbold for assistance with the calculations. Finally, we are glad to record our thanks to Hiro Kanemasu, Paul Newbold, Larry Haugh, John MacGregor and Granville Wilson for careful reading and checking of the manuscript, to Carole Leigh and Mary Esser for their care and patience in typing the manuscript and to Meg Jenkins for the initial draft of the diagrams.

G. E. P. Box, Madison, U.S.A.
G. M. Jenkins, Lancaster, U.K.
June 1969

Contents

CHAPTER 3   LINEAR STATIONARY MODELS

PART II STOCHASTIC MODEL BUILDING

CHAPTER 6 MODEL IDENTIFICATION

CHAPTER 7 MODEL ESTIMATION

CHAPTER 8   MODEL DIAGNOSTIC CHECKING

CHAPTER 9   SEASONAL MODELS

PART IV   DESIGN OF DISCRETE CONTROL SCHEMES

CHAPTER 12   DESIGN OF FEEDFORWARD AND FEEDBACK
             CONTROL SCHEMES

TIME SERIES
ANALYSIS
forecasting
and
control

1

Introduction and Summary

In this book we shall present methods for building, identifying, fitting and checking models for time series and dynamic systems. The methods discussed will be appropriate for discrete (sampled-data) systems, where observation of the system and an opportunity to take control action occur at equally spaced intervals of time.

We shall illustrate the use of these time series and dynamic models in three important areas of application.

(1) The *forecasting* of future values of a time series from current and past values.

(2) The determination of the *transfer function* of a system—the determination of a dynamic input-output model that can show the effect on the output of a system subject to inertia, of any given series of inputs.

(3) The design of simple *feed forward and feedback control schemes* by means of which potential deviations of the system output from a desired target may be compensated, so far as possible.

1.1 THREE IMPORTANT PRACTICAL PROBLEMS

1.1.1 Forecasting time series

The use at time t of available observations from a time series to forecast its value at some future time $t + l$ can provide a basis for (a) economic and business planning, (b) production planning, (c) inventory and production control, (d) control and optimization of industrial processes. As described by Holt et al. [1], Brown [2], and the Imperial Chemical Industries monograph on short term forecasting [3], forecasts are usually needed over a period known as the *lead time*, which varies with each problem. For example, the lead time in the inventory control problem is defined by Harrison [4] as a period that commences when an order to replenish stock is placed with the factory and lasts until the order is delivered into stock.

We suppose that observations are available at *discrete*, equispaced intervals of time. For example, in a sales forecasting problem, the sales z_t in the current month t and the sales $z_{t-1}, z_{t-2}, z_{t-3}, \ldots$ in previous months might be

1

used to forecast sales for lead times $l = 1, 2, 3, \ldots 12$ months ahead. Denote by $\hat{z}_t(l)$ the forecast made at *origin* t of the sales z_{t+l} at some future time $t + l$, that is at *lead time* l. The function $\hat{z}_t(l)$, $l = 1, 2, \ldots$ that provides the forecasts at origin t for all future lead times will be called the *forecast function* at origin t. Our objective is to obtain a forecast function which is such that the mean square of the deviations $z_{t+l} - \hat{z}_t(l)$ between the actual and forecasted values is as small as possible *for each lead time l.*

In addition to calculating the best forecasts, it is also necessary to specify their accuracy, so that, for example, the risks associated with decisions based upon the forecasts may be calculated. The accuracy of the forecasts may be expressed by calculating *probability limits* on either side of each forecast. These limits may be calculated for any convenient set of probabilities, for example 50% and 95%. They are such that the realized value of the time series, when it eventually occurs, will be included within these limits with the stated probability. To illustrate, Figure 1.1 shows the last 20 values of a time series culminating at time t. Also shown are forecasts made from origin t for lead times $l = 1, 2, \ldots, 13$, together with the 50% probability limits.

Methods for obtaining forecasts and estimating probability limits are discussed in detail in Chapter 5.

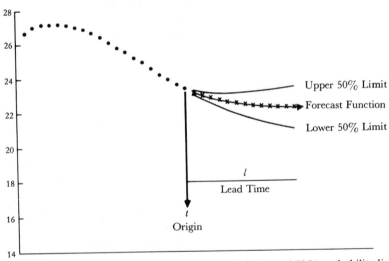

FIG. 1.1   Values of a time series with forecast function and 50% probability limits

1.1.2 *Estimation of transfer functions*

A topic of considerable industrial interest is the study of process dynamics [5], [6]. Such a study is made

   (a) to achieve better control of existing plants, and

   (b) to improve the design of new plants.

In particular, several methods have been proposed for estimating the transfer function of plant units from process records consisting of an input time series X_t and an output time series Y_t. Sections of such records are shown in Figure 1.2, where the input X_t is the rate of air supply and the output Y_t is the concentration of carbon dioxide produced in a furnace. The observations were made at nine second intervals. A hypothetical impulse response function, which determines the transfer function for the system, is also shown in the figure as a bar chart.

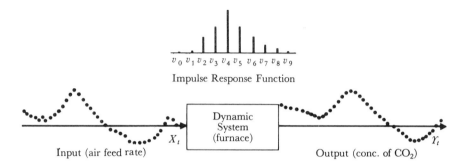

FIG. 1.2   Input and output time series in relation to a dynamic system

Classical methods for estimating transfer function models based on deterministic perturbations of the input, such as step, pulse, and sinusoidal changes, have not always been successful. This is because, for perturbations of a magnitude that are relevant and tolerable, the response of the system may be masked by uncontrollable disturbances collectively referred to as noise. Statistical methods for estimating transfer function models that make allowance for noise in the system are described in Chapter 11. The estimation of dynamic response is of considerable interest in economics, engineering, biology and many other fields.

The problem of choice of the form of statistical input to allow good estimation of the transfer function with minimal perturbation of the system is briefly discussed in Chapter 11. For many problems a good compromise is to employ as input perturbation a source of uncorrelated input changes, such as the Pseudo Random Binary Sequences (PRBS) used by Hammond [7] and others.

Another important application of transfer function models is in forecasting. If, for example, the dynamic relationship between two time series Y_t and X_t can be determined, then past values of *both* series may be used in forecasting Y_t. In some situations this approach can lead to a considerable reduction in the errors of the forecasts.

1.1.3 Design of discrete control systems

In the past, the word "control," to the statistician, has usually meant the *quality control techniques* developed originally by Shewhart [8] in the United States and by Dudding and Jennet [9] in Great Britain. Recently, the sequential aspects of quality control have been emphasized, leading to the introduction of *cumulative sum charts* by Page [10], [11] and Barnard [12] and the *geometric moving average* charts of Roberts [13].

The word "control" has a different meaning, however, to the control engineer. He thinks in terms of feed forward and feedback control loops, of the dynamics and stability of the system, and usually of particular types of hardware to carry out the control action. These control devices are *automatic* in the sense that information is fed to them automatically from instruments on the process and from them automatically to adjust the inputs to the process.

In this book we describe a statistical approach to forecasting time series and to the design of feed forward and feedback control schemes developed in previous papers [14], [15], [16], [17], [18], [19] and [20]. The control techniques discussed are closer to those of the control engineer than the standard quality control procedures developed by statisticians. This does not mean we believe that the traditional quality control chart is unimportant but rather that it performs a different function from that with which we are here concerned. An important function of standard quality control charts is to supply a continuous screening mechanism for detecting assignable causes of variation. Appropriate display of plant data ensures that changes that occur are quickly brought to the attention of those responsible for running the process. Knowing the answer to the question "*when* did a change of this particular kind occur?" we can then ask "*why* did it occur?" Hence, a continuous incentive for process improvement, often leading to new thinking about the process, can be achieved.

In contrast, the control schemes we discuss in this book are appropriate for the periodic, optimal adjustment of a manipulated variable, whose effect on some output quality characteristic is already known, so as to minimize the variation of that quality characteristic about some target value.

The reason control is necessary is that there are inherent *disturbances* or *noise* in the process. When we can measure these disturbances, it may be possible to make appropriate compensatory changes in some other variable. This is referred to as *feedforward control*. Alternatively, or in addition, we may be able to use the deviation from target or "error signal" of the output characteristic itself to calculate appropriate compensatory changes. This is called *feedback control*. Unlike feedforward control, this mode of correction can be employed even when the source of the disturbances is not accurately known or the magnitude of the disturbances is not measured. More generally, it may be desirable to use a mixture of feedforward and feedback control; feedforward control can be used to compensate for those disturbances that can be measured and feedback control for those disturbances that cannot be measured.

The approach to control adopted here is to typify the disturbance by a suitable time series or *stochastic model* and the inertial characteristics of the system by a suitable *transfer function model*. It is then possible to calculate the optimal *control equation*. This is an equation that allows the action which should be taken at any given time to be calculated given the present and previous states of the system. "Optimal" action is interpreted as that which produces the smallest mean square error at the output.

Execution of the control action called for by the control equation can be achieved in various ways corresponding to various levels of technological sophistication. The question of what means should be employed to meet a given situation is not an appropriate subject for this text. However, we may mention that at one extreme of sophistication we have so-called computer control where the action called for by the control equation is calculated by computer and is automatically effected by suitable transducers that appropriately open and close valves to adjust process variables, such as temperatures, pressures and flow rates. At an intermediate level we find automatic controllers that employ various pneumatic and electrical devices to carry out the control action. At the other extreme the action called for by the control equation may be embodied in a suitable nomogram or chart. Periodically the plant operator may make one or more readings, measurements, or determinations, consult the nomogram to calculate the action called for and then manually make the appropriate modifications.

Even in modern industry there are many situations where manual control is employed and could be improved. Chapter 12 discusses briefly the preparation of charts and nomograms that can indicate optimal control action in simple cases. For example, the upper chart of Figure 1.3 shows hourly measurements of the viscosity of a polymer made over a period of 42 hours. The viscosity is to be controlled about a target value of 90 units. As each viscosity measurement comes to hand, the process operator uses the nomogram shown in the middle of the figure to compute the optimal adjustment to be made in the manipulated variable (gas rate). The lower chart of Figure 1.3 shows the adjustments made in accordance with the nomogram.

In practice, the precise nature of the stochastic and transfer function models which are appropriate in any given situation and which are needed to design the optimal control scheme will not usually be known. Furthermore, the appropriate numerical values for the parameters contained in these models will be unknown. Therefore, as is almost always true in model building, we have to proceed *iteratively*. Whatever records and background information are available may be used to design a *pilot control scheme*, which is then run on the process for a period. The data of Figure 1.3 were generated during such a pilot scheme. It is shown in Chapter 12 how from such pilot scheme data consisting of the deviation from the target of the quality characteristic and the corresponding adjustment, a statistical analysis can then be conducted to modify models and parameter estimates, leading to an improved control scheme.

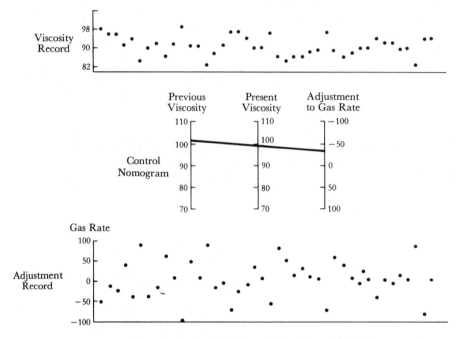

FIG. 1.3   Control of viscosity. Record of observed viscosity and of adjustments in gas rate made using nomogram

Although our applications have often been to manual control schemes, the methods we discuss are equally suited for use with an on-line process control computer as described, for example, by Oughton [21]. A large number of variables to be controlled can then be scanned, using a different control algorithm for each variable. This can also allow for the possibility of interaction between several control loops. Many computer-based, direct digital control (d.d.c) schemes today are based on replacing conventional one, two, and three term controllers with corresponding digital algorithms. The approach, which we shall describe here, can lead to control algorithms of a more general nature than have previously been used and can exploit the great flexibility of the automatic digital computer.

1.2 STOCHASTIC AND DETERMINISTIC DYNAMIC MATHEMATICAL MODELS

The idea of using a mathematical model to describe the behavior of a physical phenomenon is well established. In particular, it is sometimes possible to derive a model based on physical laws, which enables us to calculate the value of some time-dependent quantity nearly exactly at any instant of time. Thus, we might calculate the trajectory of a missile launched in a known direction with known velocity. If exact calculation were possible, such a model would be entirely *deterministic*.

Probably no phenomenon is totally deterministic, however, because unknown factors such as variable wind velocity that can throw a missile slightly off course, can occur. In many problems we have to consider a time-dependent phenomenon, such as monthly sales of newsprint, in which there are many unknown factors and for which it is not possible to write a deterministic model that allows exact calculation of the future behavior of the phenomenon. Nevertheless, it may be possible to derive a model that can be used to calculate the *probability* of a future value lying between two specified limits. Such a model is called a probability model or a *stochastic model*. The models for time series that are needed, for example to achieve optimal forecasting and control, are in fact stochastic models. It is necessary in what follows to distinguish between the probability model or *stochastic process*, as it is sometimes called, and the observed time series. Thus, a time series $z_1, z_2, \ldots, z_N$ of N successive observations is regarded as a sample realization from an infinite population of such time series that could have been generated by the stochastic process. Very often we shall omit the word stochastic from "stochastic process" and talk about the "process."

1.2.1 Stationary and nonstationary stochastic models for forecasting and control

An important class of stochastic models for describing time series, which has received a great deal of attention, is the so called *stationary* models, which assume that the process remains in *equilibrium* about a *constant mean level*. However, forecasting has been of particular importance in industry, business and economics, where many time series are often better represented as *nonstationary* and, in particular, as having no natural mean. It is not surprising, therefore, that the economic forecasting methods that have been proposed by Holt [1], [22], Winters [23], Brown [2] and the I.C.I. Monograph [3], all using exponentially weighted moving averages, can be shown to be appropriate for a particular type of *nonstationary* process. Although such methods are too narrow to deal efficiently with all time series, the fact that they give the right kind of forecast function supplies a clue to the *kind* of *nonstationary* model that might be useful in these problems.

It is not difficult to see that the autoregressive model is a special case of the linear filter model of (1.2.1). For example, we can eliminate $\tilde{z}_{t-1}$ from the right-hand side of (1.2.2) by substituting

$$\tilde{z}_{t-1} = \phi_1 \tilde{z}_{t-2} + \phi_2 \tilde{z}_{t-3} + \cdots + \phi_p \tilde{z}_{t-p-1} + a_{t-1}$$

We can likewise substitute for $\tilde{z}_{t-2}$, and so on, to yield eventually an infinite series in the a's.

Symbolically, we have

$$\phi(B)\tilde{z}_t = a_t$$

is equivalent to

$$\tilde{z}_t = \psi(B)a_t$$

with

$$\psi(B) = \phi^{-1}(B)$$

Autoregressive processes can be stationary or nonstationary. For the process to be stationary, the ϕ's must be chosen so that the weights $\psi_1, \psi_2, \ldots$ in $\psi(B) = \phi^{-1}(B)$ form a convergent series. We discuss these models in greater detail in Chapters 3 and 4.

Moving average models. The autoregressive model (1.2.2) expresses the deviation $\tilde{z}_t$ of the process as a *finite* weighted sum of p previous deviations $\tilde{z}_{t-1}, \tilde{z}_{t-2}, \ldots, \tilde{z}_{t-p}$ of the process, plus a random shock a_t. Equivalently, as we have just seen, it expresses $\tilde{z}_t$ as an *infinite* weighted sum of a's.

Another kind of model, of great practical importance in the representation of observed time series, is the so-called finite *moving average* process. Here we make $\tilde{z}_t$ linearly dependent on a *finite* number q of previous a's. Thus

$$\tilde{z}_t = a_t - \theta_1 a_{t-1} - \theta_2 a_{t-2} - \cdots - \theta_q a_{t-q} \qquad (1.2.3)$$

is called a *moving average* (MA) *process of order* q. The name "moving average" is somewhat misleading because the weights $1, -\theta_1, -\theta_2, \ldots, -\theta_q$, which multiply the a's, need not total unity nor need they be positive. However, this nomenclature is in common use, and therefore we employ it.

If we define a *moving average operator* of order q by

$$\theta(B) = 1 - \theta_1 B - \theta_2 B^2 - \cdots - \theta_q B^q$$

then the moving average model may be written economically as

$$\tilde{z}_t = \theta(B)a_t$$

It contains $q + 2$ unknown parameters $\mu, \theta_1, \ldots, \theta_q, \sigma_a^2$, which in practice have to be estimated from the data.

Mixed autoregressive–moving average models. To achieve greater flexibility in fitting of actual time series, it is sometimes advantageous to include both autoregressive and moving average terms in the model. This leads to the mixed autoregressive–moving average model

$$\tilde{z}_t = \phi_1 \tilde{z}_{t-1} + \cdots + \phi_p \tilde{z}_{t-p} + a_t - \theta_1 a_{t-1} - \cdots - \theta_q a_{t-q} \quad (1.2.4)$$

or

$$\phi(B)\tilde{z}_t = \theta(B)a_t$$

which employs $p + q + 2$ unknown parameters μ; $\phi_1, \ldots, \phi_p$; $\theta_1, \ldots, \theta_q$; σ_a^2, that are estimated from the data.

In practice, it is frequently true that adequate representation of actually occurring stationary time series can be obtained with autoregressive, moving average, or mixed models, in which p and q are not greater than 2 and often less than 2.

Nonstationary models. Many series actually encountered in industry or business (for example, stock prices) exhibit nonstationary behavior and in particular do not vary about a fixed mean. Such series may nevertheless exhibit homogeneous behavior of a kind. In particular, although the general level about which fluctuations are occurring may be different at different times, the broad behavior of the series, when differences in level are allowed for, may be similar. We show in Chapter 4 that such behavior may be represented by a generalized autoregressive operator $\varphi(B)$, in which one or more of the zeroes of the polynomial $\varphi(B)$ (that is one or more of the roots of the equation $\varphi(B) = 0$) is unity. Thus the operator $\varphi(B)$ can be written

$$\varphi(B) = \phi(B)(1 - B)^d$$

where $\phi(B)$ is a stationary operator. Thus a general model, which can represent homogeneous nonstationary behavior, is of the form

$$\varphi(B)z_t = \phi(B)(1 - B)^d z_t = \theta(B)a_t$$

that is

$$\phi(B)w_t = \theta(B)a_t \quad (1.2.5)$$

where

$$w_t = \nabla^d z_t \quad (1.2.6)$$

Homogeneous nonstationary behavior can therefore be represented by a model which calls for the d'th difference of the process to be stationary. In practice d is usually 0, 1, or at most 2.

The process defined by (1.2.5) and (1.2.6) provides a powerful model for describing stationary and nonstationary time series and is called an *autoregressive integrated moving average (ARIMA) process*, of order (p, d, q).

The process is defined by

$$w_t = \phi_1 w_{t-1} + \cdots + \phi_p w_{t-p} + a_t - \theta_1 a_{t-1} - \cdots - \theta_q a_{t-q} \quad (1.2.7)$$

with $w_t = \nabla^d z_t$. Note that if we replace w_t by $z_t - \mu$, when $d = 0$, the model (1.2.7) includes the stationary mixed model (1.2.4), as a special case, and also the pure autoregressive model (1.2.2) and the pure moving average model (1.2.3).

The reason for the inclusion of the word "integrated" (which should perhaps more appropriately be "summed") in the ARIMA title, is as follows. The relationship which is inverse to (1.2.6) is

$$z_t = S^d w_t \quad (1.2.8)$$

where it will be recalled that S is the summation operator defined by

$$Sw_t = \sum_{j=0}^{\infty} w_{t-j} = w_t + w_{t-1} + w_{t-2} + \cdots$$

Thus, the general autoregressive integrated moving average (ARIMA) process may be generated from white noise a_t by means of three filtering operations, as indicated by the block diagram of Figure 1.4(b). The first filter has input a_t, transfer function $\theta(B)$, and output e_t, where

$$e_t = a_t - \theta_1 a_{t-1} - \cdots - \theta_q a_{t-q}$$
$$= \theta(B)a_t \quad (1.2.9)$$

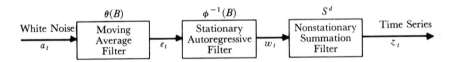

FIG. 1.4(b) Block diagram for autoregressive integrated moving average model

The second filter has input e_t, transfer function $\phi^{-1}(B)$, and output w_t, according to

$$w_t = \phi_1 w_{t-1} + \cdots + \phi_p w_{t-p} + e_t$$
$$= \phi^{-1}(B)e_t \quad (1.2.10)$$

Finally, the third filter has input w_t and output z_t, according to (1.2.8), and has transfer function S^d.

As described in Chapter 9, a special form of the model (1.2.7) can be employed to represent seasonal time series.

1.2.2 Transfer function models

An important type of dynamic relationship between a continuous input and a continuous output, for which many physical examples can be found, is that in which the *deviations* of input X and of output Y, from appropriate mean values, are related by a *linear* differential equation of the form

$$(1 + \Xi_1 D + \cdots + \Xi_R D^R)Y(t) = (H_0 + H_1 D + \cdots + H_S D^S)X(t - \tau) \quad (1.2.11)$$

where D is the differential operator d/dt, the Ξ's and H's are unknown parameters, and τ is a parameter which measures the *dead-time* or *pure delay* between input and output. The simplest example of (1.2.11) would be a system where the rate of change in the output was proportional to the difference between input and output, so that

$$\Xi \frac{dY}{dt} = X - Y$$

and hence

$$(1 + \Xi D)Y = X$$

In a similar way, for discrete data, in Chapter 10 we represent the transfer between an output Y and an input X, each measured at equispaced times, by the difference equation

$$(1 + \xi_1 \nabla + \cdots + \xi_r \nabla^r)Y_t = (\eta_0 + \eta_1 \nabla + \cdots + \eta_s \nabla^s)X_{t-b} \quad (1.2.12)$$

in which the differential operator D is replaced by the difference operator ∇. An expression of the form (1.2.12), containing only a few parameters $(r \leqslant 2, s \leqslant 2)$, may often be used as an approximation to a dynamic relationship, whose true nature is more complex.

The linear model (1.2.12) may be written equivalently in terms of past values of the input and output by substituting $B = 1 - \nabla$ in (1.2.12), that is

$$(1 - \delta_1 B - \cdots - \delta_r B^r)Y_t = (\omega_0 - \omega_1 B - \cdots - \omega_s B^s)X_{t-b} \quad (1.2.13)$$

$$= (\omega_0 B^b - \omega_1 B^{b+1} - \cdots - \omega_s B^{b+s})X_t$$

or

$$\delta(B)Y_t = \omega(B)B^b X_t$$

$$= \Omega(B)X_t$$

Alternatively, we can say that the output Y_t and input X_t are linked by a linear filter

$$Y_t = v_0 X_t + v_1 X_{t-1} + v_2 X_{t-2} + \cdots$$

$$= v(B)X_t \quad (1.2.14)$$

for which the transfer function

$$v(B) = v_0 + v_1 B + v_2 B^2 + \cdots \qquad (1.2.15)$$

can be expressed as a ratio of two polynomials

$$v(B) = \Omega(B)/\delta(B) = \delta^{-1}(B)\Omega(B)$$

The linear filter (1.2.14) is said to be stable if the series (1.2.15) converges for $|B| \leqslant 1$. The series of weights $v_0, v_1, v_2, \ldots$, which appear in the transfer function (1.2.15), is called the *impulse response function*. We note that for the model (1.2.12), the first b weights $v_0, v_1, \ldots, v_{b-1}$ are zero. A hypothetical impulse response function for the system of Figure 1.2 is shown in the center of that diagram.

The transfer function model (1.2.13) enables us to reinterpret the stochastic models (1.2.4) and (1.2.5). The disturbances occurring in some output z will often have originated elsewhere in some variable with which z is dynamically linked by an equation of the form (1.2.12). Therefore, we might expect that the complex stochastic behavior of a random variable z_t might be expressed in terms of another random variable a_t, having simpler properties, by a relationship

$$\delta(B)\tilde{z}_t = \Omega(B)a_t$$

If we allow the possibility of an unstable filter with one or more of the roots of $\delta(B) = 0$ equal to unity, then using previous notation,

$$\varphi(B)\tilde{z}_t = \theta(B)a_t \qquad (1.2.16)$$

The stochastic models we have considered are precisely of this kind, with a_t a source of *white noise*. Because (1.2.16) may be written

$$\tilde{z}_t = \varphi^{-1}(B)\theta(B)a_t$$

it is assumed that $\tilde{z}_t$ could be generated by passing white noise through a linear filter with a transfer function $\varphi^{-1}(B)\theta(B)$.

In summary:
(1) We can often represent a dynamic relationship connecting an output Y and an input X in terms of a linear filter

$$Y_t = v_0 X_t + v_1 X_{t-1} + v_2 X_{t-2} + \cdots$$
$$= v(B)X_t$$

where $v(B)$ is the transfer function of the filter.
(2) In turn $v(B)$ can frequently be represented with brevity and with sufficient accuracy by the ratio of two polynomials of low degree in B

$$v(B) = \delta^{-1}(B)\Omega(B)$$

so that the dynamic input–output equation may be written

$$\delta(B)Y_t = \Omega(B)X_t$$

(3) It is postulated that a series z_t, in which successive values are highly dependent, can be represented as the result of passing white noise a_t through such a dynamic system in which certain of the roots of $\delta(B) = 0$ could be unity. The notion yields the autoregressive integrated moving average model

$$\varphi(B)z_t = \theta(B)a_t$$

Models with superimposed noise. We have seen that the problem of estimating an appropriate model, linking an output Y_t and an input X_t, is equivalent to estimating the transfer function $v(B) = \delta^{-1}(B)\Omega(B)$. However, this problem is complicated in practice by the presence of noise N_t, which we assume corrupts the true relationship between input and output according to

$$Y_t = v(B)X_t + N_t$$

where N_t and X_t are independent. Suppose, as indicated by Figure 1.5, that the noise N_t can be described by a nonstationary stochastic model of the form (1.2.5) or (1.2.7), that is

$$N_t = \psi(B)a_t = \varphi^{-1}(B)\theta(B)a_t$$

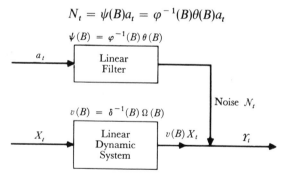

FIG. 1.5   Transfer function model for dynamic system with superimposed noise model

Then the observed relationship between output and input will be

$$\begin{aligned}
Y_t &= v(B)X_t + \psi(B)a_t \\
&= \delta^{-1}(B)\Omega(B)X_t + \varphi^{-1}(B)\theta(B)a_t
\end{aligned} \tag{1.2.17}$$

In practice, it is necessary to estimate the transfer function $\psi(B) = \varphi^{-1}(B)\theta(B)$ of the linear filter describing the noise, in addition to the transfer function $v(B) = \delta^{-1}(B)\Omega(B)$, which describes the dynamic relationship between the input and the output. Methods for doing this are discussed in Chapter 11.

1.2.3 *Models for discrete control systems*

As stated in Section 1.1.3, control is an attempt to compensate for disturbances which infect a system. Some of these disturbances are measureable;
others are not measurable and only manifest themselves as unexplained
deviations from the target of the characteristic to be controlled.

To illustrate the general principles involved, consider the special case
where unmeasured disturbances affect the output Y_t of some system, and
suppose that feedback control is employed to bring the output as close
as possible to the desired target value by adjustments applied to an input
variable X_t. This is illustrated in the block diagram of Figure 1.6. Suppose

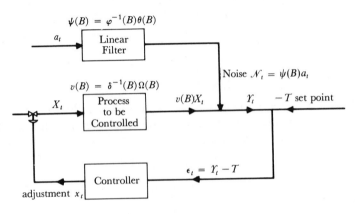

FIG. 1.6   Feedback control scheme to compensate an unmeasured disturbance N_t

that N_t represents the effect at the output of various unidentified disturbances
within the system, which in the absence of control could cause the output
to drift away from the desired target value or *set point* T. Then, in spite of
adjustments which have been made to the process, an error

$$\varepsilon_t = Y_t - T$$

$$= v(B)X_t + N_t - T$$

will occur between the output and its target value T. The object is to so
choose a control equation, that the errors ε will have the smallest possible
mean square. The control equation expresses the adjustment $x_t = X_t - X_{t-1}$
to be taken at time t, as a function of the present deviation ε_t, previous
deviations $\varepsilon_{t-1}, \varepsilon_{t-2}, \ldots$, and previous adjustments $x_{t-1}, x_{t-2}, \ldots$. The
mechanism (human, electrical, pneumatic or electronic), which carries out
the control action called for by the control equation, is called the *controller*.

It turns out that the procedure for designing an optimal scheme is equivalent to forecasting the deviation from target which would occur *if no*

control were applied, and then calculating the adjustment that would be necessary to cancel out this deviation. It follows that the forecasting and control problems are closely linked. To forecast the deviation from target that could occur if no control were applied, it is necessary to build a model

$$N_t = \psi(B)a_t = \varphi^{-1}(B)\theta(B)a_t$$

for the disturbance. Calculation of the adjustment x_t which needs to be applied to the input variable at time t to cancel out a predicted change at the output requires the building of a dynamic model with transfer function

$$v(B) = \delta^{-1}(B)\Omega(B)$$

which links the input and output. The resulting adjustment x_t will consist, in general, of a linear aggregate of previous adjustments, and current and previous control errors. Thus the control equation will be of the form

$$x_t = \zeta_1 x_{t-1} + \zeta_2 x_{t-2} + \cdots + \chi_0 \varepsilon_t + \chi_1 \varepsilon_{t-1} + \chi_2 \varepsilon_{t-2} + \cdots \quad (1.2.18)$$

where $\zeta_1, \zeta_2, \ldots, \chi_0, \chi_1, \chi_2, \ldots$ are constants. These ideas are discussed in Chapter 12.

1.3. BASIC IDEAS IN MODEL BUILDING

1.3.1 Parsimony

We have seen that the mathematical models, which we need to employ, contain certain constants or parameters whose values must be estimated from the data. It is important, in practice, that we employ the *smallest possible* number of parameters for adequate representation. The central role played by this principle of *parsimony* [25] in the use of parameters will become clearer as we proceed. As a preliminary illustration, we consider the following simple example.

Suppose we fitted a dynamic model (1.2.12) of the form

$$Y_t = (\eta_0 + \eta_1 \nabla + \eta_2 \nabla^2 + \cdots + \eta_s \nabla^s)X_t \quad (1.3.1)$$

when dealing with a system which was adequately represented by

$$(1 + \xi\nabla)Y_t = X_t \quad (1.3.2)$$

The model (1.3.2) contains only one parameter ξ but, for s sufficiently large, could be represented approximately by the model (1.3.1). Because of experimental error, we could easily fail to recognize the relationship between the coefficients in the fitted equation. Thus, we might needlessly fit a relationship like (1.3.1), containing s parameters, where the much simpler form (1.3.2), containing only one, would have been adequate. This could, for example, lead to unnecessarily poor estimation of the output Y_t for given values of the input $X_t, X_{t-1}, \ldots$

Our objective, then, must be to obtain adequate but parsimonious models. Forecasting and control procedures could be seriously deficient if these models were either inadequate or unnecessarily prodigal in the use of parameters. Care and effort is needed in selecting the model. The process of selection is necessarily iterative, that is to say, it is a process of evolution, adaptation, or trial and error.

1.3.2 Iterative stages in the selection of a model

When the physical mechanism of a phenomenon is completely understood, it may be possible to write down a mathematical expression which describes it exactly. We thus obtain a *mechanistic* or *theoretical* model. Although insufficient information may be available initially to write an adequate mechanistic model, nevertheless, an adaptive strategy [26] can sometimes lead to such a model.

In many instances the rather complete knowledge or large experimental resources needed to produce a mechanistic model are not available, and we must then resort to an empirical model. Of course the exact mechanistic model and the exclusively empirical model represent extremes. Models actually employed usually lie somewhere in between. In particular, we may use incomplete theoretical knowledge to indicate a suitable class of mathematical functions, which will then be fitted empirically; that is, the number of terms needed in the model and the numerical values of the parameters are estimated from experimental data. This is the approach that we adopt in this book. As we have indicated previously, the stochastic and dynamic models we describe can be justified, at least partially, on theoretical grounds as having the right general properties.

It is normally supposed that successive values of the time series under consideration or of the input-output data, are available for analysis. If possible, at least 50 and preferably 100 observations or more should be used. In those cases where a past history of 50 or more observations are not available, one proceeds by using experience and past information to yield a preliminary model. This model may be updated from time to time as more data becomes available.

In fitting dynamic models, a theoretical analysis can sometimes tell us not only the appropriate form for the model, but may also provide us with good estimates of the numerical values of its parameters. These values can then be checked later by analysis of plant data.

Figure 1.7 summarizes the iterative approach to model building for forecasting and control, which is employed in this book.

(1) From the interaction of theory and practice, a *useful class of models* for the purposes at hand is considered.

(2) Because this class is too extensive to be conveniently fitted directly to

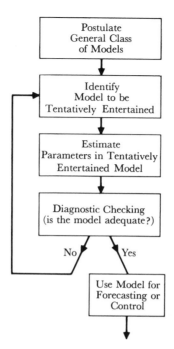

FIG. 1.7   Stages in the iterative approach to model building

data, rough methods for *identifying* subclasses of these models are developed. Such methods of model identification employ data and knowledge of the system to suggest an appropriate parsimonious subclass of models which may be tentatively entertained. In addition, the identification process can be used to yield rough preliminary estimates of the parameters in the model.

(3) The tentatively entertained model is *fitted* to data and its parameters *estimated*. The rough estimates obtained during the identification stage can now be used as starting values in more refined iterative methods for estimating the parameters.

(4) *Diagnostic checks* are applied with the object of uncovering possible lack of fit and diagnosing the cause. If no lack of fit is indicated, the model is ready to use. If any inadequacy is found, the iterative cycle of identification, estimation, and diagnostic checking is repeated until a suitable representation is found.

Identification, Estimation and Diagnostic Checking are discussed for time series models in Chapters 6, 7, 8 and 9, and for transfer function models in Chapter 11.

Part I

Stochastic Models and their Forecasting

In the first part of this book, which includes Chapters 2, 3, 4 and 5, a valuable class of stochastic models is described and its use in forecasting discussed.

A model which describes the probability structure of a sequence of observations is called a *stochastic process*. A time series of N successive observations $\mathbf{z}' = (z_1, z_2, \ldots, z_N)$ is regarded as a sample realization, from an infinite population of such samples, which could have been generated by the process. A major objective of statistical investigation is to infer properties of the population from those of the sample. For example, to make a forecast is to infer the probability distribution of a *future observation* from the population, given a sample $\mathbf{z}$ of past values. To do this we need ways of describing stochastic processes and time series, and we also need classes of stochastic models which are capable of describing practically occurring situations. An important class of stochastic processes discussed in Chapter 2 is the *stationary* processes. They are assumed to be in a specific form of statistical equilibrium, and in particular, vary about a fixed mean. Useful devices for describing the behavior of stationary processes are the *autocorrelation function* and the *spectrum*.

Particular stationary stochastic processes of value in modelling time series are the autoregressive, moving average, and mixed autoregressive moving average processes. The properties of these processes, and in particular their correlation structures are described in Chapter 3.

Because many practically occurring time series (for example stock prices and sales figures) have nonstationary characteristics, the stationary models introduced in Chapter 3 are further developed in Chapter 4 to give a useful class of nonstationary processes called autoregressive integrated moving average (ARIMA) models. The use of all these models in forecasting time series is discussed in Chapter 5 and is illustrated with examples.

2

The Autocorrelation Function and Spectrum

A central feature in the development of time series models is an assumption of some form of *statistical equilibrium.* A particular assumption of this kind (an unduly restrictive one as we shall see later) is that of *stationarity.* Usually a stationary time series can be usefully described by its mean, variance, and *autocorrelation function*, or equivalently by its mean, variance, and *spectral density function.* In this chapter we consider the properties of these functions and, in particular, the properties of the autocorrelation function which is used extensively in the chapters that follow.

2.1 AUTOCORRELATION PROPERTIES OF STATIONARY MODELS

2.1.1 Time series and stochastic processes

Time series. A time series is a set of observations generated sequentially in time. If the set is continuous, the time series is said to be *continuous.* If the set is discrete, the time series is said to be *discrete.* Thus, the observations from a discrete time series made at times $\tau_1, \tau_2, \ldots, \tau_t, \ldots, \tau_N$ may be denoted by $z(\tau_1), z(\tau_2), \ldots, z(\tau_t), \ldots, z(\tau_N)$. In this book we consider only discrete time series where observations are made at some fixed interval h. When we have N successive values of such a series available for analysis, we write $z_1, z_2, \ldots, z_t, \ldots, z_N$ to denote observations made at equidistant time intervals $\tau_0 + h, \tau_0 + 2h, \ldots, \tau_0 + th, \ldots, \tau_0 + Nh$. For many purposes the values of τ_0 and h are unimportant, but if the observation times need to be defined exactly, these two values can be specified. If we adopt τ_0 as the origin and h as the unit of time, we can regard z_t as the observation *at time t.*

Discrete time series may arise in two ways.

(1) By *sampling* a continuous time series; for example, in the situation shown in Figure 1.2, where the continuous input and output from a gas furnace was sampled at intervals of nine seconds.

(2) By *accumulating* a variable over a period of time; examples are rainfall, which is usually accumulated over a period such as a day or a month,

and the yield from a batch process, which is accumulated over the batch time. For example, Figure 2.1 shows a time series consisting of the yields from 70 consecutive batches of a chemical process.

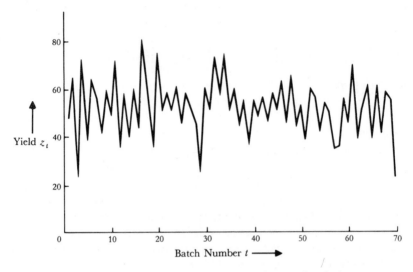

FIG. 2.1   Yields of 70 consecutive batches from a chemical process

Deterministic and statistical time series.   If future values of a time series are exactly determined by some mathematical function such as

$$z_t = \cos(2\pi f t)$$

the time series is said to be *deterministic*. If the future values can be described only in terms of a probability distribution, the time series is said to be non-deterministic or simply a *statistical time series*. The batch data of Figure 2.1 is an example of a statistical time series. Thus, although there is a well-defined, high-low pattern in the series, it is impossible to exactly forecast the yield for the next batch. It is with such statistical time series that we shall be concerned in this book.

Stochastic processes.   A statistical phenomenon that evolves in time according to probabilistic laws is called a *stochastic process*. We shall often refer to it simply as a *process*, omitting the word stochastic. The time series to be analyzed may then be thought of as one particular *realization*, produced by the underlying probability mechanism, of the system under study. In other words, *in analyzing a time series we regard it as a realization of a stochastic process.*

For example, to analyze the batch data in Figure 2.1, we can imagine other sets of observations (other realizations of the underlying stochastic process), which might have been generated by the same chemical system, in the same $n = 70$ batches. Thus, Figure 2.2 shows the yields from batches $t = 21$ to $t = 30$ (thick line), together with other time series which *might* have been obtained from the population of time series defined by the underlying stochastic process. It follows that we can regard the observation z_t at a given time t, say $t = 25$, as a realization of a random variable z_t with probability density function $p(z_t)$. Similarly, the observations at any two times, say $t_1 = 25$ and $t_2 = 27$, may be regarded as realizations of two random variables z_{t_1} and z_{t_2} with joint probability density function $p(z_{t_1}, z_{t_2})$. For example, Figure 2.3 shows contours of constant density for such a joint distribution, together with the marginal distribution at time t_1. In general, the observations making up an equispaced time series can be described by an N dimensional, random variable $(z_1, z_2, \ldots, z_N)$ with probability distribution $p(z_1, z_2, \ldots, z_N)$.

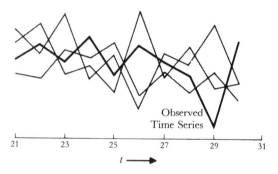

Observed
Time Series

21   23   25   27   29   31

$t \longrightarrow$

FIG. 2.2   An observed time series (thick line) with other time series representing realizations of the same stochastic process

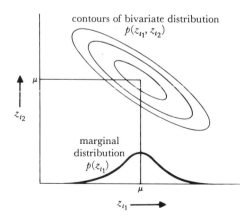

contours of bivariate distribution
$p(z_{t_1}, z_{t_2})$

μ

z_{t_2}

marginal
distribution
$p(z_{t_1})$

μ

$z_{t_1} \longrightarrow$

FIG. 2.3   Contours of constant density of a bivariate probability distribution describing a stochastic process at two times t_1, t_2, together with the marginal distribution at time t_1

2.1.2 Stationary stochastic processes

A very special class of stochastic processes, called stationary processes, is based on the assumption that the process is in a particular state of *statistical equilibrium*. A stochastic process is said to be *strictly stationary* if its properties are unaffected by a change of time origin; that is, if the joint probability distribution associated with m observations $z_{t_1}, z_{t_2}, \ldots, z_{t_m}$, made at *any* set of times $t_1, t_2, \ldots, t_m$, is the same as that associated with m observations $z_{t_1+k}, z_{t_2+k}, \ldots, z_{t_m+k}$, made at times $t_1 + k, t_2 + k, \ldots, t_m + k$. Thus for a discrete process to be strictly stationary, the joint distribution of any set of observations must be unaffected by shifting all the times of observation forward or backward by any integer amount k.

Mean and variance of a stationary process.   When $m = 1$, the stationarity assumption implies that the probability distribution $p(z_t)$ is the same for all times t and may be written $p(z)$. Hence, the stochastic process has a constant mean

$$\mu = E[z_t] = \int_{-\infty}^{\infty} zp(z)\, dz \tag{2.1.1}$$

which defines the level about which it fluctuates, and a constant variance

$$\sigma_z^2 = E[(z_t - \mu)^2] = \int_{-\infty}^{\infty} (z - \mu)^2 p(z)\, dz \tag{2.1.2}$$

which measures its *spread* about this level. Since the probability distribution $p(z)$ is the same for all times t, its shape can be inferred by forming the histogram of the observations $z_1, z_2, \ldots, z_N$, making up the observed time series. In addition, the mean μ of the stochastic process can be estimated by the mean

$$\bar{z} = \frac{1}{N} \sum_{t=1}^{N} z_t \tag{2.1.3}$$

of the time series, and the variance σ_z^2, of the stochastic process, can be estimated by the variance

$$\hat{\sigma}_z^2 = \frac{1}{N} \sum_{t=1}^{N} (z_t - \bar{z})^2 \tag{2.1.4}$$

of the time series.

Autocovariance and autocorrelation coefficients.   The stationarity assumption also implies that the joint probability distribution $p(z_{t_1}, z_{t_2})$ is the same for all times t_1, t_2, which are a constant interval apart. It follows that the nature of this joint distribution can be inferred by plotting a scatter diagram using pairs of values (z_t, z_{t+k}), of the time series, separated by a constant interval or *lag k*. For the batch data, Figure 2.4(a) is a scatter diagram for lag $k = 1$, obtained by plotting z_{t+1} versus z_t, and Figure 2.4(b) a scatter diagram for lag $k = 2$, obtained by plotting z_{t+2} versus z_t. We see that neighboring values of the time series are correlated; the correlation between

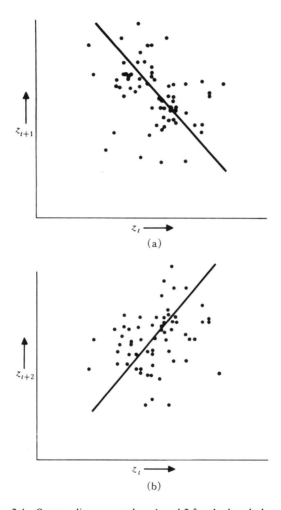

(a)

(b)

Fɪɢ. 2.4   Scatter diagrams at lags 1 and 2 for the batch data of Figure 2.1

z_t and z_{t+1} appearing to be negative and the correlation between z_t and z_{t+2} positive. The covariance between z_t and its value z_{t+k}, separated by k intervals of time, is called the *autocovariance* at lag k and is defined by

$$\gamma_k = \text{cov}\,[z_t, z_{t+k}] = E[(z_t - \mu)(z_{t+k} - \mu)] \qquad (2.1.5)$$

Similarly the *autocorrelation* at lag k is

$$\rho_k = \frac{E[(z_t - \mu)(z_{t+k} - \mu)]}{\sqrt{E[(z_t - \mu)^2]E[(z_{t+k} - \mu)^2]}}$$

$$= \frac{E[(z_t - \mu)(z_{t+k} - \mu)]}{\sigma_z^2}$$

since, for a stationary process, the variance $\sigma_z^2 = \gamma_0$ is the same at time $t + k$ as at time t.

Thus, the autocorrelation at lag k is

$$\rho_k = \frac{\gamma_k}{\gamma_0} \tag{2.1.6}$$

which implies that $\rho_0 = 1$.

2.1.3 *Positive definiteness and the autocovariance matrix*

The covariance matrix associated with a stationary process for observations $(z_1, z_2, \ldots, z_n)$, made at n successive times is

$$\Gamma_n = \begin{bmatrix} \gamma_0 & \gamma_1 & \gamma_2 & \cdots & \gamma_{n-1} \\ \gamma_1 & \gamma_0 & \gamma_1 & \cdots & \gamma_{n-2} \\ \gamma_2 & \gamma_1 & \gamma_0 & \cdots & \gamma_{n-3} \\ \vdots & \vdots & \vdots & \cdots & \vdots \\ \gamma_{n-1} & \gamma_{n-2} & \gamma_{n-3} & \cdots & \gamma_0 \end{bmatrix}$$

$$= \sigma_z^2 \begin{bmatrix} 1 & \rho_1 & \rho_2 & \cdots & \rho_{n-1} \\ \rho_1 & 1 & \rho_1 & \cdots & \rho_{n-2} \\ \rho_2 & \rho_1 & 1 & \cdots & \rho_{n-3} \\ \vdots & \vdots & \vdots & \cdots & \vdots \\ \rho_{n-1} & \rho_{n-2} & \rho_{n-3} & \cdots & 1 \end{bmatrix} = \sigma_z^2 \mathbf{P}_n \tag{2.1.7}$$

A covariance matrix Γ_n of this form, which is symmetric with constant elements on any diagonal, will be called an *autocovariance matrix* and the corresponding correlation matrix $\mathbf{P}_n$, will be called an *autocorrelation matrix*. Now consider any linear function of the random variables z_t, $z_{t-1}, \ldots, z_{t-n+1}$

$$L_t = l_1 z_t + l_2 z_{t-1} + \cdots + l_n z_{t-n+1} \tag{2.1.8}$$

Since cov $[z_i, z_j] = \gamma_{|j-i|}$ for a stationary process, the variance of L_t is

$$\text{var}\,[L_t] = \sum_{i=1}^{n} \sum_{j=1}^{n} l_i l_j\, \gamma_{|j-i|}$$

which is necessarily greater than zero if the l's are not all zero. It follows that both an autocovariance matrix and an autocorrelation matrix are positive–definite for any stationary process.

Conditions satisfied by the autocorrelations of a stationary process. The positive definiteness of the autocorrelation matrix (2.1.7) implies that its determinant and all principal minors are greater than zero. In particular for $n = 2$

$$\begin{vmatrix} 1 & \rho_1 \\ \rho_1 & 1 \end{vmatrix} > 0$$

so that

$$1 - \rho_1^2 > 0$$

and hence

$$-1 < \rho_1 < 1$$

Similarly, for $n = 3$, we must have

$$\begin{vmatrix} 1 & \rho_1 \\ \rho_1 & 1 \end{vmatrix} > 0, \qquad \begin{vmatrix} 1 & \rho_2 \\ \rho_2 & 1 \end{vmatrix} > 0$$

$$\begin{vmatrix} 1 & \rho_1 & \rho_2 \\ \rho_1 & 1 & \rho_1 \\ \rho_2 & \rho_1 & 1 \end{vmatrix} > 0$$

which implies

$$-1 < \rho_1 < 1$$

$$-1 < \rho_2 < 1$$

$$-1 < \frac{\rho_2 - \rho_1^2}{1 - \rho_1^2} < 1$$

and so on. Since $\mathbf{P}_n$ must be positive definite for *all* values of n, the autocorrelations of a stationary process must satisfy a very large number of conditions. As will be shown in Section 2.2.3, all of these conditions can be brought together in the definition of the spectrum.

Stationarity of linear functions. It follows from the definition of stationarity, that the process L_t, obtained by performing the linear operation (2.1.8) on a stationary process z_t, is also stationary. In particular, the first difference $\nabla z_t = z_t - z_{t-1}$ and higher differences $\nabla^d z_t$, are stationary. This result is of particular importance to the discussion of nonstationary time series presented in Chapter 4.

Gaussian processes. If the probability distribution associated with *any* set of times is a multivariate Normal distribution, the process is called a *Normal* or *Gaussian* process. Since the multivariate Normal distribution is fully characterized by its moments of first and second order, the existence of a fixed mean μ and an autocovariance matrix Γ_n for all n, would be sufficient to ensure the stationarity of a Gaussian process.

Weak stationarity. We have seen that for a process to be strictly stationary, the whole probability structure must depend only on time differences. A less restrictive requirement, called *weak stationarity* of order f, is that the moments up to some order f depend only on time differences. For example, the existence of a mean μ and an autocovariance matrix Γ_n of the form (2.1.7) is sufficient to ensure stationarity up to second order. Thus, second order stationarity, plus an assumption of Normality, is sufficient to produce strict stationarity.

2.1.4 The autocovariance and autocorrelation functions

It was seen in Section 2.1.2 that the autocovariance coefficient γ_k, at lag k, measures the covariance between two values z_t and z_{t+k}, a distance k apart. The plot of γ_k versus the lag k, is called the *autocovariance function* $\{\gamma_k\}$ of the stochastic process. Similarly, the plot of the autocorrelation coefficient ρ_k as a function of the lag k, is called the *autocorrelation function* $\{\rho_k\}$ of the process. Note that the autocorrelation function is dimensionless; that is, independent of the scale of measurement of the time series. Since $\gamma_k = \rho_k \sigma_z^2$ knowledge of the autocorrelation function $\{\rho_k\}$ and of the variance σ_z^2 is equivalent to knowledge of the autocovariance function $\{\gamma_k\}$.

The autocorrelation function, shown in Figure 2.5 as a plot of the diagonals of the autocorrelation matrix, reveals how the correlation between any two values of the series changes as their separation changes. Since $\rho_k = \rho_{-k}$, the autocorrelation function is necessarily symmetric about zero, and in practice it is only necessary to plot the positive half of the function. Figure 2.6 shows the positive half of the autocorrelation function given in Figure 2.5. Henceforth, when we speak of the autocorrelation function, we shall often mean the positive half. In the past, the autocorrelation function has sometimes been called the *correlogram.*

From what has been previously shown, a *Normal* stationary process z_t is completely characterized by its mean μ and its autocovariance function $\{\gamma_k\}$, or equivalently by its mean μ, variance σ_z^2, and autocorrelation function $\{\rho_k\}$.

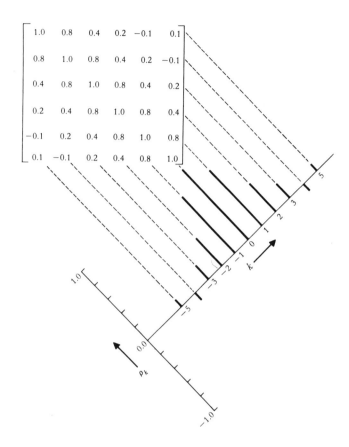

FIG. 2.5   An autocorrelation matrix and the resulting autocorrelation function

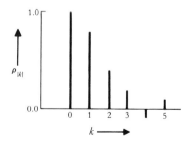

FIG. 2.6   Positive half of the autocorrelation function of Figure 2.5

2.1.5 Estimation of autocovariance and autocorrelation functions

Up to now we have only considered the theoretical autocorrelation function which describes a conceptual stochastic process. In practice, we have a finite time series $z_1, z_2, \ldots, z_N$, of N observations, from which we can only obtain *estimates* of the autocorrelations.

A number of estimates of the autocorrelation function have been suggested by statisticians and their properties are discussed in particular in [27]. It is concluded that the most satisfactory estimate of the kth lag autocorrelation ρ_k is

$$r_k = \frac{c_k}{c_0} \qquad (2.1.9)$$

where

$$c_k = \frac{1}{N} \sum_{t=1}^{N-k} (z_t - \bar{z})(z_{t+k} - \bar{z}), \qquad k = 0, 1, 2, \ldots, K \quad (2.1.10)$$

is the estimate of the autocovariance γ_k, and $\bar{z}$ is the mean of the time series. We now illustrate (2.1.10) by calculating r_1 for the first 10 values of the batch data of Figure 2.1, given in Table 2.1. The mean $\bar{z}$ of the first ten values in the table is 51 and so the deviations about the mean are $-4, 13, -28, 20, -13, 13, 4, -10, 8, -3$.

TABLE 2.1   A series of 70 consecutive yields from a batch chemical process. This series also appears as Series F in the Collection of Time Series at the end of this volume

1–15	16–30	31–45	46–60	61–70
47	44	50	62	68
64	80	71	44	38
23	55	56	64	50
71	37	74	43	60
38	74	50	52	39
64	51	58	38	59
55	57	45	59	40
41	50	54	55	57
59	60	36	41	54
48	45	54	53	23
71	57	48	49	
35	50	55	34	
57	45	45	35	
40	25	57	54	
58	59	50	45	

Thus

$$\sum_{t=1}^{9} (z_t - \bar{z})(z_{t+1} - \bar{z}) = (-4)(13) + (13)(-28) + \cdots + (8)(-3)$$

$$= -1497$$

Hence

$$c_1 = \frac{-1497}{10} = -149.7$$

Similarly we find that $c_0 = 189.6$. Hence

$$r_1 = \frac{c_1}{c_0} = \frac{-149.7}{189.6} = -0.79$$

it being sufficient for most practical purposes to round off the autocorrelation to two decimal places. The above calculation is made for illustration only. In practice, to obtain a useful estimate of the autocorrelation function, we would need at least fifty observations and the estimated autocorrelations r_k would be calculated for $k = 0, 1, \ldots, K$ where K was not larger than say $N/4$.

The first 15 values of r_k, based on the whole series of 70 observations, are given in Table 2.2 and plotted in Figure 2.7. The estimated autocorrelation function is characterized by correlations which alternate in sign and which tend to damp out with increasing lag. Autocorrelation functions of this kind are not uncommon in production data and can arise because of "carry over"

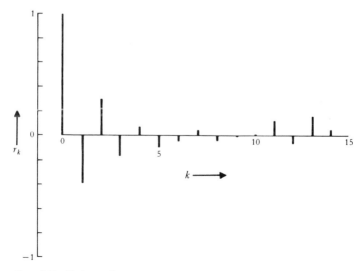

FIG. 2.7   Estimated autocorrelation function of batch data

TABLE 2.2   Estimated autocorrelation function of batch data

k	r_k	k	r_k	k	r_k
1	-0.39	6	-0.05	11	0.11
2	0.30	7	0.04	12	-0.07
3	-0.17	8	-0.04	13	0.15
4	0.07	9	-0.01	14	0.04
5	-0.10	10	0.01	15	-0.01

effects. In this particular example, a high yielding batch tended to produce tarry residues which were not entirely removed from the vessel and adversely affected the yield of the next batch.

A computer program for calculating estimates of the autocovariance function and autocorrelation function is described under Program I, in the collection of computer programs at the end of the volume.

2.1.6 Standard error of autocorrelation estimates

To identify a model for a time series, using methods to be described in Chapter 6, it is necessary to have a crude check on whether ρ_k is effectively zero beyond a certain lag. For this purpose, use can be made of the following approximate expression for the variance of the estimated autocorrelation coefficient of a stationary Normal process given by Bartlett [28].

$$\text{var}\,[r_k] \simeq \frac{1}{N} \sum_{v=-\infty}^{+\infty} \{\rho_v^2 + \rho_{v+k}\rho_{v-k} - 4\rho_k\rho_v\rho_{v-k} + 2\rho_v^2\rho_k^2\} \quad (2.1.11)$$

For example, if $\rho_k = \phi^{|k|}, (-1 < \phi < 1)$, that is the autocorrelation function damps out exponentially, (2.1.11) gives

$$\text{var}\,[r_k] \simeq \frac{1}{N} \left[\frac{(1+\phi^2)(1-\phi^{2k})}{1-\phi^2} - 2k\phi^{2k} \right] \quad (2.1.12)$$

and in particular

$$\text{var}\,[r_1] \simeq \frac{1}{N}(1-\phi^2)$$

For any process for which all the autocorrelations ρ_v are zero for $v > q$, all terms except the first appearing on the right-hand side of (2.1.11) are zero when $k > q$. Thus for the variance of the estimated autocorrelations r_k, at lags k greater than some value q *beyond which the theoretical autocorrelation function may be deemed to have "died out,"* Bartlett's approximation gives

$$\text{var}\,[r_k] \simeq \frac{1}{N}\left\{1 + 2\sum_{v=1}^{q}\rho_v^2\right\}, \qquad k > q \qquad (2.1.13)$$

For example, as k is increased, and provided ϕ is not close to unity, (2.1.12) tends quickly to its limiting form

$$\lim_{k\to\infty}\text{var}\,[r_k] \simeq \frac{1}{N}\left(\frac{1 + \phi^2}{1 - \phi^2}\right)$$

also given by (2.1.13).

To use (2.1.13) in practice, the estimated autocorrelations r_k $(k = 1, 2, \ldots, q)$ are substituted for the theoretical autocorrelations ρ_k and when this is done we shall refer to the square root of (2.1.13) as the *large-lag standard error*. On the assumption that the theoretical autocorrelations ρ_k are all essentially zero beyond some hypothesized lag $k = q$, the large lag standard error approximates the standard deviation of r_k for suitably large lags $(k > q)$.

Similar approximate expressions for the covariance between the estimated correlations r_k and r_{k+s} at two different lags k and $k + s$ have been given by Bartlett [28]. In particular, the large-lag approximation reduces to

$$\text{cov}\,[r_k, r_{k+s}] \simeq \frac{1}{N}\sum_{v=-\infty}^{\infty}\rho_v\rho_{v+s} \qquad (2.1.14)$$

Bartlett's result (2.1.14) shows that care is required in the interpretation of individual autocorrelations because large covariances can exist between neighboring values. This effect can sometimes distort the visual appearance of the autocorrelation function which may fail to damp out according to expectation [27], [29].

An example. The following estimated autocorrelations were obtained from a time series of length $N = 200$ observations, generated from a stochastic process for which it was *known* that $\rho_1 = -0.4$ and $\rho_k = 0$ for $k \geqslant 2$:

k	1	2	3	4	5
r_k	-0.38	-0.08	0.11	-0.08	0.02
k	6	7	8	9	10
r_k	0.00	0.00	0.00	0.07	-0.08

On the assumption that the series is completely random, we have that $q = 0$. Then, for *all lags*, (2.1.13) yields

$$\text{var}\,[r_k] \simeq \frac{1}{N} = \frac{1}{200} = 0.005$$

The corresponding standard error is $0.07 = (0.005)^{1/2}$. Since the value of -0.38 for r_1 is over five times this standard error, it can be concluded that ρ_1 is nonzero. Moreover, the estimated autocorrelations for lags greater than one are small. Therefore it might be reasonable to ask next whether the series was compatible with a hypothesis (whose relevance will be discussed later) whereby ρ_1 was non-zero but $\rho_k = 0$ ($k \geqslant 2$). Using (2.1.13) with $q = 1$ and substituting r_1 for ρ_1, the estimated large-lag variance under this assumption is

$$\text{var}\,[r_k] \simeq \frac{1}{200}\,\{1 + 2(-0.38)^2\} = 0.0064, \qquad k > 1$$

yielding a standard error of 0.08. Since the estimated autocorrelations for lags greater than one are small compared with this standard error, there is no reason to doubt the adequacy of the model $\rho_1 \neq 0$, $\rho_k = 0$ ($k \geqslant 2$).

2.2. SPECTRAL PROPERTIES OF STATIONARY MODELS

2.2.1 The periodogram

Another way of analyzing a time series is based on the assumption that it is made up of sine and cosine waves with different frequencies. A device which uses this idea, first introduced by Schuster [30] in 1898, is the *periodogram* (see also [31]).

The periodogram was originally used to detect and estimate the amplitude of a sine component, of known frequency, buried in noise. We shall use it later to provide a check on the randomness of a series (usually a series of residuals after fitting a particular model), where we consider the possibility that periodic components of unknown frequency may still remain in the series.

To illustrate the calculation of the periodogram, suppose that the number of observations $N = 2q + 1$ is odd. If we fit the Fourier series model

$$z_t = \alpha_0 + \sum_{i=1}^{q} (\alpha_i c_{it} + \beta_i s_{it}) + e_t \tag{2.2.1}$$

where $c_{it} = \cos 2\pi f_i t$, $s_{it} = \sin 2\pi f_i t$, and $f_i = i/N$ is the ith harmonic of the fundamental frequency $1/N$, then the least squares estimates of the coefficients α_0 and (α_i, β_i) will be

$$a_0 = \bar{z} \tag{2.2.2}$$

$$a_i = \frac{2}{N} \sum_{t=1}^{N} z_t c_{it} \tag{2.2.3}$$

$$\left.\begin{array}{c} \\ \\ \end{array}\right\} \; i = 1, 2, \ldots, q$$

$$b_i = \frac{2}{N} \sum_{t=1}^{N} z_t s_{it} \tag{2.2.4}$$

The periodogram then consists of the $q = (N - 1)/2$ values

$$I(f_i) = \frac{N}{2}(a_i^2 + b_i^2), \qquad i = 1, 2, \ldots, q \tag{2.2.5}$$

where $I(f_i)$ is called the *intensity* at frequency f_i.

When N is even, we set $N = 2q$ and (2.2.2), (2.2.3), (2.2.4), and (2.2.5), apply for $i = 1, 2, \ldots, (q - 1)$ but

$$a_q = \frac{1}{N} \sum_{t=1}^{N} (-1)^t z_t$$

$$b_q = 0$$

and

$$I(f_q) = I(0.5) = N a_q^2$$

Note that the highest frequency is 0.5 cycles per time interval because the smallest period is 2 intervals.

2.2.2 Analysis of variance

In an analysis of variance table associated with the fitted regression (2.2.1), when N is odd, we can isolate $(N - 1)/2$ pairs of degrees of freedom, after eliminating the mean. These are associated with the pairs of coefficients $(a_1, b_1), (a_2, b_2), \ldots, (a_q, b_q)$, and hence with the frequencies $1/N, 2/N, \ldots,$ q/N. The periodogram $I(f_i) = (N/2)(a_i^2 + b_i^2)$ is seen to be simply the "sum of squares" associated with the pair of coefficients (a_i, b_i), and hence with the frequency $f_i = i/N$. Thus

$$\sum_{t=1}^{N} (z_t - \bar{z})^2 = \sum_{i=1}^{q} I(f_i) \tag{2.2.6}$$

When N is even, there are $(N - 2)/2$ pairs of degrees of freedom, and a further single degree of freedom associated with the coefficient a_q.

If the series were truly random, containing no systematic sinusoidal component, that is

$$z_t = \alpha_0 + e_t$$

with α_0 the fixed mean, and the e's independent and Normal, with mean zero and variance σ^2, then each component $I(f_i)$ would have expectation $2\sigma^2$ and would be distributed* as $\sigma^2 \chi^2(2)$, independently of all the other components. By contrast, if the series contained a systematic sine component having frequency f_i, amplitude A, and phase angle F, so that

$$z_t = \alpha_0 + \alpha \cos(2\pi f_i t) + \beta \sin(2\pi f_i t) + e_t$$

---

* It is to be understood that $\chi^2(m)$ refers to a random variable having a chi-square distribution with m degrees of freedom, defined explicitly, for example, in Appendix A.7.1.

with $A \sin F = \alpha$ and $A \cos F = \beta$, then the sum of squares $I(f_i)$ would tend to be inflated, since its expected value would be $2\sigma^2 + N(\alpha^2 + \beta^2)/2 = 2\sigma^2 + NA^2/2$.

In practice, it is unlikely that the frequency f of an unknown systematic sine component would exactly match any of the frequencies f_i, for which intensities have been calculated. In this case the periodogram would show an increase in the intensities in the immediate vicinity of f.

An example. A large number of observations would usually be used in the calculation of the periodogram. However, to illustrate the process of calculation we use the set of twelve mean monthly temperatures (in degrees Celsius) for central England during 1964, given in Table 2.3.

TABLE 2.3   Mean monthly temperatures for Central England in 1964

t	1	2	3	4	5	6	7	8	9	10	11	12
z_t	3.4	4.5	4.3	8.7	13.3	13.8	16.1	15.5	14.1	8.9	7.4	3.6
c_{1t}	0.87	0.50	0.00	−0.50	−0.87	−1.00	−0.87	−0.50	0.00	0.50	0.87	1.00

The third line in Table 2.3 gives $c_{1t} = \cos(2\pi t/12)$, which is required in the calculation of a_1, obtained from

$$a_1 = \tfrac{1}{6}\{(3.4)(0.87) + \cdots + (3.6)(1.00)\}$$
$$= -5.30$$

The values of the $a_i, b_i, i = 1, 2, \ldots, 6$, are given in Table 2.4 and yield the analysis of variance of Table 2.5.

TABLE 2.4   Amplitudes of sines and cosines at different harmonics for temperature data

i	a_i	b_i
1	−5.30	−3.82
2	0.05	0.17
3	0.10	0.50
4	0.52	−0.52
5	0.09	−0.58
6	−0.30	

As would be expected, the major component of this temperature data has a period of twelve months, that is a frequency of 1/12 cycles per month.

TABLE 2.5 Analysis of variance table for temperature series

i	Frequency f_i	Period	Periodogram $I(f_i)$	Degrees of freedom	Mean square
1	1/12	12	254.96	2	127.48
2	1/6	6	0.19	2	0.10
3	1/4	4	1.56	2	0.78
4	1/3	3	3.22	2	1.61
5	5/12	12/5	2.09	2	1.05
6	1/2	2	1.08	1	1.08
	Total		263.10	11	23.92

2.2.3 The spectrum and spectral density function

For completeness, we add here a brief discussion of the spectrum and spectral density function. The use of these important tools is described more fully in [27]. We do not apply them to the analysis of time series in this book, and this section can be omitted on first reading.

The sample spectrum. The definition (2.2.5) of the periodogram assumes that the frequencies $f_i = i/N$ are harmonics of the fundamental frequency $1/N$. By way of introduction to the spectrum, we relax this assumption and allow the frequency f to vary continuously in the range 0–0.5 cycles. The definition (2.2.5) of the periodogram may be modified to

$$I(f) = \frac{2}{N}(a_f^2 + b_f^2), \qquad 0 \leqslant f \leqslant \tfrac{1}{2} \qquad (2.2.7)$$

and $I(f)$ is then referred to as the *sample spectrum* [27]. Like the periodogram, it can be used to detect and estimate the amplitude of a sinusoidal component of unknown frequency f buried in noise and is, indeed, a more appropriate tool for this purpose if it is known that the frequency f is not harmonically related to the length of the series. Moreover, it provides a starting point for the theory of spectral analysis, using a result given in Appendix A2.1. This result shows that the sample spectrum $I(f)$ and the estimate c_k of the auto-covariance function are linked by the important relation

$$I(f) = 2\left\{c_0 + 2\sum_{k=1}^{N-1} c_k \cos 2\pi f k\right\}, \qquad 0 \leqslant f \leqslant \tfrac{1}{2} \qquad (2.2.8)$$

That is, the sample spectrum is the Fourier cosine transform of the estimate of the autocovariance function.

The spectrum. The periodogram and sample spectrum are appropriate tools for analyzing time series made up of mixtures of sine and cosine waves, at *fixed* frequencies buried in noise. However, stationary time series of the

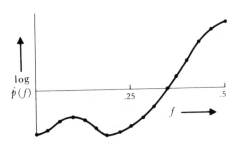

FIG. 2.8   Estimated power spectrum of batch data

2.2.4 Simple examples of autocorrelation and spectral density functions

For illustration, we show below equivalent representations of two simple stochastic processes by:
  (i)   their theoretical models,
  (ii)   their theoretical autocorrelation functions, and
  (iii)   their theoretical spectra.
Consider the two processes

$$z_t = 10 + a_t + a_{t-1} \qquad z_t = 10 + a_t - a_{t-1}$$

where $a_t, a_{t-1}, \ldots,$ are a sequence of uncorrelated, random Normal variables with mean zero and variance 1, that is, white noise. Using the definition (2.1.5)

$$\gamma_k = \operatorname{cov}[z_t, z_{t+k}] = E[(z_t - \mu)(z_{t+k} - \mu)]$$

where $E[z_t] = E[z_{t+k}] = \mu = 10$, the autocovariances of these two stochastic processes are

$$\gamma_k = \begin{cases} 2.0, k = 0 \\ 1.0, k = 1 \\ 0, k \geqslant 2 \end{cases} \qquad \gamma_k = \begin{cases} 2.0, k = 0 \\ -1.0, k = 1 \\ 0, k \geqslant 2 \end{cases}$$

Thus the theoretical autocorrelation functions are

$$\rho_k = \begin{cases} 0.5, k = 1 \\ 0.0, k \geqslant 2 \end{cases} \qquad \rho_k = \begin{cases} -0.5, k = 1 \\ 0.0, k \geqslant 2 \end{cases}$$

and using (2.2.13), the theoretical spectral density functions are

$$g(f) = 2(1 + \cos 2\pi f), \qquad g(f) = 2(1 - \cos 2\pi f)$$

The autocorrelation functions and spectral density functions are plotted in Figure 2.9, together with a sample time series from each process.

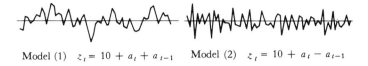

Model (1) $z_t = 10 + a_t + a_{t-1}$ Model (2) $z_t = 10 + a_t - a_{t-1}$

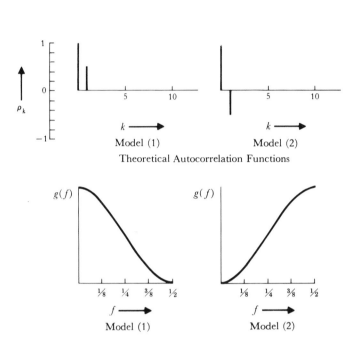

Theoretical Autocorrelation Functions

Theoretical Spectral Density Functions

FIG. 2.9 Two simple stochastic models with their corresponding theoretical autocorrelation functions and spectral density functions

(1) It should be noted that, for these two stationary processes, knowledge
of either the autocorrelation function or the spectral density function,
with the mean and variance of the process, is equivalent to knowledge
of the model (given the Normality assumption).

(2) It will be seen that the autocorrelation function reflects one aspect of
the behavior of the series. The comparatively smooth nature of the first
series is accounted for by the positive association between successive
values. The alternating tendency of the second series, in which positive
deviations usually follow negative ones, is accounted for by the negative
association between successive values.

(3) The spectral density throws light on a different, but equivalent aspect.
The predominance of low frequencies in the first series and high frequencies in the second is shown up by the spectra.

2.2.5 *Advantages and disadvantages of the autocorrelation and spectral density functions*

Because the autocorrelation function and the spectrum are transforms of each other, they are mathematically equivalent and therefore any discussion of their advantages and disadvantages turns not on mathematical questions, but on their representational value. Because, as we have seen, each sheds light on a different aspect of the data, they should be regarded not as rivals but, as allies. Each contributes something to an understanding of the stochastic process in question.

The obtaining of sample estimates of the autocorrelation function and of the spectrum are non-structural approaches, analogous to the representation of an empirical distribution function by a histogram. They are both ways of letting data from stationary series "speak for themselves" and provide a first step in the analysis of time series, just as a histogram can provide a first step in the distributional analysis of data, pointing the way to some parametric model on which subsequent analysis will be based.

Parametric time series models, such as those of Section 2.2.4, are not necessarily associated with a simple autocorrelation function or a simple spectrum. Working with either of these non-structural methods, we may be involved in the estimation of many lag correlations and many spectral ordinates, even when a parametric model containing only one or two parameters could represent the data. Each correlation and each spectral ordinate is a parameter to be estimated, so that these non-structural approaches might be very prodigal with parameters, when the approach via the model could be parsimonious. On the other hand, initially, we probably do not know what type of model may be appropriate and initial use of one or other of these non-structural approaches is necessary to *identify* the type of model which is needed. (In the same way that plotting a histogram helps to indicate which family of distributions may be appropriate). The choice between the spectrum and the autocorrelation function as a tool in model building depends upon the nature of the models which turn out to be practically useful. The models which we have found useful and which we consider in later chapters of this book are simply described in terms of the autocorrelation function and it is this tool which we employ for identification of series.

APPENDIX A2.1 LINK BETWEEN THE SAMPLE SPECTRUM AND AUTOCOVARIANCE FUNCTION ESTIMATE

Here we derive the result (2.2.8)

$$I(f) = 2\left\{c_0 + 2\sum_{k=1}^{N-1} c_k \cos 2\pi f k\right\}, \qquad 0 \leqslant f \leqslant \tfrac{1}{2}$$

which links the sample spectrum $I(f)$ and the estimate c_k of the auto-covariance function. Suppose that the least squares estimates a_f and b_f of the cosine and sine components, at frequency f, in a series are combined according to $d_f = a_f - ib_f$, where $i = \sqrt{-1}$, then

$$I(f) = \frac{N}{2}(a_f - ib_f)(a_f + ib_f)$$

$$= \frac{N}{2} d_f \, d_f^* \tag{A2.1.1}$$

where d_f^* is the complex conjugate of d_f. Then, using (2.2.3) and (2.2.4),

$$d_f = \frac{2}{N} \sum_{t=1}^{N} z_t(\cos 2\pi ft - i \sin 2\pi ft)$$

$$= \frac{2}{N} \sum_{t=1}^{N} z_t e^{-i2\pi ft}$$

$$= \frac{2}{N} \sum_{t=1}^{N} (z_t - \bar{z})e^{-i2\pi ft} \tag{A2.1.2}$$

Substituting (A.2.1.2) in (A2.1.1),

$$I(f) = \frac{2}{N} \sum_{t=1}^{N} \sum_{t'=1}^{N} (z_t - \bar{z})(z_{t'} - \bar{z})e^{-i2\pi f(t-t')} \tag{A2.1.3}$$

Since

$$c_k = \frac{1}{N} \sum_{t=1}^{N-k} (z_t - \bar{z})(z_{t+k} - \bar{z})$$

the transformation $k = t - t'$ transforms (A2.1.3) into

$$I(f) = 2 \sum_{k=-N+1}^{N-1} c_k e^{-i2\pi fk}$$

$$= 2\left\{c_0 + 2 \sum_{k=1}^{N-1} c_k \cos 2\pi fk\right\}, \qquad 0 \leqslant f \leqslant \tfrac{1}{2}$$

which is the required result.

3

Linear Stationary Models

A general linear stochastic model is described, which supposes a time series to be generated by a linear aggregation of random shocks. For practical representation, it is desirable to employ models which use parameters parsimoniously. Parsimony may often be achieved by representation of the linear process in terms of a small number of autoregressive and moving average terms. The properties of the resulting autoregressive-moving average (ARMA) models are discussed in preparation for their use in model building.

3.1. THE GENERAL LINEAR PROCESS

3.1.1 Two equivalent forms for the linear process

In Section 1.2.1 we discussed the representation of a stochastic process as the ouput from a linear filter, whose input is white noise a_t, that is

$$\tilde{z}_t = a_t + \psi_1 a_{t-1} + \psi_2 a_{t-2} + \cdots$$

$$= a_t + \sum_{j=1}^{\infty} \psi_j a_{t-j} \qquad (3.1.1)$$

where $\tilde{z}_t = z_t - \mu$ is the deviation of the process from some origin, or from its mean, if the process is stationary. The *general linear process* (3.1.1) allows us to represent $\tilde{z}_t$ as a weighted sum of present and past values of the "white noise" process a_t. Important references in the development of linear stochastic models are: [24], [28], [29], [32], [44], [92], [97], [98], [99], [100], [102] and [103]. The white noise process a_t may be regarded as a *series of shocks* which drive the system. It consists of a sequence of uncorrelated random variables with mean zero and constant variance, that is

$$E[a_t] = 0 \qquad \text{var}[a_t] = \sigma_a^2$$

Since the random variables a_t are uncorrelated, it follows that their auto-covariance function is

$$\gamma_k = E[a_t a_{t+k}] = \begin{cases} \sigma_a^2 & k = 0 \\ 0 & k \neq 0 \end{cases} \qquad (3.1.2)$$

46

Thus, the autocorrelation function of white noise has the particularly simple form

$$\rho_k = \begin{cases} 1 & k = 0 \\ 0 & k \neq 0 \end{cases} \tag{3.1.3}$$

The model (3.1.1) implies that, under suitable conditions, $\tilde{z}_t$ is a weighted sum of past values of the $\tilde{z}$'s, plus an added shock a_t, that is

$$\tilde{z}_t = \pi_1 \tilde{z}_{t-1} + \pi_2 \tilde{z}_{t-2} + \cdots + a_t$$

$$= \sum_{j=1}^{\infty} \pi_j \tilde{z}_{t-j} + a_t \tag{3.1.4}$$

The alternative form (3.1.4) may be thought of as one where the current deviation $\tilde{z}_t$, from the level μ, is "regressed" on past deviations $\tilde{z}_{t-1}, \tilde{z}_{t-2}, \cdots$ of the process.

Relationships between the ψ weights and π weights. The relationships between the ψ weights and π weights may be obtained using the previously defined *backward shift operator*

$$B z_t = z_{t-1} \qquad B^j z_t = z_{t-j}$$

Later we shall also need to use the forward shift operator $F = B^{-1}$, such that

$$F z_t = z_{t+1} \qquad F^j z_t = z_{t+j}$$

As an example of the use of the operator B, consider the model

$$\tilde{z}_t = a_t - \theta a_{t-1} = (1 - \theta B) a_t$$

in which $\psi_1 = -\theta$, $\psi_j = 0$ for $j > 1$. Expressing a_t in terms of the $\tilde{z}$'s, we obtain

$$(1 - \theta B)^{-1} \tilde{z}_t = a_t$$

Hence, for $|\theta| < 1$,

$$(1 + \theta B + \theta^2 B^2 + \theta^3 B^3 + \cdots) \tilde{z}_t = a_t$$

and the deviation $\tilde{z}_t$ expressed in terms of previous deviations, as in (3.1.4), is

$$\tilde{z}_t = -\theta \tilde{z}_{t-1} - \theta^2 \tilde{z}_{t-2} - \theta^3 \tilde{z}_{t-3} - \cdots + a_t$$

so that for this model $\pi_j = -\theta^j$.

In general, (3.1.1) may be written

$$\tilde{z}_t = \left(1 + \sum_{j=1}^{\infty} \psi_j B^j\right) a_t$$

or

$$\tilde{z}_t = \psi(B) a_t \tag{3.1.5}$$

where

$$\psi(B) = 1 + \sum_{j=1}^{\infty} \psi_j B^j = \sum_{j=0}^{\infty} \psi_j B^j$$

with $\psi_0 = 1$. As has been mentioned in Section 1.2.1, $\psi(B)$ is called the *transfer function* of the linear filter relating $\tilde{z}_t$ to a_t. It can also be regarded as the *generating function* of the ψ weights, with B now treated simply as a dummy variable whose jth power is the coefficient of ψ_j.

Similarly, (3.1.4) may be written

$$\left(1 - \sum_{j=1}^{\infty} \pi_j B^j\right) \tilde{z}_t = a_t$$

or

$$\pi(B)\tilde{z}_t = a_t \qquad (3.1.6)$$

Thus

$$\pi(B) = 1 - \sum_{j=1}^{\infty} \pi_j B^j$$

is the generating function of the π weights. After operating on both sides of (3.1.6) by $\psi(B)$, we obtain

$$\psi(B)\pi(B)\tilde{z}_t = \psi(B)a_t = \tilde{z}_t$$

Hence

$$\psi(B)\pi(B) = 1$$

that is

$$\pi(B) = \psi^{-1}(B) \qquad (3.1.7)$$

The relationship (3.1.7) may be used to derive the π weights, knowing the ψ weights, and vice versa.

3.1.2 Autocovariance generating function of a linear process

A basic data analysis tool for identifying models in Chapter 6 will be the autocorrelation function. Therefore, it is important to know the autocorrelation function of a linear process. It is shown in Appendix A3.1 that the autocovariance function of the linear process (3.1.1) is given by

$$\gamma_k = \sigma_a^2 \sum_{j=0}^{\infty} \psi_j \psi_{j+k} \qquad (3.1.8)$$

In particular, by setting $k = 0$ in (3.1.8), we find that its variance is

$$\gamma_0 = \sigma_z^2 = \sigma_a^2 \sum_{j=0}^{\infty} \psi_j^2 \tag{3.1.9}$$

It follows, that if the process is to have a finite variance, the weights ψ_j must decrease fast enough for the series on the right of (3.1.9) to converge.

A more convenient way of obtaining the autocovariances of a linear process is often via the *autocovariance generating function*

$$\gamma(B) = \sum_{k=-\infty}^{\infty} \gamma_k B^k \tag{3.1.10}$$

in which it is noted that γ_0, the variance of the process, is the coefficient of $B^0 = 1$, while γ_k, the autocovariance of lag k, is the coefficient of both B^j and of $B^{-j} = F^j$. It is shown in Appendix A3.1 that

$$\gamma(B) = \sigma_a^2 \psi(B) \psi(B^{-1}) = \sigma_a^2 \psi(B) \psi(F) \tag{3.1.11}$$

For example, suppose

$$\tilde{z}_t = a_t - \theta a_{t-1} = (1 - \theta B) a_t$$

so that $\psi(B) = 1 - \theta B$. Then, substituting in (3.1.11),

$$\gamma(B) = \sigma_a^2 (1 - \theta B)(1 - \theta B^{-1})$$
$$= \sigma_a^2 \{ -\theta B^{-1} + (1 + \theta^2) - \theta B \}$$

Comparing with (3.1.10), the autocovariances are

$$\gamma_0 = (1 + \theta^2) \sigma_a^2$$
$$\gamma_1 = -\theta \sigma_a^2$$
$$\gamma_k = 0 \qquad k \geqslant 2$$

In the development that follows, when treated as a dummy variable in a generating function, B will be supposed capable of taking complex values. In particular, it will often be necessary to consider the different situations occurring when $|B| < 1$, $|B| = 1$, or $|B| > 1$, that is, when the complex number B lies inside, on or outside the unit circle.

3.1.3 Stationarity and invertibility conditions for a linear process

Stationarity.   The convergence of the series (3.1.9) ensures that the process has a finite variance. Also, we have seen in Section 2.1.3, that the auto-covariances and autocorrelations must satisfy a set of conditions to ensure stationarity. For a linear process these conditions can be embodied in the single condition that the series $\psi(B)$, which is the generating function of the ψ weights, must converge for $|B| \leqslant 1$. That is, on or within the unit circle. This result is discussed in Appendix A3.1.

Spectrum of a linear stationary process. It is shown in Appendix A3.1 that if we substitute $B = e^{-i2\pi f}$, where $i = \sqrt{-1}$, in the autocovariance generating function (3.1.11), we obtain one half of the power spectrum. Thus the spectrum of a linear process is

$$p(f) = 2\sigma_a^2 \psi(e^{-i2\pi f})\psi(e^{i2\pi f})$$
$$= 2\sigma_a^2 |\psi(e^{-i2\pi f})|^2 \qquad 0 \leqslant f \leqslant \tfrac{1}{2} \qquad (3.1.12)$$

In fact, (3.1.12) is the well known expression [27], which relates the spectrum $p(f)$ of the output from a linear system, to the uniform spectrum $2\sigma_a^2$ of a white noise input by multiplying by the squared gain $G^2(f) = |\psi(e^{-i2\pi f})|^2$ of the system.

Invertibility. We have seen above, that the ψ weights of a linear process must satisfy the condition that $\psi(B)$ converges on or within the unit circle, if the process is to be stationary. We now consider a restriction applied to the π weights to ensure what is called "invertibility." The invertibility condition is independent of the stationarity condition and is applicable also to the non-stationary linear models, which we introduce in Chapter 4.

To illustrate the basic idea of invertibility, consider again the model

$$\tilde{z}_t = (1 - \theta B)a_t \qquad (3.1.13)$$

Expressing the a's in terms of the $\tilde{z}$'s, (3.1.13) becomes

$$a_t = (1 - \theta B)^{-1}\tilde{z}_t = (1 + \theta B + \theta^2 B^2 + \cdots + \theta^k B^k)(1 - \theta^{k+1}B^{k+1})^{-1}\tilde{z}_t$$

that is

$$\tilde{z}_t = -\theta\tilde{z}_{t-1} - \theta^2\tilde{z}_{t-2} - \cdots - \theta^k z_{t-k} + a_t - \theta^{k+1}a_{t-k-1} \qquad (3.1.14)$$

and, if $|\theta| < 1$, on letting k tend to infinity, we obtain the infinite series

$$\tilde{z}_t = -\theta\tilde{z}_{t-1} - \theta^2\tilde{z}_{t-2} - \cdots + a_t \qquad (3.1.15)$$

and the π weights of the model in the form of (3.1.4), are $\pi_j = -\theta^j$. Whatever the value of θ, (3.1.13) defines a perfectly proper stationary process. However, if $|\theta| \geqslant 1$, the current deviation $\tilde{z}_t$ in (3.1.14) depends on $\tilde{z}_{t-1}, \tilde{z}_{t-2}, \ldots, \tilde{z}_{t-k}$, with weights which increase as k increases. We avoid this situation by requiring that $|\theta| < 1$. We shall then say that the series is *invertible*. We see that this condition is satisfied if the series

$$\pi(B) = (1 - \theta B)^{-1} = \sum_{j=0}^{\infty} \theta^j B^j$$

converges for all $|B| \leqslant 1$, that is, on or within the unit circle.

In Chapter 6, where we consider questions of uniqueness of these models, we shall see that a convergent expansion for a_t is possible when $|\theta| \geqslant 1$, but only in terms of $z_t, z_{t+1}, z_{t+2}, \ldots$ (that is in terms of present and *future* values of the process). The requirement of invertibility is needed if we are interested in associating present events with *past* happenings in a sensible manner.

In general, the linear process

$$\pi(B)\tilde{z}_t = a_t$$

is invertible if the weights π_j are such that the series $\pi(B)$ converges on, or within the unit circle.

To sum up, a linear process is *stationary* if $\psi(B)$ converges on, or within the unit circle and is *invertible* if $\pi(B)$ converges on, or within the unit circle.

3.1.4 Autoregressive and moving average processes

The representations (3.1.1) and (3.1.4) of the general linear process would not be very useful in practice, if they contained an infinite number of parameters ψ_j and π_j. We now consider how to introduce parsimony and yet retain models which are representationally useful.

Autoregressive processes. Consider the special case of (3.1.4), in which only the first p of the weights are nonzero. This model may be written

$$\tilde{z}_t = \phi_1 \tilde{z}_{t-1} + \phi_2 \tilde{z}_{t-2} + \cdots + \phi_p \tilde{z}_{t-p} + a_t \qquad (3.1.16)$$

where we now use the symbols $\phi_1, \phi_2, \ldots, \phi_p$ for the *finite* set of weight parameters. The process defined by (3.1.16) is called an *autoregressive* process of order p, or more succinctly, an AR(p) process. In particular, the autoregressive processes of first order ($p = 1$), and of second order ($p = 2$),

$$\tilde{z}_t = \phi_1 \tilde{z}_{t-1} + a_t$$

$$\tilde{z}_t = \phi_1 \tilde{z}_{t-1} + \phi_2 \tilde{z}_{t-2} + a_t$$

are of considerable practical importance.

Now we can write (3.1.16) in the equivalent form

$$(1 - \phi_1 B - \phi_2 B^2 - \cdots - \phi_p B^p)\tilde{z}_t = a_t$$

or

$$\phi(B)\tilde{z}_t = a_t \qquad (3.1.17)$$

Since (3.1.17) implies

$$\tilde{z}_t = \frac{1}{\phi(B)} a_t = \phi^{-1}(B)a_t$$

the autoregressive process can be thought of as the output $\tilde{z}_t$ from a linear filter with transfer function $\phi^{-1}(B)$, when the input is white noise a_t.

Moving average processes. Consider the special case of (3.1.1) when only the first q of the ψ weights are non-zero. The process

$$\tilde{z}_t = a_t - \theta_1 a_{t-1} - \theta_2 a_{t-2} - \cdots - \theta_q a_{t-q} \qquad (3.1.18)$$

where we now use the symbols $-\theta_1, -\theta_2, \ldots, -\theta_q$ for the *finite* set of weight parameters. The process defined by (3.1.18) is called a *moving average* process* of order q, which we sometimes abbreviate to MA(q). In particular, the processes of first order ($q = 1$), and second order ($q = 2$),

$$\tilde{z}_t = a_t - \theta_1 a_{t-1}$$

$$\tilde{z}_t = a_t - \theta_1 a_{t-1} - \theta_2 a_{t-2}$$

are particularly important in practice.

We can also write (3.1.18) in the equivalent form

$$\tilde{z}_t = (1 - \theta_1 B - \theta_2 B^2 - \cdots - \theta_q B^q)a_t$$

or

$$\tilde{z}_t = \theta(B)a_t \qquad (3.1.19)$$

Hence, the moving average process can be thought of as the output $\tilde{z}_t$ from a linear filter with transfer function $\theta(B)$, when the input is white noise a_t.

Mixed autoregressive—moving average processes. We have seen in Section 3.1.1 that the *finite* moving average process

$$\tilde{z}_t = a_t - \theta_1 a_{t-1} = (1 - \theta_1 B)a_t \qquad |\theta_1| < 1$$

can be written as an *infinite* autoregressive process

$$\tilde{z}_t = -\theta_1 \tilde{z}_{t-1} - \theta_1^2 \tilde{z}_{t-2} - \cdots + a_t$$

Hence, if the process were really MA(1), we would obtain a nonparsimonious representation in terms of an autoregressive model. Conversely, an AR(1) could not be parsimoniously represented using a moving average process. In practice, to obtain a parsimonious parameterization, it will sometimes be necessary to include both autoregressive and moving average terms in the model. Thus

$$\tilde{z}_t = \phi_1 \tilde{z}_{t-1} + \cdots + \phi_p \tilde{z}_{t-p} + a_t - \theta_1 a_{t-1} - \cdots - \theta_q a_{t-q}$$

or

$$\phi(B)\tilde{z}_t = \theta(B)a_t \qquad (3.1.20)$$

---

* As we remarked in Chapter 1, the term "moving average" is somewhat misleading, since the weights do not sum to unity. However, this nomenclature is now well established and we shall use it.

is called the *mixed autoregressive–moving average* process of order (p, q), which we sometimes abbreviate to ARMA (p, q). For example, the ARMA $(1, 1)$ process is

$$\tilde{z}_t - \phi_1 \tilde{z}_{t-1} = a_t - \theta_1 a_{t-1}$$

Since (3.1.20) may be written

$$\tilde{z}_t = \phi^{-1}(B)\theta(B)a_t$$

$$= \frac{\theta(B)}{\phi(B)} a_t = \frac{1 - \theta_1 B - \cdots - \theta_q B^q}{1 - \phi_1 B - \cdots - \phi_p B^p} a_t$$

the mixed autoregressive–moving average process can be thought of as the output $\tilde{z}_t$ from a linear filter, whose transfer function is the ratio of two polynominals $\theta(B)$ and $\phi(B)$, when the input is white noise a_t.

In the following sections we discuss important characteristics of auto-regressive, moving average, and mixed models. We study their variances, autocorrelation functions, spectra, and the stationarity and invertibility conditions which must be imposed upon their parameters.

3.2. AUTOREGRESSIVE PROCESSES

3.2.1 Stationarity conditions for autoregressive processes

The set of adjustable parameters $\phi_1, \phi_2, \ldots, \phi_p$ of an AR(p) process

$$\tilde{z}_t = \phi_1 \tilde{z}_{t-1} + \cdots + \phi_p \tilde{z}_{t-p} + a_t$$

or

$$(1 - \phi_1 B - \cdots - \phi_p B^p)\tilde{z}_t = \phi(B)\tilde{z}_t = a_t$$

must satisfy certain conditions for the process to be stationary.

For illustration, the first-order autoregressive process

$$(1 - \phi_1 B)\tilde{z}_t = a_t$$

may be written

$$\tilde{z}_t = (1 - \phi_1 B)^{-1} a_t = \sum_{j=0}^{\infty} \phi_1^j a_{t-j}$$

Hence

$$\psi(B) = (1 - \phi_1 B)^{-1} = \sum_{j=0}^{\infty} \phi_1^j B^j \qquad (3.2.1)$$

We have seen in Section 3.1.3 that for stationarity, $\psi(B)$ must converge for $|B| \leqslant 1$. From (3.2.1) we see that this implies that the parameter ϕ_1, of an AR(1) process, must satisfy the condition $|\phi_1| < 1$ to ensure stationarity.

Since the root of $1 - \phi_1 B = 0$ is $B = \phi_1^{-1}$, this condition is equivalent to saying that the root of $1 - \phi_1 B = 0$ must lie *outside* the unit circle.

For the general AR(p) process $\tilde{z}_t = \phi^{-1}(B)a_t$, we obtain,

$$\phi(B) = (1 - G_1 B)(1 - G_2 B) \cdots (1 - G_p B)$$

and expanding in partial fractions,

$$\tilde{z}_t = \phi^{-1}(B)a_t = \sum_{i=1}^{p} \frac{K_i}{(1 - G_i B)} a_t$$

Hence, if $\psi(B) = \phi^{-1}(B)$ is to be a convergent series for $|B| \leq 1$, then we must have $|G_i| < 1$, where $i = 1, 2, \ldots, p$. Equivalently, the roots of $\phi(B) = 0$ must lie *outside* the unit circle. The roots of the equation $\phi(B) = 0$ may be referred to as the zeroes of the polynomial $\phi(B)$. Thus, the stationarity condition may be expressed by saying that the zeroes of $\phi(B)$ must lie *outside* the unit circle. A similar argument may be applied when the zeroes of $\phi(B)$ are not all distinct. The equation $\phi(B) = 0$ is called the *characteristic equation* for the process.

Since the series

$$\pi(B) = \phi(B) = 1 - \phi_1 B - \phi_2 B^2 - \cdots - \phi_p B^p$$

is finite, no restrictions are required on the parameters of an autoregressive process to ensure invertibility.

3.2.2 Autocorrelation function and spectrum of autoregressive processes

Autocorrelation function. An important recurrence relation for the auto-correlation function of a stationary autoregressive process is found by multiplying throughout in

$$\tilde{z}_t = \phi_1 \tilde{z}_{t-1} + \phi_2 \tilde{z}_{t-2} + \cdots + \phi_p \tilde{z}_{t-p} + a_t$$

by $\tilde{z}_{t-k}$, to obtain

$$\tilde{z}_{t-k}\tilde{z}_t = \phi_1 \tilde{z}_{t-k}\tilde{z}_{t-1} + \phi_2 \tilde{z}_{t-k}\tilde{z}_{t-2} + \cdots + \phi_p \tilde{z}_{t-k}\tilde{z}_{t-p} + \tilde{z}_{t-k}a_t \quad (3.2.2)$$

On taking expected values in (3.2.2), we obtain the difference equation

$$\gamma_k = \phi_1 \gamma_{k-1} + \phi_2 \gamma_{k-2} + \cdots + \phi_p \gamma_{k-p} \quad k > 0 \quad (3.2.3)$$

Note that the expectation $E[\tilde{z}_{t-k}a_t]$ vanishes when $k > 0$, since $\tilde{z}_{t-k}$ can only involve the shocks a_j up to time $t - k$, which are uncorrelated with a_t. On dividing throughout in (3.2.3) by γ_0, it is seen that the autocorrelation function satisfies the same form of difference equation

$$\rho_k = \phi_1 \rho_{k-1} + \phi_2 \rho_{k-2} + \cdots + \phi_p \rho_{k-p} \quad k > 0 \quad (3.2.4)$$

We note that this is analogous to the difference equation satisfied by the process $\tilde{z}_t$ itself.

Now suppose that (3.2.4) is written

$$\phi(B)\rho_k = 0$$

where $\phi(B) = 1 - \phi_1 B - \cdots - \phi_p B^p$ and B now operates on k and not t. Then, writing

$$\phi(B) = \prod_{i=1}^{p} (1 - G_i B)$$

the general solution of (3.2.4) is

$$\rho_k = A_1 G_1^k + A_2 G_2^k + \cdots + A_p G_p^k \qquad (3.2.5)$$

where $G_1^{-1}, G_2^{-1}, \ldots, G_p^{-1}$ are the roots of the *characteristic equation*

$$\phi(B) = 1 - \phi_1 B - \phi_2 B^2 - \cdots - \phi_p B^p = 0$$

For stationarity we require that $|G_i| < 1$. Thus, two situations can arise in practice if we assume that the roots G_i are distinct.

(1) A root G_i is real, in which case a term $A_i G_i^k$ in (3.2.5) geometrically decays to zero as k increases. We shall frequently refer to this as a damped exponential.

(2) A pair of roots G_i, G_j is complex, in which case they contribute a term

$$d^k \sin (2\pi f k + F)$$

to the autocorrelation function (3.2.5), which follows a damped sine wave.

In general, the autocorrelation function of a stationary autoregressive process will consist of a mixture of damped exponentials and damped sine waves.

Autoregressive parameters in terms of the autocorrelations—Yule–Walker equations. If we substitute $k = 1, 2, \ldots, p$ in (3.2.4), we obtain a set of linear equations for $\phi_1, \phi_2, \ldots, \phi_p$ in terms of $\rho_1, \rho_2, \ldots, \rho_p$, that is

$$
\begin{aligned}
\rho_1 &= \phi_1 && + \phi_2 \rho_1 && + \cdots + \phi_p \rho_{p-1} \\
\rho_2 &= \phi_1 \rho_1 && + \phi_2 && + \cdots + \phi_p \rho_{p-2} \\
&\;\;\vdots && \;\;\vdots && \quad\;\; \vdots \\
\rho_p &= \phi_1 \rho_{p-1} && + \phi_2 \rho_{p-2} && + \cdots + \phi_p
\end{aligned}
\qquad (3.2.6)
$$

These are usually called the *Yule–Walker* equations [24], [32]. We obtain *Yule–Walker estimates* of the parameters, by replacing the theoretical autocorrelations ρ_k by the estimated autocorrelations r_k. Note that if we write

$$\boldsymbol{\phi} = \begin{bmatrix} \phi_1 \\ \phi_2 \\ \vdots \\ \phi_p \end{bmatrix} \quad \boldsymbol{\rho}_p = \begin{bmatrix} \rho_1 \\ \rho_2 \\ \vdots \\ \rho_p \end{bmatrix} \quad \mathbf{P}_p = \begin{bmatrix} 1 & \rho_1 & \rho_2 & \cdots & \rho_{p-1} \\ \rho_1 & 1 & \rho_1 & \cdots & \rho_{p-2} \\ \vdots & \vdots & \vdots & \cdots & \vdots \\ \rho_{p-1} & \rho_{p-2} & \rho_{p-3} & \cdots & 1 \end{bmatrix}$$

the solution of (3.2.6) for the parameters $\boldsymbol{\phi}$ in terms of the autocorrelations may be written

$$\boldsymbol{\phi} = \mathbf{P}_p^{-1}\boldsymbol{\rho}_p \qquad (3.2.7)$$

Variance. When $k = 0$, the contribution from the term $E[z_{t-k}a_t]$, on taking expectations in (3.2.2), is $E[a_t^2] = \sigma_a^2$, since the only part of z_t which will be correlated with a_t is the most recent shock a_t. Hence, when $k = 0$

$$\gamma_0 = \phi_1\gamma_{-1} + \phi_2\gamma_{-2} + \cdots + \phi_p\gamma_{-p} + \sigma_a^2$$

On dividing throughout by $\gamma_0 = \sigma_z^2$, and substituting $\gamma_k = \gamma_{-k}$, the variance σ_z^2 may be written

$$\sigma_z^2 = \frac{\sigma_a^2}{1 - \rho_1\phi_1 - \rho_2\phi_2 - \cdots - \rho_p\phi_p} \qquad (3.2.8)$$

Spectrum. For the AR(p) process,

$$\psi(B) = \phi^{-1}(B)$$

and

$$\phi(B) = 1 - \phi_1 B - \phi_2 B^2 - \cdots - \phi_p B^p$$

Therefore, using (3.1.12), the spectrum of an autoregressive process is

$$p(f) = \frac{2\sigma_a^2}{|1 - \phi_1 e^{-i2\pi f} - \phi_2 e^{-i4\pi f} - \cdots - \phi_p e^{-i2\pi pf}|^2} \quad 0 \leqslant f \leqslant \tfrac{1}{2} \qquad (3.2.9)$$

We now discuss two particularly important autoregressive processes, namely those of first and second order.

3.2.3 The first-order autoregressive (Markov) process

The first-order autoregressive process is

$$\tilde{z}_t = \phi_1 \tilde{z}_{t-1} + a_t$$

$$= a_t + \phi_1 a_{t-1} + \phi_1^2 a_{t-2} + \cdots \qquad (3.2.10)$$

where it has been shown in Section 3.2.1 that ϕ_1 must satisfy the condition $-1 < \phi_1 < 1$ for the process to be stationary.

Autocorrelation function. Using (3.2.4), the autocorrelation function satisfies the first-order difference equation

$$\rho_k = \phi_1 \rho_{k-1} \qquad k > 0 \tag{3.2.11}$$

which, with $\rho_0 = 1$, has the solution

$$\rho_k = \phi_1^k \qquad k \geqslant 0 \tag{3.2.12}$$

As shown in Figure 3.1, the autocorrelation function decays exponentially to zero when ϕ_1 is positive, but decays exponentially to zero and oscillates

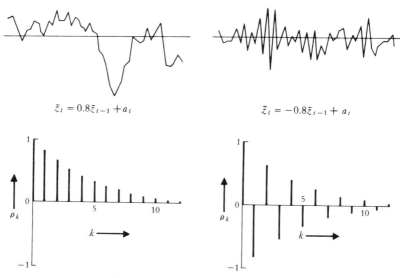

$$\tilde{z}_t = 0.8\tilde{z}_{t-1} + a_t \qquad\qquad\qquad \tilde{z}_t = -0.8\tilde{z}_{t-1} + a_t$$

Theoretical Autocorrelation Functions

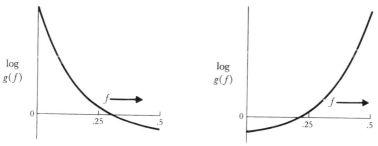

Theoretical Log Spectral Density Functions

Fig. 3.1   Realizations from first-order autoregressive processes and their corresponding theoretical autocorrelation functions and spectral density functions

in sign when ϕ_1 is negative. In particular, it will be noted that

$$\rho_1 = \phi_1 \qquad (3.2.13)$$

Variance. Using (3.2.8), the variance of the process is

$$\sigma_z^2 = \frac{\sigma_a^2}{1 - \rho_1\phi_1}$$

$$= \frac{\sigma_a^2}{1 - \phi_1^2} \qquad (3.2.14)$$

on substituting $\rho_1 = \phi_1$.

Spectrum. Finally, using (3.2.9), the spectrum is

$$p(f) = \frac{2\sigma_a^2}{|1 - \phi_1 e^{-i2\pi f}|^2}$$

$$= \frac{2\sigma_a^2}{1 + \phi_1^2 - 2\phi_1 \cos 2\pi f} \qquad 0 \le f \le \tfrac{1}{2} \qquad (3.2.15)$$

Figure 3.1 shows realizations from processes with $\phi_1 = 0.8$, $\phi_1 = -0.8$, and the corresponding theoretical autocorrelation functions and spectra. Thus, when the parameter has the large positive value $\phi_1 = 0.8$, neighboring values in the series are similar and the series exhibits marked trends. This is reflected in the autocorrelation function, which slowly decays exponentially to zero, and in the spectrum which is dominated by low frequencies. When, on the other hand, the parameter has the large negative value $\phi_1 = -0.8$, the series tends to oscillate rapidly and this is reflected in the autocorrelation function, which alternates in sign as it decays to zero, and in the spectrum which is dominated by high frequencies.

3.2.4 The second-order autoregressive process

Stationarity condition. The second-order autoregressive process may be written

$$\tilde{z}_t = \phi_1 \tilde{z}_{t-1} + \phi_2 \tilde{z}_{t-2} + a_t \qquad (3.2.16)$$

For stationarity, the roots of

$$\phi(B) = 1 - \phi_1 B - \phi_2 B^2 = 0 \qquad (3.2.17)$$

must lie outside the unit circle, which implies that the parameters ϕ_1 and ϕ_2 must lie in the triangular region

$$\phi_2 + \phi_1 < 1$$

$$\phi_2 - \phi_1 < 1 \qquad (3.2.18)$$

$$-1 < \phi_2 < 1$$

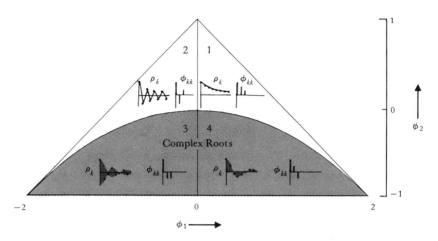

FIG. 3.2   Typical autocorrelation and partial autocorrelation functions ρ_k and ϕ_{kk} for various stationary AR(2) models

shown in Figure 3.2.

Autocorrelation function.   Using (3.2.4), the autocorrelation function satisfies the second-order difference equation

$$\rho_k = \phi_1 \rho_{k-1} + \phi_2 \rho_{k-2} \qquad k > 0 \qquad (3.2.19)$$

with starting values $\rho_0 = 1$ and $\rho_1 = \phi_1/(1 - \phi_2)$. From (3.2.5), the general solution of the difference equation (3.2.19) is

$$\rho_k = A_1 G_1^k + A_2 G_2^k$$
$$= \frac{G_1(1 - G_2^2)G_1^k - G_2(1 - G_1^2)G_2^k}{(G_1 - G_2)(1 + G_1 G_2)} \qquad (3.2.20)$$

where G_1^{-1} and G_2^{-1} are the roots of the characteristic equation (3.2.17). When the roots are real, the autocorrelation function consists of a mixture of damped exponentials. This occurs when $\phi_1^2 + 4\phi_2 \geqslant 0$ and corresponds to regions 1 and 2, which lie above the parabolic boundary in Figure 3.2 (taken from Stralkowski [33]). Specifically, in region 1, the autocorrelation function remains positive as it damps out, corresponding to a positive dominant root in (3.2.20). In region 2, the autocorrelation function alternates in sign as it damps out, corresponding to a negative dominant root.

If the roots G_1 and G_2 are complex ($\phi_1^2 + 4\phi_2 < 0$), a second-order autoregressive process displays *pseudo periodic behavior.* This behavior is reflected in the autocorrelation function, for on substituting $G_1 = de^{i2\pi f_0}$ and $G_2 = de^{-i2\pi f_0}$ in (3.2.20), we obtain

$$\rho_k = \frac{\{\text{sgn}\,(\phi_1)\}^k \, d^k \sin\,(2\pi f_0 k + F)}{\sin F} \qquad (3.2.21)$$

where sgn (ϕ_1) is $+1$ if ϕ_1 is positive and -1 if ϕ_1 is negative. In either case, we refer to (3.2.21) as a *damped sine wave* with damping factor d, *frequency* f_0, and *phase F*. These factors are related to the process parameters as follows:

$$d = \sqrt{-\phi_2} \qquad (3.2.22)$$

where the positive square root is taken,

$$\cos 2\pi f_0 = \frac{|\phi_1|}{2\sqrt{-\phi_2}} \qquad (3.2.23)$$

$$\tan F = \frac{1 + d^2}{1 - d^2} \tan 2\pi f_0 \qquad (3.2.24)$$

Again referring to Figure 3.2, the autocorrelation function is a damped sine wave in regions 3 and 4; the phase angle F being less than $90°$ in region 4 and lying between $90°$ and $180°$ in region 3. This means that the autocorrelation function starts with a positive value throughout region 4, but always switches sign from lag zero to lag 1 in region 3.

Yule–Walker equations. Substituting $p = 2$ in (3.2.6), the Yule–Walker equations are

$$\rho_1 = \phi_1 + \phi_2\rho_1$$
$$\rho_2 = \phi_1\rho_1 + \phi_2 \qquad (3.2.25)$$

which, when solved for ϕ_1 and ϕ_2, give

$$\phi_1 = \frac{\rho_1(1 - \rho_2)}{1 - \rho_1^2}$$

$$\phi_2 = \frac{\rho_2 - \rho_1^2}{1 - \rho_1^2} \qquad (3.2.26)$$

Chart B in the collection of Tables and Charts at the end of the volume, allows values of ϕ_1 and ϕ_2 to be read off for any given values of ρ_1 and ρ_2. The chart is used in Chapters 6 and 7 to obtain estimates of the ϕ's from values of the estimated autocorrelations r_1 and r_2.

Equations (3.2.25) may also be solved to express ρ_1 and ρ_2 in terms of ϕ_1 and ϕ_2, to give

$$\rho_1 = \frac{\phi_1}{1 - \phi_2}$$

$$\rho_2 = \phi_2 + \frac{\phi_1^2}{1 - \phi_2} \qquad (3.2.27)$$

which explains the starting values for (3.2.19) quoted above. The forms (3.2.20) and (3.2.21) for the autocorrelation function are useful for explaining the different types which may arise in practice. However, for computing the autocorrelations of an AR(2) process, with given values of ϕ_1 and ϕ_2, it is simplest to make direct use of the difference equation (3.2.19).

Using the stationarity conditions (3.2.18) and the expressions (3.2.27) for ρ_1 and ρ_2, it is found that the admissible values of ρ_1 and ρ_2, for a stationary AR(2) process, must lie in the region

$$-1 < \rho_1 < 1$$

$$-1 < \rho_2 < 1$$

$$\rho_1^2 < \tfrac{1}{2}(\rho_2 + 1)$$

Figure 3.3(a) shows the admissible region for the parameters ϕ_1 and ϕ_2, and Figure 3.3(b) shows the corresponding admissible region for ρ_1 and ρ_2.

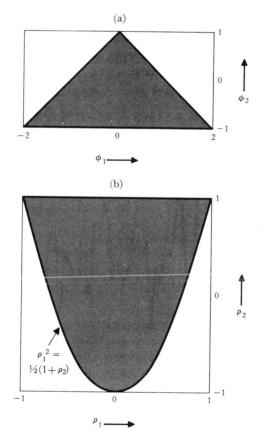

FIG. 3.3   Admissible regions for (a) ϕ_1, ϕ_2 and (b) ρ_1, ρ_2, for a stationary AR(2) process

Variance. From (3.2.8), the variance of the process is

$$\sigma_z^2 = \frac{\sigma_a^2}{1 - \rho_1 \phi_1 - \rho_2 \phi_2}$$

$$= \left(\frac{1 - \phi_2}{1 + \phi_2}\right) \frac{\sigma_a^2}{\{(1 - \phi_2)^2 - \phi_1^2\}} \qquad (3.2.28)$$

Spectrum. From (3.2.9), the spectrum is

$$p(f) = \frac{2\sigma_a^2}{|1 - \phi_1 e^{-i2\pi f} - \phi_2 e^{-i4\pi f}|^2}$$

$$= \frac{2\sigma_a^2}{\{1 + \phi_1^2 + \phi_2^2 - 2\phi_1(1 - \phi_2)\cos 2\pi f - 2\phi_2 \cos 4\pi f\}}$$

$$0 \leqslant f \leqslant \tfrac{1}{2} \quad (3.2.29)$$

The spectrum also reflects the pseudo-periodic behavior which the series exhibits when the roots of the characteristic equation are complex. For illustration, Figure 3.4 shows seventy terms of a series generated by the second-order autoregressive model

$$\tilde{z}_t = 0.75\,\tilde{z}_{t-1} - 0.50\,\tilde{z}_{t-2} + a_t$$

obtained by setting $\phi_1 = 0.75$ and $\phi_2 = -0.50$ in (3.2.16). Figure 3.5 shows

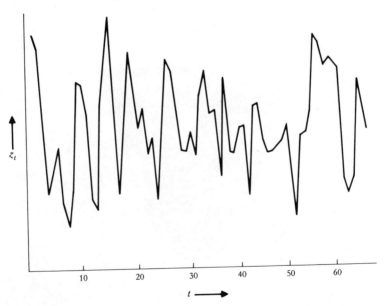

FIG. 3.4   A time series generated from a second-order autoregressive process
$\tilde{z}_t = 0.75\,\tilde{z}_{t-1} - 0.50\,\tilde{z}_{t-2} + a_t$

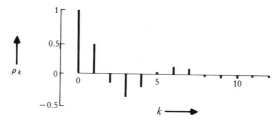

FIG. 3.5 Theoretical autocorrelation function of second-order autoregressive process $\tilde{z}_t = 0.75\,\tilde{z}_{t-1} - 0.50\,\tilde{z}_{t-2} + a_t$

the corresponding theoretical autocorrelation function calculated from (3.2.19), with starting values of $\rho_0 = 1$ and $\rho_1 = 0.75/\{1 - (-0.5)\} = 0.5$. The roots of the characteristic equation

$$1 - 0.75B + 0.5B^2 = 0$$

are complex, so that pseudo-periodic behavior which may be observed in the series is to be expected. We clearly see this behavior reflected in the theoretical autocorrelation function of Figure 3.5; the average apparent period being about 6.

The damping factor d and frequency f_0, from (3.2.22) and (3.2.23), are

$$d = \sqrt{0.50} = 0.71 \qquad f_0 = \frac{\cos^{-1} 0.5303}{2\pi} = \frac{1}{6.2}$$

Thus, the fundamental period of the autocorrelation function is 6.2.

Finally, the theoretical spectral density function in Figure 3.6, obtained from (3.2.29), shows that a large proportion of the variance of the series is accounted for by frequencies in the neighborhood of f_0.

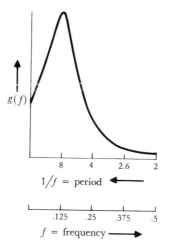

FIG. 3.6 Theoretical spectral density function of second-order autoregressive process $\tilde{z}_t = 0.75\,\tilde{z}_{t-1} - 0.50\,\tilde{z}_{t-2} + a_t$

3.2.5 The partial autocorrelation function

Initially, we may not know which order of autoregressive process to fit to an observed time series. This problem is analogous to deciding on the number of independent variables to be included in a multiple regression.

The partial autocorrelation function is a device which exploits the fact that whereas an AR(p) process has an autocorrelation function which is infinite in extent, it can by its very nature be described in terms of p non-zero *functions* of the autocorrelations. Denote by ϕ_{kj}, the jth coefficient in an autoregressive process of order k, so that ϕ_{kk} is the last coefficient. From (3.2.4), the ϕ_{kj} satisfy the set of equations

$$\rho_j = \phi_{k1}\rho_{j-1} + \cdots + \phi_{k(k-1)}\rho_{j-k+1} + \phi_{kk}\rho_{j-k} \qquad j = 1, 2, \ldots, k \qquad (3.2.30)$$

leading to the Yule–Walker equations (3.2.6), which may be written

$$\begin{bmatrix} 1 & \rho_1 & \rho_2 & \cdots & \rho_{k-1} \\ \rho_1 & 1 & \rho_1 & \cdots & \rho_{k-2} \\ \vdots & \vdots & \vdots & \cdots & \vdots \\ \rho_{k-1} & \rho_{k-2} & \rho_{k-3} & \cdots & 1 \end{bmatrix} \begin{bmatrix} \phi_{k1} \\ \phi_{k2} \\ \vdots \\ \phi_{kk} \end{bmatrix} = \begin{bmatrix} \rho_1 \\ \rho_2 \\ \vdots \\ \rho_k \end{bmatrix} \qquad (3.2.31)$$

or

$$\mathbf{P}_k \boldsymbol{\phi}_k = \boldsymbol{\rho}_k \qquad (3.2.32)$$

Solving these equations for $k = 1, 2, 3, \ldots$, successively, we obtain

$$\phi_{11} = \rho_1$$

$$\phi_{22} = \frac{\begin{vmatrix} 1 & \rho_1 \\ \rho_1 & \rho_2 \end{vmatrix}}{\begin{vmatrix} 1 & \rho_1 \\ \rho_1 & 1 \end{vmatrix}} = \frac{\rho_2 - \rho_1^2}{1 - \rho_1^2}$$

$$\qquad (3.2.33)$$

$$\phi_{33} = \frac{\begin{vmatrix} 1 & \rho_1 & \rho_1 \\ \rho_1 & 1 & \rho_2 \\ \rho_2 & \rho_1 & \rho_3 \end{vmatrix}}{\begin{vmatrix} 1 & \rho_1 & \rho_2 \\ \rho_1 & 1 & \rho_1 \\ \rho_2 & \rho_1 & 1 \end{vmatrix}}$$

In general, for ϕ_{kk}, the determinant in the numerator has the same elements as that in the denominator, but with the last column replaced by $\boldsymbol{\rho}_k$. The quantity ϕ_{kk}, regarded as a function of the lag k, is called the *partial auto-correlation* function.

For an autoregressive process of order p, the partial autocorrelation function ϕ_{kk} will be nonzero for k less than or equal to p and *zero for k greater than p*. In other words, the partial autocorrelation function of a pth order autoregressive process has a *cutoff* after lag p. For the second-order autoregressive process, partial autocorrelation functions ϕ_{kk} are shown in each of the four regions of Figure 3.2.

3.2.6 Estimation of the partial autocorrelation function

The partial autocorrelations may be estimated by fitting successively auto-regressive processes of orders 1, 2, 3, ... by least squares, as will be described in Chapter 7, and picking out the estimates $\hat{\phi}_{11}, \hat{\phi}_{22}, \hat{\phi}_{33}, \ldots$ of the last co-efficient fitted at each stage. Alternatively, if the values of the parameters are not too close to the nonstationary boundaries, approximate Yule–Walker estimates of the successive autoregressive processes may be employed. The estimated partial autocorrelations then can be obtained by substituting estimates r_j for the theoretical autocorrelations in (3.2.30), to yield

$$r_j = \hat{\phi}_{k1} r_{j-1} + \hat{\phi}_{k2} r_{j-2} + \cdots + \hat{\phi}_{k(k-1)} r_{j-k+1} + \hat{\phi}_{kk} r_{j-k}$$

$$j = 1, 2, \ldots, k \tag{3.2.34}$$

and solving the resultant equations for $k = 1, 2, \ldots$. A simple recursive method for doing this, due to Durbin [34], is given in Appendix A3.2. However, these estimates obtained from (3.2.34) become very sensitive to rounding errors and should not be used if the values of the parameters are close to the nonstationary boundaries. Programs 1 and 3 in the collection of Computer Programs at the end of the book, include routines for calculating estimates of the autocorrelation and partial autocorrelation functions.

3.2.7 Standard errors of partial autocorrelation estimates

It was shown by Quenouille [35] (see also [36] and [37]) that on the hypothesis that the process is autoregressive of order p, the estimated partial auto-correlations of order $p + 1$, and higher, are approximately independently distributed. Also if n is the number of observations used in fitting,

$$\text{var} \, [\hat{\phi}_{kk}] \simeq \frac{1}{n} \qquad k \geqslant p + 1$$

Thus the standard error (S.E.) of the estimated partial autocorrelation $\hat{\phi}_{kk}$ is

$$\text{S.E.} \, [\hat{\phi}_{kk}] = \hat{\sigma}[\hat{\phi}_{kk}] \simeq \frac{1}{\sqrt{n}} \qquad k \geqslant p + 1 \tag{3.2.35}$$

Table 3.1 shows the first 15 estimated partial autocorrelations for the batch data of Table 2.1, obtained by direct fitting* of autoregressive processes of increasing order. These partial autocorrelations are plotted in Figure 3.7

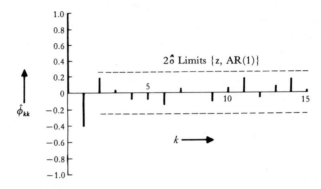

FIG. 3.7   Estimated partial autocorrelation function for batch data of Figure 2.1, together with two standard error limits calculated on the assumption that the model is AR(1)

and may be compared with the autocorrelations of Figure 2.7. The behavior of these functions resembles that associated with an AR(1) process with a negative value of ϕ_1 (see Figure 3.1), or possibly an AR(2) process with a dominant negative root (see region 2 of Figure 3.2). Also shown in Figure 3.7 by dotted lines are the 2 S.E. limits for $\hat{\phi}_{22}, \hat{\phi}_{33}, \ldots$, calculated from (3.2.35) on the assumption that the process is AR(1). Since $\hat{\phi}_{22}$ is the second biggest partial autocorrelation of those considered, the possibility that the process is AR(2) ought to be kept in mind.

TABLE 3.1   Estimated partial autocorrelation function for batch data of Table 2.1.

k	$\hat{\phi}_{kk}$	k	$\hat{\phi}_{kk}$	k	$\hat{\phi}_{kk}$
1	-0.40	6	-0.15	11	0.18
2	0.19	7	0.05	12	-0.05
3	0.01	8	0.00	13	0.09
4	-0.07	9	-0.10	14	0.18
5	-0.07	10	0.05	15	0.01

The use of the partial autocorrelation function for identifying models is discussed more fully in Chapter 6.

* Approximate values agreeing to the first decimal may be obtained by solving equations 3.2.34.

3.3 MOVING AVERAGE PROCESSES

3.3.1 Invertibility conditions for moving average processes

We now derive the conditions which the parameters $\theta_1, \theta_2, \ldots, \theta_q$ must satisfy to ensure the invertibility of the MA(q) process

$$\tilde{z}_t = a_t - \theta_1 a_{t-1} - \cdots - \theta_q a_{t-q}$$
$$= (1 - \theta_1 B - \cdots - \theta_q B^q) a_t$$
$$= \theta(B) a_t \tag{3.3.1}$$

We have already seen in Section 3.1.3 that the first-order moving average process

$$\tilde{z}_t = (1 - \theta_1 B) a_t$$

is invertible if $|\theta_1| < 1$, that is,

$$\pi(B) = (1 - \theta_1 B)^{-1} = \sum_{j=0}^{\infty} \theta_1^j B^j$$

converges on, or within the unit circle. However, this is equivalent to saying that the root, $B = \theta_1^{-1}$ of $(1 - \theta_1 B) = 0$, lies *outside* the unit circle.

The invertibility condition for higher order MA processes may be obtained by writing (3.3.1) as

$$a_t = \theta^{-1}(B) \tilde{z}_t$$

Hence, if

$$\theta(B) = \prod_{j=1}^{q} (1 - H_j B)$$

then, on expanding in partial fractions,

$$\pi(B) = \theta^{-1}(B) = \sum_{j=1}^{q} \frac{M_j}{(1 - H_j B)}$$

which converges if $|H_j| < 1$, when $j = 1, 2, \ldots, q$. Since the roots of $\theta(B) = 0$ are H_j^{-1}, it follows that the invertibility condition for a MA(q) process is that the roots of the characteristic equation

$$\theta(B) = 1 - \theta_1 B - \theta_2 B^2 - \cdots - \theta_q B^q = 0 \tag{3.3.2}$$

lie *outside* the unit circle.

Note, since the series

$$\psi(B) = \theta(B) = 1 - \theta_1 B - \theta_2 B^2 - \cdots - \theta_q B^q$$

is finite, no restrictions are needed on the parameters of the moving average process to ensure stationarity.

3.3.2 Autocorrelation function and spectrum of moving average processes

Autocorrelation function. Using (3.3.1), the autocovariance function of a MA(q) process is

$$\gamma_k = E[(a_t - \theta_1 a_{t-1} - \cdots - \theta_q a_{t-q})(a_{t-k} - \theta_1 a_{t-k-1} - \cdots - \theta_q a_{t-k-q})]$$

Hence, the variance of the process is

$$\gamma_0 = (1 + \theta_1^2 + \theta_2^2 + \cdots + \theta_q^2)\sigma_a^2 \qquad (3.3.3)$$

and

$$\gamma_k = \begin{cases} (-\theta_k + \theta_1\theta_{k+1} + \theta_2\theta_{k+2} + \cdots + \theta_{q-k}\theta_q)\sigma_a^2 & k = 1, 2, \ldots, q \\ 0 & k > q \end{cases}$$

Thus, the autocorrelation function is

$$\rho_k = \begin{cases} \dfrac{-\theta_k + \theta_1\theta_{k+1} + \cdots + \theta_{q-k}\theta_q}{1 + \theta_1^2 + \cdots + \theta_q^2} & k = 1, 2, \ldots, q \\[2mm] 0, & k > q \end{cases} \qquad (3.3.4)$$

We see that the autocorrelation function of a MA(q) process is zero, beyond the order q, of the process. In other words, the autocorrelation function of a moving average process has a *cut-off* at lag q.

Moving average parameters in terms of the autocorrelations. If ρ_1, $\rho_2, \ldots, \rho_q$ are known, the q equations (3.3.4) may be solved for the parameters $\theta_1, \theta_2, \ldots, \theta_q$. However, unlike the Yule–Walker equations (3.2.6) for an autoregressive process, which are linear, the equations (3.3.4) are nonlinear. Hence, except in the simple case where $q = 1$, which is discussed shortly, these equations have to be solved iteratively as described in Appendix A6.2. By substituting estimates r_k for ρ_k in (3.3.4) and solving the resulting equations, initial estimates of the moving average parameters may be obtained. Unlike the corresponding autoregressive estimates, obtained by a corresponding substitution in the Yule–Walker equations, the resulting estimates may not have high statistical efficiency. However, they can provide useful rough estimates at the identification stage discussed in Chapter 6. Furthermore, they provide useful starting values for an iterative procedure, discussed in Chapter 7, which converges to the efficient maximum likelihood estimates.

Spectrum. For the MA(q) process,

$$\psi(B) = \theta(B)$$

and

$$\theta(B) = 1 - \theta_1 B - \theta_2 B^2 - \cdots - \theta_q B^q$$

Therefore, using (3.1.12), the spectrum of a MA(q) process is

$$p(f) = 2\sigma_a^2|1 - \theta_1 e^{-i2\pi f} - \theta_2 e^{-i4\pi f} - \cdots - \theta_q e^{-i2\pi qf}|^2 \qquad 0 \leqslant f \leqslant \tfrac{1}{2}$$

(3.3.5)

We now discuss in greater detail the moving average processes of first and second order, which are of considerable practical importance.

3.3.3 The first-order moving average process

We have already met this process in the form

$$\tilde{z}_t = a_t - \theta_1 a_{t-1}$$
$$= (1 - \theta_1 B)a_t$$

and it has been shown in Section 3.1.3 that θ_1 must lie in the range $-1 < \theta_1 < 1$ for the process to be invertible. However, the process is of course stationary for all values of θ_1.

Autocorrelation function. Using (3.3.3), the variance of the process is

$$\gamma_0 = (1 + \theta_1^2)\sigma_a^2$$

and using (3.3.4), the autocorrelation function is

$$\rho_k = \begin{cases} \dfrac{-\theta_1}{1 + \theta_1^2} & k = 1 \\ 0 & k \geqslant 2 \end{cases}$$

(3.3.6)

From (3.3.6), with $k = 1$, we find

$$\theta_1^2 + \frac{\theta_1}{\rho_1} + 1 = 0$$

(3.3.7)

Since the product of the roots is unity, we see that if θ_1 is a solution, then so is θ_1^{-1}. Furthermore, if θ_1 satisfies the invertibility condition $|\theta_1| < 1$, the other root θ_1^{-1} will be greater than unity and will not satisfy the condition. For example, if $\rho_1 = -0.4$, (3.3.7) has two solutions, $\theta_1 = 0.5$ and $\theta_1 = 2.0$. However, only the solution $\theta_1 = 0.5$ corresponds to an invertible process. Table A, in the collection of Tables and Charts at the end of this volume, allows such solutions for θ_1 to be read off over the whole range of possible values $-0.5 < \rho_1 < 0.5$.

Spectrum. Using (3.3.5), the spectrum is

$$p(f) = 2\sigma_a^2|1 - \theta_1 e^{-i2\pi f}|^2$$
$$= 2\sigma_a^2(1 + \theta_1^2 - 2\theta_1 \cos 2\pi f) \qquad 0 \leqslant f \leqslant \tfrac{1}{2}$$

(3.3.8)

In general when θ_1 is negative, ρ_1 is positive, and the spectrum is dominated by low frequencies. Conversely, when θ_1 is positive, ρ_1 is negative, and the spectrum is dominated by high frequencies.

Partial autocorrelation function. Using (3.2.31) with $\rho_1 = -\theta_1/(1 + \theta_1^2)$ and $\rho_k = 0$, for $k > 1$, we obtain after some algebraic manipulation

$$\phi_{kk} = -\theta_1^k \{1 - \theta_1^2\}/\{1 - \theta_1^{2(k+1)}\}$$

Thus, $|\phi_{kk}| < \theta_1^k$, and the partial autocorrelation function is dominated by a damped exponential. If ρ_1 is positive, so that θ_1 is negative, the partial autocorrelations alternate in sign. If, however, ρ_1 is negative, so that θ_1 is positive, the partial autocorrelations are negative.

We now note a duality between the AR(1) and MA(1) processes. Thus, whereas the autocorrelation function of a MA(1) process has a cutoff after lag 1, the autocorrelation function of an AR(1) process tails off exponentially. Conversely, whereas the partial autocorrelation function of a MA(1) process tails off and is dominated by a damped exponential, the partial autocorrelation function of an AR(1) process has a cutoff after lag 1. It turns out that a corresponding approximate duality of this kind occurs in general.

3.3.4 The second-order moving average process

Invertibility conditions. The second-order moving average process is defined by

$$\tilde{z}_t = a_t - \theta_1 a_{t-1} - \theta_2 a_{t-2}$$

and is stationary for all values of θ_1 and θ_2. However, it is invertible only if the roots of the characteristic equation

$$1 - \theta_1 B - \theta_2 B^2 = 0 \qquad\qquad (3.3.9)$$

lie outside the unit circle, that is

$$\begin{aligned} \theta_2 + \theta_1 &< 1 \\ \theta_2 - \theta_1 &< 1 \\ -1 < \theta_2 &< 1 \end{aligned} \qquad\qquad (3.3.10)$$

These are parallel to the conditions (3.2.18) required for the *stationarity* of an AR(2) process.

Autocorrelation function. Using (3.3.3), the variance of the process is

$$\gamma_0 = \sigma_a^2(1 + \theta_1^2 + \theta_2^2)$$

and using (3.3.4), the autocorrelation function is

$$\rho_1 = \frac{-\theta_1(1 - \theta_2)}{1 + \theta_1^2 + \theta_2^2}$$

$$\rho_2 = \frac{-\theta_2}{1 + \theta_1^2 + \theta_2^2} \qquad\qquad (3.3.11)$$

$$\rho_k = 0 \qquad k \geqslant 3$$

Thus the autocorrelation function has a cut-off after lag 2.

It follows from (3.3.10) and (3.3.11) that the first two autocorrelations of an invertible MA(2) process must lie within the area bounded by segments of the curves

$$\rho_2 + \rho_1 = -0.5$$

$$\rho_2 - \rho_1 = -0.5 \qquad\qquad (3.3.12)$$

$$\rho_1^2 = 4\rho_2(1 - 2\rho_2)$$

The invertibility region (3.3.10) for the parameters is shown in Figure 3.8(a) and the corresponding region (3.3.12) for the autocorrelations in Figure 3.8(b). The latter shows whether a given pair of autocorrelations ρ_1 and ρ_2 is consistent with the hypothesis that the model is a MA(2) process. If they are consistent, the values of the parameters θ_1 and θ_2 can be obtained by solving the nonlinear equations (3.3.11). To facilitate this calculation, Chart C at the end of the book has been prepared so that the values of θ_1 and θ_2 can be read off directly, given ρ_1 and ρ_2.

Spectrum.   Using (3.3.5), the spectrum is

$$p(f) = 2\sigma_a^2|1 - \theta_1 e^{-i2\pi f} - \theta_2 e^{-i4\pi f}|^2$$

$$= 2\sigma_a^2\{1 + \theta_1^2 + \theta_2^2 - 2\theta_1(1 - \theta_2)\cos 2\pi f - 2\theta_2 \cos 4\pi f\}$$

$$0 \leqslant f \leqslant \tfrac{1}{2} \quad (3.3.13)$$

and is the reciprocal of the spectrum of a second-order autoregressive process (3.2.29), apart from the constant $2\sigma_a^2$.

Partial autocorrelation function.   The exact expression for the partial autocorrelation function of an MA(2) process is complicated, but it is dominated by the sum of two exponentials, if the roots of the characteristic equation (3.3.9) are real, and by a damped sine wave, if the roots of (3.3.9) are complex. Thus, it behaves like the autocorrelation of an AR(2) process. The autocorrelation functions (left-hand curves) and partial autocorrelation functions (right-hand curves) for various values of the parameters within the invertible region are shown in Figure 3.9, which is taken from Stralkowski [33]. Comparison of Figure 3.9 with Figure 3.2, which shows the corresponding autocorrelations and partial autocorrelations for an AR(2) process, illustrates the duality between the MA(2) and the AR(2) processes.

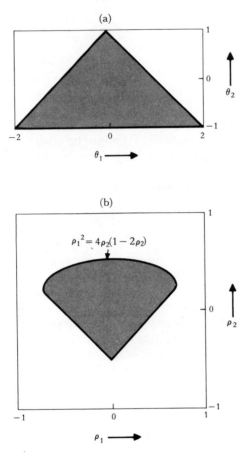

FIG. 3.8   Admissible regions for (a) θ_1, θ_2 and (b) ρ_1, ρ_2, for an invertible MA (2) process

3.3.5 Duality between autoregressive and moving average processes

The results of the previous sections have shown further aspects of the *duality* between autoregressive and finite moving average processes. As illustrated in Table 3.2 at the end of this chapter, this duality has the following consequences:

(1) In a stationary autoregressive process of order p, a_t can be represented as a *finite* weighted sum of previous $\tilde{z}$'s, or $\tilde{z}_t$ as an infinite weighted sum

$$\tilde{z}_t = \phi^{-1}(B)a_t$$

of previous a's. Also, in an invertible moving average process of order q,

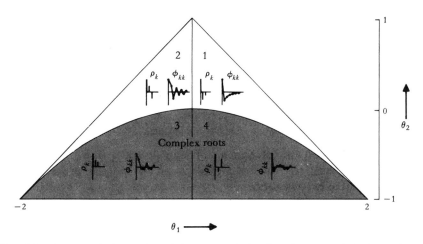

F<small>IG</small>. 3.9   Autocorrelation and partial autocorrelation functions ρ_k and ϕ_{kk} for various MA (2) models

$\tilde{z}_t$ can be represented as a finite weighted sum of previous a's, or a_t as an infinite weighted sum

$$\theta^{-1}(B)\tilde{z}_t = a_t$$

of previous $\tilde{z}$'s.

(2) The finite MA process has an autocorrelation function which is zero beyond a certain point, but since it is equivalent to an infinite AR process, its partial autocorrelation function is infinite in extent and is dominated by damped exponentials and/or damped sine waves. Conversely, the AR process has a partial autocorrelation function which is zero beyond a certain point, but its autocorrelation function is infinite in extent and consists of a mixture of damped exponentials and/or damped sine waves.

(3) For an autoregressive process of finite order p, the parameters are not required to satisfy any conditions to ensure invertibility. However, for stationarity, the roots of $\phi(B) = 0$ must lie outside the unit circle. Conversely, the parameters of the MA process are not required to satisfy any conditions to ensure stationarity. However, for invertibility of the MA process, the roots of $\theta(B) = 0$ must lie outside the unit circle.

(4) The spectrum of a moving average process has an inverse relationship to the spectrum of the corresponding autoregressive process.

3.4 MIXED AUTOREGRESSIVE—MOVING AVERAGE PROCESSES

3.4.1 Stationarity and invertibility properties

We have noted in Section 3.1.4 that to achieve parsimony it may be necessary to include both autoregressive and moving average terms. Thus, we may need to employ the mixed autoregressive-moving average (ARMA) model

$$\tilde{z}_t = \phi_1 \tilde{z}_{t-1} + \cdots + \phi_p \tilde{z}_{t-p} + a_t - \theta_1 a_{t-1} - \cdots - \theta_q a_{t-q} \quad (3.4.1)$$

that is

$$(1 - \phi_1 B - \phi_2 B^2 - \cdots - \phi_p B^p)\tilde{z}_t = (1 - \theta_1 B - \theta_2 B^2 \cdots - \theta_q B^q)a_t$$

or

$$\phi(B)\tilde{z}_t = \theta(B)a_t$$

where $\phi(B)$ and $\theta(B)$ are polynomials of degree p and q, in B.

We subsequently refer to this process as an ARMA(p, q) process. It may be thought of in two ways. Namely,

(a) as a pth order autoregressive process

$$\phi(B)\tilde{z}_t = e_t$$

with e_t following the qth order moving average process

$$e_t = \theta(B)a_t$$

(b) as a qth order moving average process

$$\tilde{z}_t = \theta(B)b_t$$

with b_t following the pth order autoregressive process

$$\phi(B)b_t = a_t$$

so that

$$\phi(B)\tilde{z}_t = \theta(B)\phi(B)b_t = \theta(B)a_t$$

It is obvious that moving average terms on the right of (3.4.1) will not affect the argument of Section 3.2.1, which establishes conditions for stationarity of an autoregressive process. Thus, $\phi(B)\tilde{z}_t = \theta(B)a_t$ will define a stationary process, provided that the characteristic equation $\phi(B) = 0$ has all its roots lying outside the unit circle. Similarly, the roots of $\theta(B) = 0$ must lie outside the unit circle if the process is to be invertible.

3.4.2 Autocorrelation function and spectrum of mixed processes

Autocorrelation function. The autocorrelation function of the mixed process may be derived by a similar method to that used for autoregressive processes in Section 3.2.2. On multiplying throughout in (3.4.1) by $\tilde{z}_{t-k}$ and taking expectations, we see that the autocovariance function satisfies the difference equation

$$\gamma_k = \phi_1 \gamma_{k-1} + \cdots + \phi_p \gamma_{k-p} + \gamma_{za}(k) - \theta_1 \gamma_{za}(k - 1) - \cdots$$
$$- \theta_q \gamma_{za}(k - q) \quad (3.4.2)$$

where $\gamma_{za}(k)$ is the cross covariance function between z and a, and is defined by $\gamma_{za}(k) = E[\tilde{z}_{t-k}a_t]$. Since z_{t-k} depends only on shocks which have occurred up to time $t - k$, it follows that

$$\gamma_{za}(k) = 0 \qquad k > 0$$

$$\gamma_{za}(k) \neq 0 \qquad k \leqslant 0$$

We see that (3.4.2) implies

$$\gamma_k = \phi_1\gamma_{k-1} + \phi_2\gamma_{k-2} + \cdots + \phi_p\gamma_{k-p} \qquad k \geqslant q + 1$$

and hence

$$\rho_k = \phi_1\rho_{k-1} + \phi_2\rho_{k-2} + \cdots + \phi_p\rho_{k-p} \qquad k \geqslant q + 1 \qquad (3.4.3)$$

or

$$\phi(B)\rho_k = 0 \qquad\qquad\qquad k \geqslant q + 1$$

Thus, for the ARMA(p, q) process, there will be q autocorrelations ρ_q, $\rho_{q-1}, \ldots, \rho_1$ whose values depend directly, through (3.4.2), on the choice of the q moving average parameters $\boldsymbol{\theta}$, as well as on the p autoregressive parameters $\boldsymbol{\phi}$. Also, the p values $\rho_q, \rho_{q-1}, \ldots, \rho_{q-p+1}$ provide the necessary starting values for the difference equation $\phi(B)\rho_k = 0$, where $k \geqslant q + 1$, which then entirely determines the autocorrelations at higher lags. If $q - p < 0$, the whole autocorrelation function ρ_j, for $j = 0, 1, 2, \ldots$, will consist of a mixture of damped exponentials and/or damped sine waves, whose nature is dictated by the polynomial $\phi(B)$ and the starting values. If, however, $q - p \geqslant 0$ there will be $q - p + 1$ initial values $\rho_0, \rho_1, \ldots, \rho_{q-p}$, which do not follow this general pattern. These facts are useful in identifying mixed series.

Variance. When $k = 0$, we have

$$\gamma_0 = \phi_1\gamma_1 + \cdots + \phi_p\gamma_p + \sigma_a^2 - \theta_1\gamma_{za}(-1) - \cdots - \theta_q\gamma_{za}(-q) \quad (3.4.4)$$

which has to be solved along with the p equations (3.4.2) for $k = 1, 2, \cdots, p$ to obtain $\gamma_0, \gamma_1, \ldots, \gamma_p$.

Spectrum. Using (3.1.12), the spectrum of a mixed process is

$$p(f) = 2\sigma_a^2 \frac{|\theta(e^{-i2\pi f})|^2}{|\phi(e^{-i2\pi f})|^2}$$

$$= 2\sigma_a^2 \frac{|1 - \theta_1 e^{-i2\pi f} - \cdots - \theta_q e^{-i2\pi qf}|^2}{|1 - \phi_1 e^{-i2\pi f} - \cdots - \phi_p e^{-i2\pi pf}|^2} \qquad 0 \leqslant f \leqslant \tfrac{1}{2} \qquad (3.4.5)$$

Partial autocorrelation function. The process (3.4.1) may be written

$$a_t = \theta^{-1}(B)\phi(B)\tilde{z}_t$$

and $\theta^{-1}(B)$ is an infinite series in B. Hence, the partial autocorrelation function of a mixed process is infinite in extent. It behaves eventually like the partial autocorrelation function of a pure moving average process, being dominated by a mixture of damped exponentials and/or damped sine waves, depending on the order of the moving average and the values of the parameters it contains.

3.4.3 The first-order autoregressive—first-order moving average process

A mixed process of considerable practical importance is the first-order autoregressive—first-order moving average ARMA(1, 1) process

$$\tilde{z}_t - \phi_1 \tilde{z}_{t-1} = a_t - \theta_1 a_{t-1} \tag{3.4.6}$$

that is

$$(1 - \phi_1 B)\tilde{z}_t = (1 - \theta_1 B)a_t$$

We now derive some of its more important properties.

Stationarity and invertibility conditions.   First, we note that the process is stationary if $-1 < \phi_1 < 1$, and invertible if $-1 < \theta_1 < 1$. Hence, the admissible parameter space is the square shown in Figure 3.10(a).

Autocorrelation function.   From (3.4.2) and (3.4.4) we obtain

$$\gamma_0 = \phi_1 \gamma_1 + \sigma_a^2 - \theta_1 \gamma_{za}(-1)$$

$$\gamma_1 = \phi_1 \gamma_0 - \theta_1 \sigma_a^2$$

$$\gamma_k = \phi_1 \gamma_{k-1} \qquad k \geqslant 2$$

On multiplying throughout (3.4.6) by a_{t-1}, and taking expectations, we obtain

$$\gamma_{za}(-1) = (\phi_1 - \theta_1)\sigma_a^2$$

Hence, the autocovariance function of the process is

$$\gamma_0 = \frac{1 + \theta_1^2 - 2\phi_1\theta_1}{1 - \phi_1^2}\sigma_a^2$$

$$\gamma_1 = \frac{(1 - \phi_1\theta_1)(\phi_1 - \theta_1)}{1 - \phi_1^2}\sigma_a^2 \tag{3.4.7}$$

$$\gamma_k = \phi_1 \gamma_{k-1} \qquad k \geqslant 2$$

Thus, the autocorrelation function decays exponentially from the starting value ρ_1, which depends on θ_1 as well as on ϕ_1*. As shown in Figure 3.11, this exponential decay is smooth if ϕ_1 is positive and alternates if ϕ_1 is negative. Furthermore, the sign of ρ_1 is determined by the sign of $(\phi_1 - \theta_1)$ and dictates from which side of zero the exponential decay takes place.

---

* By contrast the autocorrelation function for the AR(1) process decays exponentially from the starting value $\rho_0 = 1$.

From (3.4.7), the first two autocorrelations may be expressed in terms of the parameters of the process, as follows

$$\rho_1 = \frac{(1 - \phi_1\theta_1)(\phi_1 - \theta_1)}{1 + \theta_1^2 - 2\phi_1\theta_1}$$

$$\rho_2 = \phi_1\rho_1$$

(3.4.8)

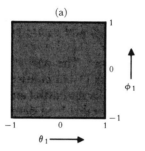
(a)

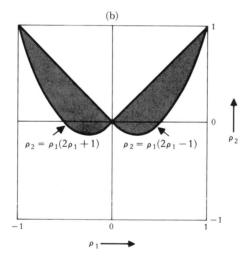
(b)

$$\rho_2 = \rho_1(2\rho_1 + 1) \qquad \rho_2 = \rho_1(2\rho_1 - 1)$$

FIG. 3.10   Admissible regions for (a) θ_1, ϕ_1 and (b) ρ_1, ρ_2, for a stationary and invertible ARMA $(1, 1)$ process

Chart D at the end of the book is so constructed, that the solution of the equations (3.4.8) for ϕ_1 and θ_1 can be read off, knowing ρ_1 and ρ_2. By substituting estimates r_1 and r_2 for ρ_1 and ρ_2, initial estimates for the parameters ϕ_1 and θ_1 can be obtained.

Using (3.4.8) and the stationarity and invertibility conditions, it may be shown that ρ_1 and ρ_2 must lie in the region

$$|\rho_2| < |\rho_1|$$

$$\rho_2 > \rho_1(2\rho_1 + 1) \qquad \rho_1 < 0 \tag{3.4.9}$$

$$\rho_2 > \rho_1(2\rho_1 - 1) \qquad \rho_1 > 0$$

Figure 3.10(b) shows the admissible space for ρ_1 and ρ_2, that is to say it indicates which combinations of ρ_1 and ρ_2 are possible for the mixed (1, 1) stationary, invertible process.

Partial autocorrelation function. The partial autocorrelation function of the mixed ARMA(1, 1) process (3.4.6) consists of a single initial value $\phi_{11} = \rho_1$. Thereafter it behaves like the partial autocorrelation function of a pure MA(1) process, and is dominated by a damped exponential. Thus, as shown in Figure 3.11, when θ_1 is positive it is dominated by a smoothly damped exponential which decays from a value of ρ_1, with sign determined by the sign of $(\phi_1 - \theta_1)$. Similarly, when θ_1 is negative, it is dominated by an exponential which oscillates as it decays from a value of ρ_1, with sign determined by the sign of $(\phi_1 - \theta_1)$.

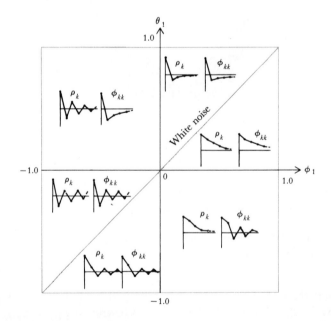

Fig. 3.11   Autocorrelation and partial autocorrelation functions ρ_k and ϕ_{kk} for various ARMA (1, 1) models

TABLE 3.2   Summary of properties of autoregressive, moving average, and mixed ARMA processes

	Autoregressive processes	Moving average processes	Mixed processes
Model in terms of previous $\tilde{z}$'s	$\phi(B)\tilde{z}_t = a_t$	$\theta^{-1}(B)\tilde{z}_t = a_t$	$\theta^{-1}(B)\phi(B)\tilde{z}_t = a_t$
Model in terms of previous a's	$\tilde{z}_t = \phi^{-1}(B)a_t$	$\tilde{z}_t = \theta(B)a_t$	$\tilde{z}_t = \phi^{-1}(B)\theta(B)a_t$
π weights	finite series	infinite series	infinite series
ψ weights	infinite series	finite series	infinite series
Stationarity condition	roots of $\phi(B) = 0$ lie outside the unit circle	always stationary	roots of $\phi(B) = 0$ lie outside the unit circle
Invertibility condition	always invertible	roots of $\theta(B) = 0$ lie outside unit circle	roots of $\theta(B) = 0$ lie outside unit circle
Autocorrelation function	infinite (damped exponentials and/or damped sine waves)	finite	infinite (damped exponentials and/or damped sine waves after first $q - p$ lags)
	tails off	cuts off	tails off
Partial autocorrelation function	finite	infinite (dominated by damped exponentials and/or sine waves)	infinite (dominated by damped exponentials and/or sine waves after first $p - q$ lags)
	cuts off	tails off	tails off

3.4.4 Summary

Figure 3.12 brings together the admissible regions for the parameters and for the correlations ρ_1, ρ_2 for AR(2), MA(2), and ARMA(1, 1) processes which are restricted to being both stationary and invertible. Table 3.2 summarizes the properties of mixed autoregressive–moving average processes, and brings together all the important results for autoregressive, moving average and mixed processes, which will be needed in Chapter 6 to identify models for observed time series. In the next chapter we extend the mixed ARMA model to produce models which can describe nonstationary behavior of the kind that is frequently met in practice.

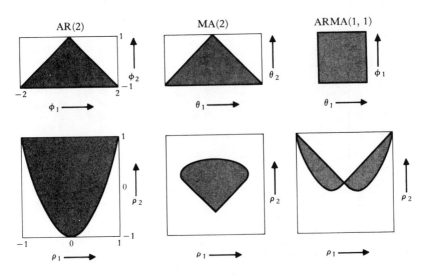

Fig. 3.12   Admissible regions for the parameters and ρ_1, ρ_2 for AR (2), MA (2), and ARMA (1, 1) processes which are restricted to being both stationary and invertible

APPENDIX   A3.1   AUTOCOVARIANCES, AUTOCOVARIANCE GENERATING FUNCTION AND STATIONARITY CONDITIONS FOR A GENERAL LINEAR PROCESS

Autocovariances.   The autocovariance at lag k of the linear process

$$\tilde{z}_t = \sum_{j=0}^{\infty} \psi_j a_{t-j}$$

with $\psi_0 = 1$, is clearly

$$\gamma_k = E[\tilde{z}_t \tilde{z}_{t+k}]$$

$$= E\left[\sum_{j=0}^{\infty} \sum_{h=0}^{\infty} \psi_j \psi_h a_{t-j} a_{t+k-h}\right]$$

$$= \sigma_a^2 \sum_{j=0}^{\infty} \psi_j \psi_{j+k} \tag{A3.1.1}$$

using the property (3.1.2) for the autocovariance function of white noise.

Autocovariance generating function. The result (A3.1.1) may be substituted in the autocovariance generating function

$$\gamma(B) = \sum_{k=-\infty}^{\infty} \gamma_k B^k \tag{A3.1.2}$$

to give

$$\gamma(B) = \sigma_a^2 \sum_{k=-\infty}^{\infty} \sum_{j=0}^{\infty} \psi_j \psi_{j+k} B^k$$

$$= \sigma_a^2 \sum_{j=0}^{\infty} \sum_{k=-j}^{\infty} \psi_j \psi_{j+k} B^k$$

since $\psi_h = 0$ for $h < 0$. Writing $j + k = h$, so that $k = h - j$,

$$\gamma(B) = \sigma_a^2 \sum_{j=0}^{\infty} \sum_{h=0}^{\infty} \psi_j \psi_h B^{h-j}$$

$$= \sigma_a^2 \sum_{h=0}^{\infty} \psi_h B^h \sum_{j=0}^{\infty} \psi_j B^{-j}$$

that is

$$\gamma(B) = \sigma_a^2 \psi(B)\psi(B^{-1}) = \sigma_a^2 \psi(B)\psi(F) \tag{A3.1.3}$$

which is the result (3.1.11) quoted in the text.

Stationarity conditions. If we substitute $B = e^{-i2\pi f}$, and $F = B^{-1} = e^{i2\pi f}$, in the autocovariance generating function (A3.1.2), we obtain half the power spectrum. Hence, the power spectrum of a linear process is

$$p(f) = 2\sigma_a^2 \psi(e^{-i2\pi f})\psi(e^{i2\pi f})$$

$$= 2\sigma_a^2 |\psi(e^{-i2\pi f})|^2 \qquad 0 \leqslant f \leqslant \tfrac{1}{2} \tag{A3.1.4}$$

It follows that the variance of the process is

$$\sigma_z^2 = \int_0^{1/2} p(f)\,df = 2\sigma_a^2 \int_0^{1/2} \psi(e^{-i2\pi f})\psi(e^{i2\pi f})\,df \tag{A3.1.5}$$

Now if the integral (A3.1.5) is to converge, it may be shown [98] that the infinite series $\psi(B)$ must converge for B on or within the unit circle.

APPENDIX A3.2   A RECURSIVE METHOD FOR CALCULATING ESTIMATES OF AUTOREGRESSIVE PARAMETERS

We now show how Yule–Walker estimates for the parameters of an AR($p + 1$) process may be obtained when the estimates for an AR(p) process, fitted to the same time series, are known. This recursive method of calculation can be used to approximate the partial autocorrelation function, as described in Section 3.2.6.

To illustrate the recursion, consider the equations (3.2.34). Yule–Walker estimates are obtained for $k = 2, 3$, from

$$r_2 = \hat{\phi}_{21} r_1 + \hat{\phi}_{22}$$
$$r_1 = \hat{\phi}_{21} + \hat{\phi}_{22} r_1$$
(A3.2.1)

and

$$r_3 = \hat{\phi}_{31} r_2 + \hat{\phi}_{32} r_1 + \hat{\phi}_{33}$$
$$r_2 = \hat{\phi}_{31} r_1 + \hat{\phi}_{32} + \hat{\phi}_{33} r_1$$
$$r_1 = \hat{\phi}_{31} + \hat{\phi}_{32} r_1 + \hat{\phi}_{33} r_2$$
(A3.2.2)

The coefficients $\hat{\phi}_{31}$ and $\hat{\phi}_{32}$ may be expressed in terms of $\hat{\phi}_{33}$, using the last two equations of (A3.2.2). The solution may be written in matrix form

$$\begin{pmatrix} \hat{\phi}_{31} \\ \hat{\phi}_{32} \end{pmatrix} = \mathbf{R}_2^{-1} \begin{pmatrix} r_2 - \hat{\phi}_{33} r_1 \\ r_1 - \hat{\phi}_{33} r_2 \end{pmatrix}$$
(A3.2.3)

where

$$\mathbf{R}_2 = \begin{bmatrix} r_1 & 1 \\ 1 & r_1 \end{bmatrix}$$

Now (A3.2.3) may be rewritten

$$\begin{bmatrix} \hat{\phi}_{31} \\ \hat{\phi}_{32} \end{bmatrix} = \mathbf{R}_2^{-1} \begin{bmatrix} r_2 \\ r_1 \end{bmatrix} - \hat{\phi}_{33} \mathbf{R}_2^{-1} \begin{bmatrix} r_1 \\ r_2 \end{bmatrix}$$
(A3.2.4)

Using the fact that (A3.2.1) may be rewritten

$$\begin{bmatrix} \hat{\phi}_{21} \\ \hat{\phi}_{22} \end{bmatrix} = \mathbf{R}_2^{-1} \begin{bmatrix} r_2 \\ r_1 \end{bmatrix}$$

it follows that (A3.2.4) becomes

$$\begin{bmatrix} \hat{\phi}_{31} \\ \hat{\phi}_{32} \end{bmatrix} = \begin{bmatrix} \hat{\phi}_{21} \\ \hat{\phi}_{22} \end{bmatrix} - \hat{\phi}_{33} \begin{bmatrix} \hat{\phi}_{22} \\ \hat{\phi}_{21} \end{bmatrix}$$

that is

$$\begin{aligned} \hat{\phi}_{31} &= \hat{\phi}_{21} - \hat{\phi}_{33}\hat{\phi}_{22} \\ \hat{\phi}_{32} &= \hat{\phi}_{22} - \hat{\phi}_{33}\hat{\phi}_{21} \end{aligned} \qquad \text{(A3.2.5)}$$

To complete the calculation of $\hat{\phi}_{31}$ and $\hat{\phi}_{32}$, we need an expression for $\hat{\phi}_{33}$. On substituting (A3.2.5) in the first of the equations (A3.2.2), we obtain

$$\hat{\phi}_{33} = \frac{r_3 - \hat{\phi}_{21}r_2 - \hat{\phi}_{22}r_1}{1 - \hat{\phi}_{21}r_1 - \hat{\phi}_{22}r_2} \qquad \text{(A3.2.6)}$$

Thus, the partial autocorrelation $\hat{\phi}_{33}$, is first calculated from $\hat{\phi}_{21}$ and $\hat{\phi}_{22}$, using (A3.2.6), and then the other two coefficients, $\hat{\phi}_{31}$ and $\hat{\phi}_{32}$, may be obtained from (A3.2.5).

In general, the recursive formulae, which are due to Durbin [34], are

$$\hat{\phi}_{p+1,j} = \hat{\phi}_{pj} - \hat{\phi}_{p+1,p+1}\hat{\phi}_{p,p-j+1} \qquad j = 1, 2, \ldots, p \qquad \text{(A3.2.7)}$$

$$\hat{\phi}_{p+1,p+1} = \frac{r_{p+1} - \sum\limits_{j=1}^{p} \hat{\phi}_{pj}r_{p+1-j}}{1 - \sum\limits_{j=1}^{p} \hat{\phi}_{pj}r_j} \qquad \text{(A3.2.8)}$$

Example. As an illustration, consider the calculations of the estimates $\hat{\phi}_{31}$, $\hat{\phi}_{32}$, and $\hat{\phi}_{33}$, of the parameters of an AR(3) process, fitted to Wölfer's sunspot numbers. The estimated autocorrelations to three decimal accuracy are $r_1 = 0.806$, $r_2 = 0.428$, $r_3 = 0.070$. Then

$$\hat{\phi}_{21} = \frac{r_1(1 - r_2)}{1 - r_1^2} = 1.316$$

$$\hat{\phi}_{22} = \frac{r_2 - r_1^2}{1 - r_1^2} = -0.632$$

Using (A3.2.6), we obtain

$$\hat{\phi}_{33} = \frac{0.070 - (1.316)(0.428) + (0.632)(0.806)}{1 - (1.316)(0.806) + (0.632)(0.428)} = 0.077$$

On substituting the values for $\hat{\phi}_{21}, \hat{\phi}_{22}, \hat{\phi}_{33}$ in (A3.2.5),

$$\hat{\phi}_{31} = 1.316 + (0.077)(0.632) = 1.365$$

$$\hat{\phi}_{32} = -0.632 - (0.077)(1.316) = -0.733$$

We remind the reader that these estimates of the partial autocorrelations differ somewhat from the maximum likelihood values obtained by fitting autoregressive processes of successively higher order. They are very sensitive to rounding errors particularly when the process approaches nonstationarity.

4

Linear Nonstationary Models

Many empirical time series (for example, stock prices) behave as though they had no fixed mean. Even so, they exhibit homogeneity in the sense that, apart from local level, or perhaps local level and trend, one part of the series behaves much like any other part. Models which describe such homogeneous nonstationary behavior can be obtained by supposing some suitable *difference* of the process to be stationary. We now consider the properties of the important class of models for which the dth difference is a stationary mixed autoregressive-moving average process. These models are called autoregressive integrated moving average (ARIMA) processes.

4.1 AUTOREGRESSIVE INTEGRATED MOVING AVERAGE PROCESSES

4.1.1 The nonstationary first-order autoregressive process

Figure 4.1 shows sections of four time series encountered in practice. These series have arisen in forecasting and control problems, and all of them exhibit behavior suggestive of nonstationarity. Series A, C, and D represent "uncontrolled" outputs (concentration, temperature and viscosity, respectively) from three different chemical processes. These series were collected to show the effect on these outputs of uncontrolled and unmeasured disturbances such as variations in feed stock and ambient temperature. The temperature series C was obtained by temporarily disconnecting the controllers on the pilot plant involved, and recording the subsequent temperature fluctuations. Both A and D were collected on full scale processes where it was necessary to maintain some output quality characteristic as close as possible to a fixed level. To achieve this control, another variable had been manipulated to approximately cancel out variations in the output. However, the effect of these manipulations on the output was in each case accurately known, so that it was possible to compensate numerically for the control action. That is to say, it was possible to calculate very nearly, the values of the series that would have been obtained if no corrective action been taken. It is these compensated values which are recorded here and referred to as

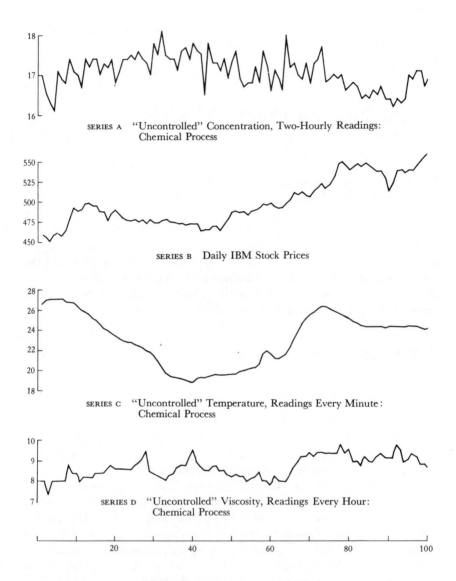

SERIES A   "Uncontrolled" Concentration, Two-Hourly Readings:
Chemical Process

SERIES B   Daily IBM Stock Prices

SERIES C   "Uncontrolled" Temperature, Readings Every Minute:
Chemical Process

SERIES D   "Uncontrolled" Viscosity, Readings Every Hour:
Chemical Process

FIG 4.1   Some typical time series arising in forecasting and control problems

the "uncontrolled" series. Series B consists of the daily I.B.M. stock prices
during a period beginning in May, 1961. A complete listing of all the series
is given in the collection of time series at the end of this volume. In Figure 4.1,
100 successive observations have been plotted from each series and the
points joined by straight lines.

There are an unlimited number of ways in which a process can be non-stationary. However, the types of economic and industrial series which we wish to analyze, frequently exhibit a particular kind of homogeneous non-stationary behavior, that can be represented by a stochastic model, which is a modified form of the ARMA model. In Chapter 3 we considered the auto-regressive-moving average model

$$\phi(B)\tilde{z}_t = \theta(B)a_t \tag{4.1.1}$$

with $\phi(B)$ and $\theta(B)$ polynomials in B, of degree p and q, respectively. To ensure stationarity, the roots of $\phi(B) = 0$ must lie outside the unit circle. A natural way of obtaining nonstationary processes would be to relax this restriction.

To gain some insight into the possibilities, consider the first-order auto-regressive model

$$(1 - \phi B)\tilde{z}_t = a_t \tag{4.1.2}$$

which is stationary for $|\phi| < 1$. Let us study the behavior of this process for $\phi = 2$; a value outside the stationary range.

Table 4.1 shows a set of unit random Normal deviates a_t and the corresponding values of the series $\tilde{z}_t$, generated by the model $\tilde{z}_t = 2\tilde{z}_{t-1} + a_t$, with $\tilde{z}_0 = 0.7$.

TABLE 4.1   First 11 values of a nonstationary first-order autoregressive process

t	0	1	2	3	4	5	6	7	8	9	10
a_t		0.1	-1.1	0.2	-2.0	-0.2	-0.8	0.8	0.1	0.1	-0.9
$\tilde{z}_t$	0.7	1.5	1.9	4.0	6.0	11.8	22.8	46.4	92.9	185.9	370.9

The series is plotted in Figure 4.2. It is seen that, after a short induction period, the series "breaks loose" and essentially follows an exponential curve with the generating a's playing almost no further part. The behavior of series generated by processes of higher order, which violate the stationarity condition, is similar. Furthermore, this behavior is essentially the same whether or not moving average terms are introduced on the right of the model.

4.1.2 A general model for a nonstationary process exhibiting homogeneity

The autoregressive integrated moving average model. Although non-stationary models, of the kind described above, are of value to represent explosive or evolutionary behavior (such as bacterial growth), the situations which we describe in this book are not of this type. So far we have seen that an ARMA process is stationary, if the roots of $\phi(B) = 0$ lie *outside* the unit

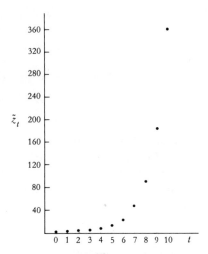

FIG. 4.2   A realization of the nonstationary first-order autoregressive process $\tilde{z}_t = 2\tilde{z}_{t-1} + a_t$ with $\sigma_a^2 = 1$

circle, and exhibits explosive nonstationary behavior if the roots lie *inside* the unit circle. The only other case open to us is that for which the roots of $\phi(B) = 0$ lie *on* the unit circle. It turns out that the resulting models are of great value in representing homogeneous nonstationary time series. In particular, nonseasonal series are often well represented by models in which one or more of these roots are *unity* and these are considered in the present chapter.*

Let us consider the model

$$\varphi(B)\tilde{z}_t = \theta(B)a_t \qquad (4.1.3)$$

where $\varphi(B)$ is a nonstationary autoregressive operator, such that d of the roots of $\varphi(B) = 0$ are unity and the remainder lie outside the unit circle. Then we can express the model (4.1.3) in the form

$$\varphi(B)\tilde{z}_t = \phi(B)(1 - B)^d \tilde{z}_t = \theta(B)a_t \qquad (4.1.4)$$

where $\phi(B)$ is a *stationary* autoregressive operator. Since $\nabla^d \tilde{z}_t = \nabla^d z_t$, for $d \geqslant 1$, we can write the model as

$$\phi(B)\nabla^d z_t = \theta(B)a_t \qquad (4.1.5)$$

Equivalently, the process is defined by the two equations

$$\phi(B)w_t = \theta(B)a_t \qquad (4.1.6)$$

---

* In Chapter 9 we consider models, capable of representing seasonality of period s, for which the characteristic equation has roots lying on the unit circle which are the sth roots of unity.

and

$$w_t = \nabla^d z_t \qquad (4.1.7)$$

Thus, we see that the model corresponds to assuming that the dth difference of the series can be represented by a stationary, invertible ARMA process. An alternative way of looking at the process for $d \geqslant 1$ results from inverting (4.1.7) to give

$$z_t = S^d w_t \qquad (4.1.8)$$

where S is the infinite summation operator defined by

$$Sx_t = \sum_{h=-\infty}^{t} x_h = (1 + B + B^2 + \cdots)x_t$$

$$= (1 - B)^{-1}x_t = \nabla^{-1}x_t$$

Thus

$$S = (1 - B)^{-1} = \nabla^{-1}$$

The operator $S^2 x_t$ is similarly defined as

$$S^2 x_t = Sx_t + Sx_{t-1} + Sx_{t-2} + \cdots$$

$$= \sum_{i=-\infty}^{t} \sum_{h=-\infty}^{i} x_h$$

Also

$$S^3 x_t = \sum_{j=-\infty}^{t} \sum_{i=-\infty}^{j} \sum_{h=-\infty}^{i} x_h$$

and so on.

Equation (4.1.8) implies that the process (4.1.5) can be obtained by summing (or "integrating") the stationary process (4.1.6) d times. Therefore, we call the process (4.1.5), an *autoregressive integrated moving average* (*ARIMA*) *process*. The ARIMA models for nonstationary time series, which have also been considered by Yaglom [38], are of fundamental importance to problems of forecasting and control [14], [15], [16], [17], [18], [19], [20]. For a further discussion of nonstationary processes, see also Zadeh and Ragazzini [39], and Kalman [40], [41]. An earlier procedure for time series analysis which employed differencing was the "variate difference method." (See Tintner [90], [91].) However, the motivation, methods and objectives of this procedure were quite different from those discussed here.

As mentioned in Chapter 1, the model (4.1.5) is equivalent to representing the process z_t as the output from a linear filter (unless $d = 0$ this is an *unstable* linear filter), whose input is white noise a_t. Alternatively, we can regard it as a device *for transforming the highly dependent, and possibly nonstationary process z_t, to a sequence of uncorrelated random variables a_t;* that is, for transforming the process to white noise.

If in (4.1.5), the autoregressive operator $\phi(B)$ is of order p, the dth difference is taken, and the moving average operator $\theta(B)$ is of order q, then we say we have an ARIMA model of order (p, d, q), or simply an ARIMA (p, d, q) process.

Two interpretations of the ARIMA model. We now show that (4.1.5) is an intuitively reasonable model for the time series we wish to describe in practice. First, we note that a basic characteristic of the first-order auto-regressive process (4.1.2), for $|\phi| < 1$ and for $|\phi| > 1$, is that the local behavior of a series generated from the model is heavily dependent upon the *level* of $\tilde{z}_t$. This is to be contrasted with the behavior of series such as those in Figure 4.1, where the local behavior of the series appears to be independent of its level.

If we are to use models for which the behavior of the process is independent of its level, then we must choose the autoregressive operator $\varphi(B)$ such that

$$\varphi(B)(\tilde{z}_t + c) = \varphi(B)\tilde{z}_t$$

where c is any constant. Thus, $\varphi(B)$ must be of the form

$$\varphi(B) = \phi_1(B)(1 - B) = \phi_1(B)\nabla$$

Therefore, a class of processes having the desired property will be of the form

$$\phi_1(B)w_t = \theta(B)a_t$$

where $w_t = \nabla\tilde{z}_t = \nabla z_t$. Required homogeneity excludes the possibility that w_t should increase explosively. This means that either $\phi_1(B)$ is a stationary autoregressive operator, or $\phi_1(B) = \phi_2(B)(1 - B)$ so that $\phi_2(B)w_t = \theta(B)a_t$, where now $w_t = \nabla^2 z_t$. In the latter case the same argument can be applied to the second difference, and so on.

Eventually we arrive at the conclusion that, for the representation of time series which are nonstationary, but nevertheless exhibit homogeneity, the operator on the left of (4.1.3) should be of the form $\phi(B)\nabla^d$, where $\phi(B)$ is a *stationary* autoregressive operator. Thus, we are led back to the model (4.1.5).

To approach the model from a somewhat different viewpoint, consider the situation where $d = 0$ in (4.1.4), so that $\phi(B)\tilde{z}_t = \theta(B)a_t$. The requirement that the zeroes of $\phi(B)$ lie outside the unit circle would ensure, not only that the process $\tilde{z}_t$ was stationary with mean zero, but also that $\nabla z_t, \nabla^2 z_t, \nabla^3 z_t, \ldots$ etc., were each stationary with mean zero. Figure 4.3(a) shows one kind of nonstationary series we would like to represent. This series is homogeneous except in level, in that, except for a vertical translation, one part of it looks much the same as another. We can represent such behavior by retaining the requirement that each of the differences be stationary with zero mean, but letting the level "go free." This we do by using the model

$$\phi(B)\nabla z_t = \theta(B)a_t$$

Figure 4.3(b) shows a second kind of nonstationarity of fairly common occurrence. The series has neither a fixed level nor a fixed slope, but its behavior is homogeneous if we allow for differences in these characteristics. We can represent such behavior by the model

$$\phi(B)\nabla^2 z_t = \theta(B)a_t$$

which ensures stationarity and zero mean, for all differences after the first and second, but allows the level and the slope to "go free."

(a) A Series Showing Nonstationarity in Level such as can be
Represented by the Model $\phi(B)\nabla z_t = \theta(B)a_t$

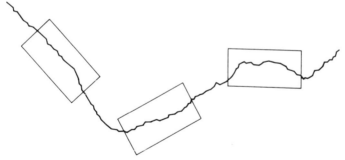

(b) A Series Showing Nonstationarity in Level and in Slope such as can be
Represented by the Model $\phi(B)\nabla^2 z_t = \theta(B)a_t$

FIG. 4.3   Two kinds of homogeneous nonstationary behavior

4.1.3 The general form of the autoregressive integrated moving average process

For reasons to be given below, it is sometimes useful to consider a slight extension of the ARIMA model (4.1.5), by adding a constant term θ_0. Thus, a rather general form of the model which we shall use to describe time series, is the autoregressive integrated moving average process

$$\varphi(B)z_t = \phi(B)\nabla^d z_t = \theta_0 + \theta(B)a_t \qquad (4.1.9)$$

where

$$\phi(B) = 1 - \phi_1 B - \phi_2 B^2 \cdots - \phi_p B^p$$

$$\theta(B) = 1 - \theta_1 B - \theta_2 B^2 \cdots - \theta_q B^q$$

In what follows:
(1) $\phi(B)$ will be called the *autoregressive operator*; it is assumed to be stationary, that is the roots of $\phi(B) = 0$ lie outside the unit circle.
(2) $\varphi(B) = \nabla^d \phi(B)$ will be called the *generalized autoregressive operator*; it is a nonstationary operator with d of the roots of $\varphi(B) = 0$ equal to unity.
(3) $\theta(B)$ will be called the *moving average operator*; it is assumed to be invertible, that is, the roots of $\theta(B) = 0$ lie outside the unit circle.

When $d = 0$, the model (4.1.9) represents a stationary process. The requirements of stationarity and invertibility apply independently, and in general, the operators $\phi(B)$ and $\theta(B)$ will not be of the same order. Examples of the stationarity regions for the simple cases of $p = 1, 2$, and the identical invertibility regions for $q = 1, 2$ have been given in Chapter 3.

Stochastic and deterministic trends. We have seen in Section 4.1.2, that when the constant term θ_0 is omitted, the model (4.1.9) is capable of representing series which have *stochastic* trends, as typified for example, by random changes in the level and slope of the series. In general, however, we may wish to include a *deterministic* function of time $f(t)$ in the model. In particular, automatic allowance for a deterministic polynomial trend, of degree d, can be made by permitting θ_0 to be nonzero. For example, when $d = 1$, we may use the model with $\theta_0 \neq 0$ to estimate a possible deterministic linear trend in the presence of nonstationary noise. Since to allow θ_0 to be nonzero is equivalent to permitting

$$E[w_t] = E[\nabla^d z_t] = \mu_w = \theta_0/(1 - \phi_1 - \phi_2 - \cdots - \phi_p)$$

to be nonzero, an alternative way of expressing this more general model (4.1.9), is in the form of a stationary invertible ARMA process in $\tilde{w}_t = w_t - \mu_w$. That is

$$\phi(B)\tilde{w}_t = \theta(B)a_t \qquad (4.1.10)$$

In many applications, where no physical reason for a deterministic component exists, the mean of w can be assumed to be zero unless such an assumption proves contrary to facts presented by the data. It is clear that, for many applications, the assumption of a stochastic trend is often more realistic than the assumption of a deterministic trend. This is of special importance in forecasting a time series, since a stochastic trend does not necessitate the series to follow the identical pattern which it has developed in the past. In what follows, when $d > 0$, we shall often assume that $\mu_w = 0$, or equivalently that $\theta_0 = 0$, unless it is clear from the data or from the nature of the problem that a nonzero mean, or more generally a deterministic component of known form, is needed.

Some important special cases of the ARIMA model. In Chapter 3 we have become acquainted with some important special cases of the model (4.1.9), corresponding to the stationary situation, $d = 0$. The following models represent some special cases of the nonstationary model $(d \geqslant 1)$, which seem to be of frequent occurrence.

(1) The (0, 1, 1) process.

$$\nabla z_t = a_t - \theta_1 a_{t-1}$$
$$= (1 - \theta_1 B)a_t$$

corresponding to $p = 0, d = 1, q = 1, \phi(B) = 1, \theta(B) = 1 - \theta_1 B$.

(2) The (0, 2, 2) process.

$$\nabla^2 z_t = a_t - \theta_1 a_{t-1} - \theta_2 a_{t-2}$$

$$= (1 - \theta_1 B - \theta_2 B^2)a_t$$

corresponding to $p = 0, d = 2, q = 2, \phi(B) = 1, \theta(B) = 1 - \theta_1 B - \theta_2 B^2$.

(3) The (1, 1, 1) process.

$$\nabla z_t - \phi_1 \nabla z_{t-1} = a_t - \theta_1 a_{t-1}$$

or

$$(1 - \phi_1 B)\nabla z_t = (1 - \theta_1 B)a_t$$

corresponding to $p = 1, d = 1, q = 1, \phi(B) = 1 - \phi_1 B, \theta(B) = 1 - \theta_1 B$.

For the representation of nonseasonal time series (seasonal models are considered in Chapter 9), we rarely seem to meet situations for which either p, d, or q need be greater than two. Frequently, values of zero or unity will be appropriate for one or more of these coefficients. For example, we show later that Series A, B, C, D, given in Figure 4.1, are reasonably well fitted* by the simple models shown in Table 4.2.

---

* As is discussed more fully later, there are certain advantages in using a nonstationary rather than a stationary model in cases of doubt. In particular, none of the fitted models above assume that z_t has a fixed mean. However, we show in Chapter 7 that it is possible in certain cases to obtain stationary models of slightly better fit.

TABLE 4.2   Summary of simple nonstationary models fitted to time series of Figure 4.1

Series	Model	Order of Model
A	$\nabla z_t = (1 - 0.7B)a_t$	$(0, 1, 1)$
B	$\nabla z_t = (1 + 0.1B)a_t$	$(0, 1, 1)$
C	$(1 - 0.8B)\nabla z_t = a_t$	$(1, 1, 0)$
D	$\nabla z_t = (1 - 0.1B)a_t$	$(0, 1, 1)$

Nonlinear transformation of z.   A considerable widening of the range of useful application of the model (4.1.9), is achieved if we allow the possibility of transformation. Thus, we may substitute $z_t^{(\lambda)}$ for z_t in (4.1.9), where $z_t^{(\lambda)}$ is some nonlinear transformation of z_t involving one or more transformation parameters λ. A suitable transformation may be suggested by the situation, or in some cases it can be estimated from the data. For example, if we were interested in the sales of a recently introduced commodity, we might find that sales volume was increasing at a rapid rate and that it was the *percentage* fluctuation which showed nonstationary stability, rather than the absolute fluctuation, supporting the analysis of the logarithm of sales. When the data cover a wide range and especially for seasonal data, estimation of the transformation using the approach of Box and Cox [42] may be helpful (Section 9.4).

4.2 THREE EXPLICIT FORMS FOR THE AUTOREGRESSIVE INTEGRATED MOVING AVERAGE MODEL

We now consider three different "explicit" forms for the general model (4.1.9). Each of these allows some special aspect to be appreciated. Thus, the current value z_t of the process can be expressed:
(a) in terms of previous values of the z's and current and previous values of the a's, by direct use of the *difference equation*;
(b) in terms of *current and previous shocks* a_{t-j} only;
(c) in terms of a weighted sum of *previous values* z_{t-j} of the process and the current shock a_t.

In this chapter we are mainly concerned with *nonstationary* models, in which $\nabla^d z_t$ is a stationary process and d is greater than zero. For such models we can, without loss of generality, omit μ from the specification or equivalently replace $\tilde{z}_t$ by z_t. The results of this chapter will, however, apply to stationary models for which $d = 0$, provided z_t is then interpreted as the *deviation* from the mean μ.

4.2.1 Difference equation form of the model

Direct use of the difference equation permits us to express the current value z_t of the process in terms of previous values of the z's and of current and previous values of the a's. Thus, if

$$\varphi(B) = \phi(B)(1 - B)^d = 1 - \varphi_1 B - \varphi_2 B^2 \cdots - \varphi_{p+d} B^{p+d}$$

the general model (4.1.9), with $\theta_0 = 0$, may be written

$$z_t = \varphi_1 z_{t-1} + \cdots + \varphi_{p+d} z_{t-p-d} - \theta_1 a_{t-1} \cdots - \theta_q a_{t-q} + a_t \quad (4.2.1)$$

For example consider the process represented by the model of order $(1, 1, 1)$

$$(1 - \phi B)(1 - B)z_t = (1 - \theta B)a_t$$

where for convenience we drop the suffix 1 on ϕ_1 and θ_1. Then this process may be written

$$\{1 - (1 + \phi)B + \phi B^2\}z_t = (1 - \theta B)a_t$$

that is

$$z_t = (1 + \phi)z_{t-1} - \phi z_{t-2} + a_t - \theta a_{t-1} \quad (4.2.2)$$

For many purposes, and in particular for calculating the forecasts in Chapter 5, the difference equation (4.2.1) is the most convenient form to employ.

4.2.2   Random shock form of the model

The model in terms of current and previous shocks.   We have seen in Section 3.1.1 that a linear model can be written as the output z_t from the linear filter

$$z_t = a_t + \psi_1 a_{t-1} + \psi_2 a_{t-2} + \cdots$$

$$= a_t + \sum_{j=1}^{\infty} \psi_j a_{t-j}$$

$$= \psi(B)a_t \quad (4.2.3)$$

whose input is white noise, or a sequence of uncorrelated shocks a_t. It is sometimes useful to express the ARIMA model in the form (4.2.3), and in particular, the ψ weights will be needed in Chapter 5 to calculate the variance of the forecasts. We now show that the ψ weights for an ARIMA process may be obtained directly from the difference equation form of the model.

General expression for the ψ weights.   If we operate on both sides of (4.2.3) with the generalized autoregressive operator $\varphi(B)$, we obtain

$$\varphi(B)z_t = \varphi(B)\psi(B)a_t$$

However, since

$$\varphi(B)z_t = \theta(B)a_t$$

it follows that

$$\varphi(B)\psi(B) = \theta(B) \quad (4.2.4)$$

Therefore, the ψ weights may be obtained by equating coefficients of B in the expansion

$$(1 - \varphi_1 B - \cdots - \varphi_{p+d} B^{p+d})(1 + \psi_1 B + \psi_2 B^2 + \cdots)$$
$$= (1 - \theta_1 B - \cdots - \theta_q B^q) \tag{4.2.5}$$

We note that for j greater than the larger of $p + d - 1$ and q, that is

$$j > p + d - 1 \quad \text{if} \quad p + d - 1 \geqslant q$$
$$j > q \qquad\qquad \text{if} \quad p + d - 1 < q$$

the ψ weights satisfy the difference equation defined by the generalized autoregressive operator, that is

$$\varphi(B)\psi_j = \phi(B)(1 - B)^d \psi_j = 0 \tag{4.2.6}$$

where B now operates on the subscript j. Thus, for sufficiently large j, the weights ψ_j are represented by a mixture of polynomials, damped exponentials, and damped sinusoids in the argument j.

   Example. To illustrate the use of (4.2.5), consider the $(1, 1, 1)$ process (4.2.2) for which

$$\varphi(B) = (1 - \phi B)(1 - B)$$
$$= 1 - (1 + \phi)B + \phi B^2$$

and

$$\theta(B) = 1 - \theta B$$

Substituting in (4.2.5) gives

$$\{1 - (1 + \phi)B + \phi B^2\}(1 + \psi_1 B + \psi_2 B^2 + \cdots) = 1 - \theta B$$

and hence

$$\psi_0 = 1$$
$$\psi_1 = A_0 + A_1 \phi$$
$$\psi_2 = A_0 + A_1 \phi^2$$
$$\vdots \qquad \vdots \qquad \vdots$$
$$\psi_j = A_0 + A_1 \phi^j \tag{4.2.7}$$

where

$$A_0 = \frac{1 - \theta}{1 - \phi} \qquad\qquad A_1 = \frac{\theta - \phi}{1 - \phi}$$

and $\psi_0 = A_0 + A_1 = 1$. Thus, we can express the model (4.2.2) in the equivalent form

$$z_t = \sum_{j=0}^{\infty} (A_0 + A_1 \phi^j) a_{t-j} \tag{4.2.8}$$

Since $|\phi| < 1$, the weights ψ_j tend to A_0 for large j, so that shocks a_{t-j}, which entered in the remote past, receive a constant weight A_0.

Truncated form of the random shock model.   For some purposes it is convenient to consider the model in a slightly different form from (4.2.3). Suppose we wish to express the current value z_t of the process in terms of the $t - k$ shocks $a_t, a_{t-1}, \ldots, a_{k+1}$, which have entered the system since some time origin $k < t$. This time origin k might, for example, be the time at which the process was first observed.

The general model

$$\varphi(B) z_t = \theta(B) a_t \tag{4.2.9}$$

is a difference equation with the solution

$$z_t = C_k(t - k) + I_k(t - k) \tag{4.2.10}$$

A short discussion of linear difference equations is given in Appendix A4.1. We remind the reader that the solution of such equations closely parallels the solution of linear differential equations. The *complementary function* $C_k(t - k)$, is the general solution of the difference equation

$$\varphi(B) C_k(t - k) = 0 \tag{4.2.11}$$

In general, this solution will consist of a *linear* combination of certain functions of time. These functions are powers t^j, real geometric (exponential) terms G^t, and complex geometric (exponential) terms $d^t \sin(2\pi f_0 t + F)$, where the constants G, f_0, F, are functions of the parameters (ϕ, θ) of the model. The *particular integral* $I_k(t - k)$ is *any* function which satisfies

$$\varphi(B) I_k(t - k) = \theta(B) a_t \tag{4.2.12}$$

It should be carefully noted that in this expression B operates on t and *not* on k. It is shown in Appendix A4.1, that this equation is satisfied for $t - k > q$ by

$$I_k(s - k) = \begin{cases} 0, & s \leqslant k \\ \displaystyle\sum_{j=k+1}^{s} \psi_{s-j} a_j = a_s + \psi_1 a_{s-1} + \cdots + \psi_{s-k-1} a_{k+1} \\ & \qquad\qquad\qquad\qquad s > k \end{cases} \tag{4.2.13}$$

For illustration, consider Figure 4.4. The above discussion implies that any observation z_t can be considered in relation to any previous time k and can be divided up into two additive parts. The first part $C_k(t - k)$ is the component of z_t, *already determined at time k*, and indicates what the observations prior to time $k + 1$ had to tell us about the value of the series at time t.

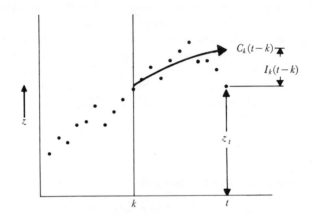

FIG. 4.4   Role of the complementary function $C_k(t - k)$ and of the particular integral $I_k(t - k)$ in describing the behavior of a time series

It represents the course which the process would take, if at time k, the source of shocks a_t had been "switched off." The second part $I_k(t - k)$ represents an additional component, *unpredictable at time k*, which embodies the entire effect of shocks entering the system after time k.

    Example.   For illustration, consider again the example

$$(1 - \phi B)(1 - B)z_t = (1 - \theta B)a_t$$

The complementary function is the solution of the difference equation

$$(1 - \phi B)(1 - B)C_k(t - k) = 0$$

that is

$$C_k(t - k) = b_0^{(k)} + b_1^{(k)}\phi^{t-k}$$

where $b_0^{(k)}$, $b_1^{(k)}$ are coefficients which depend on the past history of the process and, it will be noted, *change with the origin k*.

    Making use of the ψ weights (4.2.7), a particular integral (4.2.13) is

$$I_k(t - k) = \sum_{j=k+1}^{t} (A_0 + A_1\phi^{t-j})a_j$$

so that finally we can write the model (4.2.8) in the equivalent form

$$z_t = b_0^{(k)} + b_1^{(k)}\phi^{t-k} + \sum_{j=k+1}^{t} (A_0 + A_1\phi^{t-j})a_j \qquad (4.2.14)$$

Link between the truncated and nontruncated forms of the random shock model. Returning to the general case, we can always think of the process with reference to some infinitely remote origin, at which the complementary function $C_{-\infty}(t)$ can, without loss of generality, be taken to be zero. In that case

$$z_t = \sum_{j=-\infty}^{t} \psi_{t-j}a_j = \psi(B)a_t = I_{-\infty}(t) \qquad (4.2.15)$$

which is the nontruncated form (4.2.3) of the model.

The complementary function $C_k(t-k)$ can also be expressed in terms of the ψ weights, for on subtracting (4.2.10) from (4.2.15), we obtain for $t-k > q$

$$C_k(t-k) = I_{-\infty}(t) - I_k(t-k) = \sum_{j=-\infty}^{k} \psi_{t-j}a_j \qquad (4.2.16)$$

It may be verified that the form (4.2.16) for the complementary function does satisfy (4.2.11) since

$$\varphi(B)\{I_{-\infty}(t) - I_k(t-k)\} = \{\theta(B) - \theta(B)\}a_t = 0$$

In summary then, for the general model (4.2.9):
(1) We can express the value z_t of the process as an infinite weighted sum of current and previous shocks a_j, according to

$$z_t = \sum_{j=-\infty}^{t} \psi_{t-j}a_j = \sum_{j=0}^{\infty} \psi_j a_{t-j} = \psi(B)a_t$$

(2) For $t-k > q$, the value of z_t can be expressed as a weighted finite sum of the $t-k$ current and previous shocks occurring after some origin k, plus a complementary function $C_k(t-k)$. This finite sum consists of the first $t-k$ terms of the infinite sum, so that

$$z_t = C_k(t-k) + \sum_{j=k+1}^{t} \psi_{t-j}a_j \qquad (4.2.17)$$

Finally, the complementary function $C_k(t-k)$ can be taken to be equal to the truncated infinite sum, so that

$$C_k(t-k) = \sum_{j=-\infty}^{k} \psi_{t-j}a_j \qquad (4.2.18)$$

For illustration, consider once more the model

$$(1 - \phi B)(1 - B)z_t = (1 - \theta B)a_t$$

We can write z_t either as an infinite weighted sum of the a_j's

$$z_t = \sum_{j=-\infty}^{t} (A_0 + A_1 \phi^{t-j}) a_j$$

or in terms of the weighted finite sum as

$$z_t = C_k(t - k) + \sum_{j=k+1}^{t} (A_0 + A_1 \phi^{t-j}) a_j$$

Furthermore, the complementary function is the truncated sum

$$C_k(t - k) = \sum_{j=-\infty}^{k} (A_0 + A_1 \phi^{t-j}) a_j$$

which can be written

$$C_k(t - k) = b_0^{(k)} + b_1^{(k)} \phi^{t-k}$$

where

$$b_0^{(k)} = A_0 \sum_{j=-\infty}^{k} a_j = \frac{1 - \theta}{1 - \phi} \sum_{j=-\infty}^{k} a_j$$

$$b_1^{(k)} = A_1 \sum_{j=-\infty}^{k} \phi^{k-j} a_j = \frac{\theta - \phi}{1 - \phi} \sum_{j=-\infty}^{k} \phi^{k-j} a_j$$

Complementary function as a conditional expectation. One consequence of the relationship (4.2.18) is that for $m > 0$

$$C_k(t - k) = C_{k-m}(t - k + m) + \psi_{t-k} a_k + \psi_{t-k+1} a_{k-1} + \cdots$$
$$+ \psi_{t-k+m-1} a_{k-m+1} \qquad (4.2.19)$$

which shows how the complementary function changes as the origin k is changed. Now denote by $\underset{k}{E}[z_t]$, the *conditional expectation of z_t at time k.* That is the expectation given complete historical knowledge up to, but not beyond time k. To calculate this expectation, note that

$$\underset{k}{E}[a_j] = \begin{cases} 0 & j > k \\ a_j & j \leqslant k \end{cases}$$

That is to say, *standing at time k,* the expected values of a's that have yet to happen is zero and the expectation of those which have happened already is the value they have actually realized.

By taking conditional expectations at time k on both sides of (4.2.17), we obtain $\underset{k}{E}[z_t] = C_k(t - k)$. Thus, for $(t - k) > q$, the complementary function provides the expected value of the future value z_t of the process, *viewed from time k* and based on knowledge of the past. The particular integral shows how that expectation is modified by *subsequent* events represented by the shocks $a_{k+1}, a_{k+2}, \ldots, a_t$. In the problem of forecasting, which we discuss in Chapter 5, it will turn out that $C_k(t - k)$ is the minimum mean square error forecast of z_t made at time k. Equation (4.2.19) may be used in "updating" this forecast.

4.2.3 Inverted form of the model

The model in terms of previous z's and the current shock a_t. We have seen in Section 3.1.1 that the model

$$z_t = \psi(B)a_t$$

may also be written in the inverted form

$$\psi^{-1}(B)z_t = a_t$$

or

$$\pi(B)z_t = \left(1 - \sum_{j=1}^{\infty} \pi_j B^j\right)z_t = a_t \tag{4.2.20}$$

Thus z_t is an infinite weighted sum of previous values of z, plus a random shock

$$z_t = \pi_1 z_{t-1} + \pi_2 z_{t-2} + \cdots + a_t$$

Because of the invertibility condition, the π weights in (4.2.20) must form a convergent series, that is, $\pi(B)$ must converge on or within the unit circle.

General expression for the π weights. To derive the π weights for the general ARIMA model, we can substitute (4.2.20) in

$$\varphi(B)z_t = \theta(B)a_t$$

to obtain

$$\varphi(B)z_t = \theta(B)\pi(B)z_t$$

Hence, the π weights can be obtained explicitly by equating coefficients of B in

$$\varphi(B) = \theta(B)\pi(B) \tag{4.2.21}$$

that is

$$(1 - \varphi_1 B - \cdots - \varphi_{p+d}B^{p+d}) = (1 - \theta_1 B - \cdots - \theta_q B^q)$$
$$\times (1 - \pi_1 B - \pi_2 B^2 - \cdots) \tag{4.2.22}$$

It will be noted that for j greater than the larger of $p + d$ and q, that is for

$$j > p + d \quad \text{if} \quad p + d \geqslant q$$

$$j > q \qquad \text{if} \quad p + d < q$$

the π weights satisfy the difference equation defined by the *moving average operator*

$$\theta(B)\pi_j = 0$$

where B now operates on j. Hence, for sufficiently large j, the π weights will behave like the autocorrelation function (3.2.5) of an autoregressive process, that is, they follow a mixture of damped exponentials and damped sine waves.

Another interesting fact is that if $d \geqslant 1$, the π weights in (4.2.20) sum to unity. This may be verified by substituting $B = 1$ in (4.2.21). Thus, $\varphi(B) = \phi(B)(1 - B)^d$ is zero when $B = 1$ and $\theta(1) \neq 0$, because the roots of $\theta(B) = 0$ lie outside the unit circle. Hence, it follows from (4.2.21) that $\pi(1) = 0$, that is

$$\sum_{j=1}^{\infty} \pi_j = 1 \qquad (4.2.23)$$

Therefore, if $d \geqslant 1$, the process may be written in the form

$$z_t = \bar{z}_{t-1}(\pi) + a_t \qquad (4.2.24)$$

where

$$\bar{z}_{t-1}(\pi) = \sum_{j=1}^{\infty} \pi_j z_{t-j}$$

is a *weighted average* of previous values of the process.

Example. We again consider, for illustration, the ARIMA $(1, 1, 1)$ process

$$(1 - \phi B)(1 - B)z_t = (1 - \theta B)a_t$$

Then, using (4.2.21),

$$\pi(B) = \varphi(B)\theta^{-1}(B) = \{1 - (1 + \phi)B + \phi B^2\}(1 + \theta B + \theta^2 B^2 + \cdots)$$

so that

$$\pi_1 = \phi + (1 - \theta), \quad \pi_2 = (\theta - \phi)(1 - \theta), \quad \pi_j = (\theta - \phi)(1 - \theta)\theta^{j-2} \quad j \geqslant 3$$

For example, the first seven π weights corresponding to $\phi = -0.3$ and $\theta = 0.5$, are given in Table 4.3.

TABLE 4.3 First seven π weights for an ARIMA $(1, 1, 1)$ process with $\phi = -0.3$, $\theta = 0.5$

j	1	2	3	4	5	6	7
π_j	0.2	0.4	0.2	0.1	0.05	0.025	0.0125

Thus, the tth value of the process would be generated by a weighted average of previous values, plus an additional shock, according to

$$z_t = (0.2z_{t-1} + 0.4z_{t-2} + 0.2z_{t-3} + 0.1z_{t-4} + \cdots) + a_t$$

We notice in particular that the π weights die out as more and more remote values of z_{t-j} are involved. This is the property that results from requiring that the series be invertible (in this case by requiring that $-1 < \theta < 1$).

We mention in passing that, for statistical models representing practically occurring time series, the convergent π weights usually die out rather quickly. Thus, although z_t may be theoretically dependent on the remote past, the representation

$$z_t = \sum_{j=1}^{\infty} \pi_j z_{t-j} + a_t$$

will usually show that z_t is dependent *to an important extent* only on recent past values z_{t-j} of the time series. This is still true even though, for nonstationary models with $d > 0$, the ψ weights in the "weighted shock" representation

$$z_t = \sum_{j=0}^{\infty} \psi_j a_{t-j}$$

do not converge. What happens, of course, is that all the information which remote values of the shocks a_{t-j} supply about z_t is contained in recent values $z_{t-1}, z_{t-2}, \ldots$ of the series. In particular the expectation $E[z_t]$, in theory conditional on complete historical knowledge up to time k, can usually be computed to sufficient accuracy from *recent* values of the time series. This fact is particularly important in forecasting applications.

4.3 INTEGRATED MOVING AVERAGE PROCESSES

A nonstationary model which is useful in representing some commonly occurring series is the $(0, 1, 1)$ process

$$\nabla z_t = a_t - \theta a_{t-1}$$

The model contains only two parameters, θ and σ_a^2. Figure 4.5 shows two time series generated by this model from the same sequence of random Normal deviates a_t. For the first series, $\theta = 0.6$ and for the second, $\theta = 0$. Models of this kind have often been found useful in inventory control problems, in representing certain kinds of disturbance occurring in industrial processes and in econometrics. As we have noted in Section 4.1.3, it will be shown in Chapter 7 that this simple process can, with suitable parameter values, supply useful representations of Series A, B, and D shown in Figure

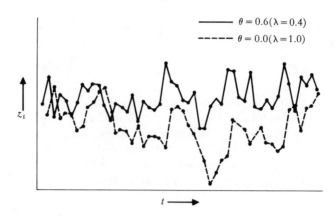

FIG 4.5   Two time series generated from an IMA (0, 1, 1) process

4.1. Another valuable model is the (0, 2, 2) process

$$\nabla^2 z_t = a_t - \theta_1 a_{t-1} - \theta_2 a_{t-2}$$

which contains three parameters, namely, θ_1, θ_2, and σ_a^2. Figure 4.6 shows two series generated from the model using the same set of Normal deviates.

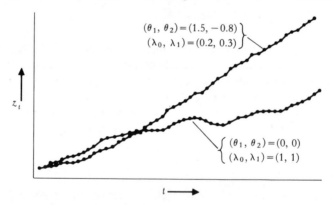

FIG. 4.6   Two time series generated from an IMA (0, 2, 2) process

For the first series the parameters $(\theta_1, \theta_2) = (0, 0)$ and for the second $(\theta_1, \theta_2) = (1.5, -0.8)$. The series tend to be much smoother than those generated by the $(0, 1, 1)$ process. The $(0, 2, 2)$ models are useful in representing disturbances (such as Series C) in systems with a large degree of inertia. Both the $(0, 1, 1)$ and $(0, 2, 2)$ models are special cases of the class

$$\nabla^d z_t = \theta(B) a_t \qquad (4.3.1)$$

We call such models as (4.3.1) *integrated moving average* (IMA) processes, of order $(0, d, q)$ and consider their properties in the following section.

4.3.1 *The integrated moving average process of order* (0, 1, 1)

Difference equation form.   The (0, 1, 1) process

$$\nabla z_t = (1 - \theta B)a_t, \qquad -1 < \theta < 1$$

possesses useful representational capability and we now study its properties in more detail. The model can be written in terms of the z's and the a's in the form

$$z_t = z_{t-1} + a_t - \theta a_{t-1} \tag{4.3.2}$$

Random shock form of model.   Alternatively, we can obtain z_t in terms of the a's alone by summing on both sides of (4.3.2). Before doing this, there is some advantage in expressing the right-hand operator in terms of ∇ rather than B. Thus we can write

$$1 - \theta B = (1 - \theta)B + (1 - B) = (1 - \theta)B + \nabla = \lambda B + \nabla$$

where $\lambda = 1 - \theta$, and the invertibility region in terms of λ is defined by $0 < \lambda < 2$. Hence

$$\nabla z_t = \lambda a_{t-1} + \nabla a_t$$

and on summation, this yields

$$z_t = \lambda S a_{t-1} + a_t \tag{4.3.3}$$

whence, in $z_t = \sum_{j=0}^{\infty} \psi_j a_{t-j}$, the weights are $\psi_0 = 1, \psi_j = \lambda$ for $j \geqslant 1$.

If we express the model in terms of a's entering the system after the time origin k, we obtain, as in (4.2.17)

$$z_t = b_0^{(k)} + \lambda \sum_{j=k+1}^{t-1} a_j + a_t \tag{4.3.4}$$

where the complementary function $C_k(t - k) = b_0^{(k)}$ (a constant b_o for each k) is the solution of the difference equation $(1 - B)C_k(t - k) = 0$. Moreover, since

$$Sa_{t-1} = \sum_{j=k+1}^{t-1} a_j + Sa_k$$

we can write

$$b_0^{(k)} = \lambda S a_k = \lambda \sum_{j=-\infty}^{k} a_j$$

For this model then, the complementary function is simply a constant (i.e. a polynomial in t of degree zero) representing the current "level" of the process and associated with the particular origin of reference k. If the origin is changed from $k - 1$ to k, then b_0 is "updated" according to

$$b_0^{(k)} = b_0^{(k-1)} + \lambda a_k$$

Inverted form of model.   Finally, we can consider the model in the form

$$\pi(B)z_t = a_t$$

or equivalently, in the form

$$z_t = \sum_{j=1}^{\infty} \pi_j z_{t-j} + a_t = \bar{z}_{t-1}(\pi) + a_t$$

where $\bar{z}_{t-1}(\pi)$ is a weighted moving average of previous values of the process. Using (4.2.21), the π weights for the IMA (0, 1, 1) process are given by

$$(1 - \theta B)\pi(B) = 1 - B$$

that is

$$\pi(B) = \frac{1 - B}{1 - \theta B} = \frac{1 - \theta B - (1 - \theta)B}{1 - \theta B}$$

$$= 1 - (1 - \theta)\{B + \theta B^2 + \theta^2 B^3 + \cdots\}$$

whence

$$\pi_j = (1 - \theta)\theta^{j-1} = \lambda(1 - \lambda)^{j-1} \quad j \geqslant 1$$

Thus, the process may be written

$$z_t = \bar{z}_{t-1}(\lambda) + a_t \tag{4.3.5}$$

The weighted moving average of previous values of the process

$$\bar{z}_{t-1}(\lambda) = \lambda \sum_{j=1}^{\infty} (1 - \lambda)^{j-1} z_{t-j} \tag{4.3.6}$$

is in this case, an *exponentially weighted moving average*, or EWMA.

This moving average (4.3.6) is said to be exponentially (or geometrically) weighted because the weights

$$\lambda \qquad \lambda(1 - \lambda) \qquad \lambda(1 - \lambda)^2 \qquad \lambda(1 - \lambda)^3 \ldots$$

fall off exponentially (that is as a geometric progression). The weight function for an IMA process of order (0, 1, 1), with $\lambda = 0.4$ ($\theta = 0.6$), is shown in Figure 4.7.

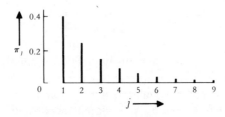

FIG. 4.7 π weights for an IMA process of order (0, 1, 1) with $\lambda = 1 - \theta = 0.4$

Although the invertibility condition is satisfied for $0 < \lambda < 2$, we are in practice most often concerned with values of λ between zero and one. We note that if λ had a value equal to 1, then the weight function would consist of a single spike ($\pi_1 = 1$, $\pi_j = 0$ for $j > 1$). As the value λ approaches zero, the exponential weights die out more and more slowly and the EWMA stretches back further into the process. Finally, with $\lambda = 0$ and $\theta = 1$, the model $(1 - B)z_t = (1 - B)a_t$ is equivalent to $z_t = \theta_0 + a_t$, with θ_0 being given by the mean of all past values.

On comparing (4.3.3) and (4.3.5), it is evident that

$$\bar{z}_{t-1}(\lambda) = \lambda S a_{t-1} \qquad (4.3.7)$$

It follows in particular, that for this process, the complementary function $b_0^{(k)} = C_k(t - k)$ in (4.3.4) is

$$b_0^{(k)} = \bar{z}_k(\lambda)$$

an exponentially weighted average of values up to the origin k. In fact, (4.3.4) may be written

$$z_t = \bar{z}_k(\lambda) + \lambda \sum_{j=k+1}^{t-1} a_j + a_t$$

We have seen that the complementary function $C_k(t - k)$ can be thought of as telling us what is known of the future value of the process at time t, based on knowledge of the past when *we are standing at time k*. For the IMA (0, 1, 1) process, this takes the form of information about the "level" or location of the process $b_0^{(k)} = \bar{z}_k(\lambda)$. At time k, our knowledge of the future behavior of the process is that it will diverge from this level in accordance with the "random walk" represented by $\lambda \sum_{j=k+1}^{t-1} a_j + a_t$, whose expectation is zero and whose behavior we cannot predict. As soon as a new observation is available, that is, as soon as we move our origin to time $k + 1$, the level will be updated to $b_0^{(k+1)} = \bar{z}_{k+1}(\lambda)$.

Important properties of the IMA (0, 1, 1) process.   Since the process is non-stationary, it possesses no mean. However, the exponentially weighted moving average $\bar{z}_t(\lambda)$, can be regarded as measuring the location or "level" of the process at time t. From its definition (4.3.6), we obtain the well-known recursion formula for the EWMA.

$$\bar{z}_t(\lambda) = \lambda z_t + (1 - \lambda)\bar{z}_{t-1}(\lambda) \qquad (4.3.8)$$

This expression shows that for the IMA (0, 1, 1) model, each new level is arrived at by interpolating between the new observation and the previous level. If λ is equal to unity, $\bar{z}_t(\lambda) = z_t$, which would ignore all evidence concerning location coming from previous observations. On the other hand, if λ had some value close to zero, $\bar{z}_t(\lambda)$ would rely heavily on the previous value $\bar{z}_{t-1}(\lambda)$, which would have weight $1 - \lambda$. Only the small weight λ would be given to the new observation.

Now consider the two equations

$$z_t = \bar{z}_{t-1}(\lambda) + a_t$$
$$\bar{z}_t(\lambda) = \bar{z}_{t-1}(\lambda) + \lambda a_t \tag{4.3.9}$$

the latter being obtained by substituting (4.3.5) in (4.3.8) and is also directly derivable from (4.3.7).

It has been pointed out by Muth [43] that these two equations (4.3.9), provide a useful way of thinking about the generation of the process. The first equation shows how, with the "level" of the system at $\bar{z}_{t-1}(\lambda)$, a shock a_t is added at time t and produces the value z_t. However, the second equation shows that only a proportion λ of the shock is actually absorbed into the level and has a lasting influence; the remaining proportion $\theta = 1 - \lambda$ of the shock being dissipated. Now a new level $\bar{z}_t(\lambda)$ having been established by the absorption of a_t, a new shock a_{t+1} enters the system at time $t + 1$. The equations (4.3.9), with subscripts increased by unity, will then show how this shock produces z_{t+1} and how a proportion λ of it is absorbed into the system to produce the new level $\bar{z}_{t+1}(\lambda)$, and so on.

The manner in which this updating occurs is also exemplified by writing the pair of equations (4.3.9) in terms of infinite sums of the a's, when they become

$$z_t = \lambda S a_{t-1} + a_t$$
$$\lambda S a_t = \lambda S a_{t-1} + \lambda a_t \tag{4.3.10}$$

The properties of the IMA $(0, 1, 1)$ process with deterministic drift

$$\nabla z_t = \theta_0 + (1 - \theta_1 B)a_t$$

are discussed in Appendix A4.2.

4.3.2 The integrated moving average process of order $(0, 2, 2)$

Difference equation form. The process

$$\nabla^2 z_t = (1 - \theta_1 B - \theta_2 B^2)a_t \tag{4.3.11}$$

possesses representational possibilities for series possessing stochastic trends (for example, see Figure 4.6) and we now study its general properties within the invertibility region

$$-1 < \theta_2 < 1 \qquad \theta_2 + \theta_1 < 1 \qquad \theta_2 - \theta_1 < 1$$

Proceeding as before, z_t can be written explicitly in terms of z's and a's as

$$z_t = 2z_{t-1} - z_{t-2} + a_t - \theta_1 a_{t-1} - \theta_2 a_{t-2}$$

Random shock form of model. Alternatively, to obtain z_t in terms of the a's, we first rewrite the right hand operator in terms of differences

$$1 - \theta_1 B - \theta_2 B^2 = (\lambda_0 \nabla + \lambda_1)B + \nabla^2$$

and on equating coefficients, we find expressions for the θ's in terms of the λ's, and vice versa, as follows

$$\left.\begin{array}{l} \theta_1 = 2 - \lambda_0 - \lambda_1 \\ \theta_2 = \lambda_0 - 1 \end{array}\right\} \qquad \left.\begin{array}{l} \lambda_0 = 1 + \theta_2 \\ \lambda_1 = 1 - \theta_1 - \theta_2 \end{array}\right\} \qquad (4.3.12)$$

The model (4.3.11) may then be rewritten

$$\nabla^2 z_t = (\lambda_0 \nabla + \lambda_1)a_{t-1} + \nabla^2 a_t \qquad (4.3.13)$$

On summing (4.3.13) twice, we find

$$z_t = \lambda_0 S a_{t-1} + \lambda_1 S^2 a_{t-1} + a_t \qquad (4.3.14)$$

so, for this process, the ψ weights are

$$\psi_0 = 1, \psi_1 = (\lambda_0 + \lambda_1), \dots, \psi_j = (\lambda_0 + j\lambda_1), \dots$$

There is an important advantage in using the forms (4.3.13) or (4.3.14) of the model, as compared with (4.3.11). This stems from the fact that if we set $\lambda_1 = 0$ in (4.3.13), we obtain

$$\nabla z_t = \{1 - (1 - \lambda_0)B\}a_t$$

which corresponds to a $(0, 1, 1)$ process, with $\theta = 1 - \lambda_0$. However, if we set $\theta_2 = 0$ in (4.3.11), we obtain

$$\nabla^2 z_t = (1 - \theta_1 B)a_t$$

It is shown later in Chapter 5, that for a series generated by the $(0, 2, 2)$ process, the optimal forecasts lie along a straight line, the level and *slope* of which are continually updated as new data becomes available. By contrast, a series generated by a $(0, 1, 1)$ process can supply no information about slope, but only about a continually updated level. It can be an important question whether a linear trend, as well as the level, can be forecasted and updated. When the choice is between these two models, this question turns on whether or not λ_1 in (4.3.14) is zero.

The invertibility region for an IMA $(0, 2, 2)$ process is the same as that given for a MA(2) process in Chapter 3. It may be written in terms of the θ's and λ's as follows:

$$\left.\begin{array}{l} \theta_2 + \theta_1 < 1 \\ \theta_2 - \theta_1 < 1 \\ -1 < \theta_2 < 1 \end{array}\right\} \qquad \left.\begin{array}{l} 0 < 2\lambda_0 + \lambda_1 < 4 \\ \lambda_1 > 0 \\ \lambda_0 > 0 \end{array}\right\} \qquad (4.3.15)$$

The triangular region for the θ's was shown in Figure 3.8 and the corresponding region for the λ's is shown in Figure 4.8.

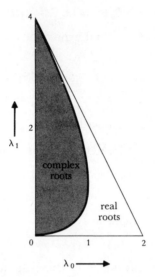

FIG. 4.8   Invertibility region for parameters λ_0 and λ_1 of an IMA (0, 2, 2) process

Truncated form of random shock model.   Using (4.3.14), the truncated form of the model (4.2.17), may be written

$$z_t = C_k(t - k) + \lambda_0 \sum_{j=k+1}^{t-1} a_j + \lambda_1 \sum_{i=k+1}^{t-1} \sum_{j=k+1}^{i} a_j + a_t \qquad (4.3.16)$$

where the complementary function is the solution of

$$(1 - B)^2 C_k(t - k) = 0$$

that is

$$C_k(t - k) = b_0^{(k)} + b_1^{(k)}(t - k)$$

which is a polynomial in $(t - k)$ of degree 1, whose coefficients depend on the location of the origin k.

It is shown in Appendix A4.3, that the complementary function may be written explicitly in terms of the a's as

$$C_k(t - k) = \{(\lambda_0 - \lambda_1)Sa_k + \lambda_1 S^2 a_k\} + \{\lambda_1 Sa_k\}(t - k) \qquad (4.3.17)$$

so that

$$b_0^{(k)} = (\lambda_0 - \lambda_1)Sa_k + \lambda_1 S^2 a_k$$
$$b_1^{(k)} = \lambda_1 Sa_k$$

Also, by considering the differences $b_0^{(k)} - b_0^{(k-1)}$ and $b_1^{(k)} - b_1^{(k-1)}$, it follows that if the origin is updated from $k - 1$ to k, then b_0 and b_1 are updated according to

$$
\begin{aligned}
b_0^{(k)} &= b_0^{(k-1)} + b_1^{(k-1)} + \lambda_0 a_k \\
b_1^{(k)} &= b_1^{(k-1)} + \lambda_1 a_k
\end{aligned}
\tag{4.3.18}
$$

We see that when this model is appropriate, our expectation of the future behavior of the series, judged from origin k, would be represented by the straight line (4.3.17), having location $b_0^{(k)}$ and slope $b_1^{(k)}$. In practice, the process will, by time t, have diverged from this line because of the influence of the random component

$$
\lambda_0 \sum_{j=k+1}^{t-1} a_j + \lambda_1 \sum_{i=k+1}^{t-1} \sum_{j=k+1}^{i} a_j + a_t
$$

which at time k is unpredictable. Moreover, on moving from origin $k - 1$ to origin k, the intercept and slope are updated according to (4.3.18).

Inverted form of model. Finally, we consider the model in the inverted form

$$
z_t = \sum_{j=1}^{\infty} \pi_j z_{t-j} + a_t = \bar{z}_{t-1}(\pi) + a_t
$$

Using (4.2.22), we find on equating coefficients in

$$
1 - 2B + B^2 = (1 - \theta_1 B - \theta_2 B^2)(1 - \pi_1 B - \pi_2 B^2 - \cdots)
$$

that the π weights of the IMA $(0, 2, 2)$ process are

$$
\begin{aligned}
\pi_1 &= 2 - \theta_1 = \lambda_0 + \lambda_1 \\
\pi_2 &= \theta_1(2 - \theta_1) - (1 + \theta_2) = \lambda_0 + 2\lambda_1 - (\lambda_0 + \lambda_1)^2 \\
(1 - \theta_1 B - \theta_2 B^2)\pi_j &= 0 \qquad j \geqslant 3
\end{aligned}
\tag{4.3.19}
$$

where B now operates on j.

If the roots of the characteristic equation $1 - \theta_1 B - \theta_2 B^2 = 0$ are real, the π weights applied to previous z's are a mixture of two damped exponentials. If the roots are complex, the weights follow a damped sine wave. Figure 4.9 shows the weights for a process with $\theta_1 = 0.9$ and $\theta_2 = -0.5$, that is $\lambda_0 = 0.5$ and $\lambda_1 = 0.6$. We see from Figures 3.9 and 4.8, that for these values the characteristic equation has complex roots (the discriminant $\theta_1^2 + 4\theta_2 = -1.19$, is less than zero). Hence, the weights in Figure 4.9 would be expected to follow a damped sine wave, as they do.

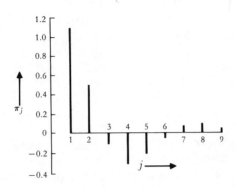

FIG. 4.9 π weights for an IMA process of order $(0, 2, 2)$ with $\lambda_0 = 0.5$, $\lambda_1 = 0.6$

4.3.3 The general integrated moving average process of order $(0, d, q)$

Difference equation form.   The general integrated moving average process of order $(0, d, q)$ is

$$\nabla^d z_t = (1 - \theta_1 B - \theta_2 B^2 - \cdots - \theta_q B^q)a_t = \theta(B)a_t \qquad (4.3.20)$$

where the zeros of $\theta(B)$ must lie outside the unit circle for the process to be invertible. The model (4.3.20) may be written explicitly in terms of past z's and a's in the form

$$z_t = dz_{t-1} - \tfrac{1}{2}d(d-1)z_{t-2} + \cdots + (-1)^{d+1}z_{t-d} + a_t - \theta_1 a_{t-1} - \cdots - \theta_q a_{t-q}$$

Random shock form of model.   To obtain z_t in terms of the a's, we write the right-hand operator in (4.3.20) in terms of $\nabla = 1 - B$. In this way we obtain

$$(1 - \theta_1 B - \cdots - \theta_q B^q) = (\lambda_{d-q}\nabla^{q-1} + \cdots + \lambda_0 \nabla^{d-1} + \cdots + \lambda_{d-1})B + \nabla^d$$
$$(4.3.21)$$

where, as before, the λ's may be written explicitly in terms of the θ's, by equating coefficients of B.

On substituting (4.3.21) in (4.3.20) and summing d times, we obtain

$$z_t = (\lambda_{d-q}\nabla^{q-d-1} + \cdots + \lambda_0 S + \cdots + \lambda_{d-1}S^d)a_{t-1} + a_t \quad (4.3.22)$$

From (4.3.22), if $q > d$, we notice that in addition to the d sums, we pick up $q - d$ additional terms $\nabla^{q-d-1}a_{t-1}$ etc., involving $a_{t-1}, a_{t-2}, \ldots, a_{t+d-q}$.

If we write this solution in terms of finite sums of a's entering the system after some origin k, we obtain the same form of equation, but with an added complementary function which is the solution of

$$\nabla^d C_k(t - k) = 0$$

that is the polynomial

$$C_k(t - k) = b_0^{(k)} + b_1^{(k)}(t - k) + b_2^{(k)}(t - k)^2 + \cdots + b_{d-1}^{(k)}(t - k)^{d-1}$$

As before, the complementary function $C_k(t - k)$ represents that aspect of the finite behavior of the process, which is predictable at time k. Similarly, the coefficients $b_j^{(k)}$ may be expressed in terms of the infinite sums up to origin k, that is, $Sa_k, S^2a_k, \ldots, S^da_k$. Accordingly, we can discover how the coefficients $b_j^{(k)}$ change as the origin is changed, from $k - 1$ to k.

Inverted form of model. Finally, the model can be expressed in the inverted form

$$\pi(B)z_t = a_t$$

or

$$z_t = \bar{z}_{t-1}(\pi) + a_t$$

The π weights may be obtained by equating coefficients in (4.2.22), that is

$$(1 - B)^d = (1 - \theta_1 B - \theta_2 B^2 - \cdots - \theta_q B^q)(1 - \pi_1 B - \pi_2 B^2 - \cdots) \quad (4.3.23)$$

For a given model, they are best obtained by substituting numerical values in (4.3.23), rather than by deriving a general formula. We note that (4.3.23) implies that, for j greater than the larger of d and q, the π weights satisfy the difference equation

$$\theta(B)\pi_j = 0$$

defined by the moving average operator. Hence, for sufficiently large j, the weights π_j follow a mixture of damped exponentials and sine waves.

IMA process of order $(0, 2, 3)$. One final special case of sufficient interest to merit comment is the IMA process of order $(0, 2, 3)$

$$\nabla^2 z_t = (1 - \theta_1 B - \theta_2 B^2 - \theta_3 B^3)a_t$$

Proceeding as before, this can be written in the integrated form as

$$z_t = \lambda_{-1}a_{t-1} + \lambda_0 Sa_{t-1} + \lambda_1 S^2 a_{t-1} + a_t$$

where the relations between the λ's and θ's are

$$\left.\begin{array}{l} \theta_1 = 2 - \lambda_{-1} - \lambda_0 - \lambda_1 \\ \theta_2 = \lambda_0 - 1 + 2\lambda_{-1} \\ \theta_3 = -\lambda_{-1} \end{array}\right\} \qquad \left.\begin{array}{l} \lambda_{-1} = -\theta_3 \\ \lambda_0 = 1 + \theta_2 + 2\theta_3 \\ \lambda_1 = 1 - \theta_1 - \theta_2 - \theta_3 \end{array}\right\}$$

Alternatively, it can be written in the truncated form as

$$z_t = b_0^{(k)} + b_1^{(k)}(t - k) + \lambda_{-1}a_{t-1} + \lambda_0 \sum_{j=k+1}^{t-1} a_j + \lambda_1 \sum_{i=k+1}^{t-1} \sum_{j=k+1}^{i} a_j + a_t$$

Finally, the invertibility region is defined by

$$
\left.
\begin{aligned}
\theta_1 + \theta_2 + \theta_3 &< 1 \\
-\theta_1 + \theta_2 - \theta_3 &< 1 \\
\theta_3(\theta_3 - \theta_1) - \theta_2 &< 1 \\
|\theta_3| &< 1
\end{aligned}
\right\}
\qquad
\left.
\begin{aligned}
\lambda_1 &> 0 \\
2\lambda_0 + \lambda_1 &< 4(1 - \lambda_{-1}) \\
\lambda_0(1 + \lambda_{-1}) &> -\lambda_1\lambda_{-1} \\
|\lambda_{-1}| &< 1
\end{aligned}
\right\}
$$

and is shown in Figure 4.10.

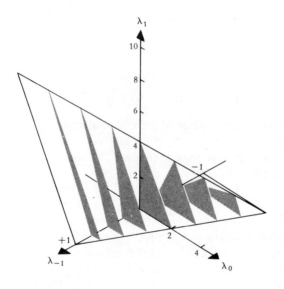

Fig. 4.10   Invertibility region for parameters λ_{-1}, λ_0 and λ_1 of an IMA $(0, 2, 3)$ process

In Chapter 5 we show how future values of a time series can be forecast in an optimal manner when the model is an **ARIMA** process. In studying these forecasts we make considerable use of the various model forms discussed in this chapter.

APPENDIX A4.1   LINEAR DIFFERENCE EQUATIONS

In this book we are often concerned with linear difference equations. In particular, the **ARIMA** model relates an output z_t to an input a_t in terms of the difference equation

$$z_t - \varphi_1 z_{t-1} - \varphi_2 z_{t-2} - \cdots - \varphi_{p'} z_{t-p'} = a_t - \theta_1 a_{t-1} - \theta_2 a_{t-2} - \cdots$$
$$- \theta_q a_{t-q} \tag{A4.1.1}$$

where $p' = p + d$.

Alternatively, we may write (A4.1.1) as

$$\varphi(B) z_t = \theta(B) a_t$$

where

$$\varphi(B) = 1 - \varphi_1 B - \varphi_2 B^2 - \cdots - \varphi_{p'} B^{p'}$$
$$\theta(B) = 1 - \theta_1 B - \theta_2 B^2 - \cdots - \theta_q B^q$$

We now derive an expression for the general solution of the difference equation (A4.1.1) relative to an origin $k < t$.

(1) We show that the general solution may be written as

$$z_t = C_k(t - k) + I_k(t - k)$$

where $C_k(t - k)$ is the complementary function and $I_k(t - k)$ a "particular integral."

(2) We then derive a general expression for the complementary function $C_k(t - k)$.

(3) Finally we derive a general expression for a particular integral $I_k(t - k)$.

The general solution.   The argument is identical to that for the solution of linear differential or linear algebraic equations. Suppose that z'_t is any particular solution of

$$\varphi(B) z_t = \theta(B) a_t \tag{A4.1.2}$$

that is, it satisfies

$$\varphi(B) z'_t = \theta(B) a_t \tag{A4.1.3}$$

On subtracting (A4.1.3) from (A4.1.2), we obtain

$$\varphi(B)(z_t - z'_t) = 0$$

Thus $z''_t = z_t - z'_t$ satisfies

$$\varphi(B) z''_t = 0 \tag{A4.1.4}$$

Now

$$z_t = z'_t + z''_t$$

and hence the general solution of (A4.1.2) is the sum of the complementary function z''_t, which is the general solution of the homogeneous difference equation (A4.1.4), and a particular integral z'_t, which is any particular solution of (A4.1.2). Relative to any origin $k < t$ we denote the complementary function z_t'' by $C_k(t - k)$ and the particular integral z_t' by $I_k(t - k)$.

Evaluation of the complementary function—Distinct roots. Consider the homogeneous difference equation

$$\varphi(B)z_t = 0 \qquad (A4.1.5)$$

where

$$\varphi(B) = (1 - G_1 B)(1 - G_2 B)\cdots(1 - G_{p'}B) \qquad (A4.1.6)$$

and where we assume in the first instance that $G_1, G_2, \ldots, G_{p'}$ are *distinct*. Then it is shown below that the general solution of (A4.1.5) at time t, when the series is referred to an origin at time k, is

$$z_t = A_1 G_1^{t-k} + A_2 G_2^{t-k} + \cdots + A_{p'} G_{p'}^{t-k} \qquad (A4.1.7)$$

where the A_i's are constants. Thus, a real root of $\varphi(B) = 0$ contributes a damped exponential term G^{t-k} to the complementary function. A pair of complex roots contributes a damped sine wave term $d^{t-k}\sin(2\pi f_0 t + F)$.

To see that (A4.1.7) does satisfy (A4.1.5), we can substitute (A4.1.7) in (A4.1.5) to give

$$\varphi(B)\{A_1 G_1^{t-k} + A_2 G_2^{t-k} + \cdots + A_{p'} G_{p'}^{t-k}\} = 0 \qquad (A4.1.8)$$

Now consider

$$\varphi(B)G_i^{t-k} = (1 - \varphi_1 B - \varphi_2 B^2 - \cdots - \varphi_{p'}B^{p'})G_i^{t-k}$$
$$= G_i^{t-k-p'}(G_i^{p'} - \varphi_1 G_i^{p'-1} - \cdots - \varphi_{p'})$$

We see that $\varphi(B)G_i^{t-k}$ vanishes for each value of i if

$$G_i^{p'} - \varphi_1 G_i^{p'-1} - \cdots - \varphi_{p'} = 0$$

that is, if $B = 1/G_i$ is a root of $\varphi(B) = 0$. Now, since (A4.1.6) implies that the roots of $\varphi(B) = 0$ are $B = 1/G_i$, it follows that $\varphi(B)G_i^{t-k}$ is zero for all i and hence (A4.1.8) holds, confirming that (A4.1.7) is a general solution of (A.4.1.5).

To prove (A4.1.7) directly, consider the special case of the second-order equation

$$(1 - G_1 B)(1 - G_2 B)z_t = 0$$

which we can write as

$$(1 - G_1 B)y_t = 0 \qquad (A4.1.9)$$

where

$$y_t = (1 - G_2 B)z_t \qquad (A4.1.10)$$

Now (A4.1.9) implies

$$y_t = G_1 y_{t-1} = G_1^2 y_{t-2} = \cdots = G_1^{t-k}y_k$$

and hence

$$y_t = D_1 G_1^{t-k}$$

where $D_1 = y_k$ is a constant determined by the starting value y_k. Hence (A4.1.10) may be written

$$
\begin{aligned}
z_t &= G_2 z_{t-1} + D_1 G_1^{t-k} \\
&= G_2(G_2 z_{t-2} + D_1 G_1^{t-k-1}) + D_1 G_1^{t-k} \\
&\quad \vdots \qquad \vdots \qquad \vdots \qquad \vdots \qquad \vdots \\
&= G_2^{t-k} z_k + D_1 \{ G_1^{t-k} + G_2 G_1^{t-k-1} + \cdots + G_2^{t-k-1} G_1 \} \quad \text{(A4.1.11)} \\
&= G_2^{t-k} z_k + \frac{D_1}{1 - G_2/G_1} \{ G_1^{t-k} - G_2^{t-k} \} \\
&= A_1 G_1^{t-k} + A_2 G_2^{t-k}
\end{aligned}
$$

where A_1, A_2 are constants determined by the starting values of the series. By an extension of the above argument, it may be shown that the general solution of (A4.1.5), when the roots of $\varphi(B) = 0$ are distinct, is given by (A4.1.7).

Equal roots. Suppose that $\varphi(B) = 0$ has d equal roots G_0^{-1}, so that $\varphi(B)$ contains a factor $(1 - G_0 B)^d$. In particular, consider the solution (A4.1.11) for the second-order equation when both G_1 and G_2 are equal to G_0. Then (A4.1.11) reduces to

$$z_t = G_0^{t-k} z_k + D_1 G_0^{t-k}(t - k)$$

or

$$z_t = \{ A_0 + A_1(t - k) \} G_0^{t-k}$$

In general, if there are d equal roots G_0, it may be verified by direct substitution in (A4.1.5) that the general solution is

$$z_t = \{ A_0 + A_1(t - k) + A_2(t - k)^2 + \cdots + A_{d-1}(t - k)^{d-1} \} G_0^{t-k} \quad \text{(A4.1.12)}$$

In particular, when the equal roots G_0 are all equal to unity as in the IMA $(0, d, q)$ process, the solution is

$$z_t = A_0 + A_1(t - k) + \cdots + A_{d-1}(t - k)^{d-1} \quad \text{(A4.1.13)}$$

that is, a polynomial in $t - k$ of degree $d - 1$.

In general, when $\varphi(B)$ factors according to

$$(1 - G_1 B)(1 - G_2 B) \cdots (1 - G_p B)(1 - G_0 B)^d$$

the complementary function is

$$C_k(t - k) = G_0^{t-k} \sum_{j=0}^{d-1} A_j(t - k)^j + \sum_{i=1}^{p} D_i G_i^{t-k} \quad \text{(A4.1.14)}$$

Thus, in general, the complementary function consists of a mixture of damped exponential terms G^{t-k}, polynomial terms $(t-k)^j$, damped sine wave terms $d^{t-k} \sin(2\pi f_0 t + F)$, and combinations of these functions.

Evaluation of the "Particular Integral." We now show that a particular integral $I_k(s-k)$, satisfying

$$\varphi(B)I_k(t-k) = \theta(B)a_t, \quad t-k > q \tag{A4.1.15}$$

is a function defined as follows

$$
\left.
\begin{aligned}
&I_k(s-k) = 0, \quad s \leqslant k \\
&\quad I_k(1) = a_{k+1} \\
&\quad I_k(2) = a_{k+2} + \psi_1 a_{k+1} \\
&\quad \vdots \qquad \vdots \qquad \vdots \qquad \vdots \\
&I_k(t-k) = a_t + \psi_1 a_{t-1} + \psi_2 a_{t-2} + \cdots + \psi_{t-k-1} a_{k+1}, \quad t > k
\end{aligned}
\right\} \tag{A4.1.16}
$$

where the ψ weights are those appearing in the form (4.2.3) of the model. Thus, the ψ weights satisfy

$$\varphi(B)\psi(B)a_t = \theta(B)a_t \tag{A4.1.17}$$

Now the terms on the left-hand side of (A4.1.17) may be set out as follows:

$$
\begin{array}{l|l}
a_t + \psi_1 a_{t-1} + \psi_2 a_{t-2} + \cdots + \psi_{t-k-1} a_{k+1} & +\psi_{t-k} a_k + \cdots \\
-\varphi_1(\qquad a_{t-1} + \psi_1 a_{t-2} + \cdots + \psi_{t-k-2} a_{k+1} & +\psi_{t-k-1} a_k + \cdots \\
-\varphi_2(\qquad \cdots \quad \cdots \quad \cdots \quad \cdots & \cdots \quad \cdots \quad \cdots \\
\quad \vdots \qquad \qquad \vdots \quad \vdots \quad \vdots \quad \vdots & \vdots \quad \vdots \quad \vdots \\
-\varphi_{p'}(\qquad a_{t-p'} + \cdots + \psi_{t-k-p'-1} a_{k+1} & +\psi_{t-k-p'} a_k + \cdots
\end{array} \tag{A4.1.18}
$$

Since the right-hand side of (A4.1.17) is

$$a_t - \theta_1 a_{t-1} - \cdots - \theta_q a_{t-q}$$

it follows that the first $q+1$ columns in this array sum to $a_t, -\theta_1 a_{t-1}, \ldots, -\theta_q a_{t-q}$. Now the left hand term in (A4.1.15), where $I_k(s-k)$ is given by (A4.1.16), is equal to the sum of the terms in the first $(t-k)$ columns of the array, that is, those to the left of the vertical line. Therefore, if $t-k > q$, that is, the vertical line is drawn after $q+1$ columns, the sum of all terms up to the vertical line is equal to $\theta(B)a_t$. This shows that (A4.1.16) is a "particular integral" of the difference equation.

Example. Consider the IMA $(0, 1, 1)$ process

$$z_t - z_{t-1} = a_t - \theta a_{t-1} \tag{A4.1.19}$$

for which $\psi_j = 1 - \theta$ for $j \geqslant 1$. Then

$$
\left.
\begin{aligned}
I_k(0) &= 0 \\
I_k(1) &= a_{k+1} \\
\vdots \quad &\quad \vdots \\
I_k(t - k) &= a_t + (1 - \theta) \sum_{j=k+1}^{t-1} a_j \quad t - k > 1
\end{aligned}
\right\}
\tag{A4.1.20}
$$

Now, if $z_t = I_k(t - k)$ is a solution of (A4.1.19) then

$$
I_k(t - k) - I_k(t - k - 1) = a_t - \theta a_{t-1}
$$

and, as is easily verified, while this is not satisfied by (A4.1.20) for $t - k = 1$, it is satisfied by (A4.1.20) for $t - k > 1$, that is, for $t - k > q$.

APPENDIX A4.2 THE IMA (0, 1, 1) PROCESS WITH DETERMINISTIC DRIFT

The general model $\phi(B)\nabla^d z_t = \theta_0 + \theta(B)a_t$ can also be written

$$
\phi(B)\nabla^d z_t = \theta(B)\varepsilon_t
$$

with the shocks ε_t having a nonzero mean $\xi = \theta_0/(1 - \theta_1 - \cdots - \theta_q)$. For example, the IMA (0, 1, 1) model is then

$$
\nabla z_t = (1 - \theta B)\varepsilon_t
$$

with $E[\varepsilon_t] = \xi = \theta_0/(1 - \theta)$. In this form, z_t could represent, for example, the outlet temperature from a reactor when heat was being supplied from a heating element at a fixed rate. Now if

$$
\varepsilon_t = \xi + a_t
\tag{A4.2.1}
$$

where a_t is white noise with zero mean, then with reference to a time origin k, the integrated form for the model is

$$
z_t = b_0^{(k)} + \lambda \sum_{j=k+1}^{t-1} \varepsilon_j + \varepsilon_t
\tag{A4.2.2}
$$

with $\lambda = 1 - \theta$. Substituting for (A4.2.1) in (A4.2.2), the model written in terms of the a's is

$$
z_t = b_0^{(k)} + \lambda\xi(t - k - 1) + \xi + \lambda \sum_{j=k+1}^{t-1} a_j + a_t
\tag{A4.2.3}
$$

Thus, we see that z_t contains a deterministic slope or drift due to the term $\lambda\xi(t - k - 1)$. Moreover, if we denote the "level" of the process at time $t - 1$ by l_{t-1}, where

$$
z_t = l_{t-1} + a_t
$$

we see that the level is changed from time $t - 1$ to time t, according to

$$l_t = l_{t-1} + \lambda \xi + \lambda a_t$$

The change in the level thus contains a deterministic component $\lambda \xi$, as well as a stochastic component λa_t.

APPENDIX A4.3 PROPERTIES OF THE FINITE SUMMATION OPERATOR

Relation between finite and infinite sums. It may be verified directly, in the following simple cases, that the relation between the infinite and the finite sum with first term having subscript $k + 1$, where $t \geqslant k + 1$, is

$$\left. \begin{aligned} Sx_t &= \sum_{h=k+1}^{t} x_h + Sx_k \\ S^2 x_t &= \sum_{i=k+1}^{t} \sum_{h=k+1}^{i} x_h + S^2 x_k + (t - k) Sx_k \end{aligned} \right\} \qquad (A4.3.1)$$

Now, for example, the last equality may be written

$$S^2 x_t = \sum_{i=k+1}^{t} \sum_{h=k+1}^{i} x_h + b_0^{(k)} + b_1^{(k)}(t - k)$$

In general, the d-fold multiple infinite sum will be equal to the d-fold finite sum, plus a polynomial of degree $(d - 1)$ in t,

$$b_0^{(k)} + b_1^{(k)}(t - k) + b_2^{(k)} \binom{t - k + 1}{2} + \cdots + b_{d-1}^{(k)} \binom{t - k + d - 2}{d - 1}$$

Application to the IMA $(0, 2, 2)$ process. The integrated form $(4.3.14)$ of the IMA $(0, 2, 2)$ process is

$$z_t = \lambda_0 Sa_{t-1} + \lambda_1 S^2 a_{t-1} + a_t$$

which, on using $(A4.3.1)$, may be written

$$z_t = \lambda_0 \left(\sum_{h=k+1}^{t-1} a_h + Sa_k \right) + \lambda_1 \left\{ \sum_{i=k+1}^{t-1} \sum_{h=k+1}^{i} a_h + S^2 a_k + (t - k - 1) Sa_k \right\} + a_t$$

$$= \{(\lambda_0 - \lambda_1) Sa_k + \lambda_1 S^2 a_k\} + \lambda_1 Sa_k(t - k)$$

$$+ \lambda_0 \sum_{h=k+1}^{t-1} a_h + \lambda_1 \sum_{i=k+1}^{t-1} \sum_{h=k+1}^{i} a_h + a_t \qquad (A4.3.2)$$

Now the truncated form $(4.3.16)$, of the random shock model is

$$z_t = C_k(t - k) + \lambda_0 \sum_{h=k+1}^{t-1} a_h + \lambda_1 \sum_{i=k+1}^{t-1} \sum_{h=k+1}^{i} a_h + a_t \qquad (A4.3.3)$$

so that on equating (A4.3.2) and (A4.3.3), the complementary function can be regarded as a function of infinite sums of random shocks as follows:

$$C_k(t - k) = \{(\lambda_0 - \lambda_1)Sa_k + \lambda_1 S^2 a_k\} + \{\lambda_1 Sa_k\}(t - k)$$

which is the result (4.3.17) quoted in the text.

APPENDIX A4.4 ARIMA PROCESSES WITH ADDED NOISE

In this appendix, we consider the effect of adding noise (for example measurement error) to a general ARIMA (p, d, q) process.

A4.4.1 The sum of two independent moving average processes

As a necessary preliminary to what follows, consider a stochastic process w_t, which is the sum of two *independent* moving average processes of orders q_1 and q_2 respectively. That is

$$w_t = \theta_1(B)a_t + \theta_2(B)b_t \tag{A4.4.1}$$

where $\theta_1(B)$ and $\theta_2(B)$ are polynomials in B, of order q_1 and q_2, and the white noise processes a_t and b_t have zero means, and are mutually independent. Suppose $q = \max(q_1, q_2)$; then it is clear that the autocovariance function γ_j for w_t must be zero for $j > q$. It follows that there exists a representation of w_t as a single moving average process of order q

$$w_t = \theta_3(B)u_t \tag{A4.4.2}$$

where u_t is a white noise process with mean zero.

Thus, the sum of two independent moving average processes is another moving average process, whose order is the same as that of the component process of higher order.

A4.4.2 Effect of added noise on the general model

Correlated noise. Consider the general nonstationary model of order (p, d, q)

$$\phi(B)\nabla^d z_t = \theta(B)a_t \tag{A4.4.3}$$

Suppose that we cannot observe z_t itself, but only $Z_t = z_t + b_t$, where b_t represents some extraneous noise (for example measurement error) and may be correlated. We wish to determine the nature of the observed process Z_t. In general we have

$$\phi(B)\nabla^d Z_t = \theta(B)a_t + \phi(B)\nabla^d b_t$$

If the noise follows a stationary ARMA process of order $(p_1, 0, q_1)$

$$\phi_1(B)b_t = \theta_1(B)\alpha_t \tag{A4.4.4}$$

where α_t is a white noise process independent of the a_t process, then

$$\underbrace{\phi_1(B)\phi(B)\nabla^d Z_t}_{p_1 + p + d} = \underbrace{\phi_1(B)\theta(B)a_t}_{p_1 + q} + \underbrace{\phi(B)\theta_1(B)\nabla^d \alpha_t}_{p + q_1 + d} \tag{A4.4.5}$$

where the numbers below the brackets indicate the degrees of the various polynomials in B. Now the right-hand side of (A4.4.5) is of the form (A4.4.1). Let $P = p_1 + p$ and Q be equal to whichever of $(p_1 + q)$ and $(p + q_1 + d)$ is the larger. Then we can write

$$\phi_2(B)\nabla^d Z_t = \theta_2(B)u_t$$

with u_t, a white noise process, and the Z_t process is seen to be of order (P, d, Q).

Added white noise. If, as might be true in some applications, the added noise is white, then $\phi_1(B) = \theta_1(B) = 1$ in (A4.4.4), and we obtain

$$\phi(B)\nabla^d Z_t = \theta_2(B)u_t \tag{A4.4.6}$$

with

$$\theta_2(B)u_t = \theta(B)a_t + \phi(B)\nabla^d b_t$$

which is of order (p, d, Q), where Q is the larger of q and $(p + d)$. If $p + d \leqslant q$, then the order of the process with error is the same as that of the original process. The only effect of the added white noise, is to change the values of the θ's (but not the ϕ's).

Effect of added white noise on an integrated moving average process. In particular, an IMA process of order $(0, d, q)$, with white noise added, remains an IMA of order $(0, d, q)$ if $d \leqslant q$; otherwise it becomes an IMA of order $(0, d, d)$. In either case, the parameters of the process are changed by the addition of noise. The nature of these changes can be determined by equating the autocovariances of the process, with added noise, to those of a simple IMA process. The procedure will now be illustrated with an example.

A4.4.3 Example for an IMA (0, 1, 1) process with added white noise

Consider the properties of the process $Z_t = z_t + b_t$ when

$$z_t = \lambda \sum_{j=1}^{\infty} a_{t-j} + a_t \tag{A4.4.7}$$

and the b_t and a_t are mutually independent white noise processes. The Z_t process has first difference $W_t = Z_t - Z_{t-1}$, given by

$$W_t = \{1 - (1 - \lambda)B\}a_t + (1 - B)b_t \tag{A4.4.8}$$

The autocovariances for the first differences W_t are

$$\left. \begin{aligned} \gamma_0 &= \sigma_a^2\{1 + (1 - \lambda)^2\} + 2\sigma_b^2 \\ \gamma_1 &= -\sigma_a^2(1 - \lambda) - \sigma_b^2 \\ \gamma_j &= 0 \qquad j \geqslant 2 \end{aligned} \right\} \qquad \text{(A4.4.9)}$$

The fact that the γ_j are zero beyond the first, confirms that the process with added noise is, as expected, an IMA process of order $(0, 1, 1)$. To obtain explicitly the parameters of the IMA which represents the noise process, we suppose it can be written

$$Z_t = \Lambda \sum_{j=1}^{\infty} u_{t-j} + u_t \qquad \text{(A4.4.10)}$$

with u_t a white noise process. The process (A4.4.10) has autocovariances

$$\left. \begin{aligned} \gamma_0 &= \sigma_u^2\{1 + (1 - \Lambda)^2\} \\ \gamma_1 &= -\sigma_u^2(1 - \Lambda) \\ \gamma_j &= 0 \qquad j \geqslant 2 \end{aligned} \right\} \qquad \text{(A4.4.11)}$$

Equating (A4.4.9) and (A4.4.11), we can solve for Λ and for σ_u^2 explicitly. Thus

$$\frac{\Lambda^2}{1 - \Lambda} = \frac{\lambda^2}{1 - \lambda + \sigma_b^2/\sigma_a^2} \qquad \text{(A4.4.12)}$$

$$\sigma_u^2 = \sigma_a^2 \frac{\lambda^2}{\Lambda^2}$$

Suppose, for example, that the original series has $\lambda = 0.5$ and $\sigma_b^2 = \sigma_a^2$, then $\Lambda = 0.33$ and $\sigma_u^2 = 2.25\,\sigma_a^2$.

A4.4.4 Relation between the IMA $(0, 1, 1)$ process and a random walk

The process

$$z_t = \sum_{j=0}^{\infty} a_{t-j} = \sum_{j=1}^{\infty} a_{t-j} + a_t \qquad \text{(A4.4.13)}$$

which is an IMA $(0, 1, 1)$ process, with $\lambda = 1$, is sometimes called a *random walk*. If the a_t are steps taken forward or backwards at time t, then z_t will represent the position of the walker at time t.

Any IMA $(0, 1, 1)$ process can be thought of as a random walk buried in white noise b_t, uncorrelated with the shocks a_t associated with the random walk process. If the noisy process is $Z_t = z_t + b_t$, where z_t is defined by (A4.4.13), then using (A4.4.12)

$$Z_t = \Lambda \sum_{j=1}^{\infty} u_{t-j} + u_t$$

with

$$\frac{\Lambda^2}{1 - \Lambda} = \frac{\sigma_a^2}{\sigma_b^2} \qquad \sigma_u^2 = \frac{\sigma_a^2}{\Lambda^2} \qquad\qquad (A4.4.14)$$

A4.4.5 Autocovariance function of the general model with added correlated noise

Suppose the basic process is an ARIMA process of order (p, d, q)

$$\phi(B)\nabla^d z_t = \theta(B)a_t$$

and that $Z_t = z_t + b_t$ is observed, where the stationary b_t process, which has autocovariance function $\gamma_j(b)$, is independent of the a_t process, and hence of z_t. Suppose that $\gamma_j(w)$ is the autocovariance function for $w_t = \nabla^d z_t = \phi^{-1}(B)\theta(B)a_t$ and that $W_t = \nabla^d Z_t$. We require the autocovariance function for W_t. Now

$$\nabla^d(Z_t - b_t) = \phi^{-1}(B)\theta(B)a_t$$

$$W_t = w_t + v_t$$

where

$$v_t = \nabla^d b_t = (1 - B)^d b_t$$

Hence

$$\gamma_j(W) = \gamma_j(w) + \gamma_j(v)$$

$$\gamma_j(v) = (1 - B)^d(1 - F)^d \gamma_j(b)$$

$$= (-1)^d(1 - B)^{2d}\gamma_{j+d}(b)$$

and

$$\gamma_j(W) = \gamma_j(w) + (-1)^d(1 - B)^{2d}\gamma_{j+d}(b) \qquad\qquad (A4.4.15)$$

For example, suppose correlated noise b_t is added to an IMA $(0, 1, 1)$ process defined by $w_t = \nabla z_t = (1 - \theta B)a_t$. Then the autocovariances of the first difference W_t of the "noisy" process will be

$$\gamma_0(W) = \sigma_a^2(1 + \theta^2) + 2\{\gamma_0(b) - \gamma_1(b)\}$$

$$\gamma_1(W) = -\sigma_a^2\theta + \{2\gamma_1(b) - \gamma_0(b) - \gamma_2(b)\}$$

$$\gamma_j(W) = \{2\gamma_j(b) - \gamma_{j-1}(b) - \gamma_{j+1}(b)\} \qquad j \geqslant 2$$

In particular, if b_t was first-order autoregressive, so that

$$b_t = \phi b_{t-1} + \alpha_t$$

$$\gamma_0(W) = \sigma_a^2(1 + \theta^2) + 2\sigma_b^2(1 - \phi)$$

$$\gamma_1(W) = -\sigma_a^2\theta - \sigma_b^2(1 - \phi)^2$$

$$\gamma_j(W) = -\sigma_b^2\phi^{j-1}(1 - \phi)^2 \quad j \geqslant 2$$

In fact, from (A4.4.5), the resulting noisy process $Z_t = z_t + b_t$ is in this case defined by

$$(1 - \phi B)\nabla Z_t = (1 - \phi B)(1 - \theta B)a_t + (1 - B)\alpha_t$$

which is of order (1, 1, 2).

5

Forecasting

Having considered in Chapter 4 some of the properties of ARIMA models, we now show how they may be used to forecast future values of an observed time series. In Part II of the book we consider the problem of fitting the model to actual data. For the present however, we proceed as if the model were known *exactly*, bearing in mind that estimation errors in the parameters will not seriously affect the forecasts unless the number of data points, used to fit the model, is small.

In this chapter we consider nonseasonal time series. The forecasting of seasonal time series is described in Chapter 9. We show how minimum mean square error forecasts may be generated directly from the *difference equation* form of the model. A further recursive calculation yields the probability limits for the forecasts. It is to be emphasized that, for practical computation of the forecasts, this approach via the difference equation is the simplest and most elegant. However, to provide insight into the nature of the forecasts, we also consider them from other viewpoints.

5.1 MINIMUM MEAN SQUARE ERROR FORECASTS AND THEIR PROPERTIES

In Section 4.2 we discussed three explicit forms for the general ARIMA model

$$\varphi(B)z_t = \theta(B)a_t \qquad (5.1.1)$$

where $\varphi(B) = \phi(B)\nabla^d$. We begin by recalling these three forms, since each one throws light on a different aspect of the forecasting problem.

We shall be concerned with forecasting a value z_{t+l}, $l \geqslant 1$, when we are currently standing at time t. This forecast is said to be made at *origin t* for *lead time l*. We now summarize the results of Section 4.2, but writing $t + l$ for t and t for k.

The three explicit forms for the model. An observation z_{t+l} generated by the process (5.1.1) may be expressed:

(1) Directly in terms of the difference equation by

$$z_{t+l} = \varphi_1 z_{t+l-1} + \cdots + \varphi_{p+d} z_{t+l-p-d} - \theta_1 a_{t+l-1} - \cdots$$
$$- \theta_q a_{t+l-q} + a_{t+l} \qquad (5.1.2)$$

(2) As an infinite weighted sum of current and previous shocks a_j,

$$z_{t+l} = \sum_{j=-\infty}^{t+l} \psi_{t+l-j} a_j = \sum_{j=0}^{\infty} \psi_j a_{t+l-j} \qquad (5.1.3)$$

where $\psi_0 = 1$ and, as in (4.2.5), the ψ weights may be obtained by equating coefficients in

$$\varphi(B)(1 + \psi_1 B + \psi_2 B^2 + \cdots) = \theta(B) \qquad (5.1.4)$$

Equivalently, for positive $l > q$, the model may be written in the truncated form

$$z_{t+l} = C_t(l) + a_{t+l} + \psi_1 a_{t+l-1} + \cdots + \psi_{l-1} a_{t+1} \qquad (5.1.5)$$

where the complementary function $C_t(l)$ is equal to the truncated infinite sum

$$C_t(l) = \sum_{j=-\infty}^{t} \psi_{t+l-j} a_j = \sum_{j=0}^{\infty} \psi_{l+j} a_{t-j} \qquad (5.1.6)$$

(3) As an infinite weighted sum of previous observations, plus a random shock

$$z_{t+l} = \sum_{j=1}^{\infty} \pi_j z_{t+l-j} + a_{t+l} \qquad (5.1.7)$$

Also, if $d \geqslant 1$

$$\bar{z}_{t+l-1}(\pi) = \sum_{j=1}^{\infty} \pi_j z_{t+l-j} \qquad (5.1.8)$$

will be a weighted average, since then $\sum_{j=1}^{\infty} \pi_j = 1$.
  As in (4.2.22), the π weights may be obtained from

$$\varphi(B) = (1 - \pi_1 B - \pi_2 B^2 - \cdots)\theta(B) \qquad (5.1.9)$$

5.1.1 Derivation of the minimum mean square error forecasts

Now suppose, standing at origin t, we are to make a forecast $\hat{z}_t(l)$ of z_{t+l} which is to be a linear function of current and previous observations z_t, $z_{t-1}, z_{t-2}, \ldots$. Then it will also be a linear function of current and previous shocks $a_t, a_{t-1}, a_{t-2}, \ldots$.
  Suppose then, that the best forecast is

$$\hat{z}_t(l) = \psi_l^* a_t + \psi_{l+1}^* a_{t-1} + \psi_{l+2}^* a_{t-2} + \cdots$$

where the weights ψ_l^*, $\psi_{l+1}^*, \ldots$ are to be determined. Then, using (5.1.3), the mean square error of the forecast is

$$E[z_{t+l} - \hat{z}_t(l)]^2 = (1 + \psi_1^2 + \cdots + \psi_{l-1}^2)\sigma_a^2 + \sum_{j=0}^{\infty} \{\psi_{l+j} - \psi_{l+j}^*\}^2\sigma_a^2 \quad (5.1.10)$$

which is minimized by setting $\psi_{l+j}^* = \psi_{l+j}$, a conclusion which is a special case of more general results in prediction theory due to Wold [44], Kolmogoroff [45], [46], [47], Wiener [48] and Whittle [49]. We have then

$$z_{t+l} = (a_{t+l} + \psi_1 a_{t+l-1} + \cdots + \psi_{l-1} a_{t+1}) + (\psi_l a_t + \psi_{l+1} a_{t-1} + \cdots)$$
$$(5.1.11)$$

$$= e_t(l) + \hat{z}_t(l) \quad (5.1.12)$$

where $e_t(l)$ is the error of the forecast $\hat{z}_t(l)$ at lead time l.

Certain important facts emerge. As before denote $E[z_{t+l}|z_t, z_{t-1}, \ldots]$, the conditional expectation of z_{t+l} given knowledge of all the z's up to time t, by $\underset{t}{E}[z_{t+l}]$. Then

(1) $$\hat{z}_t(l) = \psi_l a_t + \psi_{l+1} a_{t-1} + \cdots = \underset{t}{E}[z_{t+l}] \quad (5.1.13)$$

Thus, the minimum mean square error forecast at origin t, for lead time l, is the conditional expectation of z_{t+l}, at time t. When $\hat{z}_t(l)$ is regarded as a function of l for fixed t, it will be called the *forecast function* for origin t.

(2) The forecast error for lead time l is

$$e_t(l) = a_{t+l} + \psi_1 a_{t+l-1} + \cdots + \psi_{l-1} a_{t+1} \quad (5.1.14)$$

Since

$$\underset{t}{E}[e_t(l)] = 0 \quad (5.1.15)$$

the forecast is unbiased. Also, the variance of the forecast error is

$$V(l) = \text{var } [e_t(l)] = (1 + \psi_1^2 + \psi_2^2 + \cdots + \psi_{l-1}^2)\sigma_a^2 \quad (5.1.16)$$

(3) It is readily shown that, not only is $\hat{z}_t(l)$ the minimum mean square error forecast of z_{t+l}, but that any linear function $\sum_{l=1}^{L} w_l \hat{z}_t(l)$, of the forecasts, is a minimum mean square error forecast of the corresponding linear function $\sum_{l=1}^{L} w_l z_{t+l}$ of the future observations. For example, suppose that using (5.1.13) we have obtained, from monthly data, minimum mean square error forecasts $\hat{z}_t(1)$, $\hat{z}_t(2)$, and $\hat{z}_t(3)$ of the sales of a product, one, two, and three months ahead. Then it is true that $\hat{z}_t(1) + \hat{z}_t(2) + \hat{z}_t(3)$ is the minimum mean square error forecast of the sales $z_{t+1} + z_{t+2} + z_{t+3}$, during the next quarter.

(4) *The residuals as one step ahead forecast errors.* Using (5.1.14), the one step ahead forecast error is

$$e_t(1) = z_{t+1} - \hat{z}_t(1) = a_{t+1} \qquad (5.1.17)$$

Hence, the residuals a_t which generate the process, and which so far we have introduced merely as a set of independent random variables or shocks, turn out to be the *one step ahead forecast errors.*

It follows that, for a minimum mean square error forecast, the one step ahead forecast errors must be uncorrelated. This is eminently sensible, for if one step ahead errors were correlated, then the forecast error a_{t+1} could, to some extent, be predicted from available forecast errors $a_t, a_{t-1}, a_{t-2}, \ldots$ If the prediction so obtained was $\hat{a}_{t+1}$, then $\hat{z}_t(1) + \hat{a}_{t+1}$ would be a better forecast of z_{t+1} than was $\hat{z}_t(1)$.

(5) *Correlation between the forecast errors.* Although the optimal forecast errors at lead time 1 will be uncorrelated, the forecast errors for longer lead times in general will be correlated. In Section A5.1.1 we derive a general expression for the correlation between the forecast errors $e_t(l)$ and $e_{t-j}(l)$, made at the *same* lead time l from *different* origins t and $t-j$.

Now it is also true that the forecast errors $e_t(l)$ and $e_t(l+j)$, made at different lead times from the same origin t, are correlated. One consequence of this is that there will often be a tendency for the forecast function to lie either wholly above or below the values of the series when they eventually come to hand. In Section A5.1.2, we give a general expression for the correlation between the forecast errors $e_t(l)$ and $e_t(l+j)$, made from the same origin.

5.1.2 Three basic forms for the forecast

We have seen that the minimum mean square error forecast $\hat{z}_t(l)$ for lead time l is the conditional expectation $E[z_{t+l}]$, of z_{t+l}, at origin t.

Using this fact, we can write down expressions for the forecast in any one of three different ways, corresponding to the three ways of expressing the model, summarized earlier in this section. For simplicity in notation, we will temporarily adopt the convention that squared brackets imply that the conditional expectation, at time t, is to be taken. Thus

$$[a_{t+l}] = E_t[a_{t+l}] \qquad [z_{t+l}] = E_t[z_{t+l}]$$

For $l > 0$, the three different ways of expressing the forecasts are:

Forecasts from difference equation. Taking conditional expectations at time t in (5.1.2), we obtain

$$[z_{t+l}] = \hat{z}_t(l) = \varphi_1[z_{t+l-1}] + \cdots + \varphi_{p+d}[z_{t+l-p-d}] - \theta_1[a_{t+l-1}] \cdots$$
$$- \theta_q[a_{t+l-q}] + [a_{t+l}] \qquad (5.1.18)$$

Forecasts in integrated form.   Using (5.1.3),

$$[z_{t+l}] = \hat{z}_t(l) = \psi_1[a_{t+l-1}] + \cdots + \psi_{l-1}[a_{t+1}] + \psi_l[a_t]$$
$$+ \psi_{l+1}[a_{t-1}] + \cdots + [a_{t+l}] \tag{5.1.19}$$

yielding the form (5.1.13) that we have met already. Alternatively, using the truncated form of the model (5.1.5), for positive $l > q$

$$[z_{t+l}] = \hat{z}_t(l) = C_t(l) + [a_{t+l}] + \psi_1[a_{t+l-1}] + \cdots + \psi_{l-1}[a_{t+1}] \tag{5.1.20}$$

where $C_t(l)$ is the complementary function at origin t.

Forecasts as a weighted average of previous observations and forecasts made at previous lead times from the same origin.   Finally, taking conditional expectations in (5.1.7),

$$[z_{t+l}] = \hat{z}_t(l) = \sum_{j=1}^{\infty} \pi_j[z_{t+l-j}] + [a_{t+l}] \tag{5.1.21}$$

It is to be noted that the minimum mean square error forecast is defined in terms of the conditional expectation

$$[z_{t+l}] = \underset{t}{E}[z_{t+l}] = E[z_{t+l}|z_t, z_{t-1}, \ldots]$$

which theoretically requires knowledge of the z's stretching back into the infinite past. However, the requirement of invertibility, which we have imposed on the general ARIMA model, ensures that the π weights in (5.1.21) form a convergent series. Hence, for the computation of a forecast to a given degree of accuracy, for some k, the dependence on z_{t-j} for $j > k$ can be ignored. In practice, the π weights usually decay rather quickly, so that whatever form of the model is employed in the computation, only a moderate length of series $z_t, z_{t-1}, \ldots, z_{t-k}$ is needed to calculate the forecasts to sufficient accuracy.

To calculate the conditional expectations which occur in the expressions (5.1.18) to (5.1.21), we note that if j is a nonnegative integer,

$$\left.\begin{aligned}
[z_{t-j}] &= \underset{t}{E}[z_{t-j}] = z_{t-j} & j = 0, 1, 2, \ldots \\
[z_{t+j}] &= \underset{t}{E}[z_{t+j}] = \hat{z}_t(j) & j = 1, 2, \ldots \\
[a_{t-j}] &= \underset{t}{E}[a_{t-j}] = a_{t-j} = z_{t-j} - \hat{z}_{t-j-1}(1) & j = 0, 1, 2, \ldots \\
[a_{t+j}] &= \underset{t}{E}[a_{t+j}] = 0 & j = 1, 2, \ldots
\end{aligned}\right\} \tag{5.1.22}$$

Therefore, to obtain the forecast $\hat{z}_t(l)$, one writes down the model for z_{t+l} in any one of the above three explicit forms and treats the terms on the right according to the following rules:

The z_{t-j} $(j = 0, 1, 2, \ldots)$, which have already happened at origin t, are left unchanged.

The z_{t+j} $(j = 1, 2, \ldots)$, which have not yet happened, are replaced by their forecasts $\hat{z}_t(j)$ at origin t.

The a_{t-j} $(j = 0, 1, 2, \ldots)$, which have happened, are available from $z_{t-j} - \hat{z}_{t-j-1}(1)$.

The a_{t+j} $(j = 1, 2, \ldots)$, which have not yet happened, are replaced by zeroes. For routine calculation it is by far the simplest to work directly with the difference equation form (5.1.18).

Example: Forecasting using the difference equation form. It will be shown in Chapter 7 that Series C is closely represented by the model

$$(1 - 0.8B)(1 - B)z_{t+l} = a_{t+l}$$

that is

$$(1 - 1.8B + 0.8B^2)z_{t+l} = a_{t+l}$$

or

$$z_{t+l} = 1.8z_{t+l-1} - 0.8z_{t+l-2} + a_{t+l}$$

The forecasts at origin t are given by

$$\left.\begin{aligned}
\hat{z}_t(1) &= 1.8z_t - 0.8z_{t-1} \\
\hat{z}_t(2) &= 1.8\hat{z}_t(1) - 0.8z_t \\
\hat{z}_t(l) &= 1.8\hat{z}_t(l - 1) - 0.8\hat{z}_t(l - 2) \quad l = 3, 4, 5, \ldots
\end{aligned}\right\} \qquad (5.1.23)$$

It is seen that the forecasts are readily generated recursively in the order $\hat{z}_t(1), \hat{z}_t(2), \ldots$.

In the above example, there happen to be no moving average terms in the model; such terms produce no added difficulties. Thus, we consider later in this chapter, a series arising in a control problem, for which the model at time $t + l$ is

$$\nabla^2 z_{t+l} = (1 - 0.9B + 0.5B^2)a_{t+l}$$

Then

$$z_{t+l} = 2z_{t+l-1} - z_{t+l-2} + a_{t+l} - 0.9a_{t+l-1} + 0.5a_{t+l-2}$$

$$\hat{z}_t(1) = 2z_t - z_{t-1} - 0.9a_t + 0.5a_{t-1}$$

$$\hat{z}_t(2) = 2\hat{z}_t(1) - z_t + 0.5a_t$$

$$\hat{z}_t(l) = 2\hat{z}_t(l - 1) - \hat{z}_t(l - 2) \quad l = 3, 4, \ldots$$

In these expressions we remember that $a_t = z_t - \hat{z}_{t-1}(1)$, $a_{t-1} = z_{t-1} - \hat{z}_{t-2}(1)$, and the forecasting process may be started off initially by setting unknown a's equal to their unconditional expected values of zero.

In general, if the moving average operator $\theta(B)$ is of degree q, the forecast equations for $\hat{z}_t(1), \hat{z}_t(2), \ldots, \hat{z}_t(q)$ will depend directly on the a's but forecasts at longer lead times will not. It should not, of course, be thought that the influence of these a's is not contained in forecasts at longer lead times. In the above example, for instance, $\hat{z}_t(3)$ depends on $\hat{z}_t(2)$ and $\hat{z}_t(1)$, which in turn depend on a_t and a_{t-1}.

5.2 CALCULATING AND UPDATING FORECASTS

5.2.1 A convenient format for the forecasts

It is frequently the case that forecasts are needed for several lead times, say at $1, 2, 3, \ldots, L$ steps ahead. The forecasts are very easily built up one from the other in the calculation scheme illustrated in Table 5.1, which shows the forecasts made at origin $t = 20$, for lead times $l = 1, 2, 3, \ldots, 14$ for Series C.

The diagonal arrangement allows each forecast to appear opposite the value it forecasts. Thus, $\hat{z}_{20}(6) = 22.51$, and this forecast made at origin 20 is a forecast of z_{26} and so appears opposite that value. The actual values of z_t at $t = 21, 22$, etc., are shown in italic type as a reminder that these values would not actually be available when the forecast was made.

The calculations are easily performed using (5.1.23). For example,

$$\hat{z}_{20}(1) = (1.8 \times 23.4) - (0.8 \times 23.7) = 23.16$$

$$\hat{z}_{20}(2) = (1.8 \times 23.16) - (0.8 \times 23.4) = 22.97$$

and so on. As soon as the new piece of data z_{21} became available, we could immediately generate a new set of forecasts which would fill a diagonal immediately below that shown. Using the fact that $a_t = z_t - \hat{z}_{t-1}(1)$, each a_t is computed, as each new piece of data z_t comes to hand, as the difference of entries on its immediate right and immediate left. Thus, as soon as $z_{21} = 23.1$ is available, we could insert the entry $-0.06 = 23.1 - 23.16$ for a_{21}.

5.2.2 Calculation of the ψ weights

Suppose forecasts at lead times $1, 2, \ldots, L$ are required. To obtain probability limits for these forecasts and also to allow new forecasts to be calculated by a process of updating the old, it is necessary to calculate the weights $\psi_1, \psi_2, \ldots, \psi_{L-1}$. This is accomplished using (5.1.4), namely

$$\varphi(B)\psi(B) = \theta(B) \tag{5.2.1}$$

That is, by equating coefficients of powers of B in

$$(1 - \varphi_1 B - \cdots - \varphi_{p+d} B^{p+d})(1 + \psi_1 B + \psi_2 B^2 + \cdots)$$
$$= (1 - \theta_1 B - \theta_2 B^2 - \cdots - \theta_q B^q) \tag{5.2.2}$$

TABLE 5.1 A convenient format for the forecasts

			Lead times l →													
		Coefficients ψ_l →	1	2	3	4	5	6	7	8	9	10	11	12	13	14
			1.80	2.44	2.95	3.36	3.69	3.95	4.16	4.33	4.46	4.57	4.65	4.72	4.78	4.82
		Nature of forecast →	$\hat{z}_{t-1}(1)$	$\hat{z}_{t-2}(2)$	$\hat{z}_{t-3}(3)$	$\hat{z}_{t-4}(4)$	$\hat{z}_{t-5}(5)$	$\hat{z}_{t-6}(6)$	$\hat{z}_{t-7}(7)$	$\hat{z}_{t-8}(8)$	$\hat{z}_{t-9}(9)$	$\hat{z}_{t-10}(10)$	$\hat{z}_{t-11}(11)$	$\hat{z}_{t-12}(12)$	$\hat{z}_{t-13}(13)$	$\hat{z}_{t-14}(14)$
		95% Limits ±	0.26	0.55	0.84	1.15	1.46	1.75	2.04	2.32	2.59	2.84	3.09	3.32	3.58	3.77
		50% Limits ±	0.09	0.19	0.29	0.39	0.50	0.60	0.70	0.79	0.88	0.97	1.05	1.13	1.22	1.29
t	z_t	a_t														
19	**23.7**															
Origin 20	**23.4**															
21	23.1	-0.06	**23.16**													
22	22.9			**22.97**												
23	22.8				**22.81**											
24	22.7					**22.69**										
25	22.6						**22.59**									
26	22.4							**22.51**								
27	22.2								**22.45**							
28	22.0									**22.40**						
29	21.8										**22.36**					
30	21.4											**22.32**				
31	20.9												**22.30**			
32	20.3													**22.28**		
33	19.7														**22.27**	
34	19.4															**22.25**
35	19.3															

Knowing the values of the φ's and the θ's, the ψ's may be obtained by equating coefficients of B as follows:

$$\left.\begin{aligned}
\psi_1 &= \varphi_1 - \theta_1 \\
\psi_2 &= \varphi_1\psi_1 + \varphi_2 - \theta_2 \\
\;\vdots\quad &\quad\;\vdots\qquad\;\;\vdots\qquad\vdots \\
\psi_j &= \varphi_1\psi_{j-1} + \cdots + \varphi_{p+d}\psi_{j-p-d} - \theta_j
\end{aligned}\right\}\qquad(5.2.3)$$

where $\psi_0 = 1$, $\psi_j = 0$ for $j < 0$ and $\theta_j = 0$ for $j > q$. If K is the greater of the integers $p + d - 1$ and q, then for $j > K$ the ψ's satisfy the difference equation

$$\psi_j = \varphi_1\psi_{j-1} + \varphi_2\psi_{j-2} + \cdots + \varphi_{p+d}\psi_{j-p-d}\qquad(5.2.4)$$

Thus the ψ's are easily calculated recursively. For example, for the model $(1 - 1.8B + 0.8B^2)z_t = a_t$, appropriate to Series C, we have

$$(1 - 1.8B + 0.8B^2)(1 + \psi_1 B + \psi_2 B^2 + \cdots) = 1$$

Either by directly equating coefficients of B^j or by using (5.2.3) and (5.2.4) with $\varphi_1 = 1.8$ and $\varphi_2 = -0.8$, we obtain

$$\psi_0 = 1$$

$$\psi_1 = 1.8$$

$$\psi_j = 1.8\psi_{j-1} - 0.8\psi_{j-2}\qquad j = 2, 3, 4, \ldots$$

Thus

$$\psi_2 = (1.8 \times 1.8) - (0.8 \times 1.0) = 2.44$$

$$\psi_3 = (1.8 \times 2.44) - (0.8 \times 1.8) = 2.95$$

and so on. The ψ's for this example are displayed in the second row of Table 5.1.

5.2.3 Use of the ψ weights in updating the forecasts

It is interesting to consider yet another way of generating the forecasts. Using (5.1.13), we can express the forecasts $\hat{z}_{t+1}(l)$ and $\hat{z}_t(l + 1)$, of the future observation z_{t+l+1}, made at origins $t + 1$ and t, as

$$\hat{z}_{t+1}(l) = \psi_l a_{t+1} + \psi_{l+1}a_t + \psi_{l+2}a_{t-1} + \cdots$$

$$\hat{z}_t(l + 1) = \qquad\qquad \psi_{l+1}a_t + \psi_{l+2}a_{t-1} + \cdots$$

On subtraction, it follows that

$$\hat{z}_{t+1}(l) = \hat{z}_t(l + 1) + \psi_l a_{t+1}\qquad(5.2.5)$$

Explicitly, the t-origin forecast of z_{t+l+1} can be updated to become the $t + 1$ origin forecast of the same z_{t+l+1}, by adding a constant multiple of the one step ahead forecast error a_{t+1}, with multiplier ψ_l.

This leads to a rather remarkable conclusion. Suppose that we currently have forecasts at origin t for lead times $1, 2, \ldots, L$. Then, as soon as z_{t+1} becomes available, we can calculate $a_{t+1} = z_{t+1} - \hat{z}_t(1)$ and proportionally update to obtain forecasts $\hat{z}_{t+1}(l) = \hat{z}_t(l + 1) + \psi_l a_{t+1}$ at origin $t + 1$, for lead times $1, 2, \ldots, L - 1$. The new forecast $\hat{z}_{t+1}(L)$, for lead time L, cannot be calculated by this means, but is easily obtained from the forecasts at shorter lead times, using the difference equation.

Referring once more to the forecasting of Series C, in Table 5.2, the forecasts at origin $t = 21$ have been added to those previously shown in Table 5.1. These can either be obtained directly, as was done to obtain the forecasts at origin $t = 20$, or we can employ the updating equation (5.2.5). The values of the ψ's are located in the second row of the table so that this can be done conveniently.

Specifically, as soon as we know $z_{21} = 23.1$, we can calculate $a_{21} = 23.1 - 23.16 = -0.06$. Then

$$\hat{z}_{21}(1) = 22.86 = 22.97 + (1.8)(-0.06)$$

$$\hat{z}_{21}(2) = 22.67 = 22.81 + (2.44)(-0.06)$$

$$\hat{z}_{21}(3) = 22.51 = 22.69 + (2.95)(-0.06)$$

and so on.

5.2.4. Calculation of the probability limits of the forecasts at any lead time

The expression (5.1.16) shows that, in general, the variance of the l steps ahead forecast error for any origin t is the expected value of

$$e_t^2(l) = \{z_{t+l} - \hat{z}_t(l)\}^2$$

and is given by

$$V(l) = \left\{ 1 + \sum_{j=1}^{l-1} \psi_j^2 \right\} \sigma_a^2$$

For example, using the weights given in Table 5.1, the function $V(l)/\sigma_a^2$ for Series C is shown in Table 5.3.

Assuming that the a's are Normal, it follows that, given information up to time t, the conditional probability distribution $p(z_{t+l}|z_t, z_{t-1}, \ldots)$ of a future value z_{t+l} of the process will be Normal with mean $\hat{z}_t(l)$ and standard deviation $\{1 + \sum_{j=1}^{l-1} \psi_j^2\}^{1/2} \sigma_a$. Figure 5.1 shows the conditional probability distributions of future values z_{21}, z_{22}, z_{23} for Series C, given information up to origin $t = 20$.

TABLE 5.2   Updating of forecasts

			1	2	3	4	5	6	7	8	9	10	11	12	13	14
Lead times $l \rightarrow$			1	2	3	4	5	6	7	8	9	10	11	12	13	14
Coefficients $\psi_l \rightarrow$			1.8	2.44	2.95	3.36	3.69	3.95	4.16	4.33	4.46	4.57	4.65	4.72	4.78	4.82
Nature of forecast $\rightarrow$			$\hat{z}_{t-1}(1)$	$\hat{z}_{t-2}(2)$	$\hat{z}_{t-3}(3)$	$\hat{z}_{t-4}(4)$	$\hat{z}_{t-5}(5)$	$\hat{z}_{t-6}(6)$	$\hat{z}_{t-7}(7)$	$\hat{z}_{t-8}(8)$	$\hat{z}_{t-9}(9)$	$\hat{z}_{t-10}(10)$	$\hat{z}_{t-11}(11)$	$\hat{z}_{t-12}(12)$	$\hat{z}_{t-13}(13)$	$\hat{z}_{t-14}(14)$
95% Limits ±			0.26	0.55	0.84	1.15	1.46	1.75	2.04	2.32	2.59	2.84	3.09	3.32	3.58	3.77
50% Limits ±			0.09	0.19	0.29	0.39	0.50	0.60	0.70	0.79	0.88	0.97	1.05	1.13	1.22	1.29
t	z_t	a_t														
19	**23.7**															
20	**23.4**															
21	**23.1**	**-0.06**	*23.16* **22.86**													
22	22.9			*22.97* **22.67**													
23	22.8				*22.81* **22.51**												
24	22.7					*22.69* **22.39**											
25	22.6						*22.59* **22.29**										
26	22.4							*22.51* **22.21**									
27	22.2								*22.45* **22.15**								
28	22.0									*22.40* **22.10**							
29	21.8										*22.36* **22.06**						
30	21.4											*22.32* **22.03**					
31	20.9												*22.30* **22.00**				
32	20.3													*22.28* **21.99**			
33	19.7														*22.27* **21.96**		
34	19.4															*22.25* **21.95**	
35	19.3																

TABLE 5.3   Variance function for series C

l	1	2	3	4	5	6	7	8	9	10
$V(l)/\sigma_a^2$	1.00	4.24	10.19	18.96	30.24	43.86	59.46	76.79	95.52	115.41

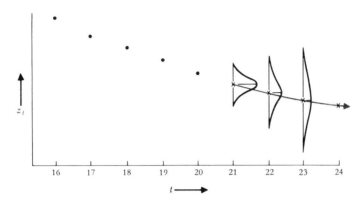

FIG. 5.1   Conditional probability distributions of future values z_{21}, z_{22} and z_{23} for
Series C, given information up to origin $t = 20$

We shall show in Chapter 7 how an estimate s_a^2, of the variance σ_a^2, may
be obtained from time series data. When the number of observations on
which such an estimate is based is, say, at least 50, s_a may be substituted for
σ_a and approximate $1 - \varepsilon$ probability limits $z_{t+l}(-)$ and $z_{t+l}(+)$ for z_{t+l}
will be given by

$$z_{t+l}(\pm) = \hat{z}_t(l) \pm u_{\varepsilon/2}\left\{1 + \sum_{j=1}^{l-1} \psi_j^2\right\}^{1/2} s_a \qquad (5.2.6)$$

where $u_{\varepsilon/2}$ is the deviate exceeded by a proportion $\varepsilon/2$ of the unit Normal
distribution.

It is shown in Table 7.13, that for Series C, $s_a = 0.134$; hence the 50%
and 95% limits, for $\hat{z}_t(2)$, for example, are given by

50% limits: $\hat{z}_t(2) \pm (0.674)(1 + 1.8^2)^{1/2}(0.134) = \hat{z}_t(2) \pm 0.19$

95% limits: $\hat{z}_t(2) \pm 1.96(1 + 1.8^2)^{1/2}(0.134) = \hat{z}_t(2) \pm 0.55$

The quantities to be added and subtracted from the forecast to obtain the
50% and 95% limits are shown in the fourth and fifth rows of the headings
of Tables 5.1 and 5.2. They apply to the forecasts which are immediately
below them.

In Figure 5.2, a section of Series C is shown, together with the several steps ahead forecasts (indicated by crosses) from origins $t = 20$ and $t = 67$.

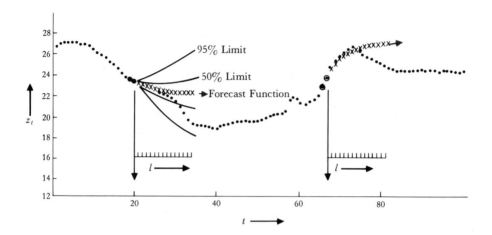

FIG. 5.2   Forecasts for Series C and probability limits

Also shown are the 50 % and 95 % probability limits for z_{20+l}, for $l = 1$ to 14. The interpretation of the limits $z_{t+l}(-)$ and $z_{t+l}(+)$ should be carefully noted. These limits are such that, *given the information available at origin t*, there is a probability of $1 - \varepsilon$, that the actual value z_{t+l}, when it occurs, will be within them, that is

$$\Pr\{z_{t+l}(-) < z_{t+l} < z_{t+l}(+)\} = 1 - \varepsilon$$

It should also be explained that the probabilities quoted apply to *individual* forecasts and not jointly to the forecasts at all the different lead times. For example, it is true that with 95 % probability, the limits for lead time 10 will include the value z_{t+10} when it occurs. It is not true that the series can be expected to remain within *all* the limits simultaneously at this level of probability.

5.3 THE FORECAST FUNCTION AND FORECAST WEIGHTS

Forecasts are calculated most simply in the manner just described—by direct use of the difference equation. From the purely *computational* standpoint, the other model forms are less convenient. However, from the point of view of studying the nature of the forecasts, it is profitable to consider in greater detail the alternative forms discussed in Section 5.1.2 and, in particular, to consider the explicit form of the forecast function.

5.3.1 The eventual forecast function determined by the autoregressive operator

At time $t + l$ the ARIMA model may be written

$$z_{t+l} - \varphi_1 z_{t+l-1} - \cdots - \varphi_{p+d} z_{t+l-p-d} = a_{t+l} - \theta_1 a_{t+l-1}$$
$$- \cdots - \theta_q a_{t+l-q} \tag{5.3.1}$$

Taking conditional expectations at time t in (5.3.1), we have, for $l > q$,

$$\hat{z}_t(l) - \varphi_1 \hat{z}_t(l-1) - \cdots - \varphi_{p+d} \hat{z}_t(l-p-d) = 0 \qquad l > q \tag{5.3.2}$$

where it is understood that $\hat{z}_t(-j) = z_{t-j}$ for $j \geqslant 0$. The difference equation (5.3.2) has the solution

$$\hat{z}_t(l) = b_0^{(t)} f_0(l) + b_1^{(t)} f_1(l) + \cdots + b_{p+d-1}^{(t)} f_{p+d-1}(l) \tag{5.3.3}$$

for $l > q - p - d$. Note that the forecast $\hat{z}_t(l)$ is the complementary function introduced in Chapter 4. In (5.3.3), $f_0(l), f_1(l), \ldots, f_{p+d-1}(l)$, are functions of the lead time l. In general, they could include polynomials, exponentials, sines and cosines, and products of these functions. For a *given origin* t, the coefficients $b_j^{(t)}$ are constants applying for all lead times l, but they change from one origin to the next, *adapting* themselves appropriately to the particular part of the series being considered. From now on we shall call the function defined by (5.3.3) the *eventual forecast function*; "eventual" because when it occasionally happens that $q > p + d$, it supplies the forecasts only for lead times $l > q - p - d$.

We see from (5.3.2) that it is the general autoregressive operator $\varphi(B)$ which determines the mathematical form of the forecast function, that is, the nature of the f's in (5.3.3). Specifically, it determines whether the forecast function is to be a polynomial, a mixture of sines and cosines, a mixture of exponentials, or some combination of these functions.

5.3.2 Role of the moving average operator in fixing the initial values

While the autoregressive operator decides the nature of the eventual forecast function, the moving average operator is influential in determining how that function is to be "fitted" to the data and hence how the coefficients $b_0^{(t)}$, $b_1^{(t)}, \ldots, b_{p+d-1}^{(t)}$ are to be calculated and updated.

For example, consider the IMA $(0, 2, 3)$ process

$$z_{t+l} - 2z_{t+l-1} + z_{t+l-2} = a_{t+l} - \theta_1 a_{t+l-1} - \theta_2 a_{t+l-2} - \theta_3 a_{t+l-3}$$

Using the conditional expectation argument of Section 5.1.2, the forecast function is defined by

$$\hat{z}_t(1) = 2z_t - z_{t-1} - \theta_1 a_t - \theta_2 a_{t-1} - \theta_3 a_{t-2}$$

$$\hat{z}_t(2) = 2\hat{z}_t(1) - z_t - \theta_2 a_t - \theta_3 a_{t-1}$$

$$\hat{z}_t(3) = 2\hat{z}_t(2) - \hat{z}_t(1) - \theta_3 a_t$$

$$\hat{z}_t(l) = 2\hat{z}_t(l-1) - \hat{z}_t(l-2) \qquad l > 3$$

Therefore, the eventual forecast function is the unique straight line

$$\hat{z}_t(l) = b_0^{(t)} + b_1^{(t)} l \qquad l > 1$$

which passes through $\hat{z}_t(2)$ and $\hat{z}_t(3)$ as shown in Figure 5.3. However, note that if the θ_3 term had been omitted, then $q - p - d = 0$, and the forecast would have been given at *all lead times* by the straight line passing through $\hat{z}_t(1)$ and $\hat{z}_t(2)$.

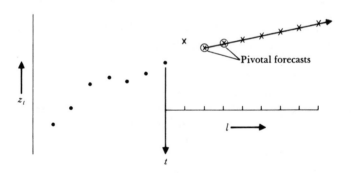

FIG. 5.3   Eventual forecast function for an IMA $(0, 2, 3)$ process

In general, since only one function of the form (5.3.3) can pass through $p + d$ points, the eventual forecast function is that unique curve of the form required by $\varphi(B)$, which passes through the $p + d$ "pivotal" values $\hat{z}_t(q)$, $\hat{z}_t(q - 1), \ldots, \hat{z}_t(q - p - d + 1)$, where $\hat{z}_t(-j) = z_{t-j}$ ($j = 0, 1, 2, \ldots$). In the extreme case where $q = 0$, so that the model is of the purely autoregressive form $\varphi(B)z_t = a_t$, the curve passes through the points $z_t, z_{t-1}, \ldots, z_{t-p-d+1}$. Thus, the pivotal values can consist of forecasts or of actual values of the series; they are indicated in the figures by circled points.

The moving average terms, which appear in the model form, help to decide the way in which we "reach back" into the series to fit the forecast function determined by the autoregressive operator $\varphi(B)$. Figure 5.4 illustrates the situation for the model of order $(1, 1, 3)$ given by $(1 - \phi B)\nabla z_t = (1 - \theta_1 B - \theta_2 B^2 - \theta_3 B^3)a_t$. The (hypothetical) weight functions indicate the linear functional dependence of the three forecasts $\hat{z}_t(1)$, $\hat{z}_t(2)$ and $\hat{z}_t(3)$ on $z_t, z_{t-1}, z_{t-2}, \ldots$. Since the forecast function contains $p + d = 2$ coefficients, it is uniquely determined by the forecasts $\hat{z}_t(3)$ and $\hat{z}_t(2)$, that is by $\hat{z}_t(q)$ and $\hat{z}_t(q - 1)$. We next consider how the forecast weight functions, referred to above, are determined.

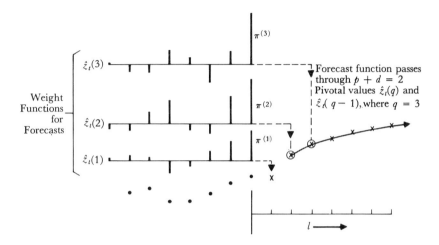

FIG. 5.4   Dependence of forecast function on observations for $(1, 1, 3)$ process $(1 - \phi B)\nabla z_t = (1 - \theta_1 B - \theta_2 B^2 - \theta_3 B^3)a_t$

5.3.3 The lead-l forecast weights

The fact that the general model may also be written in inverted form

$$a_t = \pi(B)z_t = (1 - \pi_1 B - \pi_2 B^2 - \pi_3 B^3 - \cdots)z_t \qquad (5.3.4)$$

results in our being able to write the forecast as in (5.1.21).

On substituting for the conditional expectations in (5.1.21), we obtain

$$\hat{z}_t(l) = \sum_{j=1}^{\infty} \pi_j \hat{z}_t(l - j) \qquad (5.3.5)$$

where, as before, $\hat{z}_t(-h) = z_{t-h}$ for $h = 0, 1, 2, \ldots$. Thus, in general,

$$\hat{z}_t(l) = \pi_1 \hat{z}_t(l - 1) + \cdots + \pi_{l-1}\hat{z}_t(1) + \pi_l z_t + \pi_{l+1} z_{t-1} + \cdots \qquad (5.3.6)$$

and in particular,

$$\hat{z}_t(1) = \pi_1 z_t + \pi_2 z_{t-1} + \pi_3 z_{t-2} + \cdots$$

The forecasts for higher lead times may also be expressed directly as linear functions of the observations $z_t, z_{t-1}, z_{t-2}, \ldots$. For example, the lead-two forecast at origin t is

$$\hat{z}_t(2) = \pi_1 \hat{z}_t(1) + \pi_2 z_t + \pi_3 z_{t-1} + \cdots$$

$$= \pi_1 \sum_{j=1}^{\infty} \pi_j z_{t-j+1} + \sum_{j=1}^{\infty} \pi_{j+1} z_{t-j+1}$$

$$= \sum_{j=1}^{\infty} \pi_j^{(2)} z_{t-j+1}$$

where

$$\pi_j^{(2)} = \pi_1 \pi_j + \pi_{j+1} \qquad j = 1, 2, \ldots \tag{5.3.7}$$

Proceeding in this way, it is readily shown that

$$\hat{z}_t(l) = \sum_{j=1}^{\infty} \pi_j^{(l)} z_{t-j+1} \tag{5.3.8}$$

where

$$\pi_j^{(l)} = \pi_{j+l-1} + \sum_{h=1}^{l-1} \pi_h \pi_j^{(l-h)} \qquad j = 1, 2, \ldots \tag{5.3.9}$$

and $\pi_j^{(1)} = \pi_j$. Alternative methods for computing these weights are given in Appendix A5.2.

As we have seen in (4.2.22) and (5.1.9), the π_j's themselves may be obtained explicitly by equating coefficients in

$$\theta(B)(1 - \pi_1 B - \pi_2 B - \cdots) = \varphi(B)$$

Given these values, the $\pi_j^{(l)}$'s may be readily obtained, if so desired, using (5.3.9), or the results of Appendix A5.2.

As an example, consider again the model

$$\nabla^2 z_t = (1 - 0.9B + 0.5B^2)a_t$$

which was fitted to a series, a part of which is shown in Figure 5.5. Equating coefficients in

$$(1 - 0.9B + 0.5B^2)(1 - \pi_1 B - \pi_2 B^2 \ldots) = 1 - 2B + B^2$$

yields the weights $\pi_j = \pi_j^{(1)}$, from which the weights $\pi_j^{(2)}$ may be computed using (5.3.7). The two sets of weights are given for $j = 1, 2, \ldots, 12$ in Table 5.4.

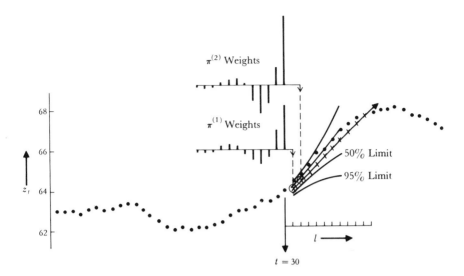

FIG. 5.5   Part of a series fitted by $\nabla^2 z_t = (1 - 0.9B + 0.5B^2)a_t$ with forecast function for origin $t = 30$, forecast weights and probability limits

TABLE 5.4 π weights for the model $\nabla^2 z_t = (1 - 0.9B + 0.5B^2)a_t$.

j	$\pi_j = \pi_j^{(1)}$	$\pi_j^{(2)}$
1	1.100	1.700
2	0.490	0.430
3	−0.109	−0.463
4	−0.343	−0.632
5	−0.254	−0.336
6	−0.057	0.013
7	0.076	0.181
8	0.097	► 0.156
9	0.049	0.050
10	−0.004	−0.032
11	−0.028	−0.054
12	−0.023	−0.026

In this example, the lead-one and lead-two forecasts, expressed in terms of the observations $z_t, z_{t-1}, \ldots,$ are

$$\hat{z}_t(1) = 1.10z_t + 0.49z_{t-1} - 0.11z_{t-2} - 0.34z_{t-3} - 0.25z_{t-4} - \cdots$$

and

$$\hat{z}_t(2) = 1.70z_t + 0.43z_{t-1} - 0.46z_{t-2} - 0.63z_{t-3} - 0.34z_{t-4} + \cdots$$

In fact, the weights follow damped sine waves as shown in Figure 5.5.

5.4 EXAMPLES OF FORECAST FUNCTIONS AND THEIR UPDATING

The forecast functions for some special cases of the general ARIMA model will now be considered. We shall exhibit these in the three different forms discussed in Section 5.1.2. As mentioned earlier, the forecasts are most easily computed from the difference equation itself. The other forms are useful because they provide insight into the nature of the forecast function in particular cases.

5.4.1 Forecasting an IMA (0, 1, 1) process

The model is $\nabla z_t = (1 - \theta B)a_t$.

   Difference equation approach.   At time $t + l$ the model may be written

$$z_{t+l} = z_{t+l-1} + a_{t+l} - \theta a_{t+l-1}$$

Taking conditional expectations at origin t,

$$\left.\begin{aligned}
\hat{z}_t(1) &= z_t - \theta a_t \\
\hat{z}_t(l) &= \hat{z}_t(l - 1) \qquad l \geqslant 2
\end{aligned}\right\} \tag{5.4.1}$$

Hence, for all lead times, the forecasts at origin t will follow a straight line parallel to the time axis. Using the fact that $z_t = \hat{z}_{t-1}(1) + a_t$, we can write (5.4.1) in either of two useful forms.

   The first of these is

$$\hat{z}_t(l) = \hat{z}_{t-1}(l) + \lambda a_t \tag{5.4.2}$$

where $\lambda = 1 - \theta$. This implies that, having seen that our previous forecast $\hat{z}_{t-1}(l)$ falls short of the realized value by a_t, we adjust it by an amount λa_t. It will be recalled from Section 4.3.1 that λ measures the proportion of any given shock a_t, which is permanently absorbed by the "level" of the process. Therefore, it is reasonable to increase the forecast by that part λa_t of a_t, which we expect to be absorbed.

   The second way of rewriting (5.4.1) is

$$\hat{z}_t(l) = \lambda z_t + (1 - \lambda)\hat{z}_{t-1}(l) \tag{5.4.3}$$

This implies that the new forecast is a linear interpolation at argument λ between old forecast and new observation. The form (5.4.3) makes it clear that if λ is very small, we shall be relying principally on a weighted average of past data and heavily discounting the new observation z_t. By contrast, if $\lambda = 1$, the evidence of past data is completely ignored, $\hat{z}_t(1) = z_t$, and the forecast for all future time is the current value. With $\lambda > 1$, we induce an extrapolation rather than an interpolation between $\hat{z}_{t-1}(l)$ and z_t. The forecast error must now be *magnified* in (5.4.2) to indicate the change in the forecast.

Forecast function in integrated form. The eventual forecast function is the solution of $(1 - B)\hat{z}_t(l) = 0$. Thus, $\hat{z}_t(l) = b_0^{(t)}$, and since $q - p - d = 0$, it provides the forecast for all lead times, that is

$$\hat{z}_t(l) = b_0^{(t)} \qquad l > 0 \qquad\qquad (5.4.4)$$

For *any fixed origin*, $b_0^{(t)}$ is a constant, and the forecasts for all lead times will follow a straight line parallel to the time axis. However, the coefficient $b_0^{(t)}$ will be updated as a new observation becomes available and the origin advances. Thus, the forecast function can be thought of as a polynomial of degree zero in the lead time l, with a coefficient which is adaptive with respect to the origin t.

Since the integrated form for the model is

$$z_t = \lambda S a_{t-1} + a_t$$

it follows that

$$\hat{z}_t(l) = b_0^{(t)} = \lambda S a_t$$

Also, $\psi_j = \lambda (j = 1, 2, \ldots)$ and hence the adaptive coefficient $b_0^{(t)}$ can be updated from origin t to origin $t + 1$ according to

$$b_0^{(t+1)} = b_0^{(t)} + \lambda a_{t+1} \qquad\qquad (5.4.5)$$

The forecast as a weighted average of previous observations. Since, for this process, the $\pi_j^{(l)}$ weights of (5.3.8) are also the weights for the one step ahead forecast, we can also write using (4.3.6),

$$\hat{z}_t(l) = b_0^{(t)} = \lambda z_t + \lambda(1 - \lambda)z_{t-1} + \lambda(1 - \lambda)^2 z_{t-2} + \cdots \qquad (5.4.6)$$

Thus, for the IMA $(0, 1, 1)$ model, the forecast for all future time is an *exponentially weighted moving average* of current and past z's.

An example: Forecasting Series A. It will be shown in Chapter 7 that Series A is closely fitted by the model

$$(1 - B)z_t = (1 - 0.7B)a_t$$

In Figure 5.6 the forecasts at origins $t = 39, 40, 41, 42$ and 43 and also at origin $t = 79$ are shown for lead times $1, 2, \ldots, 20$. The weights π_j, which for this model are forecast weights for any lead time, are given in Table 5.5.

These weights are shown diagrammatically in their appropriate positions for the forecast $\hat{z}_{39}(l)$ on Figure 5.6.

Variance functions. Since, for this model, $\psi_j = \lambda$ $(j = 1, 2, \ldots)$, the expression (5.1.16) for the variance of the lead l forecasts is

$$V(l) = \sigma_a^2 \{1 + (l - 1)\lambda^2\} \qquad\qquad (5.4.7)$$

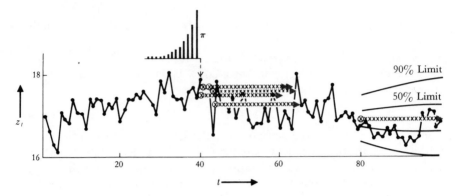

FIG. 5.6   Part of Series A with forecasts at origins $t = 39, 40, 41, 42, 43$ and at $t = 79$

TABLE 5.5   Forecast weights applied to previous z's
for any lead time used in forecasting
series A with model $\nabla z_t = (1 - 0.7B)a_t$

j	π_j	j	π_j
1	0.300	7	0.035
2	0.210	8	0.025
3	0.147	9	0.017
4	0.103	10	0.012
5	0.072	11	0.008
6	0.050	12	0.006

Using the estimate $s_a^2 = 0.101$, appropriate for Series A, in (5.4.7), 50%
and 95% probability limits were calculated and are shown in Figure 5.6
for origin $t = 79$.

5.4.2 Forecasting an IMA (0, 2, 2) process

The model is $\nabla^2 z_t = (1 - \theta_1 B - \theta_2 B^2)a_t$.
   Difference equation approach.   At time $t + l$, the model may be written

$$z_{t+l} = 2z_{t+l-1} - z_{t+l-2} + a_{t+l} - \theta_1 a_{t+l-1} - \theta_2 a_{t+l-2}$$

On taking conditional expectations at time t

$$\hat{z}_t(1) = 2z_t - z_{t-1} - \theta_1 a_t - \theta_2 a_{t-1}$$

$$\hat{z}_t(2) = 2\hat{z}_t(1) - z_t - \theta_2 a_t$$

$$\hat{z}_t(l) = 2\hat{z}_t(l - 1) - \hat{z}_t(l - 2) \qquad l \geqslant 3$$

from which the forecasts may be calculated. Forecasting of the series of
Figure 5.5 in this way, was illustrated in Section 5.1.2. An alternative way

of generating the first $L - 1$ of L forecasts, is via the updating formula (5.2.5),

$$\hat{z}_{t+1}(l) = \hat{z}_t(l + 1) + \psi_l a_{t+1} \qquad (5.4.8)$$

The integrated model is

$$z_t = \lambda_0 S a_{t-1} + \lambda_1 S^2 a_{t-1} + a_t \qquad \lambda_0 = 1 + \theta_2 \qquad \lambda_1 = 1 - \theta_1 - \theta_2$$

$$(5.4.9)$$

so that $\psi_j = \lambda_0 + j\lambda_1 (j = 1, 2, \ldots)$. Therefore, the updating function for this model is

$$\hat{z}_{t+1}(l) = \hat{z}_t(l + 1) + (\lambda_0 + l\lambda_1) a_{t+1} \qquad (5.4.10)$$

Forecast in integrated form.　The eventual forecast function is the solution of $(1 - B)^2 \hat{z}_t(l) = 0$, that is, $\hat{z}_t(l) = b_0^{(t)} + b_1^{(t)} l$. Since $q - p - d = 0$, the eventual forecast function provides the forecast for all lead times, that is

$$\hat{z}_t(l) = b_0^{(t)} + b_1^{(t)} l \qquad l > 0 \qquad (5.4.11)$$

Thus, the forecast function is a linear function of the lead time l, with coefficients which are adaptive with respect to the origin t. The stochastic model in integrated form is

$$z_{t+l} = a_{t+l} + \lambda_0 S a_{t+l-1} + \lambda_1 S^2 a_{t+l-1}$$

and taking expectations at origin t

$$\hat{z}_t(l) = \{\lambda_0 S a_t + \lambda_1 S^2 a_{t-1}\} + \{\lambda_1 S a_t\} l$$

The adaptive constants may thus be identified as

$$\left. \begin{array}{l} b_0^{(t)} = \lambda_0 S a_t + \lambda_1 S^2 a_{t-1} \\ b_1^{(t)} = \lambda_1 S a_t \end{array} \right\} \qquad (5.4.12)$$

whence their updating formulae are

$$\left. \begin{array}{l} b_0^{(t)} = b_0^{(t-1)} + b_1^{(t-1)} + \lambda_0 a_t \\ b_1^{(t)} = b_1^{(t-1)} + \lambda_1 a_t \end{array} \right\} \qquad (5.4.13)$$

The additional slope term $b_1^{(t-1)}$, which occurs in the updating formula for $b_0^{(t)}$, is an adjustment to change the location parameter b_0 to a value appropriate to the new origin.

It will be noticed also, that λ_0 and λ_1 are the fractions of the shock a_t, which are transmitted to the location parameter and the slope respectively.

Forecasts as a weighted average of previous observations.　For this model then, the forecast function is the straight line that passes through the forecasts $\hat{z}_t(1)$ and $\hat{z}_t(2)$. This is illustrated for the series in Figure 5.5 which shows the forecasts made at origin $t = 30$, with appropriate weight functions. It will be seen how dependence of the whole forecast function on previous z's in the series is a reflection of the dependence of $\hat{z}_t(1)$ and $\hat{z}_t(2)$ on these values. The weight functions for $\hat{z}_t(1)$ and $\hat{z}_t(2)$, plotted in the figure, have been given in Table 5.4.

The example illustrates once more, that while the AR operator $\varphi(B)$ determines the form of function to be used (a straight line in this case), the MA operator is of importance in determining the way in which that function is "fitted" to previous data.

Dependence of the adaptive coefficients in the forecast function on previous z's. Since, for the general model, the values of the adaptive coefficients in the forecast function are determined by $\hat{z}_t(q), \hat{z}_t(q-1), \ldots, \hat{z}_t(q-p-d+1)$, which can be expressed as functions of the observations, it follows that the adaptive coefficients themselves may be so expressed.

For instance, in the case of the model $\nabla^2 z_t = (1 - 0.9B + 0.5B^2)a_t$ of Figure 5.5,

$$\hat{z}_t(1) = b_0^{(t)} + b_1^{(t)} = \sum_{j=1}^{\infty} \pi_j^{(1)} z_{t-j+1}$$

$$\hat{z}_t(2) = b_0^{(t)} + 2b_1^{(t)} = \sum_{j=1}^{\infty} \pi_j^{(2)} z_{t-j+1}$$

so that

$$b_0^{(t)} = 2\hat{z}_t(1) - \hat{z}_t(2) = \sum_{j=1}^{\infty} \{2\pi_j^{(1)} - \pi_j^{(2)}\} z_{t-j+1}$$

and

$$b_1^{(t)} = \hat{z}_t(2) - \hat{z}_t(1) = \sum_{j=1}^{\infty} \{\pi_j^{(2)} - \pi_j^{(1)}\} z_{t-j+1}$$

These weight functions are plotted in Figure 5.7.

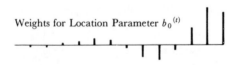

Weights for Location Parameter $b_0{}^{(t)}$

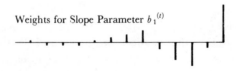

Weights for Slope Parameter $b_1{}^{(t)}$

FIG 5.7   Weights applied to previous z's determining location and slope for the model $\nabla^2 z_t = (1 - 0.9B + 0.5B^2)a_t$

The variance of the forecast error. Using (5.1.16), and the fact that $\psi_j = \lambda_0 + j\lambda_1$, the variance of the lead-l forecast is

$$V(l) = \sigma_a^2\{1 + (l-1)\lambda_0^2 + \tfrac{1}{6}l(l-1)(2l-1)\lambda_1^2 + \lambda_0\lambda_1 l(l-1)\} \quad (5.4.14)$$

Using the estimate $s_a^2 = 0.032$, $\lambda_0 = 0.5$, and $\lambda_1 = 0.6$, the 50% and 95% limits are shown in Figure 5.5 for the forecast at origin $t = 30$.

5.4.3. Forecasting a general $IMA\,(0, d, q)$ process

As an example, consider the process of order $(0, 1, 3)$

$$(1 - B)z_{t+l} = (1 - \theta_1 B - \theta_2 B^2 - \theta_3 B^3)a_{t+l}$$

Taking conditional expectations at time t, we obtain

$$\hat{z}_t(1) - z_t = -\theta_1 a_t - \theta_2 a_{t-1} - \theta_3 a_{t-2}$$

$$\hat{z}_t(2) - \hat{z}_t(1) = -\theta_2 a_t - \theta_3 a_{t-1}$$

$$\hat{z}_t(3) - \hat{z}_t(2) = -\theta_3 a_t$$

$$\hat{z}_t(l) - \hat{z}_t(l-1) = 0 \qquad l = 4, 5, 6, \ldots$$

Hence, $\hat{z}_t(l) = \hat{z}_t(3) = b_0^{(t)}$ for all $l > 2$, as expected, since $q - p - d = 2$. As shown in Figure 5.8, the forecast function makes two initial "jumps," depending on previous a's, before levelling out to the eventual forecast function.

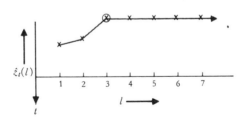

FIG. 5.8   Forecast function for an IMA $(0, 1, 3)$ process

For the IMA $(0, d, q)$ process, the eventual forecast function satisfies $(1 - B)^d \hat{z}_t(l) = 0$, and has for its solution, a polynomial in l of degree $d - 1$,

$$\hat{z}_t(l) = b_0^{(t)} + b_1^{(t)}l + b_2^{(t)}l^2 + \cdots + b_{d-1}^{(t)}l^{d-1}$$

This will provide the forecasts $\hat{z}_t(l)$ for $l > q - d$. The coefficients $b_0^{(t)}$, $b_1^{(t)}$, $\ldots, b_{d-1}^{(t)}$, must be progressively updated as the origin advances. The forecast for origin t will make $q - d$ initial "jumps," which depend upon a_t, a_{t-1}, $\ldots, a_{t-q+1}$ and after this, will follow the above polynomial.

5.4.4 Forecasting autoregressive processes

Consider a process of order $(p, d, 0)$

$$\varphi(B)z_t = a_t$$

The eventual forecast function is the solution of $\varphi(B)\hat{z}_t(l) = 0$. It applies for all lead times and passes through the last $p + d$ available values of the series. For example, the model for the IBM stock series (Series B) is very nearly

$$(1 - B)z_t = a_t$$

so that

$$\hat{z}_t(l) \approx z_t$$

The best forecast for all future time is very nearly the current value of the stock. The weight function for $\hat{z}_t(l)$ is a spike at time t and there is no averaging over past history.

Stationary autoregressive models. The process $\phi(B)\tilde{z}_t = a_t$ of order $(p, 0, 0)$, where $\phi(B)$ is a stationary operator, and $\tilde{z}_t = z_t - \mu$, with $E[z_t] = \mu$, will in general produce a forecast function which is a mixture of exponentials and damped sines.

In particular, for $p = 1$ the model of order $(1, 0, 0)$

$$(1 - \phi B)\tilde{z}_t = a_t, \qquad -1 < \phi < 1$$

has a forecast function, which for all $l > 0$, is the solution of $(1 - \phi B)\hat{\tilde{z}}_t(l) = 0$. Thus

$$\hat{\tilde{z}}_t(l) = b_0^{(t)}\phi^l \qquad l > 0 \tag{5.4.15}$$

Also, $\hat{\tilde{z}}_t(1) = \phi\tilde{z}_t$, so that $b_0^{(t)} = \tilde{z}_t$ and

$$\hat{\tilde{z}}_t(l) = \tilde{z}_t\phi^l$$

Hence the minimum mean square error forecast predicts the current deviation from the mean decaying exponentially to zero. In Figure 5.9(a) a time series is shown which is generated from the process $(1 - 0.5B)\tilde{z}_t = a_t$, with the forecast function at origin $t = 14$. The course of this function is seen to be determined entirely by the single deviation $\tilde{z}_{14}$. Similarly, the minimum mean square error forecast for a second-order autoregressive process, is such that the current deviation from the mean decays to zero via a damped sine wave or a mixture of two exponentials. Figure 5.9(b) shows a time series generated from the process $(1 - 0.75B + 0.50B^2)\tilde{z}_t = a_t$, and the forecast at origin $t = 14$. Here the course of the forecast function at origin t is determined entirely by the last two deviations $\tilde{z}_{14}$ and $\tilde{z}_{13}$.

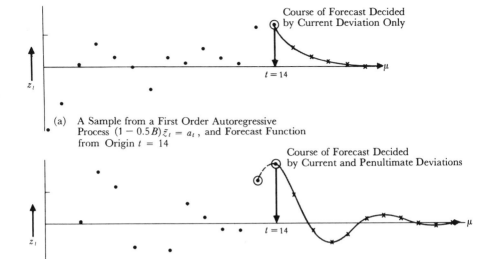

(a) A Sample from a First Order Autoregressive
     Process $(1 - 0.5B)\tilde{z}_t = a_t$, and Forecast Function
     from Origin $t = 14$

(b) A Sample from a Second Order Autoregressive
     Process $(1 - 0.75\,B + 0.5\,B^2)\tilde{z}_t = a_t$, and Forecast Function
     from Origin $t = 14$

FIG. 5.9   Forecast functions for first- and second-order autoregressive processes

Variance function for the forecast from a $(1, 0, 0)$ process.   As a further
illustration of the use of (5.1.16), we derive the variance function for a first-
order autoregressive process. Since the model at time $t + l$ may be written

$$\tilde{z}_{t+l} = a_{t+l} + \phi a_{t+l-1} + \cdots + \phi^{l-1}a_{t+1} + \phi^l \tilde{z}_t$$

it follows from (5.4.15) that

$$e_t(l) = \tilde{z}_{t+l} - \hat{z}_t(l) = a_{t+l} + \phi a_{t+l-1} + \cdots + \phi^{l-1}a_{t+1}$$

Hence

$$V(l) = \text{var}\,[e_t(l)] = \sigma_a^2(1 + \phi^2 + \cdots + \phi^{2l-2})$$

$$= \frac{\sigma_a^2(1 - \phi^{2l})}{1 - \phi^2} \qquad (5.4.16)$$

We see that, for this stationary process, as l tends to infinity the variance
increases to a constant value $\sigma_a^2/(1 - \phi^2)$, associated with the variation of
the process about the ultimate forecast μ. This is in contrast to the behavior
of forecast variance functions for nonstationary models which "blow up"
for large lead times.

Nonstationary autoregressive models of order $(p, d, 0)$. For the model

$$\phi(B)\nabla^d z_t = a_t$$

it will be the dth difference of the process which decays back to its mean when projected several steps ahead. The mean of $\nabla^d z_t$ will usually be assumed to be zero unless contrary evidence is available. When needed, it is possible, as discussed in Chapter 4, to introduce a nonzero mean by replacing $\nabla^d z_t$ by $\nabla^d z_t - \mu$ in the model. For example, consider the model,

$$(1 - \phi B)(\nabla z_t - \mu) = a_t \tag{5.4.17}$$

After substituting $t + j$ for t and taking conditional expectations at origin t, we readily obtain (compare with (5.4.15) et seq.)

$$\hat{z}_t(j) - \hat{z}_t(j - 1) - \mu = \phi^j(z_t - z_{t-1} - \mu)$$

which shows how the forecasted *difference* decays exponentially from the initial value $z_t - z_{t-1}$ to its mean value μ. On summing this expression from $j = 1$ to $j = l$, we obtain the forecast function

$$\hat{z}_t(l) = z_t + \mu l + (z_t - z_{t-1} - \mu)\frac{\phi(1 - \phi^l)}{1 - \phi} \qquad l \geqslant 1$$

which approaches asymptotically the straight line

$$f(l) = z_t + \mu l + (z_t - z_{t-1} - \mu)\frac{\phi}{1 - \phi}$$

Figure 5.10 shows forecasts for the two cases $\phi = 0.8$, $\mu = 0$ and $\phi = 0.8$, $\mu = 0.2$. We show in Chapter 7 that the model (5.4.17), with $\phi = 0.8$, $\mu = 0$, closely represents Series C and forecasts based on this model have already been illustrated in Figures 5.1 and 5.2. We now consider the forecasting of some important mixed models.

5.4.5 Forecasting a $(1, 0, 1)$ process

Difference equation approach. Consider the stationary model

$$(1 - \phi B)\tilde{z}_t = (1 - \theta B)a_t$$

The forecasts are readily obtained from

$$\left.\begin{array}{l} \hat{\tilde{z}}_t(1) = \phi\tilde{z}_t - \theta a_t \\ \hat{\tilde{z}}_t(l) = \phi\hat{\tilde{z}}_t(l - 1) \qquad l \geqslant 2 \end{array}\right\} \tag{5.4.18}$$

The forecasts decay geometrically to the mean, as in the first-order autoregressive process, but with a lead-one forecast modified by a factor depending

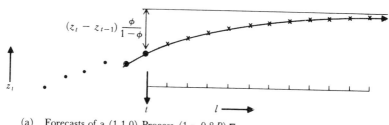

(a)   Forecasts of a (1,1,0) Process $(1 - 0.8\,B)\, \nabla z_t = a_t$

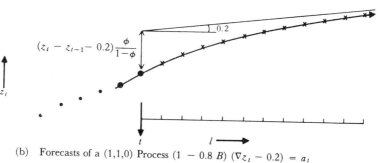

(b)   Forecasts of a (1,1,0) Process $(1 - 0.8\,B)\, (\nabla z_t - 0.2) = a_t$

FIG. 5.10   Forecast functions for two (1, 1, 0) processes

on $a_t = z_t - \hat{z}_{t-1}(1)$. The ψ weights are

$$\psi_j = (\phi - \theta)\phi^{j-1} \qquad j = 1, 2, \ldots$$

and hence, using (5.2.5), the updated forecasts for lead times $1, 2, \ldots, L-1$ could be obtained from previous forecasts for lead times $2, 3, \ldots, L$ according to

$$\hat{z}_{t+1}(l) = \hat{z}_t(l+1) + (\phi - \theta)\phi^{l-1} a_{t+1}$$

Integrated form. The eventual forecast function for all $l > 0$, is the solution of $(1 - \phi B)\hat{z}_t(l) = 0$, that is

$$\hat{z}_t(l) = b_0^{(t)}\phi^l \qquad l > 0$$

However,

$$\hat{z}_t(1) = b_0^{(t)}\phi = \phi z_t - \theta a_t = \left\{\left(1 - \frac{\theta}{\phi}\right)z_t + \frac{\theta}{\phi}\hat{z}_{t-1}(1)\right\}\phi$$

Thus

$$\hat{z}_t(l) = \left\{\left(1 - \frac{\theta}{\phi}\right)z_t + \frac{\theta}{\phi}\hat{z}_{t-1}(1)\right\}\phi^l \tag{5.4.19}$$

Hence, the forecasted deviation at lead l decays exponentially from an initial value, which is a linear interpolation between the previous lead-1 forecasted deviation and the current deviation. When ϕ is equal to unity, the forecast for all lead times becomes the familiar exponentially weighted moving average and (5.4.19) becomes equal to (5.4.3).

Weights applied to previous observations. The π weights, and hence the weights applied to previous observations to obtain the lead one forecasts, are

$$\pi_j = (\phi - \theta)\theta^{j-1} \qquad j = 1, 2, \ldots$$

Note that the weights for this stationary process sum to $(\phi - \theta)/(1 - \theta)$ and not to unity. If ϕ were equal to 1, the process would become a non-stationary IMA $(0, 1, 1)$ process, the weights would then sum to unity, and the behavior of the generated series would be independent of the level of z_t.

For example, Series A is later fitted to a $(1, 0, 1)$ process with $\phi = 0.9$ and $\theta = 0.6$, and hence the weights are $\pi_1 = 0.30$, $\pi_2 = 0.18$, $\pi_3 = 0.11$, $\pi_4 = 0.07, \ldots$, which sum to 0.75. The forecasts (5.4.19) decay very slowly to the mean, and for short lead times, are practically indistinguishable from the forecasts obtained from the alternative IMA $(0, 1, 1)$ model $\nabla z_t = a_t - 0.7a_{t-1}$, for which the weights are $\pi_1 = 0.30$, $\pi_2 = 0.21$, $\pi_3 = 0.15$, $\pi_4 = 0.10$, etc and sum to unity. The latter model has the advantage that it does not tie the process to a fixed mean.

Variance function. Since the ψ weights are given by

$$\psi_j = (\phi - \theta)\phi^{j-1} \qquad j = 1, 2, \ldots$$

it follows that the variance function is

$$V(l) = \sigma_a^2 \left\{ 1 + (\phi - \theta)^2 \left[\frac{1 - \phi^{2l-2}}{1 - \phi^2} \right] \right\} \qquad (5.4.20)$$

which increases asymptotically to the value $\sigma_a^2(1 - 2\phi\theta + \theta^2)/(1 - \phi^2)$.

5.4.6 Forecasting a $(1, 1, 1)$ process

Another important mixed model is the nonstationary $(1, 1, 1)$ process

$$(1 - \phi B)\nabla z_t = (1 - \theta B)a_t$$

Difference equation approach. At time $t + l$, the model may be written

$$z_{t+l} = (1 + \phi)z_{t+l-1} - \phi z_{t+l-2} + a_{t+l} - \theta a_{t+l-1}$$

On taking conditional expectations,

$$\left. \begin{array}{l} \hat{z}_t(1) = (1 + \phi)z_t - \phi z_{t-1} - \theta a_t \\ \hat{z}_t(l) = (1 + \phi)\hat{z}_t(l - 1) - \phi\hat{z}_t(l - 2) \qquad l > 1 \end{array} \right\} \qquad (5.4.21)$$

Integrated form. Since $q < p + d$, the eventual forecast function for all $l > 0$ is the solution of $(1 - \phi B)(1 - B)\hat{z}_t(l) = 0$, which is

$$\hat{z}_t(l) = b_0^{(t)} + b_1^{(t)}\phi^l$$

Substituting for $\hat{z}_t(1)$ and $\hat{z}_t(2)$ in (5.4.21), we find explicitly that

$$b_0^{(t)} = z_t + \frac{\phi}{1 - \phi}(z_t - z_{t-1}) - \frac{\theta}{1 - \phi}a_t$$

$$b_1^{(t)} = \frac{\theta a_t - \phi(z_t - z_{t-1})}{1 - \phi}$$

Thus finally

$$\hat{z}_t(l) = z_t + \phi\frac{(1 - \phi^l)}{1 - \phi}(z_t - z_{t-1}) - \theta\frac{(1 - \phi^l)}{1 - \phi}a_t \qquad (5.4.22)$$

Evidently for large l, the forecast tends to $b_0^{(t)}$.

Weights applied to previous observations. Eliminating a_t from (5.4.22), we obtain the alternative form for the forecast in terms of previous z's

$$\hat{z}_t(l) = \left\{1 - \frac{\theta - \phi}{1 - \phi}(1 - \phi^l)\right\}z_t + \left\{\frac{\theta - \phi}{1 - \phi}(1 - \phi^l)\right\}\bar{z}_{t-1}(\theta) \quad (5.4.23)$$

where $\bar{z}_{t-1}(\theta)$ is an exponentially weighted moving average with parameter θ, that is, $\bar{z}_{t-1}(\theta) = (1 - \theta)\sum_{j=1}^{\infty}\theta^{j-1}z_{t-j}$. Thus the π weights for the process consist of a "spike" at time t and an EWMA starting at time $t - 1$. If we refer to $(1 - \alpha)x + \alpha y$ as a linear interpolation between x and y at argument α, then the forecast (5.4.23) is a linear interpolation between z_t and $\bar{z}_{t-1}(\theta)$. The argument for lead time one is $\theta - \phi$, but as the lead time is increased, the argument approaches $(\theta - \phi)/(1 - \phi)$. For example, when $\theta = 0.9$ and $\phi = 0.5$, the lead-1 forecast is

$$\hat{z}_t(1) = 0.6z_t + 0.4\bar{z}_{t-1}(\theta)$$

and for long lead times, the forecast approaches

$$\hat{z}_t(\infty) = 0.2z_t + 0.8\bar{z}_{t-1}(\theta)$$

5.5 SUMMARY

The results of this chapter may be summarized as follows: Let $\bar{z}_t$ be the deviation of an observed time series from any known deterministic function of time $f(t)$. In particular, for a stationary series, $f(t)$ could be equal to μ, the mean of the series, or it could be equal to zero, so that $\bar{z}_t$ was the observed series. Then consider the general ARIMA model

$$\phi(B)\nabla^d\bar{z}_t = \theta(B)a_t$$

or

$$\varphi(B)\tilde{z}_t = \theta(B)a_t$$

Minimum mean square error forecast. Given knowledge of the series up to some origin t, the minimum mean square error forecast $\hat{z}_t(l)$ $(l > 0)$ of $\tilde{z}_{t+l}$, is the conditional expectation

$$\hat{z}_t(l) = [\tilde{z}_{t+l}] = E[\tilde{z}_{t+l}|\tilde{z}_t, \tilde{z}_{t-1}, \ldots]$$

Lead-1 forecast errors. A necessary consequence is that the lead-1 forecast errors are the generating a's in the model, and are uncorrelated.

Calculation of the forecasts. It is usually simplest in practice to compute the forecasts directly from the difference equation to give

$$\hat{z}_t(l) = \varphi_1[\tilde{z}_{t+l-1}] + \cdots + \varphi_{p+d}[\tilde{z}_{t+l-p-d}] + [a_{t+l}] - \theta_1[a_{t+l-1}]$$
$$- \cdots - \theta_q[a_{t+l-q}] \tag{5.5.1}$$

The conditional expectations in (5.5.1) are evaluated by inserting actual $\tilde{z}$'s when these are known, forecasted $\tilde{z}$'s for future values, actual a's when these are known, and zeroes for future a's. The forecasting process may be initiated by approximating unknown a's by zeroes and, in practice, the appropriate form for the model and suitable estimates for the parameters are obtained by methods set out in Chapters 6, 7, and 8.

Probability limits for forecasts. These may be obtained

(a) by first calculating the ψ weights from

$$\psi_0 = 1$$
$$\psi_1 = \varphi_1 - \theta_1$$
$$\psi_2 = \varphi_1\psi_1 + \varphi_2 - \theta_2 \tag{5.5.2}$$
$$\vdots \qquad \vdots \qquad \vdots \qquad \vdots$$
$$\psi_j = \varphi_1\psi_{j-1} + \cdots + \varphi_{p+d}\psi_{j-p-d} - \theta_j$$

where, $\theta_j = 0, j > q$

(b) for each desired level of probability ε, and for each lead time l, substituting in

$$\tilde{z}_{t+l}(\pm) = \hat{z}_t(l) \pm u_{\varepsilon/2}\left(1 + \sum_{j=1}^{l-1} \psi_j^2\right)^{1/2}\sigma_a \tag{5.5.3}$$

where in practice σ_a is replaced by an estimate s_a, of the standard deviation of the white noise process a_t, and $u_{\varepsilon/2}$ is the deviate exceeded by a proportion $\varepsilon/2$ of the unit Normal distribution.

Updating the forecasts. When a new deviation $\tilde{z}_{t+1}$ comes to hand, the forecasts may be updated to origin $t + 1$, by calculating the new forecast error $a_{t+1} = \tilde{z}_{t+1} - \hat{\tilde{z}}_t(1)$ and using the difference equation (5.5.1) with $t + 1$ replacing t. However, an *alternative* method is to use the forecasts $\hat{\tilde{z}}_t(1), \hat{\tilde{z}}_t(2), \ldots, \hat{\tilde{z}}_t(L)$ at origin t, to obtain the first $L - 1$ forecasts $\hat{\tilde{z}}_{t+1}(1)$, $\hat{\tilde{z}}_{t+1}(2), \ldots, \hat{\tilde{z}}_{t+1}(L - 1)$ at origin $t + 1$, from

$$\hat{\tilde{z}}_{t+1}(l) = \hat{\tilde{z}}_t(l + 1) + \psi_l a_{t+1} \tag{5.5.4}$$

and then generate the last forecast $\hat{\tilde{z}}_{t+1}(L)$ using the difference equation (5.5.1).

Other ways of expressing the forecasts. The above is all that is needed for *practical* utilization of the forecasts. However, the following alternative forms provide theoretical insight into the nature of the forecasts generated by different models:

(1) *Forecasts in integrated form:* For $l > q - p - d$, the forecasts lie on the unique curve

$$\hat{\tilde{z}}_t(l) = b_0^{(t)} f_0(l) + b_1^{(t)} f_1(l) + \cdots + b_{p+d-1}^{(t)} f_{p+d-1}(l) \tag{5.5.5}$$

determined by the "pivotal" values $\hat{\tilde{z}}_t(q), \hat{\tilde{z}}_t(q - 1), \ldots, \hat{\tilde{z}}_t(q - p - d + 1)$, where $\hat{\tilde{z}}_t(-j) = \tilde{z}_{t-j}(j = 0, 1, 2, \ldots)$. If $q > p + d$, the first $q - p - d$ forecasts do not lie on this curve. In general, the stationary autoregressive operator contributes damped exponential and damped sine wave terms to (5.5.5) and the nonstationary operator ∇^d polynomial terms.

The adaptive coefficients $b_j^{(t)}$ in (5.5.5) may be updated from origin t to $t + 1$ by amounts depending on the last lead-1 forecast error a_{t+1}, according to the general formula

$$\mathbf{b}^{(t+1)} = \mathbf{L}'\mathbf{b}^{(t)} + \mathbf{g}a_{t+1} \tag{5.5.6}$$

given in Appendix A5.3.

(2) *Forecasts as a weighted sum of past observations:* It is instructive from a theoretical point of view to express the forecasts as a weighted sum of past observations. Thus, if the model is written in inverted form

$$a_t = \pi(B)\tilde{z}_t = (1 - \pi_1 B - \pi_2 B^2 - \cdots)\tilde{z}_t$$

the lead-1 forecast is

$$\hat{\tilde{z}}_t(1) = \pi_1 \tilde{z}_t + \pi_2 \tilde{z}_{t-1} + \cdots \tag{5.5.7}$$

and the forecasts for longer lead times may be obtained from

$$\hat{\tilde{z}}_t(l) = \pi_1[\tilde{z}_{t+l-1}] + \pi_2[\tilde{z}_{t+l-2}] + \cdots \tag{5.5.8}$$

where the conditional expectations in (5.5.8) are evaluated by replacing $\tilde{z}$'s by actual values, when known, and forecasted values when unknown.

Alternatively, the forecast for any lead time may be written as a linear function of the available observations. Thus

$$\hat{z}_t(l) = \sum_{j=1}^{\infty} \pi_j^{(l)} z_{t+j-1}$$

where the $\pi_j^{(l)}$ are functions of the π_j's.

APPENDIX A5.1 CORRELATIONS BETWEEN FORECAST ERRORS

A5.1.1 Autocorrelation function of forecast errors at different origins

While it is true that, for an optimal forecast, the forecast errors for lead time 1 will be uncorrelated, this will not generally be true of forecasts at longer lead times. Consider forecasts for lead times l, made at origins t and $t - j$ respectively, where j is a positive integer. Then, if $j = l, l + 1, l + 2, \ldots$, the forecast errors will contain no common component, but for $j = 1, 2, \ldots,$ $l - 1$, certain of the a's will be included in both forecast errors. Specifically,

$$e_t(l) = z_{t+l} - \hat{z}_t(l) = a_{t+l} + \psi_1 a_{t+l-1} + \cdots + \psi_{l-1} a_{t+1}$$

$$e_{t-j}(l) = z_{t+l-j} - \hat{z}_{t-j}(l) = a_{t-j+l} + \psi_1 a_{t-j+l-1} + \cdots + \psi_{l-1} a_{t-j+1}$$

and for $j < l$, the lag j autocovariance of the forecast errors for lead time l is

$$E[e_t(l)e_{t-j}(l)] = \sigma_a^2 \sum_{i=j}^{l-1} \psi_i \psi_{i-j} \qquad (A5.1.1)$$

where $\psi_0 = 1$. The corresponding autocorrelations are

$$\rho[e_t(l), e_{t-j}(l)] = \begin{cases} \dfrac{\displaystyle\sum_{i=j}^{l-1} \psi_i \psi_{i-j}}{\displaystyle\sum_{i=0}^{l-1} \psi_i^2} & 0 \leqslant j < l \\[4mm] 0 & j \geqslant l \end{cases} \qquad (A5.1.2)$$

We show in Chapter 7 that Series C of Figure 4.1 is well fitted by the $(1, 1, 0)$ model $(1 - 0.8B)\nabla z_t = a_t$. To illustrate (A5.1.2), we calculate the auto-correlation function of the forecast errors at lead time 6 for this model. It is shown in Section 5.2.2 that the ψ weights $\psi_1, \psi_2, \ldots, \psi_5$ for this model, are 1.80, 2.44, 2.95, 3.36, 3.69. Thus, for example, the lag 1 autocovariance is,

$$E[e_t(6)e_{t-1}(6)] = \sigma_a^2\{(1.80 \times 1.00) + (2.44 \times 1.80) + \cdots + (3.69 \times 3.36)\}$$

$$= 35.70\sigma_a^2$$

On dividing by $E[e_t^2(6)] = 43.86\sigma_a^2$, we obtain $\rho[e_t(6)e_{t-1}(6)] = 0.81$. The first six autocorrelations are shown in Table A5.1 and plotted in Figure A5.1(a). As expected, the autocorrelations beyond the fifth are zero.

TABLE A5.1   Autocorrelations of forecast errors at lead 6 for Series C

j	0	1	2	3	4	5	6
$\rho[e_t(6), e_{t-j}(6)]$	1.00	0.81	0.61	0.41	0.23	0.08	0.00

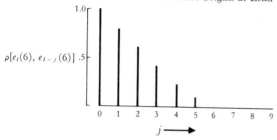

(a) Autocorrelations of Forecast Errors for SERIES C from Different Origins at Lead Time $l = 6$

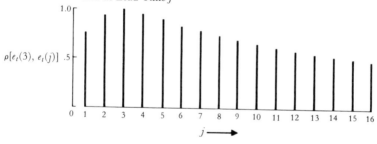

(b) Correlation Between Forecast Errors for SERIES C from Same Origin at Lead Time 3 and at Lead Time j

FIG. A5.1   Correlations between various forecast errors for Series C

A5.1.2 Correlation between forecast errors at the same origin with different lead times

Suppose we make a series of forecasts for different lead times from the *same* fixed origin t. Then the errors for these forecasts will be correlated. We have for $j = 1, 2, 3, \ldots,$

$$e_t(l) = z_{t+l} - \hat{z}_t(l) = a_{t+l} + \psi_1 a_{t+l-1} + \cdots + \psi_{l-1} a_{t+1}$$

$$e_t(l + j) = z_{t+l+j} - \hat{z}_t(l + j) = a_{t+l+j} + \psi_1 a_{t+l+j-1} + \cdots + \psi_j a_{t+l}$$

$$+ \psi_{j+1} a_{t+l-1} + \cdots + \psi_{l+j-1} a_{t+1}$$

so that the covariance between the t-origin forecasts at lead times l and $l+j$ is $\sigma_a^2 \sum_{i=0}^{l-1} \psi_i \psi_{j+i}$, where $\psi_0 = 1$.

Thus the correlation coefficient between the t-origin forecast errors at lead times l and $l+j$ is

$$\rho[e_t(l), e_t(l+j)] = \frac{\sum_{i=0}^{l-1} \psi_i \psi_{j+i}}{\left\{\sum_{h=0}^{l-1} \psi_h^2 \sum_{g=0}^{l+j-1} \psi_g^2\right\}^{1/2}} \tag{A5.1.3}$$

To illustrate (A5.1.3), we compute, for forecasts made from the same origin, the correlation between the forecast error at lead time 3 and the forecast errors at lead times $j = 1, 2, 3, 4, \ldots, 16$ for Series C. For example, using (A5.1.3) and the ψ weights given in Section 5.2.2,

$$E[e_t(3)e_t(5)] = \sigma_a^2\{(1.00 \times 2.44) + (1.80 \times 2.95) + (2.44 \times 3.36)\}$$

$$= 15.94\sigma_a^2$$

The correlations for lead times $j = 1, 2, \ldots, 16$ are shown in Table A5.2 and plotted in Figure A5.1(b).

TABLE A5.2   Correlation between forecast errors at lag 3 and at lag j made from a fixed origin (Series C)

j	1	2	3	4	5	6	7	8
$\rho[e_t(3), e_t(j)]$	0.76	0.94	1.00	0.96	0.91	0.85	0.80	0.75
j	9	10	11	12	13	14	15	16
$\rho[e_t(3), e_t(j)]$	0.71	0.67	0.63	0.60	0.57	0.54	0.52	0.50

As is to be expected, forecasts made from the same origin at different lead times are highly correlated.

APPENDIX A5.2 FORECAST WEIGHTS FOR ANY LEAD TIME

In this appendix we consider an alternative procedure for calculating the forecast weights $\pi_j^{(l)}$ applied to previous z's for any lead time l. To derive this result, we make use of the identity (3.1.7), namely

$$(1 + \psi_1 B + \psi_2 B^2 + \cdots)(1 - \pi_1 B - \pi_2 B^2 - \cdots) = 1$$

from which the π weights may be obtained in terms of the ψ weights and vice versa.

On equating coefficients, we find, for $j \geqslant 1$,

$$\psi_j = \sum_{i=1}^{j} \psi_{j-i} \pi_i \qquad (\psi_0 = 1) \tag{A5.2.1}$$

Thus, for example,

$$\psi_1 = \pi_1 \qquad\qquad\qquad \pi_1 = \psi_1$$

$$\psi_2 = \psi_1 \pi_1 + \pi_2 \qquad\qquad \pi_2 = \psi_2 - \psi_1 \pi_1$$

$$\psi_3 = \psi_2 \pi_1 + \psi_1 \pi_2 + \pi_3 \qquad \pi_3 = \psi_3 - \psi_2 \pi_1 - \psi_1 \pi_2$$

Now from (5.3.6)

$$\hat{z}_t(l) = \pi_1 \hat{z}_t(l-1) + \pi_2 \hat{z}_t(l-2) + \cdots + \pi_{l-1} \hat{z}_t(1) + \pi_l z_t + \pi_{l+1} z_{t-1} + \cdots \tag{A5.2.2}$$

Since each of the forecasts in (A5.2.2) is itself a function of the observations $z_t, z_{t-1}, z_{t-2}, \ldots$, we can write

$$\hat{z}_t(l) = \pi_1^{(l)} z_t + \pi_2^{(l)} z_{t-1} + \pi_3^{(l)} z_{t-2} + \cdots$$

where the lead l forecast weights may be calculated from the lead one forecast weights $\pi_j^{(1)} = \pi_j$. We now show that the weights $\pi_j^{(l)}$ can be obtained using the identity

$$\pi_j^{(l)} = \sum_{i=1}^{l} \psi_{l-i} \pi_{i+j-1} \tag{A5.2.3}$$

For example, the weights for the forecast at lead time 3 are

$$\pi_1^{(3)} = \pi_3 + \psi_1 \pi_2 + \psi_2 \pi_1$$

$$\pi_2^{(3)} = \pi_4 + \psi_1 \pi_3 + \psi_2 \pi_2$$

$$\pi_3^{(3)} = \pi_5 + \psi_1 \pi_4 + \psi_2 \pi_3$$

and so on. To derive (A5.2.3), we write

$$\hat{z}_t(l) = \qquad\qquad\qquad\qquad \psi_l a_t + \psi_{l+1} a_{t-1} + \cdots$$

$$\hat{z}_{t+l-1}(1) = \psi_1 a_{t+l-1} + \cdots + \psi_l a_t + \psi_{l+1} a_{t-1} + \cdots$$

On subtraction

$$\hat{z}_t(l) = \hat{z}_{t+l-1}(1) - \psi_1 a_{t+l-1} - \psi_2 a_{t+l-2} - \cdots - \psi_{l-1} a_{t+1}$$

Hence

$$\hat{z}_t(l) = \pi_1 z_{t+l-1} + \pi_2 z_{t+l-2} + \cdots + \pi_{l-1} z_{t+1} + \pi_l z_t + \pi_{l+1} z_{t-1} + \cdots$$
$$+ \psi_1\{-z_{t+l-1} + \pi_1 z_{t+l-2} + \cdots + \pi_{l-2} z_{t+1} + \pi_{l-1} z_t + \pi_l z_{t-1} + \cdots\}$$
$$+ \psi_2\{-z_{t+l-2} + \cdots + \pi_{l-3} z_{t+1} + \pi_{l-2} z_t + \pi_{l-1} z_{t-1} + \cdots\}$$
$$+ \cdots$$
$$+ \psi_{l-1}\{-z_{t+1} + \pi_1 z_t + \pi_2 z_{t-1} + \cdots\}$$

Using the relations (A5.2.1), each one of the coefficients of $z_{t+l-1}, \ldots, z_{t+1}$ is seen to vanish, as they should, and on collecting terms, we obtain the required result (A5.2.3). Alternatively, we may use the formula in the recursive form

$$\pi_j^{(l)} = \pi_{j+1}^{(l-1)} + \psi_{l-1} \pi_j \qquad (A5.2.4)$$

Using the model $\nabla^2 z_t = (1 - 0.9B + 0.5B^2) a_t$ for illustration, we calculate the weights for lead time 2. Equation (A5.2.4) gives

$$\pi_j^{(2)} = \pi_{j+1} + \psi_1 \pi_j$$

and using the weights in Table 5.4 with $\psi_1 = 1.1$ we have, for example,

$$\pi_1^{(2)} = \pi_2 + \psi_1 \pi_1 = 0.490 + (1.1)(1.1) = 1.700$$

$$\pi_2^{(2)} = \pi_3 + \psi_1 \pi_2 = -0.109 + (1.1)(0.49) = 0.430$$

and so on. The first 12 weights have been given in Table 5.4.

APPENDIX A5.3 FORECASTING IN TERMS OF THE GENERAL INTEGRATED FORM

A5.3.1 A general method of obtaining the integrated form

We emphasize once more that, for practical computation of the forecasts, the difference equation procedure is by far the simplest. The following general treatment of the integrated form is given only to elaborate further on the forecasts obtained. In this treatment, rather than solve explicitly for the forecast function as we did in the examples given in Section 5.4, it will be appropriate to write down the general form of the eventual forecast function involving $p + d$ adaptive coefficients. We then show how the eventual forecast function needs to be modified to deal with the first $q - p - d$ forecasts if $q > p + d$. Finally, we show how to update the adaptive coefficients from origin t to origin $t + 1$.

If it is understood that $\hat{z}_t(-j) = z_{t-j}$ for $j = 0, 1, 2, \ldots$, then using the conditional expectation argument of Section 5.1.2, the forecasts satisfy the difference equation

$$\hat{z}_t(1) - \varphi_1\hat{z}_t(0) - \cdots \qquad - \varphi_{p+d}\hat{z}_t(1 - p - d) = -\theta_1 a_t - \cdots - \theta_q a_{t-q+1}$$

$$\hat{z}_t(2) - \varphi_1\hat{z}_t(1) - \cdots \qquad - \varphi_{p+d}\hat{z}_t(2 - p - d) = -\theta_2 a_t - \cdots - \theta_q a_{t-q+2}$$

$$\vdots \qquad \vdots \qquad \cdots \qquad \vdots \qquad \qquad \vdots \qquad \cdots \qquad (A5.3.1)$$

$$\hat{z}_t(q) - \varphi_1\hat{z}_t(q - 1) - \cdots - \varphi_{p+d}\hat{z}_t(q - p - d) = -\theta_q a_t$$

$$\hat{z}_t(l) - \varphi_1\hat{z}_t(l - 1) \cdots \qquad - \varphi_{p+d}\hat{z}_t(l - p - d) = 0 \qquad l > q$$

The eventual forecast function is the solution of the last equation, and may be written

$$\hat{z}_t(l) = b_0^{(t)}f_0(l) + b_1^{(t)}f_1(l) + \cdots + b_{p+d-1}^{(t)}f_{p+d-1}(l) = \sum_{i=0}^{p+d-1} b_i^{(t)}f_i(l)$$

$$l > q - p - d \qquad (A5.3.2)$$

When q is less than, or equal to $p + d$, the eventual forecast function will provide forecasts $\hat{z}_t(1), \hat{z}_t(2), \hat{z}_t(3), \ldots$ for all lead times $l \geq 1$.

As an example of such a model with $q \leq p + d$, suppose

$$(1 - B)(1 - \sqrt{3}B + B^2)^2 z_t = (1 - 0.5B)a_t$$

so that $p + d = 5$ and $q = 1$. Then

$$(1 - B)(1 - \sqrt{3}B + B^2)^2 \hat{z}_t(l) = 0 \qquad l = 2, 3, 4, \ldots,$$

where B now operates on l and not on t. Solution of this difference equation yields the forecast function

$$\hat{z}_t(l) = b_0^{(t)} + b_1^{(t)}\cos\frac{2\pi l}{12} + b_2^{(t)}l\cos\frac{2\pi l}{12} + b_3^{(t)}\sin\frac{2\pi l}{12} + b_4^{(t)}l\sin\frac{2\pi l}{12}$$

$$l = 1, 2, \ldots,$$

If q is greater than $p + d$, then for lead times $l \leq q - p - d$, the forecast function will have additional terms containing a's. Thus

$$\hat{z}_t(l) = \sum_{i=0}^{p+d-1} b_i^{(t)}f_i(l) + \sum_{i=0}^{j} d_{li}a_{t-i} \qquad l \leq q - p - d \qquad (A5.3.3)$$

where $j = q - p - d - l$ and the d's may be obtained explicitly by substituting (A5.3.3) in (A5.3.1). For example, consider the stochastic model

$$\nabla^2 z_t = (1 - 0.8B + 0.5B^2 - 0.4B^3 + 0.1B^4)a_t$$

in which $p + d = 2$, $q = 4$, $q - p - d = 2$ and $\varphi_1 = 2$, $\varphi_2 = -1$, $\theta_1 = 0.8$, $\theta_2 = -0.5$, $\theta_3 = 0.4$ and $\theta_4 = -0.1$. Using the recurrence relation (5.2.3), we obtain $\psi_1 = 1.2$, $\psi_2 = 1.9$, $\psi_3 = 2.2$, $\psi_4 = 2.6$. Now, from (A5.3.3),

$$
\left.
\begin{aligned}
\hat{z}_t(1) &= b_0^{(t)} + b_1^{(t)} + d_{10}a_t + d_{11}a_{t-1} \\
\hat{z}_t(2) &= b_0^{(t)} + 2b_1^{(t)} + d_{20}a_t \\
\hat{z}_t(l) &= b_0^{(t)} + b_1^{(t)}l \qquad l > 2
\end{aligned}
\right\}
\qquad (A5.3.4)
$$

Using (A5.3.1) gives

$$
\hat{z}_t(4) - 2\hat{z}_t(3) + \hat{z}_t(2) = 0.1a_t
$$

so that from (A5.3.4)

$$
d_{20}a_t = 0.1a_t
$$

and hence $d_{20} = 0.1$. Similarly, from (A5.3.1)

$$
\hat{z}_t(3) - 2\hat{z}_t(2) + \hat{z}_t(1) = -0.4a_t + 0.1a_{t-1}
$$

and hence, using (A5.3.4)

$$
-0.2a_t + d_{10}a_t + d_{11}a_{t-1} = -0.4a_t + 0.1a_{t-1}
$$

yielding

$$
d_{10} = -0.2 \qquad d_{11} = 0.1
$$

Hence the forecast function is

$$
\begin{aligned}
\hat{z}_t(1) &= b_0^{(t)} + b_1^{(t)} - 0.2a_t + 0.1a_{t-1} \\
\hat{z}_t(2) &= b_0^{(t)} + 2b_1^{(t)} + 0.1a_t \\
\hat{z}_t(l) &= b_0^{(t)} + b_1^{(t)}l \qquad l > 2
\end{aligned}
$$

A5.3.2 *Updating the general integrated form*

Updating formulae for the coefficients may be obtained using the identity (5.2.5) with $t + 1$ replaced by t

$$
\hat{z}_t(l) = \hat{z}_{t-1}(l + 1) + \psi_l a_t
$$

Then, for $l > q - p - d$,

$$
\sum_{i=0}^{p+d-1} b_i^{(t)} f_i(l) = \sum_{i=0}^{p+d-1} b_i^{(t-1)} f_i(l + 1) + \psi_l a_t \qquad (A5.3.5)
$$

By solving $p + d$ such equations for different values of l, we obtain the required updating formula for the individual coefficients, in the form

$$
b_i^{(t)} = \sum_{j=0}^{p+d-1} g_{ij} b_j^{(t-1)} + v_i a_t
$$

Note that the updating of each of the coefficients of the forecast function depends only on the lead one forecast error $a_t = z_t - \hat{z}_{t-1}(1)$. As an example, consider again the stochastic model

$$(1 - B)(1 - \sqrt{3}B + B^2)^2 z_t = (1 - 0.5B)a_t$$

We showed in Section A5.3.1 that the forecast function for this example is

$$\hat{z}_t(l) = b_0^{(t)} + (b_1^{(t)} + lb_2^{(t)}) \cos \frac{2\pi l}{12} + (b_3^{(t)} + lb_4^{(t)}) \sin \frac{2\pi l}{12}$$

It is convenient to use the updating identity (A5.3.5) for the particular values $l = 6, 12, 24, 3, 9$. Using the recurrence relations (5.2.3), the ψ weights corresponding to these values are $\psi_6 = 25 + 9\sqrt{3}$, $\psi_{12} = 317 + 184\sqrt{3}$, $\psi_{24} = 611 + 346\sqrt{3}$, $\psi_3 = 4 + 7\sqrt{3}$, $\psi_9 = 134 + 90\sqrt{3}$. Hence, using (A5.3.5), we obtain the matrix equation

$$
\begin{bmatrix}
1 & -1 & -6 & 0 & 0 \\
1 & 1 & 12 & 0 & 0 \\
1 & 1 & 24 & 0 & 0 \\
1 & 0 & 0 & 1 & 3 \\
1 & 0 & 0 & -1 & -9
\end{bmatrix}
\begin{bmatrix}
b_0^{(t)} \\
b_1^{(t)} \\
b_2^{(t)} \\
b_3^{(t)} \\
b_4^{(t)}
\end{bmatrix}
=
$$

$$
\begin{bmatrix}
1 & -\dfrac{\sqrt{3}}{2} & -\dfrac{7\sqrt{3}}{2} & -\dfrac{1}{2} & -\dfrac{7}{2} \\
1 & \dfrac{\sqrt{3}}{2} & \dfrac{13\sqrt{3}}{2} & \dfrac{1}{2} & \dfrac{13}{2} \\
1 & \dfrac{\sqrt{3}}{2} & \dfrac{25\sqrt{3}}{2} & \dfrac{1}{2} & \dfrac{25}{2} \\
1 & -\dfrac{1}{2} & -2 & \dfrac{\sqrt{3}}{2} & 2\sqrt{3} \\
1 & \dfrac{1}{2} & 5 & -\dfrac{\sqrt{3}}{2} & -5\sqrt{3}
\end{bmatrix}
\begin{bmatrix}
b_0^{(t-1)} \\
b_1^{(t-1)} \\
b_2^{(t-1)} \\
b_3^{(t-1)} \\
b_4^{(t-1)}
\end{bmatrix}
+
\begin{bmatrix}
25 + 9\sqrt{3} \\
317 + 184\sqrt{3} \\
611 + 346\sqrt{3} \\
4 + 7\sqrt{3} \\
134 + 90\sqrt{3}
\end{bmatrix}
a_t
$$

The first three equations are easily solved for $b_0^{(t)}$, $b_1^{(t)}$, $b_2^{(t)}$ and the fourth and fifth for $b_3^{(t)}$, $b_4^{(t)}$. The final solution is

$$
\begin{bmatrix} b_0^{(t)} \\ b_1^{(t)} \\ b_2^{(t)} \\ b_3^{(t)} \\ b_4^{(t)} \end{bmatrix} = \begin{bmatrix} 1 & 0 & 0 & 0 & 0 \\ 0 & \dfrac{\sqrt{3}}{2} & \dfrac{\sqrt{3}}{2} & \dfrac{1}{2} & \dfrac{1}{2} \\ 0 & 0 & \dfrac{\sqrt{3}}{2} & 0 & \dfrac{1}{2} \\ 0 & -\dfrac{1}{2} & -\dfrac{1}{2} & \dfrac{\sqrt{3}}{2} & \dfrac{\sqrt{3}}{2} \\ 0 & 0 & -\dfrac{1}{2} & 0 & \dfrac{\sqrt{3}}{2} \end{bmatrix} \begin{bmatrix} b_0^{(t-1)} \\ b_1^{(t-1)} \\ b_2^{(t-1)} \\ b_3^{(t-1)} \\ b_4^{(t-1)} \end{bmatrix} + \begin{bmatrix} \dfrac{195 + 112\sqrt{3}}{2} \\ \dfrac{-149 - 68\sqrt{3}}{2} \\ \dfrac{49 + 27\sqrt{3}}{2} \\ \dfrac{-244 - 113\sqrt{3}}{2} \\ \dfrac{19 + 5\sqrt{3}}{2} \end{bmatrix} a_t
$$

yielding the required updating formulae for each of the five coefficients, $b_0^{(t)}, b_1^{(t)}, \ldots, b_4^{(t)}$.

A5.3.3 Comparison with the discounted least squares method of R. G. Brown.

Although to work with the integrated form is an unnecessarily complicated way of computing forecasts, it allows us to compare the present mean square error forecast with another type of forecast which has received considerable attention.

Let us write

$$
\mathbf{F}_l = \begin{bmatrix} f_0(l) & f_1(l) & \cdots & f_{p+d-1}(l) \\ f_0(l+1) & f_1(l+1) & \cdots & f_{p+d-1}(l+1) \\ & & \cdots & \\ f_0(l+p+d-1) & f_1(l+p+d-1) & \cdots & f_{p+d-1}(l+p+d-1) \end{bmatrix}
$$

$$
\mathbf{b}^{(t)} = \begin{bmatrix} b_0^{(t)} \\ b_1^{(t)} \\ b_{p+d-1}^{(t)} \end{bmatrix} \qquad \boldsymbol{\psi}_l = \begin{bmatrix} \psi_l \\ \psi_{l+1} \\ \psi_{l+p+d} \end{bmatrix}
$$

Then, using (A5.3.5) for $l, l+1, \ldots, l+p+d$, we obtain for $l > q - p - d$,

$$
\mathbf{F}_l \mathbf{b}^{(t)} = \mathbf{F}_{l+1} \mathbf{b}^{(t-1)} + \boldsymbol{\psi}_l a_t
$$

yielding

$$
\mathbf{b}^{(t)} = (\mathbf{F}_l^{-1} \mathbf{F}_{l+1}) \mathbf{b}^{(t-1)} + (\mathbf{F}_l^{-1} \boldsymbol{\psi}_l) a_t
$$

or

$$\mathbf{b}^{(t)} = \mathbf{L}'\mathbf{b}^{(t-1)} + \mathbf{g}a_t \qquad (A5.3.6)$$

(A5.3.6) is of the same algebraic *form* as the updating function given by the "discounted least squares" procedure of Brown [2], [50]. For comparison, if we denote the forecast error given by that method by e_t, then Brown's updating formula may be written

$$\boldsymbol{\beta}^{(t)} = \mathbf{L}'\boldsymbol{\beta}^{(t-1)} + \mathbf{h}e_t \qquad (A5.3.7)$$

where $\boldsymbol{\beta}^{(t)}$ is his vector of adaptive coefficients. The same matrix $\mathbf{L}$ appears in (A5.3.6) and (A5.3.7). This is inevitable, for this first factor merely allows for changes in the coefficients arising from translation to the new origin and would have to occur in any such formula. For example, consider the straight line forecast function

$$\hat{z}_{t-1}(l) = b_0^{(t-1)} + b_1^{(t-1)}l$$

where $b_0^{(t-1)}$ is the ordinate at time $t-1$, the origin of the forecast. This can equally well be written

$$\hat{z}_{t-1}(l) = (b_0^{(t-1)} + b_1^{(t-1)}) + b_1^{(t-1)}(l-1)$$

where now $(b_0^{(t-1)} + b_1^{(t-1)})$ is the ordinate at time t. Obviously, if we update the forecast to origin t, the coefficient b_0 must be suitably adjusted even if the forecast function were to remain unchanged.

In general, the matrix $\mathbf{L}$ does not change the forecast function, it merely re-locates it. The actual updating is done by the vector of coefficients $\mathbf{g}$ and $\mathbf{h}$. We shall see that the coefficients $\mathbf{g}$, which yield the minimum mean square error forecasts, and the coefficients $\mathbf{h}$ given by Brown are in general completely different.

Brown's method of forecasting.

(1) A forecast function is selected from the general class of linear combinations and products of polynomials, exponentials, sines and cosines.
(2) The selected forecast function is fitted to past values by a "discounted least squares" procedure. In this procedure, the coefficients are estimated and updated so that the sum of squares of weighted discrepancies

$$S_\omega = \sum_{j=0}^{\infty} \omega_j \{z_{t-j} - \hat{z}_t(-j)\}^2 \qquad (A5.3.8)$$

between past values of the series and the value given by the forecast function at the corresponding past time are minimized. The weight function ω_j is chosen arbitrarily to fall off geometrically, so that $\omega_j = (1 - \alpha)^j$, where the constant α, usually called the *smoothing constant*, is (again arbitrarily) set equal to a value in the range 0.1–0.3.

Difference between the minimum mean square error forecasts and those of Brown.

To illustrate these comments, consider the forecasting of IBM stock prices, discussed by Brown [2, page 141]. In this study he used a quadratic model which would be, in the present notation,

$$\hat{\hat{z}}_t(l) = \beta_0^{(t)} + \beta_1^{(t)}l + \tfrac{1}{2}\beta_2^{(t)}l^2$$

With this model he employed his method of discounted least squares to forecast stock prices three days ahead. The results obtained from this method are shown for a section of the IBM series in Fig. A5.2, where they are compared with the minimum mean square error forecasts.

The discounted least squares method can be criticized on the following grounds:

(1) The nature of the forecast function ought to be decided by the auto-regressive operator $\varphi(B)$ in the stochastic model, and not arbitrarily. In particular, it cannot be safely chosen by visual inspection of the time series itself. For example, consider the IBM stock prices plotted in Figure A5.2. It will be seen that a quadratic function might well be used to *fit* short pieces of this series to values already available. If such fitting were relevant to forecasting, we might conclude, as did Brown, that a polynomial forecast function of degree 2 was indicated.

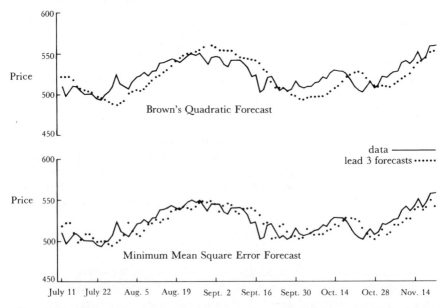

Fig. A5.2 IBM stock price series with comparison of lead 3 forecasts obtained from best IMA $(0, 1, 1)$ process and Brown's quadratic forecast for a period beginning July 11, 1960

The most general linear process for which a quadratic function would produce *minimum mean square error* forecasts at every lead time $l = 1, 2, \ldots$ is defined by the $(0, 3, 3)$ model

$$\nabla^3 z_t = (1 - \theta_1 B - \theta_2 B^2 - \theta_3 B^3) a_t$$

which, arguing as in Section 4.3.3., can be written

$$\nabla^3 z_t = \nabla^3 a_t + \lambda_0 \nabla^2 a_{t-1} + \lambda_1 \nabla a_{t-1} + \lambda_2 a_{t-1}$$

However, we shall show in Chapter 7 that, if this model is correctly fitted, the least square estimates of the parameters are $\lambda_1 = \lambda_2 = 0$, and $\lambda_0 \simeq 1.0$. Thus, $\nabla z_t = (1 - \theta B) a_t$, with $\theta = 1 - \lambda_0$ close to zero, is the appropriate stochastic model, and the appropriate forecasting polynomial is $\hat{z}_t(l) = \beta_0^{(t)}$, which is of degree zero in l and not of degree two.

(2) The choice of the weight function ω_j in (A5.3.8) must correspondingly be decided by the stochastic model, and not arbitrarily. The use of the discounted least squares fitting procedure would produce minimum mean square error forecasts in the very restricted case where:

    (a) the process was of order $(0, 1, 1)$, so that $\nabla z_t = (1 - \theta B) a_t$;

    (b) a polynomial of degree zero was fitted;

    (c) the smoothing constant α was set equal to our $\lambda = 1 - \theta$.

In the present example, even if the correct polynomial model of degree zero had been chosen, the value $\alpha = \lambda = 0.1$, actually used by Brown, would have been quite inappropriate. The correct value λ for this series is close to unity.

(3) The exponentially discounted weighted least squares procedure forces all the $p + d$ coefficients in the updating vector $\mathbf{h}$ to be functions of the single smoothing parameter α. In fact they should be functions of the $p + q$ independent parameters $(\boldsymbol{\phi}, \boldsymbol{\theta})$.

Thus, the differences between the two methods are not trivial, and it is interesting to compare their performances on the IBM data. The minimum mean square error forecast is $\hat{z}_t(l) = b_0^{(t)}$, with updating $b_0^{(t)} = b_0^{(t-1)} + \lambda a_t$, where $\lambda \simeq 1.0$. If λ is taken to be exactly equal to unity, this is equivalent to using

$$\hat{z}_t(l) = z_t$$

which implies that the best forecast of the stock price for all future time is the present price*. The suggestion that stock prices behave in this way is of course not new and goes back to Bachelier [51]. Since $z_t = Sa_t$ when $\lambda = 1$, this implies that z_t is a random walk.

To compare the minimum mean square error (MSE) forecast with Brown's quadratic forecasts, a direct comparison was made using the IBM stock

* This result is approximately true supposing that no relevant information except past values of the series itself is available and that fairly short forecasting periods are being considered. For longer periods, growth and inflationary factors would become important.

price series from July 11, 1960 to February 10, 1961 (150 observations). For this stretch of the series the minimum MSE forecast is obtained using the model $\nabla z_t = a_t - \theta a_{t-1}$, with $\theta = 0.1$, or $\lambda = 1 - \theta = 0.9$. Figure A5.2 shows the minimum mse forecasts for lead time 3 and the corresponding values of Brown's quadratic forecasts. It is seen that the minimum mse forecasts, which are virtually equivalent to using today's price to predict that 3 days ahead, are considerably better than those obtained using Brown's much more complicated procedure.

The mean square errors for the forecast at various lead times, computed by direct comparison of the value of the series and their lead l forecasts, are shown in Table A5.3 for the two types of forecasts. It is seen that Brown's quadratic forecasts have mean square errors which are much larger than those obtained by the minimum mean square error method.

TABLE A5.3   Comparison of mean square error of forecasts obtained at various lead times using best IMA (0, 1, 1) process and Brown's quadratic forecasts.

Lead Time l	1	2	3	4	5	6	7	8	9	10
MSE (Brown)	102	158	218	256	363	452	554	669	799	944
MSE ($\lambda = 0.9$)	42	91	136	180	282	266	317	371	427	483

Part II

Stochastic Model Building

We have seen that an ARIMA process of order (p, d, q) provides a class of models capable of representing time series which, although not necessarily stationary, are homogeneous and in statistical equilibrium.

The ARIMA process is defined by the equation

$$\phi(B)(1 - B)^d z_t = \theta_0 + \theta(B)a_t$$

where $\phi(B)$ and $\theta(B)$ are operators in B of degree p and q, respectively, whose zeroes lie outside the unit circle.

We have noted that the model is very general, subsuming autoregressive models, moving average models, mixed autoregressive-moving average models, and the integrated forms of all three.

Iterative approach to model building

The relating of a model of this kind to data is usually best achieved by a three stage iterative procedure based on identification, estimation, and diagnostic checking.

By *identification* we mean the use of the data, and of any information on how the series was generated, to suggest a subclass of parsimonious models worthy to be entertained.

By *estimation* we mean efficient use of the data to make inferences about parameters conditional on the adequacy of the entertained model.

By *diagnostic checking* we mean checking the fitted model in its relation to the data with intent to reveal model inadequacies and so to achieve model improvement.

In Chapter 6, which follows, we discuss identification, in Chapter 7 estimation, and in Chapter 8 diagnostic checking. In Chapter 9 all these techniques are illustrated by applying them to modelling seasonal time series.

6

Model Identification

In this chapter we discuss methods for identifying non-seasonal time series models. Identification methods are rough procedures applied to a set of data to indicate the kind of representational model which is worthy of further investigation. The specific aim here is to obtain some idea of the values of p, d, and q needed in the general linear ARIMA model and to obtain initial guesses for the parameters. The tentative model so obtained provides a starting point for the application of the more formal and efficient estimation methods described in Chapter 7.

6.1 OBJECTIVES OF IDENTIFICATION

6.1.1 Stages in the identification procedure

It should first be said that identification and estimation necessarily overlap. Thus, we may estimate the parameters in a model, which is more elaborate than that which we expect to find, so as to decide *at what point* simplification is possible. Here we employ the estimation procedure to carry out part of the identification. It should also be explained that identification is necessarily inexact. It is inexact because the question of what types of models occur in practice and in what circumstances, is a property of the behavior of the physical world and cannot, therefore, be decided by purely mathematical argument. Furthermore, because at the identification stage no precise formulation of the problem is available, statistically "inefficient" methods must necessarily be used. It is a stage at which graphical methods are particularly useful and judgment must be exercised. However, it should be borne in mind that preliminary identification commits us to nothing except to tentatively entertaining a class of models which will later be efficiently fitted and checked.

Our task then, is to identify an appropriate subclass of models from the general ARIMA family

$$\phi(B)\nabla^d z_t = \theta_0 + \theta(B)a_t \tag{6.1.1}$$

which may be used to represent a given time series. Our approach will be

(a) to difference z_t as many times as is needed to produce stationarity, hopefully reducing the process under study to the mixed autoregressive-moving average process

$$\phi(B)w_t = \theta_0 + \theta(B)a_t$$

where

$$w_t = (1 - B)^d z_t = \nabla^d z_t$$

(b) to identify the resulting ARMA process. Our principal tools for putting (a) and (b) into effect will be the autocorrelation function and the partial autocorrelation function. They are used not only to help guess the form of the model, but also to obtain approximate estimates of the parameters. Such approximations are often useful at the estimation stage to provide starting values for iterative procedures employed at that stage.

6.2 IDENTIFICATION TECHNIQUES

6.2.1 *Use of the autocorrelation and partial autocorrelation functions in identification*

Identifying the degree of differencing. We have seen in Section 3.4.2 that, for a stationary mixed autoregressive-moving average process of order $(p, 0, q)$, $\phi(B)\tilde{z}_t = \theta(B)a_t$, the autocorrelation function satisfies the difference equation

$$\phi(B)\rho_k = 0 \qquad k > q$$

Also, if $\phi(B) = \prod_{i=1}^{p} (1 - G_i B)$, then the solution of this difference equation for the kth autocorrelation is, assuming distinct roots, of the form

$$\rho_k = A_1 G_1^k + A_2 G_2^k + \cdots + A_p G_p^k \qquad k > q - p \qquad (6.2.1)$$

The stationarity requirement that the zeroes of $\phi(B)$ lie outside the unit circle, implies that the roots $G_1, G_2, \ldots, G_p$ lie inside the unit circle.

Inspection of (6.2.1) shows that, in the case of a stationary model in which none of the roots lie close to the boundary of the unit circle, the autocorrelation function will quickly "die out" for moderate and large k. Suppose now that a single real root, say G_1, approaches unity, so that

$$G_1 = 1 - \delta$$

where δ is some small positive quantity. Then, since for k large

$$\rho_k \simeq A_1(1 - k\delta)$$

the autocorrelation function will not die out quickly and will fall off slowly and very nearly linearly. A similar argument may be applied if more than one of the roots approaches unity.

Therefore, a tendency for the autocorrelation function not to die out quickly is taken as an indication that a root close to unity may exist. The estimated autocorrelation function tends to follow the behavior of the theoretical autocorrelation function. Therefore, failure of the estimated autocorrelation function to die out rapidly might logically suggest that we should treat the underlying stochastic process as nonstationary in z_t, but possibly as stationary in ∇z_t, or in some higher difference.

It turns out that it is failure of the estimated autocorrelation function to die out rapidly that suggests nonstationarity. It need not happen that the estimated correlations are extremely high even at low lags. This is illustrated in Appendix A.6.1, where the expected behavior of the estimated autocorrelation function is considered for the nonstationary (0, 1, 1) process $\nabla z_t = (1 - \theta B)a_t$. The ratio $E[c_k]/E[c_0]$ of expected values falls off only slowly, but initially depends on the value of θ and on the number of observations in the series, and need not be close to unity if θ is close to one. We shall illustrate this point again in Section 6.3.4 for Series A of Figure 4.1.

For the reasons given, it is assumed that the degree of differencing d, necessary to achieve stationarity, has been reached when the autocorrelation function of $w_t = \nabla^d z_t$ dies out fairly quickly. In practice d is normally either 0, 1, or 2 and it is usually sufficient to inspect the first 20 or so estimated autocorrelations of the original series and of its first and second differences.

Identification of resultant stationary ARMA process. Having tentatively decided what d should be, we next study the general appearance of the estimated autocorrelation and partial autocorrelation functions of the appropriately differenced series, to provide clues about the choice of the orders p and q for the autoregressive and moving average operators. In doing so, we recall the characteristic behavior of the theoretical autocorrelation function and of the theoretical partial autocorrelation function for moving average, autoregressive, and mixed processes, discussed in Chapter 3.

Briefly, whereas the autocorrelation function of an autoregressive process of order p tails off, its partial autocorrelation function has a cutoff after lag p. Conversely, the autocorrelation function of a moving average process of order q has a cutoff after lag q, while its partial autocorrelation tails off. If both the autocorrelations and partial autocorrelations tail off, a mixed process is suggested. Furthermore, the autocorrelation function for a mixed process, containing a pth order autoregressive component and a qth order moving average component, is a mixture of exponentials and damped sine waves after the first $q - p$ lags. Conversely, the partial autocorrelation function for a mixed process is dominated by a mixture of exponentials and damped sine waves after the first $p - q$ lags (see Table 3.2).

In general, autoregressive (moving average) behavior, as measured by the autocorrelation function, tends to mimic moving average (autoregressive) behavior as measured by the partial autocorrelation function. For example, the autocorrelation function of a first-order autoregressive process decays exponentially, while the partial autocorrelation function cuts off after the first lag. Correspondingly, for a first-order moving average process, the autocorrelation function cuts off after the first lag. The partial autocorrelation function, while not precisely exponential, is dominated by exponential terms and has the general appearance of an exponential.

Of particular importance are the autoregressive and moving average processes of first- and second-order, and the simple mixed $(1, d, 1)$ process. The properties of the theoretical autocorrelation and partial autocorrelation functions for these processes are summarized in Table 6.1, which requires careful study and which provides a convenient reference table. The reader

TABLE 6.1 Behavior of the autocorrelation functions for the *dth difference* of an ARIMA process of order (p, d, q). (Table A and Charts B, C and D are included at the end of this volume to facilitate the calculation of approximate estimates of the parameters for first-order moving average, second-order autoregressive, second-order moving average, and for the mixed (ARMA) $(1, 1)$ process)

Order	$(1, d, 0)$	$(0, d, 1)$
Behavior of ρ_k	decays exponentially	only ρ_1 nonzero
Behavior of ϕ_{kk}	only ϕ_{11} nonzero	exponential dominates decay
Preliminary estimates from	$\phi_1 = \rho_1$	$\rho_1 = \dfrac{-\theta_1}{1 + \theta_1^2}$
Admissible region	$-1 < \phi_1 < 1$	$-1 < \theta_1 < 1$

Order	$(2, d, 0)$	$(0, d, 2)$
Behavior of ρ_k	mixture of exponentials or damped sine wave	only ρ_1 and ρ_2 nonzero
Behavior of ϕ_{kk}	only ϕ_{11} and ϕ_{22} nonzero	dominated by mixture of exponentials or damped sine wave
Preliminary estimates from	$\phi_1 = \dfrac{\rho_1(1 - \rho_2)}{1 - \rho_1^2}$ $\phi_2 = \dfrac{\rho_2 - \rho_1^2}{1 - \rho_1^2}$	$\rho_1 = \dfrac{-\theta_1(1 - \theta_2)}{1 + \theta_1^2 + \theta_2^2}$ $\rho_2 = \dfrac{-\theta_2}{1 + \theta_1^2 + \theta_2^2}$
Admissible region	$\begin{cases} -1 < \phi_2 < 1 \\ \phi_2 + \phi_1 < 1 \\ \phi_2 - \phi_1 < 1 \end{cases}$	$\begin{cases} -1 < \theta_2 < 1 \\ \theta_2 + \theta_1 < 1 \\ \theta_2 - \theta_1 < 1 \end{cases}$

TABLE 6.1 continued

Order	$(1, d, 1)$
Behavior of ρ_k	decays exponentially from first lag
Behavior of ϕ_{kk}	dominated by exponential decay from first lag
Preliminary estimates from	$\rho_1 = \dfrac{(1 - \theta_1 \phi_1)(\phi_1 - \theta_1)}{1 + \theta_1^2 - 2\phi_1 \theta_1} \qquad \rho_2 = \rho_1 \phi_1$
Admissible region	$-1 < \phi_1 < 1 \qquad -1 < \theta_1 < 1$

should also refer again to Figures 3.2, 3.9, and 3.11, which show typical behavior of the autocorrelation function and the partial autocorrelation function for the second-order autoregressive process, the second-order moving average process, and the simple mixed $(1, d, 1)$ process.

Relation between estimated and theoretical autocorrelations. Estimated autocorrelations can have rather large variances and can be highly auto-correlated with each other. For this reason, as was emphasized by M. G. Kendall [29], *detailed* adherence to the theoretical autocorrelation function cannot be expected in the estimated function. In particular, moderately large estimated autocorrelations can occur after the theoretical autocorrelation function has damped out, and apparent ripples and trends can occur in the estimated function which have no basis in the theoretical function. In employing the estimated autocorrelation function as a tool for identification, it is usually possible to be fairly sure about broad characteristics, but more subtle indications may or may not represent real effects, and two or more related models may need to be entertained and investigated further at the estimation and diagnostic checking stages of model building.

6.2.2 Standard errors for estimated autocorrelations and partial autocorrelations

Since we do not know the theoretical correlations and since the estimated values which we compute will differ somewhat from their theoretical counterparts, it is important to have some indication of how far an estimated value may differ from the corresponding theoretical value. In particular, we need some means for judging whether the autocorrelations and partial auto-correlations are effectively zero after some specific lag q. For *larger lags*, we can compute standard errors of estimated autocorrelations from the simplified form of Bartlett's formula (2.1.13), with sample estimates replacing theoretical autocorrelations. Thus

$$\hat{\sigma}\,[r_k] \simeq \frac{1}{n^{1/2}}\{1 + 2(r_1^2 + r_2^2 + \cdots + r_q^2)\}^{1/2} \qquad k > q \qquad (6.2.2)$$

For the partial correlations we use the result quoted in (3.2.35) that, on the hypothesis that the process is autoregressive of order p, the standard error for partial correlations of order $p + 1$ and higher is

$$\hat{\sigma}\,[\hat{\phi}_{kk}] \simeq \frac{1}{(n)^{1/2}} \qquad k > p \tag{6.2.3}$$

It was shown by Anderson [52] that, for moderate n, the distribution of an estimated autocorrelation coefficient, whose theoretical value is zero, is approximately Normal. Thus, on the hypothesis that the theoretical autocorrelation ρ_k is zero, the estimate r_k divided by its standard error will be approximately distributed as a unit Normal deviate. A similar result is true for the partial autocorrelations. These facts may be used to provide an informal guide as to whether theoretical autocorrelations and partial autocorrelations beyond a particular lag are essentially zero. It is usually sufficient to remember that, for the Normal distribution, deviations exceeding one standard error in either direction have a probability of about one third, while deviations in either direction exceeding two standard deviations have a probability of about one twentieth.

6.2.3 *Identification of some actual time series*

In this section the techniques referred to above are applied to six time series, designated A, B, ..., F. Series A–D are plotted in Figure 4.1, Series E in Figure 6.1, and F in Figure 2.1. The data for all these series are listed in the Collection of Time Series in Part V at the end of this volume. Series A, B, C, and D have been described in Chapter 4 and Series F in Chapter 2. Series E

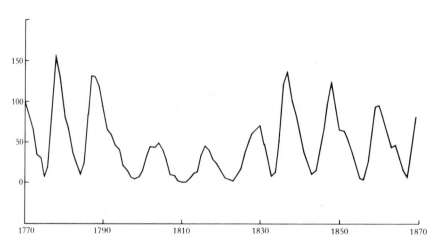

FIG. 6.1   Series E: Wölfer annual sunspot numbers (1770-1869)

is the series of annual Wölfer sunspot numbers and measures the average number of sunspots on the sun during each year. As we have remarked in Chapter 4, we might expect Series A, C, and D to possess nonstationary characteristics, since they represent the "uncontrolled" behavior of certain process outputs. Similarly, we would expect the stock price Series B to have no fixed level. On the other hand, we would expect Series F to be stationary, because it represents the variation in the yields of batches processed under uniformly controlled conditions. Similarly, we would expect the sunspot series to remain in equilibrium over long periods.

The estimated autocorrelations of z, ∇z, and $\nabla^2 z$ for Series A–F are shown in Table 6.2. Table 6.3 shows the corresponding estimated partial autocorrelations. Plotting of the correlation functions greatly assists their understanding, and for illustration, autocorrelations and partial autocorrelations are plotted in Figures 6.2 and 6.3 for Series A, and in Figures 6.4 and 6.5 for Series C.

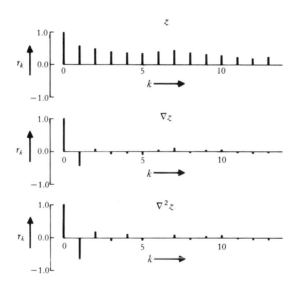

FIG. 6.2   Estimated autocorrelations of various differences of Series A

For Series A the autocorrelations for ∇z are small after the first lag. This suggests that this time series might be described by an IMA $(0, 1, 1)$ process. However, from the autocorrelation function of z, it is seen that *after lag 1* the correlations do decrease fairly regularly. Therefore, an alternative possibility is that the series is a mixed ARMA of order $(1, 0, 1)$. The partial autocorrelation function for z tends to support this possibility. We shall see later that the two possibilities result in virtually the same model.

TABLE 6.2   Estimated autocorrelations of Series A–F

SERIES A   Chemical Process Concentration Readings: Every Two Hours

197 Observations

		Autocorrelations									
		1	2	3	4	5	6	7	8	9	10
z	Lags 1–10	0.57	0.50	0.40	0.36	0.33	0.35	0.39	0.32	0.30	0.26
	11–20	0.19	0.16	0.20	0.24	0.14	0.18	0.20	0.20	0.14	0.18
∇z	Lags 1–10	−0.41	0.02	−0.07	−0.01	−0.07	−0.02	0.15	−0.07	0.04	0.02
	11–20	−0.05	−0.06	−0.01	0.16	−0.17	0.03	0.01	0.08	−0.12	0.15
$\nabla^2 z$	Lags 1–10	−0.65	0.18	−0.04	0.04	−0.04	−0.04	0.13	−0.11	0.04	0.02
	11–20	−0.02	−0.02	−0.04	0.18	−0.19	0.08	−0.03	0.09	−0.17	0.20

SERIES B   IBM Common Stock Closing Prices: Daily, 17th May 1961–2nd November 1962

369 Observations

		Autocorrelations									
		1	2	3	4	5	6	7	8	9	10
z	Lags 1–10	0.99	0.99	0.98	0.97	0.96	0.96	0.95	0.94	0.93	0.92
	11–20	0.91	0.91	0.90	0.89	0.88	0.87	0.86	0.85	0.84	0.83
∇z	Lags 1–10	0.09	0.00	−0.05	−0.04	−0.02	0.12	0.07	0.04	−0.07	0.02
	11–20	0.08	0.05	−0.05	0.07	−0.07	0.12	0.12	0.05	0.05	0.07
$\nabla^2 z$	Lags 1–10	−0.45	−0.02	−0.04	0.00	−0.07	0.11	−0.01	0.04	−0.10	0.02
	11–20	0.04	0.04	−0.12	0.13	−0.17	0.10	0.05	−0.04	−0.01	0.09

SERIES C   Chemical Process Temperature Readings: Every Minute

226 Observations

		Autocorrelations									
		1	2	3	4	5	6	7	8	9	10
z	Lags 1–10	0.98	0.94	0.90	0.85	0.80	0.75	0.69	0.64	0.58	0.52
	11–20	0.47	0.41	0.36	0.30	0.25	0.20	0.15	0.10	0.05	0.00
∇z	Lags 1–10	0.80	0.65	0.53	0.44	0.38	0.32	0.26	0.19	0.14	0.14
	11–20	0.10	0.09	0.07	0.07	0.07	0.07	0.09	0.05	0.04	0.04
$\nabla^2 z$	Lags 1–10	−0.08	−0.07	−0.12	−0.06	0.01	−0.02	0.05	−0.05	−0.12	0.12
	11–20	−0.12	0.07	−0.08	0.03	−0.01	−0.06	0.17	−0.10	−0.01	−0.02

TABLE 6.2 continued

SERIES D Chemical Process Viscosity Readings: Every Hour

310 Observations

		1	2	3	4	5	6	7	8	9	10
z	Lags 1–10	0.86	0.74	0.62	0.53	0.46	0.41	0.35	0.31	0.27	0.24
	11–20	0.22	0.20	0.18	0.15	0.14	0.13	0.16	0.19	0.21	0.23
∇z	Lags 1–10	−0.05	−0.06	−0.07	−0.08	−0.06	0.00	−0.02	−0.02	−0.03	−0.06
	11–20	−0.01	0.04	0.02	−0.07	−0.03	−0.09	−0.02	0.05	−0.01	0.06
$\nabla^2 z$	Lags 1–10	−0.50	0.00	0.00	−0.01	−0.02	0.04	−0.01	0.00	0.01	−0.04
	11–20	0.00	0.04	0.03	−0.06	0.04	−0.06	0.00	0.06	−0.06	0.06

Autocorrelations

SERIES E Wölfer Sunspot Numbers: Yearly

100 Observations

		1	2	3	4	5	6	7	8	9	10
z	Lags 1–10	0.81	0.43	0.07	−0.17	−0.27	−0.21	−0.04	0.16	0.33	0.41
	11–20	0.39	0.29	0.14	0.02	−0.06	−0.10	−0.14	−0.18	−0.17	−0.10
∇z	Lags 1–10	0.55	−0.02	−0.30	−0.40	−0.40	−0.33	−0.20	0.04	0.26	0.31
	11–20	0.29	0.16	−0.03	−0.12	−0.10	−0.09	−0.09	−0.12	−0.14	−0.05
$\nabla^2 z$	Lags 1–10	0.15	−0.31	−0.20	−0.11	−0.09	−0.02	−0.11	−0.04	0.19	0.05
	11–20	0.13	0.09	−0.10	−0.11	0.04	0.01	0.00	−0.03	−0.10	−0.04

Autocorrelations

SERIES F Yields from a Batch Chemical Process

70 Observations

		1	2	3	4	5	6	7	8	9	10
z	Lags 1–10	−0.39	0.30	−0.17	0.07	−0.10	−0.05	0.04	−0.04	−0.01	0.01
	11–20	0.11	−0.07	0.15	0.04	−0.01	0.17	−0.11	0.02	−0.05	0.02
∇z	Lags 1–10	−0.74	0.43	−0.27	0.16	−0.10	0.01	0.05	−0.05	0.04	−0.05
	11–20	0.11	−0.16	0.12	−0.01	−0.08	0.16	−0.14	0.08	−0.07	0.03
$\nabla^2 z$	Lags 1–10	−0.83	0.54	−0.33	0.21	−0.12	0.03	0.04	−0.06	0.06	−0.07
	11–20	0.12	−0.16	0.11	0.00	−0.10	0.16	−0.15	0.10	−0.07	0.03

Autocorrelations

TABLE 6.3 Estimated partial autocorrelations* of Series A–F

SERIES A Chemical Process Concentration Readings: Every Two Hours

197 Observations

		Partial autocorrelations									
		1	2	3	4	5	6	7	8	9	10
z	Lags 1–10	0.57	0.25	0.08	0.09	0.07	0.15	0.19	−0.03	0.01	−0.01
	11–20	−0.09	−0.04	0.04	0.08	−0.15	0.06	0.13	0.09	−0.06	0.07
∇z	Lags 1–10	−0.41	−0.19	−0.17	−0.14	−0.20	−0.23	0.01	−0.04	−0.01	0.06
	11–20	0.02	−0.07	−0.10	0.13	−0.09	−0.15	−0.11	0.04	−0.08	0.12
$\nabla^2 z$	Lags 1–10	−0.66	−0.43	−0.33	−0.23	−0.20	−0.36	−0.23	−0.21	−0.23	−0.16
	11–20	−0.07	−0.04	−0.25	0.00	0.04	−0.02	−0.16	−0.03	−0.22	−0.03

SERIES B IBM Common Stock Closing Prices: Daily, 17th May 1961–2nd November 1962

369 Observations

		Partial autocorrelations									
		1	2	3	4	5	6	7	8	9	10
z	Lags 1–10	0.996	−0.09	0.01	0.05	0.02	0.02	−0.12	−0.05	−0.02	0.06
	11–20	−0.05	−0.09	−0.03	0.07	−0.08	0.06	−0.14	−0.10	−0.01	−0.08
∇z	Lags 1–10	0.09	−0.01	−0.05	−0.03	−0.02	0.13	0.05	0.02	−0.06	0.05
	11–20	0.09	0.03	−0.08	0.08	−0.06	0.14	0.10	0.00	0.07	0.08
$\nabla^2 z$	Lags 1–10	−0.45	−0.28	−0.24	−0.20	−0.29	−0.17	−0.13	−0.03	−0.14	−0.16
	11–20	−0.09	0.02	−0.13	0.01	−0.19	−0.13	−0.03	−0.10	−0.10	0.06

SERIES C Chemical Process Temperature Readings: Every Minute

226 Observations

		Partial autocorrelations									
		1	2	3	4	5	6	7	8	9	10
z	Lags 1–10	0.99	−0.81	−0.03	−0.02	−0.10	−0.07	−0.01	−0.03	0.04	−0.04
	11–20	−0.15	0.10	−0.14	0.01	−0.10	−0.02	−0.07	−0.11	0.11	−0.13
∇z	Lags 1–10	0.81	−0.01	−0.01	0.06	0.03	−0.03	−0.01	−0.08	0.00	0.10
	11–20	−0.14	0.10	−0.05	0.05	0.02	0.06	0.06	−0.17	0.09	0.00
$\nabla^2 z$	Lags 1–10	−0.08	−0.08	−0.14	−0.10	−0.03	−0.05	0.02	−0.06	−0.16	0.09
	11–20	−0.14	0.01	−0.09	−0.02	−0.05	−0.09	0.13	−0.13	−0.03	−0.05

* Obtained by fitting autoregressive processes of increasing order, using least squares.

TABLE 6.3   continued

SERIES D   Chemical Process Viscosity Readings: Every Hour

310 Observations

				Partial autocorrelations							
		1	2	3	4	5	6	7	8	9	10
z	Lags 1–10	0.86	−0.02	0.00	0.01	0.03	0.03	−0.02	0.01	0.00	0.01
	11–20	0.05	0.01	−0.04	−0.03	0.07	0.04	0.10	0.06	0.00	0.06
∇z	Lags 1–10	−0.05	−0.06	−0.07	−0.09	−0.08	−0.03	−0.05	−0.05	−0.05	−0.09
	11–20	−0.05	0.01	−0.01	−0.10	−0.07	−0.13	−0.09	−0.02	−0.08	0.00
$\nabla^2 z$	Lags 1–10	−0.50	−0.32	−0.24	−0.20	−0.22	−0.16	−0.14	−0.11	−0.07	−0.12
	11–20	−0.15	−0.12	−0.02	−0.06	−0.01	−0.07	−0.12	−0.06	−0.13	−0.07

SERIES E   Wölfer Sunspot Numbers: Yearly

100 Observations

				Partial autocorrelations							
		1	2	3	4	5	6	7	8	9	10
z	Lags 1–10	0.81	−0.71	0.21	−0.15	0.10	0.10	0.18	0.23	0.01	0.00
	11–20	0.14	−0.16	0.12	0.03	−0.08	−0.14	−0.06	−0.12	0.00	0.05
∇z	Lags 1–10	0.57	−0.48	−0.06	−0.27	−0.22	−0.26	−0.29	−0.05	−0.02	−0.16
	11–20	0.13	−0.15	−0.04	0.06	0.12	0.02	0.07	−0.06	−0.09	−0.06
$\nabla^2 z$	Lags 1–10	0.15	−0.35	−0.10	−0.21	−0.16	−0.17	−0.36	−0.26	−0.09	−0.33
	11–20	−0.02	−0.13	−0.20	−0.21	−0.10	−0.13	0.00	0.03	−0.01	−0.08

SERIES F   Yields from a Batch Chemical Process

70 Observations

				Partial autocorrelations							
		1	2	3	4	5	6	7	8	9	10
z	Lags 1–10	−0.40	0.19	0.01	−0.07	−0.07	−0.15	0.05	0.00	−0.10	0.05
	11–20	0.18	−0.05	0.09	0.18	0.01	0.43	0.01	−0.14	0.11	0.18
∇z	Lags 1–10	−0.76	−0.32	−0.19	−0.16	−0.09	−0.24	−0.15	−0.06	−0.18	−0.28
	11–20	−0.02	−0.16	−0.24	−0.06	−0.44	−0.02	0.12	−0.12	−0.17	−0.24
$\nabla^2 z$	Lags 1–10	−0.83	−0.52	−0.38	−0.33	−0.15	−0.24	−0.26	−0.14	−0.09	−0.31
	11–20	−0.12	−0.09	−0.26	0.08	−0.38	−0.39	−0.07	−0.05	−0.03	−0.30

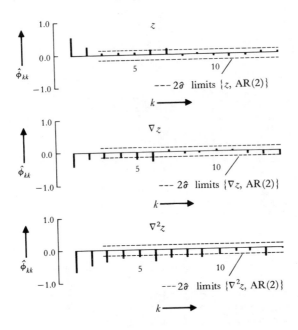

FIG. 6.3   Estimated partial autocorrelations of various differences of Series A

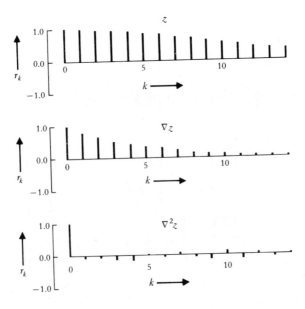

FIG. 6.4   Estimated autocorrelations of various differences of Series C

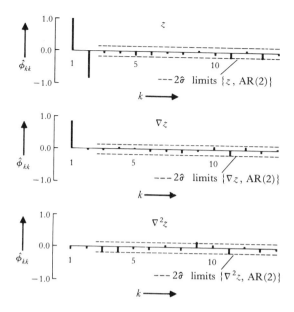

FIG. 6.5   Estimated partial autocorrelations of various differences of Series C

The autocorrelations of Series C, shown in Figure 6.4, suggest that at least one differencing is necessary. The roughly exponential fall-off in the correlations for the first difference suggests a process of order $(1, 1, 0)$, with an autoregressive parameter ϕ of about 0.8. Alternatively, it will be noticed that the autocorrelations of $\nabla^2 z$ are small, suggesting an IMA $(0, 2, 0)$ process. The same conclusions are reached by considering the partial autocorrelation function. That for z points to a generalized autoregressive operator of degree 2. That for ∇z points to a first order autoregressive process in ∇z_t, with ϕ equal to about 0.8, and that for $\nabla^2 z_t$ to uncorrelated noise. Thus the possibilities are

$$(1 - 0.8B)(1 - B)z_t = a_t$$

$$(1 - B)^2 z_t = a_t$$

The second model is very similar to the first, differing only in the choice of 0.8 rather than 1 for one of the autoregressive coefficients.

In assessing the estimated correlation functions, it is very helpful to plot "control" lines about zero at $\pm\hat{\sigma}$ or $\pm 2\hat{\sigma}$. An indication of the hypothesis in mind should accompany the limits. Thus, in Figure 6.5, the annotation "$2\hat{\sigma}$ limits $\{z, \mathrm{AR}\,(2)\}$" on the limit lines indicates that the value $\hat{\sigma}$ employed, is approximately correct on the hypothesis that the process is second-order autoregressive. Tentative identification for each of the Series A–F is given in Table 6.4.

TABLE 6.4 Tentative identification of models for Series A–F

Series		Degree of Differencing	Apparent nature of differenced series	Identification for z_t
A	either	0	mixed first-order AR with first-order MA	$(1, 0, 1)$
	or	1	first-order MA	$(0, 1, 1)$
B		1	first-order MA	$(0, 1, 1)$
C	either	1	first-order AR	$(1, 1, 0)$
	or	2	uncorrelated noise	$(0, 2, 2)$*
D	either	0	first-order AR	$(1, 0, 0)$
	or	1	uncorrelated noise	$(0, 1, 1)$*
E	either	0	second-order AR	$(2, 0, 0)$
	or	0	third-order AR	$(3, 0, 0)$
F		0	second-order AR	$(2, 0, 0)$

* The order of the moving average operator appears to be zero, but the more general form is retained for subsequent consideration.

Three other points concerning this identification procedure need to be mentioned:

(1) Simple differencing of the kind we have used will not produce stationarity in series containing seasonal components. In Chapter 9 we discuss the appropriate modifications for such seasonal time series.

(2) As discussed in Chapter 4, a nonzero value for θ_0 in (6.1.1) implies the existence of a systematic polynomial trend of degree d. For the nonstationary models in Table 6.4, a value of $\theta_0 = 0$ can perfectly well account for the behavior of the series. Occasionally, however, there will be some real physical phenomenon requiring the provision of such a component. In other cases it might be uncertain whether such a provision should be made or not. Some indication of the evidence supplied by the data, for the inclusion of θ_0 in the model, can be obtained at the identification stage by comparing the mean $\bar{w}$ of $w_t = \nabla^d z_t$ with its approximate standard error, using $\sigma^2(\bar{w}) = n^{-1}\sigma_w^2(1 + 2r_1(w) + 2r_2(w) + \cdots)$.

(3) It was noted in Section 3.4.2 that, for any ARMA (p, q) process with $p - q > 0$, the *whole positive half* of the autocorrelation function will be a mixture of damped sine waves and exponentials. This does not, of course, prevent us from tentatively identifying q, because: (a) the partial autocorrelation function will show $p - q$ "anomalous" values before

behaving like that of an MA(q) process; (b) q must be such that the autocorrelation function could take, as starting values following the general pattern, ρ_q back to $\rho_{-(p-q-1)}$.

6.3 INITIAL ESTIMATES FOR THE PARAMETERS

6.3.1 Uniqueness of estimates obtained from the autocovariance function

While it is true that a given ARMA model has a unique covariance structure, the converse is not true. At first sight this would seem to rule out the use of the estimated autocovariances as a means of identification. We show later in Section 6.4 that the estimated autocovariance function may be so used. The reason is that, although there exists a multiplicity of ARMA models possessing the same autocovariance function, there exists only one which expresses the current value of $\nabla^d z_t = w_t$, exclusively in terms of previous history and in stationary invertible form.

6.3.2 Initial estimates for moving average processes

It has been shown (3.3.4) that the first q autocorrelations of a MA (q) process are nonzero and can be written in terms of the parameters of the model as

$$\rho_k = \frac{-\theta_k + \theta_1\theta_{k+1} + \theta_2\theta_{k+2} + \cdots + \theta_{q-k}\theta_q}{(1 + \theta_1^2 + \theta_2^2 + \cdots + \theta_q^2)} \quad k = 1, 2, \ldots, q \qquad (6.3.1)$$

The expression (6.3.1) for $\rho_1, \rho_2, \ldots, \rho_q$, in terms of $\theta_1, \theta_2, \ldots, \theta_q$, supplies q equations in q unknowns. Preliminary estimates of the θ's can be obtained by substituting the estimates r_k for ρ_k in (6.3.1) and solving the resulting nonlinear equations. A preliminary estimate of σ_a^2 may then be obtained from

$$\gamma_0 = \sigma_a^2(1 + \theta_1^2 + \cdots + \theta_q^2)$$

by substituting the preliminary estimates of the θ's and replacing $\gamma_0 = \sigma^2$ by its estimate c_0.

Preliminary estimates for a $(0, d, 1)$ process. Table A, at the end of the volume, relates ρ_1 to θ_1, and by substituting $r_1(w)$ for ρ_1, can be used to provide initial estimates for any $(0, d, 1)$ process $w_t = (1 - \theta_1 B)a_t$, where $w_t = \nabla^d z_t$.

Preliminary estimates for a $(0, d, 2)$ process. Chart C, at the end of the volume, relates ρ_1 and ρ_2 to θ_1 and θ_2, and by substituting $r_1(w)$ and $r_2(w)$ for ρ_1 and ρ_2, can be used to provide initial estimates for any $(0, d, 2)$ process.

In obtaining preliminary estimates in this way, the following points should be borne in mind:

(1) The autocovariances are second moments of the joint distribution of the w's. Thus, the parameter estimates are obtained by equating sample

moments to their theoretical values. It is well known that the method of moments is not necessarily efficient, and it can be demonstrated that it lacks efficiency in these particular cases. However, the rough estimates obtained can be useful in obtaining fully efficient estimates, because they supply an approximate idea of "where in the parameter space to look" for the most efficient estimates.

(2) In general, the equations (6.3.1), obtained by equating moments, will have multiple solutions. For instance, when $q = 1$

$$\rho_1 = \frac{-\theta_1}{1 + \theta_1^2} \tag{6.3.2}$$

and hence both

$$\theta_1 = -\frac{1}{2\rho_1} + \left\{ \frac{1}{(2\rho_1)^2} - 1 \right\}^{1/2}$$

and

$$\theta_1 = -\frac{1}{2\rho_1} - \left\{ \frac{1}{(2\rho_1)^2} - 1 \right\}^{1/2} \tag{6.3.3}$$

are possible solutions. Thus, from Table 6.2, the first lag autocorrelation, of the first difference of Series A, is about -0.4. Substitution in (6.3.3) yields the pair of solutions $\theta_1 \simeq 0.5$ and $\theta_1' \simeq 2.0$. However, the chosen value $\theta_1 \simeq 0.5$ is the only value that lies within the invertibility interval $-1 < \theta_1 < 1$. In fact, it is shown in Section 6.4.1, that it is always true that only one of the multiple solutions can satisfy the invertibility condition.

Examples. Series A, B, and D have all been identified in Table 6.4 as possible IMA processes of order $(0, 1, 1)$. We have seen in Section 4.3.1 that this model may be written in the alternative forms

$$\nabla z_t = (1 - \theta_1 B)a_t$$

$$z_t = \lambda_0 S a_{t-1} + a_t \qquad (\lambda_0 = 1 - \theta_1)$$

$$z_t = \lambda_0 \sum_{j=1}^{\infty} (1 - \lambda_0)^{j-1} z_{t-j} + a_t$$

Using Table A at the end of this volume, the approximate estimates of the parameters shown in Table 6.5 were obtained.

Series C has been tentatively identified in Table 6.4 as an IMA $(0, 2, 2)$ process

$$\nabla^2 z_t = (1 - \theta_1 B - \theta_2 B^2)a_t$$

TABLE 6.5   Initial estimates of parameters for Series
A, B, and D

Series	r_1	$\hat{\theta}_1$	$\hat{\lambda}_0 = 1 - \hat{\theta}_1$
A	-0.41	0.5	0.5
B	0.09	-0.1	1.1
D	-0.05	0.1	0.9

or equivalently,

$$z_t = \lambda_0 S a_{t-1} + \lambda_1 S^2 a_{t-1} + a_t$$

Since the first two autocorrelations of $\nabla^2 z_t$, given in Table 6.2, are approximately zero, then using Chart C at the end of the volume, $\hat{\theta}_1 = 0$, $\hat{\theta}_2 = 0$,
so that $\hat{\lambda}_0 = 1 + \hat{\theta}_2 = 1$ and $\hat{\lambda}_1 = 1 - \hat{\theta}_1 - \hat{\theta}_2 = 1$. On this basis the series
would be represented by

$$\nabla^2 z_t = a_t$$

or

$$z_t = S a_{t-1} + S^2 a_{t-1} + a_t \qquad (6.3.4)$$

This would mean that the second difference $\nabla^2 z_t$, was very nearly a random
series.

6.3.3 Initial estimates for autoregressive processes

For an assumed AR process of order 1 or 2, initial estimates for ϕ_1 and ϕ_2
can be calculated by substituting estimates r_j for the theoretical autocorrelations ρ_j in the formulae of Table 6.1, which are obtained from the
Yule–Walker equations (3.2.6). In particular, for an AR(1), $\hat{\phi}_{11} = r_1$ and for
an AR(2)

$$\hat{\phi}_{21} = \frac{r_1(1 - r_2)}{1 - r_1^2}$$

$$\hat{\phi}_{22} = \frac{r_2 - r_1^2}{1 - r_1^2} \qquad (6.3.5)$$

where ϕ_{pj} denotes the jth autoregressive parameter in a process of order p.
The corresponding formulae given by the Yule–Walker equations, for
higher-order schemes may be obtained by substituting the r_j for the ρ_j in
(3.2.7). Thus

$$\hat{\boldsymbol{\phi}} = \mathbf{R}_p^{-1} \mathbf{r}_p \qquad (6.3.6)$$

where $\mathbf{R}_p$ is an estimate of the $p \times p$ matrix of correlations up to order $p - 1$, and $\mathbf{r}_p'$ is the vector $(r_1, r_2, \ldots, r_p)$. For example, if $p = 3$, (6.3.6) becomes

$$
\begin{bmatrix} \hat{\phi}_{31} \\ \hat{\phi}_{32} \\ \hat{\phi}_{33} \end{bmatrix} = \begin{bmatrix} 1 & r_1 & r_2 \\ r_1 & 1 & r_1 \\ r_2 & r_1 & 1 \end{bmatrix}^{-1} \begin{bmatrix} r_1 \\ r_2 \\ r_3 \end{bmatrix} \tag{6.3.7}
$$

A recursive method of obtaining the estimates for an $AR(p)$ from those of an $AR(p - 1)$ has been given in Appendix A3.2.

It will be shown in Chapter 7 that, by contrast to the situation for MA processes, the autoregressive parameters obtained from (6.3.6) approximate the fully efficient maximum likelihood estimates.

Example. Series E behaves in its undifferenced form like* an autoregressive process of second-order

$$(1 - \phi_1 B - \phi_2 B^2)\tilde{z}_t = a_t$$

Substituting the estimates $r_1 = 0.81$ and $r_2 = 0.43$, obtained from Table 6.2, in (6.3.5) we have $\hat{\phi}_1 = 1.32$ and $\hat{\phi}_2 = -0.63$.

As a second example, consider again Series C identified as either of order $(1, 1, 0)$, or possibly $(0, 2, 2)$. The first possibility would give

$$(1 - \phi_1 B)\nabla z_t = a_t$$

with $\hat{\phi}_1 = 0.8$, since r_1 for ∇z is 0.81.

This example is especially interesting, because it makes clear that the two alternative models which have been identified for this series are closely related. On the supposition that the series is of order $(0, 2, 2)$, we have suggested a model

$$(1 - B)(1 - B)z_t = a_t \tag{6.3.8}$$

The alternative

$$(1 - 0.8B)(1 - B)z_t = a_t \tag{6.3.9}$$

is very similar.

6.3.4 Initial estimates for mixed autoregressive-moving average processes

It will often be found, either initially or after suitable differencing, that $w_t = \nabla^d z_t$ is most economically represented by a mixed ARMA process

$$\phi(B)w_t = \theta(B)a_t$$

---

* The sunspot series has been the subject of much investigation (see, for example, Schuster [53], Yule [24], and Moran [54]). The series is almost certainly not adequately fitted by a second-order autoregressive process. A model related to the actual mechanism at work would, of course, be the most satisfactory. Recent unpublished work has suggested that empirically, a second-order autoregressive model would provide a better fit if a suitable transformation were first applied to z.

It was noted in Section 6.2.1 that, a mixed process is indicated if both the autocorrelation and partial autocorrelation functions tail off. Another fact (3.4.3) of help in identifying the mixed process, is that after lag $q - p$ the theoretical autocorrelations of the mixed process behave like the autocorrelations of the pure autoregressive process $\phi(B)w_t = a_t$. In particular, if the autocorrelation function of the dth difference appeared to be falling off exponentially from an aberrant first value r_1 we would suspect that we had a process of order $(1, d, 1)$, that is

$$(1 - \phi_1 B)w_t = (1 - \theta_1 B)a_t \tag{6.3.10}$$

where $w_t = \nabla^d z_t$.

Approximate values for the parameters of the process (6.3.10) are obtained by substituting the estimates $r_1(w)$ and $r_2(w)$ for ρ_1 and ρ_2 in the expressions (3.4.8) and also given in Table 6.1. Thus

$$r_1 = \frac{(1 - \hat{\theta}_1 \hat{\phi}_1)(\hat{\phi}_1 - \hat{\theta}_1)}{1 + \hat{\theta}_1^2 - 2\hat{\phi}_1 \hat{\theta}_1}$$

$$r_2 = r_1 \hat{\phi}_1$$

Chart D at the end of the volume relates ρ_1 and ρ_2 to ϕ_1 and θ_1, and can be used to provide initial estimates of the parameters for any $(1, d, 1)$ process.

For example, using Fig. 6.2, Series A was identified as of order $(0, 1, 1)$, with θ_1 about 0.5. Looking at the autocorrelation function of z_t, rather than that of $w_t = \nabla z_t$, we see that from r_1 onwards the autocorrelations decay roughly exponentially, although slowly. Thus, an alternative identification of Series A is that it is generated by a stationary process of order $(1, 0, 1)$. The estimated autocorrelations and the corresponding initial estimates of the parameters are then

$$r_1 = 0.57 \qquad r_2 = 0.50 \qquad \hat{\phi}_1 \simeq 0.87 \qquad \hat{\theta}_1 \simeq 0.48$$

This identification yields the approximate model of order $(1, 0, 1)$

$$(1 - 0.9B)\tilde{z}_t = (1 - 0.5B)a_t$$

whereas the previously identified model of order $(0, 1, 1)$, given in Table 6.5, is

$$(1 - B)z_t = (1 - 0.5B)a_t$$

Again we see that the "alternative" models are nearly the same.

A more general method for obtaining initial estimates of the parameters for a mixed autoregressive-moving average process is given in Appendix A6.2.

Compensation between autoregressive and moving average operators. The alternative models identified above are even more alike than they appear. This is because small changes in the autoregressive operator of a mixed

model can be nearly compensated by corresponding changes in the moving
average operator. In particular, if we have a model

$$\{1 - (1 - \delta)B\}\tilde{z}_t = (1 - \theta B)a_t$$

where δ is small and positive, then we can write

$$(1 - B)\tilde{z}_t = \{1 - (1 - \delta)B\}^{-1}(1 - B)(1 - \theta B)a_t$$

$$= [1 - \delta B\{1 + (1 - \delta)B + (1 - \delta)^2 B^2 + \cdots\}](1 - \theta B)a_t$$

$$= \{1 - (\theta + \delta)B\}a_t + \text{terms in } a_{t-2}, a_{t-3}, \ldots \text{ of order } \delta$$

6.3.5 Choice between stationary and nonstationary models in doubtful cases

The apparent ambiguity displayed in Table 6.4 in identifying models A, C,
and D is, of course, more apparent than real. It arises whenever the roots of
$\phi(B) = 0$ approach unity. When this happens it becomes less and less important whether a root near unity is included in $\phi(B)$, or an additional difference
is included corresponding to a unit root. A more precise evaluation is possible
using the estimation procedures discussed in Chapter 7, but the following
should be borne in mind:
(1) From time series which are necessarily of finite length, it is not possible
    to ever *prove* that a zero of the autoregressive operator is exactly equal
    to unity.
(2) There is, of course, no sudden transition from stationary behavior to
    nonstationary behavior. This can be understood by considering the
    behavior of the simple mixed model

$$(1 - \phi_1 B)(z_t - \mu) = (1 - \theta_1 B)a_t$$

Series generated by such a model behave in a more and more nonstationary
manner as ϕ_1 increases toward unity. For example, a series with $\phi_1 = 0.99$
can wander away from its mean μ and not return for very long periods.
It is as if the attraction which the mean exerts in the series becomes less and
less as ϕ_1 approaches unity, and finally, when ϕ_1 is equal to unity, the behavior
of the series is completely independent of μ.

In doubtful cases there may be advantage in employing the nonstationary
model rather than the stationary alternative (for example, in treating a ϕ_1,
whose estimate is close to unity, as being *equal* to unity). This is particularly
true in forecasting and control problems. Where ϕ_1 is close to unity, we do
not really know whether the mean of the series has meaning or not. Therefore, it may be advantageous to employ the nonstationary model which does
not include a mean μ. If we use such a model, forecasts of future behavior
will not in any way depend on an estimated mean, calculated from a previous
period, which may have no relevance to the future level of the series.

For comparison with the more efficient methods of estimation to be described in Chapter 7, it is interesting to see how much additional information about the model can be extracted at the identification stage. We have already shown how to obtain initial estimates of the parameters $(\hat{\phi}, \hat{\theta})$ in the ARMA model, identified for an appropriate difference $w_t = \nabla^d z_t$ of the series. To complete the picture, we now show how to obtain preliminary estimates of the residual variance σ_a^2 and an approximate standard error for the mean of the appropriately differenced series.

6.3.6 Initial estimate of residual variance

An initial estimate of the residual variance may be obtained by substituting an estimate c_0 in the expressions for the variance γ_0 given in Chapter 3. Thus, substituting in (3.2.8), an initial estimate of σ_a^2 for an AR process may be obtained from

$$\hat{\sigma}_a^2 = c_0(1 - \hat{\phi}_1 r_1 - \hat{\phi}_2 r_2 - \cdots - \hat{\phi}_p r_p) \tag{6.3.11}$$

Similarly, from (3.3.3), an initial estimate for a MA process may be obtained from

$$\hat{\sigma}_a^2 = \frac{c_0}{1 + \hat{\theta}_1^2 + \cdots + \hat{\theta}_q^2} \tag{6.3.12}$$

The form of the estimate for a mixed process is more complicated and is most easily obtained as described in Appendix A6.2. For the important ARMA $(1, 1)$ process, it takes the form (see 3.4.7)

$$\hat{\sigma}_a^2 = \frac{1 - \hat{\phi}_1^2}{1 + \hat{\theta}_1^2 - 2\hat{\phi}_1\hat{\theta}_1} c_0 \tag{6.3.13}$$

For example, consider the $(1, 0, 1)$ model identified for Series A. Using (6.3.13) with $\hat{\phi}_1 = 0.87, \hat{\theta}_1 = 0.48$ and $c_0 = 0.1586$, we obtain the estimate

$$\hat{\sigma}_a^2 = \frac{1 - (0.87)^2}{1 + (0.48)^2 - 2(0.87)(0.48)} 0.1586 = 0.098$$

6.3.7 An approximate standard error for $\bar{w}$

The general ARIMA, for which the mean μ_w of $w_t = \nabla^d z_t$ is not necessarily zero, may be written in any one of the three forms

$$\phi(B)(w_t - \mu_w) = \theta(B)a_t \tag{6.3.14}$$

$$\phi(B)w_t = \theta_0 + \theta(B)a_t \tag{6.3.15}$$

$$\phi(B)w_t = \theta(B)(a_t + \xi) \tag{6.3.16}$$

where

$$\mu_w = \frac{\theta_0}{1 - \phi_1 - \phi_2 - \cdots - \phi_p} = \frac{(1 - \theta_1 - \theta_2 - \cdots - \theta_q)\xi}{1 - \phi_1 - \phi_2 - \cdots - \phi_p}$$

Hence, if $1 - \phi_1 - \phi_2 - \cdots - \phi_p \neq 0$ and $1 - \theta_1 - \theta_2 - \cdots - \theta_q \neq 0$, $\mu_w = 0$ implies that $\theta_0 = 0$ and that $\xi = 0$. Now, in general, when $d = 0$, μ_z will not be zero. However, consider the eventual forecast function associated with the general model (6.3.14) when $d > 0$. With $\mu_w = 0$, this forecast function already contains an adaptive polynomial component of degree $d - 1$. The effect of allowing μ_w to be nonzero, is to introduce a *fixed* polynomial term into this function, of degree d. For example, if $d = 2$ and μ_w is nonzero, then the forecast function $\hat{z}_t(l)$ includes a quadratic component in l, in which the coefficient of the quadratic term is fixed and does not adapt to the series. Because models of this kind are often inapplicable when $d > 0$, the hypothesis that $\mu_w = 0$ will frequently not be contradicted by the data. Indeed, as we have indicated, we usually assume that $\mu_w = 0$ unless evidence to the contrary presents itself.

At this, the identification stage of model building, an indication of whether or not a nonzero value for μ_w is needed, may be obtained by comparison of $\bar{w} = \Sigma w/n$ with its approximate standard error. With $n = N - d$ differences available,

$$\sigma^2(\bar{w}) = n^{-1}\gamma_0 \sum_{-\infty}^{+\infty} \rho_j = n^{-1} \sum_{-\infty}^{+\infty} \gamma_j$$

that is

$$\sigma^2(\bar{w}) = n^{-1}\gamma(1) \qquad (6.3.17)$$

where $\gamma(B)$ is the autocovariance generating function defined in (3.1.10) and $\gamma(1)$ is its value when $B = B^{-1} = 1$ is substituted.

For illustration, consider the process of order $(1, d, 0)$

$$(1 - \phi B)(w_t - \mu_w) = a_t$$

with $w_t = \nabla^d z_t$. From (3.1.11), we obtain

$$\gamma(B) = \frac{\sigma_a^2}{(1 - \phi B)(1 - \phi F)}$$

so that

$$\sigma^2(\bar{w}) = n^{-1}(1 - \phi)^{-2}\sigma_a^2$$

But $\sigma_a^2 = \sigma_w^2(1 - \phi^2)$, so that

$$\sigma^2(\bar{w}) = \frac{\sigma_w^2}{n}\frac{1 - \phi^2}{(1 - \phi)^2} = \frac{\sigma_w^2}{n}\frac{1 + \phi}{1 - \phi}$$

and

$$\sigma(\overline{w}) = \sigma_w \left\{ \frac{1 + \phi}{n(1 - \phi)} \right\}^{1/2}$$

Now σ_w^2 and ϕ are estimated by c_0 and r_1, respectively, as defined in (2.1.9) and (2.1.10). Thus for a $(1, d, 0)$ process, the required standard error is given by

$$\hat{\sigma}(\overline{w}) = \left\{ \frac{c_0(1 + r_1)}{n(1 - r_1)} \right\}^{1/2}$$

Proceeding in this way, the expressions for $\sigma(\overline{w})$ given in Table 6.6 may be obtained.

TABLE 6.6 Approximate standard error for $\overline{w}$, where $w_t = \nabla^d z_t$ and z_t is an ARIMA process of order (p, d, q)

$(1, d, 0)$	$(0, d, 1)$
$\left\{ \dfrac{c_0(1 + r_1)}{n(1 - r_1)} \right\}^{1/2}$	$\left\{ \dfrac{c_0(1 + 2r_1)}{n} \right\}^{1/2}$
$(2, d, 0)$	$(0, d, 2)$
$\left\{ \dfrac{c_0(1 + r_1)(1 - 2r_1^2 + r_2)}{n(1 - r_1)(1 - r_2)} \right\}^{1/2}$	$\left\{ \dfrac{c_0(1 + 2r_1 + 2r_2)}{n} \right\}^{1/2}$

$$(1, d, 1)$$

$$\left\{ \frac{c_0}{n} \left(1 + \frac{2r_1^2}{r_1 - r_2} \right) \right\}^{1/2}$$

Tentative identification of models A–F: Table 6.7 summarizes the models tentatively identified for Series A–F, with the preliminary parameter estimates inserted. These models are used as initial guesses for the more efficient estimation methods to be described in Chapter 7.

6.4 MODEL MULTIPLICITY

6.4.1 Multiplicity of autoregressive-moving average models

With the Normal assumption, knowledge of the first and second moments of a probability distribution implies complete knowledge of the distribution. In particular, knowledge of the mean of $w_t = \nabla^d z_t$ and of its autocovariance function uniquely determines the probability structure for w_t. We now show, that although this unique probability structure can be represented by a *multiplicity* of linear models, nevertheless, uniqueness *is* achieved in the model

TABLE 6.7   Summary of models identified for Series A–F, with initial estimates inserted

Series	Differencing	*$\bar{w} \pm \hat{\sigma}(\bar{w})$	$\hat{\sigma}_w^2 = c_0$	Identified Model	$\hat{\sigma}_a^2$
A	either 0	17.06 ± 0.10	0.1586	$z_t - 0.87z_{t-1} = 2.45 + a_t$ $- 0.48a_{t-1}$	0.098
	or 1	0.002 ± 0.011	0.1364	$\nabla z_t = a_t - 0.53a_{t-1}$	0.107
B	1	-0.28 ± 0.41	52.54	$\nabla z_t = a_t + 0.09a_{t-1}$	52.2
C	either 1	-0.035 ± 0.047	0.0532	$\nabla z_t - 0.81\nabla z_{t-1} = a_t$	0.019
	or 2	-0.003 ± 0.008	0.0198	$\nabla^2 z_t = a_t - 0.09a_{t-1}$ $- 0.07a_{t-2}$	0.020
D	either 0	9.13 ± 0.04	0.3620	$z_t - 0.86z_{t-1} = 1.32 + a_t$	0.093
	or 1	0.004 ± 0.017	0.0965	$\nabla z_t = a_t - 0.05a_{t-1}$	0.096
E	either 0	46.9 ± 5.4	1382.2	$z_t - 1.32z_{t-1} + 0.63z_{t-2}$ $= 14.9 + a_t$	289
	or 0	46.9 ± 5.4	1382.2	$z_t - 1.37z_{t-1} + 0.74z_{t-2}$ $- 0.08z_{t-3} = 13.7 + a_t$	287
F	0	51.1 ± 1.1	139.80	$z_t + 0.32z_{t-1} - 0.18z_{t-2}$ $= 58.3 + a_t$	115

* When $d = 0$, read z for w.

when we introduce the appropriate stationarity and invertibility restrictions.

Suppose that w_t, having covariance generating function $\gamma(B)$, is represented by the linear model

$$\phi(B)w_t = \theta(B)a_t \qquad (6.4.1)$$

where the zeroes of $\phi(B)$ and of $\theta(B)$ lie outside the unit circle. Then this linear model may also be written

$$\prod_{i=1}^{p} (1 - G_i B)w_t = \prod_{j=1}^{q} (1 - H_j B)a_t \qquad (6.4.2)$$

where the G_i^{-1} are the roots of $\phi(B) = 0$ and H_j^{-1} are the roots of $\theta(B) = 0$, and G_i, H_j lie inside the unit circle. Using (3.1.11), the covariance generating function for w is

$$\gamma(B) = \prod_{i=1}^{p} (1 - G_i B)^{-1}(1 - G_i F)^{-1} \prod_{j=1}^{q} (1 - H_j B)(1 - H_j F)\sigma_a^2$$

Multiple choice of moving average parameters

Since

$$(1 - H_j B)(1 - H_j F) = H_j^2(1 - H_j^{-1}B)(1 - H_j^{-1}F)$$

it follows that any one of the stochastic models

$$\prod_{i=1}^{p} (1 - G_i B) w_t = \prod_{j=1}^{q} (1 - H_j^{\pm 1} B) k a_t$$

can have the same covariance generating function, if the constant k is appropriately chosen. In the above, it is understood that for complex roots, reciprocals of both members of the congugate pair will be taken. However, if a real root H is inside the unit circle, H^{-1} will lie outside, or if a complex pair, say H_1 and H_2, are inside, then the pair H_1^{-1} and H_2^{-1} will lie outside. It follows that there will be only *one stationary invertible* model of the form (6.4.2), which has a given autocovariance function.

Backward representations

Now $\gamma(B)$ also remains unchanged if in (6.4.2) we replace $1 - G_i B$ by $1 - G_i F$ or $1 - H_j B$ by $1 - H_j F$. Thus, all the stochastic models

$$\prod_{i=1}^{p} (1 - G_i B^{\pm 1}) w_t = \prod_{j=1}^{q} (1 - H_j B^{\pm 1}) a_t$$

have identical covariance structure. However, representations containing the operator $B^{-1} = F$ refer to future w's and/or future a's, so that although stationary and invertible representations exist in which w_t is expanded in terms of future w's and a's, only one such representation, (6.4.2), exists which relates w_t entirely to *past* history.

A model form which, somewhat surprisingly, is of practical interest is that in which *all* B's are replaced by F's in (6.4.1), so that

$$\phi(F) w_t = \theta(F) e_t$$

where e_t is a sequence of independently distributed random variables having mean zero and variance $\sigma_e^2 = \sigma_a^2$. This then is a stationary invertible representation in which w_t is expressed *entirely* in terms of future w's and e's. We refer to it as the *backward* form of the process, or more simply as the *backward process*.

Equation (6.4.2) is not the most general form of stationary invertible linear model having the covariance generating function $\gamma(B)$. For example, the model (6.4.2) may be multiplied on both sides by any factor $1 - QB$. Thus the process

$$(1 - QB) \prod_{i=1}^{p} (1 - G_i B) w_t = (1 - QB) \prod_{j=1}^{q} (1 - H_j B) a_t$$

has the same covariance structure as (6.4.2). This fact will present no particular difficulty where identification is concerned, since we will be naturally led to choose the simplest representation. However, we shall find in Chapter 7 that we will need to be alert to the possibility of factoring of the operators when fitting the process.

Finally, we reach the conclusion then that a stationary-invertible model, in which a current value w_t is expressed in terms only of *previous* history, and which contains no common factors, is uniquely determined by the covariance structure.

Proper understanding of model multiplicity is of importance for a number of reasons:

(1) We are reassured by the above argument that the covariance function can logically be used to identify a linear stationary-invertible model which expresses w_t in terms of previous history.
(2) The nature of the multiple solutions for moving average parameters obtained by equating moments is clarified.
(3) The backward process

$$\phi(F)w_t = \theta(F)e_t$$

obtained by replacing B by F in the linear model, is useful in estimating values of the series which have occurred before the first observation was made.

Now we consider (2) and (3) in greater detail.

6.4.2 Multiple moment solutions for moving average parameters

In estimating the q parameters $\theta_1, \theta_2, \ldots, \theta_q$ in the MA model by equating covariances, we have seen in Section 3.3 that multiple solutions are obtained. To each combination of roots there will correspond a linear representation, but to only one such combination will there be an invertible representation in terms of past history.

For example, consider the MA(1) process in w_t

$$w_t = (1 - \theta_1 B)a_t$$

and suppose $\gamma_0(w)$ and $\gamma_1(w)$ are known and we want to deduce the values of θ_1 and σ_a^2. Since

$$\gamma_0 = (1 + \theta_1^2)\sigma_a^2 \qquad \gamma_1 = -\theta_1\sigma_a^2 \qquad \gamma_k = 0 \qquad k > 1 \qquad (6.4.3)$$

then

$$-\frac{\gamma_0}{\gamma_1} = \theta_1^{-1} + \theta_1$$

and if $(\theta_1 = \theta, \sigma_a^2 = \sigma^2)$ is a solution for given γ_0 and γ_1, then so is $(\theta_1 = \theta^{-1}, \sigma_a^2 = \theta^2\sigma^2)$. Apparently then, for given values of γ_0 and γ_1, there are a *pair* of possible processes

$$w_t = (1 - \theta B)a_t$$

and

$$w_t = (1 - \theta^{-1}B)\alpha_t \qquad (6.4.4)$$

with $\sigma_\alpha^2 = \sigma^2\theta^2$. If $-1 < \theta < 1$, then (6.4.4) is not an invertible representation. However, this model may be written

$$w_t = \{(1 - \theta^{-1}B)(-\theta F)\}\{-\theta^{-1}B\alpha_t\}$$

Thus after setting

$$e_t = -\alpha_{t-1}/\theta$$

the model becomes

$$w_t = (1 - \theta F)e_t \qquad (6.4.5)$$

where e_t has the same variance as a_t. Thus (6.4.5) is simply the "backward" process, which is dual to the forward process

$$w_t = (1 - \theta B)a_t \qquad (6.4.6)$$

Just as the shock a_t in (6.4.6) is expressible as a convergent sum of current and *previous* values of w

$$a_t = w_t + \theta w_{t-1} + \theta^2 w_{t-2} + \cdots$$

so the shock e_t in (6.4.5) is expressible as a convergent sum of current and future values of w

$$e_t = w_t + \theta w_{t+1} + \theta^2 w_{t+2} + \cdots$$

Thus the root θ^{-1} *does* produce an invertible process, but only if a representation of the shock e_t in terms of future values of w is permissible.

The invertibility regions shown in Table 6.1 delimit acceptable values of the parameters, *given* that we express the shock in terms of *previous* history.

6.4.3 Use of the backward process to determine starting values

Suppose a time series $w_1, w_2, \ldots, w_n$, is available from a process

$$\phi(B)w_t = \theta(B)a_t \qquad (6.4.7)$$

In Chapter 7, problems arise where we need to estimate the values w_0, w_{-1}, w_{-2}, etc., of the series which occurred *before* the first observation was made. This happens because "starting values" are needed for certain basic recursive calculations used for estimating the parameters in the model. Now, suppose we require to estimate w_{-l}, given $w_1, \ldots, w_n$. The discussion of Section 6.4.1 shows that the probability structure of $w_1, \ldots, w_n$ is equally explained by the forward model (6.4.7), or by the backward model

$$\phi(F)w_t = \theta(F)e_t \qquad (6.4.8)$$

The value w_{-l} thus bears exactly the same probability relationship to the sequence $w_1, w_2, \ldots, w_n$, as does the value w_{n+l+1} to the sequence $w_n, w_{n-1},$ $w_{n-2}, \ldots, w_1$. Thus, to estimate a value $l + 1$ periods before observations started, we can first consider what would be the optimal estimate or forecast $l + 1$ periods after the series ended, and then apply this procedure to the *reversed* series. In other words, we "forecast" the reversed series. We call this "back forecasting."

APPENDIX A6.1 EXPECTED BEHAVIOR OF THE ESTIMATED AUTOCORRELATION FUNCTION FOR A NON-STATIONARY PROCESS

Suppose that a series of N observations $z_1, z_2, \ldots, z_N$ is generated by a nonstationary $(0, 1, 1)$ process

$$\nabla z_t = (1 - \theta B)a_t$$

and the estimated autocorrelations r_k are computed, where

$$r_k = \frac{c_k}{c_0} = \frac{\sum\limits_{t=1}^{N-k} (z_t - \bar{z})(z_{t+k} - \bar{z})}{\sum\limits_{t=1}^{N} (z_t - \bar{z})^2}$$

Some idea of the behavior of these estimated autocorrelations may be obtained by deriving expected values for the numerator and denominator of this expression and considering the ratio. We will write, following [108],

$$\mathcal{E}[r_k] = E[c_k]/E[c_0]$$

$$= \sum\limits_{t=1}^{N-k} E[(z_t - \bar{z})(z_{t+k} - \bar{z})] \bigg/ \sum\limits_{t=1}^{N} E[(z_t - \bar{z})^2]$$

After straightforward, but tedious algebra we find

$$\mathcal{E}[r_k] = \frac{(N - k)\{(1 - \theta)^2(N^2 - 1 + 2k^2 - 4kN) - 6\theta\}}{N(N - 1)\{(N + 1)(1 - \theta)^2 + 6\theta\}} \qquad \text{(A6.1.1)}$$

For θ close to zero, $\mathcal{E}[r_k]$ will be close to unity, but for large values of θ it can be considerably smaller than unity, even for small values of k. Figure A6.1 illustrates this fact and shows $\mathcal{E}[r_k]$ for $\theta = 0.8$ and for $N = 100$ and $N = 200$. Although, as anticipated for a nonstationary process, the ratios $\mathcal{E}[r_k]$ of expected values fail to damp out quickly, it will be seen that they do not approach the value 1 even for small lags.

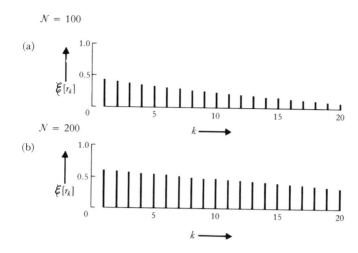

FIG. A6.1 $\xi[r_k] = E[c_k]/E[c_0]$ for series generated by $\nabla z_t = (1 - 0.8B)a_t$

Similar effects may be demonstrated whenever the parameters approach values where cancellation on both sides of the model would produce a stationary process. For instance, in the above example we can write the model as

$$(1 - B)z_t = \{(1 - B) + \delta B\}a_t$$

where $\delta = 0.2$. As δ tends to zero, the behavior of the process would be expected to come closer and closer to that of the white noise process $z_t = a_t$, for which the autocorrelation function is zero for lags $k > 0$.

APPENDIX A6.2 A GENERAL METHOD FOR OBTAINING INITIAL ESTIMATES OF THE PARAMETERS OF A MIXED AUTOREGRESSIVE-MOVING AVERAGE PROCESS

In Section 6.3 we have shown how to derive initial estimates for the para-meters of simple ARMA models. In particular, Charts B, C and D, in the collection of tables and charts at the end of the volume, have been prepared to enable the initial estimates to be read off quickly for the AR (2), MA (2), and ARMA (1, 1) processes. In this appendix we give a general method, which can be programmed on a computer, for obtaining initial estimates for a general ARMA (p, q) process $\phi(B)w_t = \theta(B)a_t$. Such a computer program is described under Program 2 in the collection of computer programs at the end of this volume.

In the general case, the calculation of the initial estimates of an ARMA (p, q) process is based on the first $p + q + 1$ autocovariances $c_j [j = 0, 1, \ldots,$ $(p + q)]$ of $w_t = \nabla^d z_t$, and proceeds in three stages.

(1) The autoregressive parameters $\phi_1, \phi_2, \ldots, \phi_p$ are estimated from the autocovariances $c_{q-p+1}, \ldots, c_{q+1}, c_{q+2}, \ldots, c_{q+p}$.

(2) Using the estimates $\hat{\phi}$ obtained in (1), the first $q + 1$ autocovariances $c'_j(j = 0, 1, \ldots, q)$ of the derived series

$$w'_t = w_t - \hat{\phi}_1 w_{t-1} - \cdots - \hat{\phi}_p w_{t-p}$$

are calculated.

(3) Finally, the autocovariances $c'_0, c'_1, \ldots, c'_q$ are used in an iterative calculation to compute initial estimates of the moving average parameters $\theta_1, \theta_2, \ldots, \theta_q$ and of the residual variance σ_a^2.

Initial estimates of autoregressive parameters. Making use of the result (3.4.3), initial estimates of the autoregressive parameters may be obtained by solving the p linear equations

$$c_{q+1} = \hat{\phi}_1 c_q + \hat{\phi}_2 c_{q-1} + \cdots + \hat{\phi}_p c_{q-p+1}$$

$$c_{q+2} = \hat{\phi}_1 c_{q+1} + \hat{\phi}_2 c_q + \cdots + \hat{\phi}_p c_{q-p+2} \qquad \text{(A6.2.1)}$$

$$c_{q+p} = \hat{\phi}_1 c_{q+p-1} + \hat{\phi}_2 c_{q+p-2} + \cdots + \hat{\phi}_p c_q$$

Autocovariances of derived moving average process. We now write $w'_t = \phi(B)w_t$, and then treat the process as a moving average process

$$w'_t = \theta(B)a_t \qquad \text{(A6.2.2)}$$

First, we need to express the autocovariances c'_j of w'_t in terms of the autocovariances c_j of w_t. It may be shown that

$$c'_j = \sum_{i=0}^{p} \phi_i^2 c_j + \sum_{i=1}^{p} (\phi_0\phi_i + \phi_1\phi_{i+1} + \cdots + \phi_{p-i}\phi_p)d_j \qquad \text{(A6.2.3)}$$

where

$$j = 0, 1, \ldots, q$$

$$d_j = c_{j+i} + c_{j-i}$$

$$\phi_0 = -1$$

Initial estimates of the moving average parameters. Using the autovariance estimates c'_j, we can obtain initial estimates of the moving average parameters in the derived process (A6.2.2) by one of two iterative processes.

(i) *Linearly convergent process:* Using the expressions

$$\gamma_0 = (1 + \theta_1^2 + \cdots + \theta_q^2)\sigma_a^2$$

$$\gamma_k = (-\theta_k + \theta_1\theta_{k+1} + \cdots + \theta_{q-k}\theta_q)\sigma_a^2 \qquad k \geq 1$$

for the autocovariance function of a MA (q) process, given in Section 3.3.2, we can compute estimates of the parameters $\sigma_a^2, \theta_q, \theta_{q-1}, \ldots, \theta_1$ in this precise order, using the iteration

$$\sigma_a^2 = \frac{c_0'}{1 + \theta_1^2 + \cdots + \theta_q^2}$$

$$\theta_j = -\left(\frac{c_j'}{\sigma_a^2} - \theta_1\theta_{j+1} - \theta_2\theta_{j+2} - \cdots - \theta_{q-j}\theta_q\right)$$

(A6.2.4)

with the convention that $\theta_0 = 0$. The parameters $\theta_1, \theta_2, \ldots, \theta_q$ are set equal to zero to start the iteration and the values of the θ_j and σ_a^2, to be used in any subsequent calculation, are the most up to date values available. For example, in the case $q = 2$, the equations (A6.2.4) are

$$\sigma_a^2 = \frac{c_0'}{1 + \theta_1^2 + \theta_2^2}$$

$$\theta_2 = -\frac{c_2'}{\sigma_a^2}$$

$$\theta_1 = -\left(\frac{c_1'}{\sigma_a^2} - \theta_1\theta_2\right)$$

(ii) *A quadratically convergent process:* A Newton–Raphson algorithm, which has superior convergence properties to method (i), has been given by Wilson [55]. We denote $\boldsymbol{\tau}' = (\tau_0, \tau_1, \ldots, \tau_q)$, where

$$\tau_0^2 = \sigma_a^2 \qquad \theta_j = -\tau_j/\tau_0 \qquad j = 1, 2, \ldots, q \qquad \text{(A6.2.5)}$$

Then, if $\boldsymbol{\tau}^i$ is the estimate of $\boldsymbol{\tau}$ obtained at the ith iteration, the new values at the $(i + 1)$ iteration are obtained from

$$\boldsymbol{\tau}^{i+1} = \boldsymbol{\tau}^i - (\mathbf{T}^i)^{-1}\mathbf{f}_i \qquad \text{(A6.2.6)}$$

where $\mathbf{f}' = (f_0, f_1, \ldots, f_q)$, $f_j = \sum_{i=0}^{q-j} \tau_i\tau_{i+j} - c_j'$ and

$$
\mathbf{T} = \begin{bmatrix}
\tau_0 & \tau_1 \cdots \tau_{q-2} & \tau_{q-1} & \tau_q \\
\tau_1 & \tau_2 \cdots \tau_{q-1} & \tau_q & 0 \\
\tau_2 & \tau_3 \cdots \tau_q & 0 & 0 \\
\vdots & \vdots \quad \vdots & \vdots & \vdots \\
\tau_q & 0 \cdots 0 & 0 & 0
\end{bmatrix}
+
\begin{bmatrix}
\tau_0 & \tau_1 & \tau_2 \cdots & \tau_q \\
0 & \tau_0 & \tau_1 \cdots \tau_{q-1} \\
0 & 0 & \tau_0 \cdots \tau_{q-2} \\
\vdots & \vdots & \vdots \quad \vdots \\
0 & 0 & 0 \cdots & \tau_0
\end{bmatrix}
$$

Knowing the values of the τ's at each iteration, the values of the parameters may be obtained from (A6.2.5).

Example.   Consider the estimation of ϕ and θ in the ARMA model

$$(1 - \phi B)\tilde{z}_t = (1 - \theta B)a_t$$

using values $c_0 = 1.25$, $c_1 = 0.50$, and $c_2 = 0.40$, which correspond to a process with $\phi = 0.8$, $\theta = 0.5$ and $\sigma_a^2 = 1.000$. The estimate of ϕ is obtained by substituting $p = q = 1$ in (A6.2.1) giving

$$c_2 = \phi c_1$$

so that $\hat{\phi} = 0.8$. Hence, using (A6.2.3), the first two covariances of the derived series

$$w_t' = \tilde{z}_t - \phi\tilde{z}_{t-1}$$

are

$$c_0' = (1 + \phi^2)c_0 - 2\phi c_1 = 1.25$$
$$c_1' = (1 + \phi^2)c_1 - \phi(c_2 + c_0) = -0.50$$

Substituting these values in (A6.2.4), the iterative process of method (i) is based upon

$$\sigma_a^2 = \frac{1.25}{1 + \theta^2}$$

$$\theta = \frac{0.5}{\sigma_a^2}$$

Similarly, substituting in (A6.2.6), the iterative process of method (ii) is based upon

$$\begin{bmatrix} \tau_0^{i+1} \\ \tau_1^{i+1} \end{bmatrix} = \begin{bmatrix} \tau_0^i \\ \tau_1^i \end{bmatrix} - (\mathbf{T}^i)^{-1} \begin{bmatrix} f_0^i \\ f_1^i \end{bmatrix}$$

where

$$\begin{bmatrix} f_0^i \\ f_1^i \end{bmatrix} = \begin{bmatrix} (\tau_0^i)^2 + (\tau_1^i)^2 - c_0' \\ \tau_0^i\tau_1^i \qquad - c_1' \end{bmatrix}$$

and

$$\mathbf{T}^i = \begin{bmatrix} \tau_0^i & \tau_1^i \\ \tau_1^i & 0 \end{bmatrix} + \begin{bmatrix} \tau_0^i & \tau_1^i \\ 0 & \tau_0^i \end{bmatrix} = \begin{bmatrix} 2\tau_0^i & 2\tau_1^i \\ \tau_1^i & \tau_0^i \end{bmatrix}$$

The variance σ_a^2 and θ may then be calculated from $\sigma_a^2 = \tau_0^2$, $\theta = -\tau_1/\tau_0$.

Table A6.1 shows how the iteration converged for methods (i) and (ii). Program 2 in the Collection of Computer Programs at the end of this volume may be used to calculate initial estimates of the parameters for any ARIMA (p, d, q) process.

TABLE A6.1   Convergence of preliminary estimates of σ_a^2 and θ_1 for a MA (1) process

Iteration	Method (i)		Method (ii)	
	σ_a^2	θ	σ_a^2	θ
0	—	0.000	1.250	0.000
1	1.250	0.400	2.250	0.667
2	1.077	0.464	1.210	0.545
3	1.029	0.486	1.012	0.503
4	1.011	0.494	1.000	0.500
5	1.004	0.498	—	—
6	1.002	0.499	—	—
7	1.001	0.500	—	—
8	1.000	0.500	—	—

APPENDIX A6.3 THE FORWARD AND BACKWARD IMA PROCESSES OF ORDER $(0, 1, 1)$

In Section 6.4 we encountered forward and backward models for stationary processes. It is also of interest to consider the corresponding dual models for nonstationary processes. As an example we consider the IMA $(0, 1, 1)$ process. With a_t and e_t random noise sequences with $\sigma_a^2 = \sigma_e^2$, the following moving average processes have identical autocovariance structure:

$$w_t = (1 - \theta B)a_t \qquad w_t = (1 - \theta F)\varepsilon_t$$

After performing the relabelling $\varepsilon_t = -e_{t-1}$ we can write the backward process as

$$-w_t = (1 - \theta F)e_{t-1}$$

Now suppose observations are actually made of z_t whose first difference is w_t, so that

$$w_t = (1 - B)z_t = -(1 - F)z_{t-1}$$

The model can then be written in terms of the z's in either of two forms with parallel development as follows:

$$(1 - B)z_t = (1 - \theta B)a_t \qquad\qquad (1 - F)z_t = (1 - \theta F)e_t$$

$$a_t = \frac{1 - B}{1 - \theta B}z_t \qquad\qquad\qquad e_t = \frac{1 - F}{1 - \theta F}z_t$$

$$= \left\{1 - \frac{(1 - \theta)B}{1 - \theta B}\right\}z_t \qquad\qquad = \left\{1 - \frac{(1 - \theta)F}{1 - \theta F}\right\}z_t$$

$$a_t = z_t - (1 - \theta)(z_{t-1}$$
$$+ \theta z_{t-2} + \theta^2 z_{t-3} + \cdots)$$
$$= z_t - \overset{\text{\tiny J}}{z}_{t-1}$$

where $\overset{\text{\tiny J}}{z}_{t-1}$ is the *backward* exponentially weighted average, using values of z *before* time t. Hence

$$\overset{\text{\tiny J}}{z}_{t-1} - \overset{\text{\tiny J}}{z}_{t-2} = (1 - \theta)a_{t-1}$$
$$\overset{\text{\tiny J}}{z}_{t-1} = (1 - \theta)z_{t-1} + \theta\overset{\text{\tiny J}}{z}_{t-2}$$

$$e_t = z_t - (1 - \theta)(z_{t+1}$$
$$+ \theta z_{t+2} + \theta^2 z_{t+3} + \cdots)$$
$$= z_t - \overset{\text{\tiny L}}{z}_{t+1}$$

where $\overset{\text{\tiny L}}{z}_{t+1}$ is the *forward* exponentially weighted average using values of z *after* time t. Hence

$$\overset{\text{\tiny L}}{z}_{t+1} - \overset{\text{\tiny L}}{z}_{t+2} = (1 - \theta)e_{t+1}$$
$$\overset{\text{\tiny L}}{z}_{t+1} = (1 - \theta)z_{t+1} + \theta\overset{\text{\tiny L}}{z}_{t+2}$$

Suppose that we have values of z extending on either side of the current time t, the covariances of whose differences satisfy (6.4.3). From this series we can construct two sets of random deviates,

$$\ldots a_{t-2}, a_{t-1}, a_t, a_{t+1}, a_{t+2}, \ldots \quad \text{and} \quad \ldots e_{t-2}, e_{t-1}, e_t, e_{t+1}, e_{t+2} \ldots,$$

where, for example, a_t is the difference between z_t and $\overset{\text{\tiny J}}{z}_{t-1}$, the backward EWMA computed from the *previous* values $z_{t-1}, z_{t-2}, z_{t-3} \ldots$, whereas e_t is the difference between z_t and $\overset{\text{\tiny L}}{z}_{t+1}$, the forward EWMA computed from $z_{t+1}, z_{t+2}, z_{t+3} \ldots$.

Relation between the a's and e's. To obtain the relationship between the a's and the e's, we can write

$$a_t = \frac{(\theta - B)}{1 - \theta B}e_t$$

$$= \theta e_t - (1 + \theta)\frac{(1 - \theta)B}{1 - \theta B}e_t$$

$$= \theta e_t - (1 + \theta)\overset{\text{\tiny J}}{e}_{t-1} \tag{A6.3.1}$$

Conversely

$$e_t = \frac{(\theta - F)}{1 - \theta F}a_t$$

$$= \theta a_t - (1 + \theta)\overset{\text{\tiny L}}{a}_{t+1} \tag{A6.3.2}$$

where $\overset{\text{\tiny J}}{e}_t$ and $\overset{\text{\tiny L}}{a}_t$ are exponentially weighted moving averages (EWMA's) as previously defined.

It is easily confirmed that if, for example, the a's are a sequence of independent random variables with mean zero and variance σ^2, then so are the e's generated from them by (A6.3.2).

Although the e's are independent of other e's, the a's and e's are cross correlated. We find, using (A6.3.1), that

$$\gamma_{ae}(k) = E[a_t e_{t+k}] = \begin{cases} -(1 - \theta^2)\theta^{-(k+1)}\sigma^2 & k < 0 \\ \theta\sigma^2 & k = 0 \\ 0 & k > 0 \end{cases}$$

Figure A6.2 illustrates the situation. At the top of the figure are shown some observations from a (0, 1, 1) process with $\theta = 0.5$, the associated backward EWMA's, and the resulting a's. Below is shown the same series with forward EWMA's and the resulting e's. At the bottom of the figure is shown the cross-covariance function $\gamma_{ae}(k)$ between a_t and e_{t+k}.

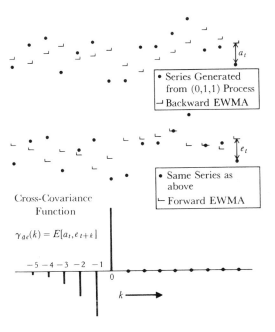

FIG. A6.2   Forward and backward exponentially weighted moving averages

7

Model Estimation

The identification process having led to a tentative formulation for the model, we then need to obtain efficient estimates of the parameters. After the parameters have been estimated, the fitted model will be subjected to diagnostic checks and tests of goodness of fit. As pointed out by R. A. Fisher, for tests of goodness of fit to be relevant, it is necessary that efficient use of the data should have been made in the fitting process. If this is not so, inadequacy of fit may simply arise because of the inefficient fitting and not because the form of the model is inadequate. This chapter contains a general account of likelihood and Bayesian methods for estimation of the parameters in the stochastic model. Throughout the chapter, bold type is used to denote vectors and matrices. Thus $\mathbf{X} = \{x_{ij}\}$ is a matrix with x_{ij} an element in the ith row and jth column and $\mathbf{X}'$ is the transpose of the matrix.

7.1 STUDY OF THE LIKELIHOOD AND SUM OF SQUARES FUNCTIONS

7.1.1 The likelihood function

Suppose we have a sample of N observations $\mathbf{z}$ with which we associate an N-dimensional random variable, whose known probability distribution $p(\mathbf{z}|\boldsymbol{\xi})$ depends on some unknown parameters $\boldsymbol{\xi}$. We use the vector $\boldsymbol{\xi}$ to denote a general set of parameters and, in particular, it could refer to the $p + q + 1$ parameters $(\boldsymbol{\phi}, \boldsymbol{\theta}, \sigma)$ of the ARIMA model.

Before the data are available, $p(\mathbf{z}|\boldsymbol{\xi})$ will associate a density with each different outcome $\mathbf{z}$, of the experiment, for fixed $\boldsymbol{\xi}$. After the data have come to hand, we are led to contemplate the various values of $\boldsymbol{\xi}$ which might have given rise to the fixed set of observations $\mathbf{z}$, actually obtained. The appropriate function for this purpose is the *likelihood function* $L(\boldsymbol{\xi}|\mathbf{z})$, which is of the same form as $p(\mathbf{z}|\boldsymbol{\xi})$, but in which $\mathbf{z}$ is now fixed, but $\boldsymbol{\xi}$ is variable. It is only the relative value of $L(\boldsymbol{\xi}|\mathbf{z})$ which is of interest, so that the likelihood function is usually regarded as containing an *arbitrary multiplicative constant*.

It is often convenient to work with the log likelihood function $\ln L(\boldsymbol{\xi}|\mathbf{z}) = l(\boldsymbol{\xi}|\mathbf{z})$, which contains an *arbitrary additive constant*. One reason that the likelihood function is of fundamental importance in estimation theory, is because of the "likelihood principle," urged on somewhat different grounds by Fisher [56], Barnard [57], and Birnbaum [58]. This principle says that (given that the assumed model is correct) all that the *data* has to tell us about the parameters is contained in the likelihood function, all other aspects of the data being irrelevant. From a Bayesian point of view, the likelihood function is equally important, since it is the component in the posterior distribution of the parameters which comes from the data.

For a complete understanding of the estimation situation, it is necessary to make a thorough analytical and graphical study of the likelihood function, or in the Bayesian framework, the posterior distribution of the parameters which, in the situations we consider, is dominated by the likelihood. In many examples, for moderate and large samples, the log-likelihood function will be unimodal and can be adequately approximated over a sufficiently extensive region near the maximum by a quadratic function. In such cases, the log-likelihood function can be described by its maximum and its second derivatives at the maximum. The values of the parameters which maximize the likelihood function, or equivalently the log-likelihood function, are called *maximum likelihood (ML) estimates*. The second derivatives of the log likelihood function provide measures of "spread" of the likelihood function and can be used to calculate approximate standard errors for the estimates.

The limiting properties of Maximum Likelihood estimates are usually established for independent observations [59]. But, as was shown by Whittle [87], they may be extended to cover stationary time series.

In what follows, we shall assume that the reader is familiar with certain basic ideas in estimation theory. Appendices A7.1 and A7.2 summarize some important results in Normal distribution theory and linear least squares, which are needed for this chapter. Some of the important earlier work on estimation of the parameters of time series models will be found in references [34], [78], [88], [89], [93], [94], [95], [97], [98], [99], [100] and [101].

7.1.2 The conditional likelihood for an ARIMA process

Let us suppose that the $N = n + d$ original observations $\mathbf{z}$, form a time series which we denote by $z_{-d+1}, \ldots, z_0, z_1, z_2, \ldots, z_n$. We assume that this series is generated by an ARIMA model of order (p, d, q). From these observations we can generate a series $\mathbf{w}$ of $n = N - d$ differences $w_1, w_2, \ldots, w_n$, where $w_t = \nabla^d z_t$. Thus the general problem of fitting the parameters $\boldsymbol{\phi}$ and $\boldsymbol{\theta}$ of the ARIMA model (6.1.1) is equivalent to fitting to the w's, the stationary, invertible* ARMA (p, q) model which may be written

$$a_t = \tilde{w}_t - \phi_1 \tilde{w}_{t-1} - \phi_2 \tilde{w}_{t-2} - \cdots - \phi_p \tilde{w}_{t-p} + \theta_1 a_{t-1} + \theta_2 a_{t-2} + \cdots$$
$$+ \theta_q a_{t-q} \tag{7.1.1}$$

where $w_t = \nabla^d z_t$ and $\tilde{w}_t = w_t - \mu$ with $E[w_t] = \mu$.

*Special care is needed to ensure that estimates lie in the invertible region. See Appendix A7.6.

For $d > 0$ it would often be appropriate to assume that $\mu = 0$ (see the discussion in Sections 4.1.3, 6.2.3, 6.3.5, and 6.3.7). When this is not appropriate, we suppose that $\bar{w} = \sum_{t=1}^{n} w_t/n$ is substituted for μ. For the sample sizes normally considered in time series analysis, this approximation will be adequate. However, if desired, μ may be included as an additional parameter to be estimated. The procedures we describe may then be used for simultaneous estimation of μ, along with the other parameters.

The w's cannot be substituted immediately in (7.1.1) to calculate the a's, because of the difficulty of starting up the difference equation. However, suppose that the p values $\mathbf{w}_*$ of the w's and the q values $\mathbf{a}_*$ of the a's prior to the commencement of the w series were given. Then the values of $a_1, a_2, \ldots, a_n$, conditional on this choice, could be calculated in turn from (7.1.1).

Thus, for any given choice of parameters $(\boldsymbol{\phi}, \boldsymbol{\theta})$ and of the starting values $(\mathbf{w}_*, \mathbf{a}_*)$, we could calculate successively a set of values $a_t(\boldsymbol{\phi}, \boldsymbol{\theta}|\mathbf{w}_*, \mathbf{a}_*, \mathbf{w})$, $t = 1, 2, \ldots, n$. Now, assuming that the a's are Normally distributed,

$$p(a_1, a_2, \ldots, a_n) \propto \sigma_a^{-n} \exp\left\{ -\left(\sum_{t=1}^{n} a_t^2/2\sigma_a^2 \right) \right\}$$

Given a particular set of data $\mathbf{w}$, the log likelihood associated with the parameter values $(\boldsymbol{\phi}, \boldsymbol{\theta}, \sigma_a)$, *conditional* on the choice of $(\mathbf{w}_*, \mathbf{a}_*)$, would then be

$$l_*(\boldsymbol{\phi}, \boldsymbol{\theta}, \sigma_a) = -n \ln \sigma_a - \frac{S_*(\boldsymbol{\phi}, \boldsymbol{\theta})}{2\sigma_a^2} \tag{7.1.2}$$

where, following the previous discussion, no additive constant term need be included, and

$$S_*(\boldsymbol{\phi}, \boldsymbol{\theta}) = \sum_{t=1}^{n} a_t^2(\boldsymbol{\phi}, \boldsymbol{\theta}|\mathbf{w}_*, \mathbf{a}_*, \mathbf{w}) \tag{7.1.3}$$

In the above, star subscripts, on the likelihood and sum of squares functions, are used to emphasize that they are conditional on the choice of the starting values. We notice that the conditional likelihood l_* involves the data only through the conditional *sum of squares function*. It follows that contours of l_* for any fixed value of σ_a in the space of $(\boldsymbol{\phi}, \boldsymbol{\theta}, \sigma_a)$ are contours of S_*, that these maximum likelihood estimates are the same as the least squares estimates, and that in general we can, on the Normal assumption, study the behavior of the conditional likelihood by studying the conditional sum of squares function. In particular for any fixed σ_a, l_* is a linear function of S_*.

7.1.3 *Choice of starting values for conditional calculation*

We shall discuss shortly the calculation of the unconditional likelihood, which, strictly, is what we need for parameter estimation.

For some purposes, when n is moderate or large, a sufficient approximation to the unconditional likelihood is obtained by using the conditional likelihood with suitable values substituted for the elements of $\mathbf{w}_*$ and $\mathbf{a}_*$ in (7.1.3). One procedure is to set the elements of $\mathbf{w}_*$ and of $\mathbf{a}_*$ equal to their unconditional expectations. The unconditional expectations of the elements of $\mathbf{a}_*$ are zero, and if the model contains no deterministic part, and in particular if $\mu = 0$, the unconditional expectations of the elements of $\mathbf{w}_*$ will also be zero.* However, this approximation can be poor if some of the roots of $\phi(B) = 0$ are close to the boundary of the unit circle, so that the process approaches nonstationarity. In this case, the initial data value w_1 could deviate considerably from its unconditional expectation, and the introduction of starting values of this sort could introduce a large transient, which would be slow to die out. In fitting a model of order (p, d, q), a more reliable approximation procedure, and one we shall employ sometimes, is to use (7.1.1) to calculate the a's *from a_{p+1} onwards*, setting previous a's equal to zero. Thus, actually occurring values are used for the w's throughout.

Using this method, we can sum the squares of only $n - p = N - p - d$ values of a_t, but the slight loss of information will be unimportant for long series. In cases where there are no autoregressive terms the two procedures are equivalent. For seasonal series, discussed in Chapter 9, the conditional approximation is not very satisfactory and the unconditional calculation becomes even more necessary.

We now illustrate the recursive calculation of the conditional sum of squares S_* with a simple example.

Calculation of the conditional sum of squares for a $(0, 1, 1)$ process.   Series B has been tentatively identified in Table 6.4 as an IMA $(0, 1, 1)$ process:

$$\nabla z_t = (1 - \theta B)a_t \qquad -1 < \theta < 1 \tag{7.1.4}$$

that is

$$a_t = w_t + \theta a_{t-1}$$

where $w_t = \nabla z_t$ and $E[w_t] = 0$. It will be recalled that in Chapter 6 a preliminary moment estimate (Table 6.5) was obtained which suggested that, for this data, θ was close to zero.

The calculation of the first few a's is set out in Table 7.1 for the particular parameter value $\theta = 0.5$. The a's are calculated recursively from $a_t = w_t + 0.5a_{t-1}$, to single decimal accuracy. In accordance with the above discussion, to start up the process, a_0 is set equal to zero. This value is shown in italic type. Proceeding in this way we find that

$$S_*(0.5) = \sum_{t=1}^{368} a_t^2(\theta = 0.5|a_0 = 0) = 27,694$$

* If the assumption $E[w_t] = \mu \neq 0$ is appropriate, we can substitute $\bar{w}$ for each of the elements of $\mathbf{w}_*$.

TABLE 7.1   Recursive calculation of a's for first 10 values
of Series B, using $\theta = 0.5$

t	z_t	$w_t = \nabla z_t$	$a_t = w_t + 0.5a_{t-1}$
0	460		*0*
1	457	-3	-3.0
2	452	-5	-6.5
3	459	7	3.8
4	462	3	4.9
5	459	-3	-0.6
6	463	4	3.7
7	479	16	17.8
8	493	14	22.9
9	490	-3	8.4

The recursive calculation is particularly well suited for use on an electronic computer. Using values from $\theta = -0.5$ to $\theta = +0.5$ in steps of 0.1, the values for the conditional sum of squares $S_*(\theta)$ (given that $a_0 = 0$) are shown in the third row of Table 7.2.

TABLE 7.2   Sum of squares functions for model $\nabla z_t = (1 - \theta B)a_t$ fitted to Series B

θ	-0.5	-0.4	-0.3	-0.2	-0.1	0.0
$\lambda = (1 - \theta)$	1.5	1.4	1.3	1.2	1.1	1.0
$S_*(\theta)$	23,929	21,595	20,222	19,483	19,220	19,363
$S(\theta)$	23,929	21,594	20,222	19,483	19,220	19,363

θ	0.1	0.2	0.3	0.4	0.5
$\lambda = (1 - \theta)$	0.9	0.8	0.7	0.6	0.5
$S_*(\theta)$	19,896	20,851	22,315	24,471	27,694
$S(\theta)$	19,896	20,849	22,315	24,478	27,691

7.1.4   The unconditional likelihood—the sum of squares function—least squares estimates

It is shown in Appendix A7.4 that, corresponding to the $N = n + d$ observations assumed to be generated by an ARIMA model, the unconditional log-likelihood is given by

$$l(\boldsymbol{\phi}, \boldsymbol{\theta}, \sigma_a) = f(\boldsymbol{\phi}, \boldsymbol{\theta}) - n \ln \sigma_a - \frac{S(\boldsymbol{\phi}, \boldsymbol{\theta})}{2\sigma_a^2} \qquad (7.1.5)$$

where $f(\boldsymbol{\phi}, \boldsymbol{\theta})$ is a function of $\boldsymbol{\phi}$ and $\boldsymbol{\theta}$. The *unconditional sum of squares function* is given by

$$S(\boldsymbol{\phi}, \boldsymbol{\theta}) = \sum_{t = -\infty}^{n} [a_t | \boldsymbol{\phi}, \boldsymbol{\theta}, \mathbf{w}]^2 \qquad (7.1.6)$$

where $[a_t | \boldsymbol{\phi}, \boldsymbol{\theta}, \mathbf{w}] = E[a_t | \boldsymbol{\phi}, \boldsymbol{\theta}, \mathbf{w}]$ denotes the expectation of a_t conditional on $\boldsymbol{\phi}, \boldsymbol{\theta}$ and $\mathbf{w}$. When the meaning is clear from the context we shall further abbreviate this conditional expectation to $[a_t]$.

Usually, $f(\boldsymbol{\phi}, \boldsymbol{\theta})$ is of importance only for small n. For moderate and large values of n, (7.1.5) is dominated by $S(\boldsymbol{\phi}, \boldsymbol{\theta})/2\sigma_a^2$ and thus the contours of the unconditional sum of squares function in the space of the parameters $(\boldsymbol{\phi}, \boldsymbol{\theta})$ are very nearly contours of likelihood and of log likelihood. It follows, in particular, that the parameter estimates obtained by minimizing the sum of squares (7.1.6), which we call *least squares estimates*, will usually provide very close approximations to the maximum likelihood estimates. From a Bayesian viewpoint, on assumptions discussed in Section 7.4, for all AR (p) and MA (q), essentially the posterior density is a function only of $S(\boldsymbol{\phi}, \boldsymbol{\theta})$. Hence, very nearly the least squares estimates are those with maximum posterior density. In the remainder of this section, and in Section 7.1.5, our main emphasis will be on the calculation, study, and use of the unconditional sum of squares function $S(\boldsymbol{\phi}, \boldsymbol{\theta})$, as defined in (7.1.6), and on calculating least squares estimates.

In the calculation of the unconditional sum of squares, the $[a]$'s are computed recursively by taking conditional expectations in (7.1.1). A preliminary back-calculation provides the values $[w_{-j}]$, $j = 0, 1, 2, \ldots$ (i.e., the back-forecasts) needed to start off the forward recursion.

Calculation of the unconditional sum of squares for a moving average process. For illustration, we reconsider the IBM stock price example, again using the first ten* values of the series given in Table 7.1. With $w_t = \nabla z_t$, we have seen in Section 6.4.3 that the model of order (0, 1, 1) we have fitted, may be written in either the forward or backward forms

$$w_t = (1 - \theta B)a_t \qquad w_t = (1 - \theta F)e_t$$

and where again $\mu = E[w_t]$ is assumed equal to zero. Hence we can write

$$[e_t] = [w_t] + \theta[e_{t+1}] \qquad (7.1.7)$$

$$[a_t] = [w_t] + \theta[a_{t-1}] \qquad (7.1.8)$$

where $[w_t] = w_t$ for $t = 1, 2, \ldots, n$ and is the back forecast of w_t for $t \leqslant 0$.

* In practice, of course, useful parameter estimates could not be obtained from as few as ten observations. We utilize this data subset merely to illustrate the calculations.

These are the two basic equations we need in the computations. A convenient format for the calculations is shown in Table 7.3.

TABLE 7.3   Calculation of the $[a]$'s from the first 10 values of Series B, using $\theta = 0.5$

t	z_t	$[a_t]$	$0.5[a_{t-1}]$	$[w_t]$	$0.5[e_{t+1}]$	$[e_t]$
−1	[458.4]	0	0	0	0	0
0	460	1.6	0	1.6	−1.6	0
1	457	−2.2	0.8	−3.0	−0.1	−3.1
2	452	−6.1	−1.1	−5.0	4.8	−0.2
3	459	4.0	−3.0	7.0	2.6	9.6
4	462	5.0	2.0	3.0	2.3	5.3
5	459	−0.5	2.5	−3.0	7.6	4.6
6	463	3.8	−0.2	4.0	11.1	15.1
7	479	17.9	1.9	16.0	6.2	22.2
8	493	23.0	9.0	14.0	−1.5	12.5
9	490	8.5	11.5	−3.0	*0*	−3.0

We commence by entering in the table what we know. These are:
(a) the data values $z_0, z_1, \ldots, z_9$, from which we can calculate the first differences $w_1, w_2, \ldots, w_9$;
(b) the values $[e_0], [e_{-1}], \ldots$ which are zero, since $e_0, e_{-1}, \ldots$ are distributed independently of $\mathbf{w}$;
(c) the values $[a_{-1}], [a_{-2}], \ldots$ which are zero, because for any MA (q) process, $a_{-q}, a_{-q-1}, \ldots$ are distributed independently of $\mathbf{w}$. However, note that in general $[a_0], [a_{-1}], \ldots, [a_{-q+1}]$ will be nonzero and are obtained by back forecasting. Thus, in the present example, $[a_0]$ is so obtained.

Beginning at the end of the series, (7.1.7) is now used to compute the $[e_t]$'s for $t = 9, 8, 7, \ldots, 1$. We start this backward process by making the same approximation as was described previously for the calculation of the conditional sum of squares. In the present instance, this amounts to setting $[e_{10}] = 0$. In general, the effect of this approximation will be to introduce a transient into the system which, because $\phi(B)$ and $\theta(B)$ are stationary operators, will for series of moderate length, almost certainly be negligible by the time the beginning of the series is reached and thus will not affect the calculation of the a's. As we see later, if desired, the adequacy of this approximation can be checked in any given case by performing a second iterative cycle.

Thus, to start the recursion in Table 7.3, in the row corresponding to $t = 9$, we enter a zero (shown in italic type) in column 6 for the unknown value

$0.5[e_{10}]$. Then, using (7.1.7), we obtain

$$[e_9] = [w_9] + 0.5[e_{10}]$$
$$= w_9 + 0 = -3$$

and so $0.5[e_9] = -1.5$ can be entered in the line $t = 8$, which enables us to compute $[e_8]$ and so on. Finally we obtain

$$[e_0] = [w_0] + \theta[e_1]$$

that is

$$0 = [w_0] - 1.6$$

which gives $[w_0] = 1.6$, and thereafter $[w_{-h}] = 0$, $h = 1, 2, 3, \ldots$.
Using (7.1.8) with $t = 0$, we obtain

$$[a_0] = [w_0] + \theta[a_{-1}] = 1.6 + (0.5)(0) = 1.6$$

and we can then continue the forward calculations of the remaining $[a]$'s. Comparison of the values of a_t given in Tables 7.1 and 7.3 shows that, in this particular example, the transient introduced by the change in the starting value has little effect for $t > 5$. Proceeding in this way with the whole series, we find for the unconditional sum of squares

$$S(0.5) = \sum_{t=0}^{368} [a_t|0.5, \mathbf{w}]^2 = 27{,}691$$

which for this particular example is very close to the conditional value $S_*(0.5) = 27{,}694$.

The unconditional sums of squares $S(\theta)$, for values of θ between -0.5 and $+0.5$, are given in the bottom row of Table 7.2 and are very close to the conditional values $S_*(\theta)$ for this particular example.

7.1.5 General procedure for calculating the unconditional sum of squares

In the above example, w_t was a first order moving average process, with zero mean. It followed that all forecasts for lead times greater than 1 were zero and consequently that, only one preliminary value (the back forecast $[w_0] = 1.6$) was required to start the recursive calculation. For a qth order moving average process, q nonzero preliminary values $[w_0], [w_{-1}], \ldots, [w_{1-q}]$ would be needed. Special procedures, which we discuss later in Section 7.3.1, are available for estimating autoregressive parameters. However, we show in Appendix A7.4 that, the procedure described in this section can supply the unconditional sum of squares to any desired degree of approximation, for any ARIMA model.

Specifically, suppose the w_t's are assumed to be generated by the stationary forward model

$$\phi(B)\tilde{w}_t = \theta(B)a_t \tag{7.1.9}$$

where $\nabla^d z_t = w_t$ and $\tilde{w}_t = w_t - \mu$. Then they could equally well have been generated by the backward model

$$\phi(F)\tilde{w}_t = \theta(F)e_t \tag{7.1.10}$$

As before, we first employ (7.1.10) to supply back forecasts $[\tilde{w}_{-j}|\phi, \theta, w]$. Theoretically, the presence of the autoregressive operator ensures a series of such estimates which is infinite in extent. However, because of the stationary character of this operator, in practice, the estimates $[\tilde{w}_t]$ at and beyond some point $t = -Q$ with Q of moderate size, become essentially equal to zero.

Thus, to a sufficient approximation we can write

$$\tilde{w}_t = \phi^{-1}(B)\theta(B)a_t = \sum_{j=0}^{\infty} \psi_j a_{t-j} \simeq \sum_{j=0}^{Q} \psi_j a_{t-j}$$

This means that the original mixed process can be replaced by a moving average process of order Q and the procedure for moving averages already outlined in Section 7.1.4 may be used.

Thus, in general, the dual set of equations for generating the conditional expectations $[a_t|\phi, \theta, w]$ is obtained by taking conditional expectations in (7.1.10) and (7.1.9), that is

$$\phi(F)[\tilde{w}_t] = \theta(F)[e_t] \tag{7.1.11}$$

is first used to generate the backward forecasts and then

$$\phi(B)[\tilde{w}_t] = \theta(B)[a_t] \tag{7.1.12}$$

is used to generate the $[a_t]$'s. If we find that the forecasts are negligible in magnitude beyond some lead time Q, then the recursive calculation goes forward with

$$\begin{aligned}[e_{-j}|\phi, \theta, w] &= 0 \qquad j = 0, 1, 2, \ldots \\ [a_{-j}|\phi, \theta, w] &= 0 \qquad j > Q - 1\end{aligned} \tag{7.1.13}$$

Calculation of the unconditional sum of squares for a mixed autoregressive-moving average process. For illustration, consider the following $n = 12$ successive values of $w_t = \nabla^d z_t$

t	1	2	3	4	5	6	7	8	9	10	11	12
w_t	2.0	0.8	−0.3	−0.3	−1.9	0.3	3.2	1.6	−0.7	3.0	4.3	1.1

Suppose we wish to compute the unconditional sum of squares $S(\phi, \theta)$ associated with the ARIMA $(1, d, 1)$ process

$$(1 - \phi B)w_t = (1 - \theta B)a_t$$

with $\nabla^d z_t = w_t$, or equivalently, with the backward process

$$(1 - \phi F)w_t = (1 - \theta F)e_t$$

it being assumed that the w's have mean zero.

Of course, estimates based on twelve observations would be almost valueless, but nevertheless, this short series serves to explain the nature of the calculation. We illustrate with the parameter values $\phi = 0.3$, $\theta = 0.7$. Then (7.1.11) and (7.1.12) may be written

$$[e_t] = [w_t] - 0.3[w_{t+1}] + 0.7[e_{t+1}] \tag{7.1.14}$$

$$[a_t] = [w_t] - 0.3[w_{t-1}] + 0.7[a_{t-1}] \tag{7.1.15}$$

where $[w_t] = w_t$ $(t = 1, 2, \ldots, n)$.

The layout of the calculation is shown in Table 7.4. The data are first entered in the center column and then the known zero values for $[e_0]$, $[e_{-1}]$, $[e_{-2}]$, ... are entered in the last column. The backward equation (7.1.14), is now started off in precisely the manner described in Section 7.1.3 for the conditional calculation. Thus, because we have a *first*-order autoregressive operator in the model, we commence *one* step from the end of the series and substitute 0 in line $t = 11$ for the unknown value of $0.7[e_{12}]$. The recursive calculation now goes ahead using (7.1.14). In Tables 7.4 and 7.5 the data are given to one decimal and the calculations are taken to two decimals. The back forecasts $[w_{-j}]$ $(j = 0, 1, 2, \ldots)$ die out quickly and to the accuracy with which we are working are equal to zero for $j > 4$. Using (7.1.13), the estimates $[a_{-j}]$ are taken as being equal to zero for $j > 4$. The forward recursion is now begun using (7.1.15) and the $[a_t]$ computed.

The unconditional sum of squares $S(\phi, \theta)$ is obtained by summing the squares of all the calculated $[a_t]$'s. Thus

$$S(0.3, 0.7) = \sum_{t=-4}^{12} [a_t]^2 = 89.2$$

In practice, a second iterative cycle would almost never be needed. However, to show the rapidity of convergence of this process, even for this impractically short series of 12 observations, a second cycle of the iteration is performed in Table 7.5. Here the value for a_{12} of 3.99, computed in the previous iteration, is used to start off a new iteration to calculate the forward forecasts of the w's using (7.1.15). These may then be substituted in the backward equations (7.1.14) to eventually obtain new back forecasts $[w_0]$,

TABLE 7.4   Calculation of $[a]$'s and of $S(0.3, 0.7)$ from 12 values of a series assumed to be generated by the process $(1 - 0.3B)w_t = (1 - 0.7B)a_t$

t	$[a_t]$	$0.7[a_{t-1}]$	$-0.3[w_{t-1}]$	$[w_t]$	$-0.3[w_{t+1}]$	$0.7[e_{t+1}]$	$[e_t]$
−4	−0.01	0.00	0.00	−0.01	0.01	0	0
−3	−0.04	−0.01	0.00	−0.03	0.03	0	0
−2	−0.11	−0.03	0.01	−0.09	0.09	0	0
−1	−0.36	−0.08	0.03	−0.31	0.31	0	0
0	−1.20	−0.25	0.09	−1.04	−0.60	1.64	0
1	1.47	−0.84	0.31	2.0	−0.24	0.58	2.34
2	1.23	1.03	−0.60	0.8	0.09	−0.06	0.83
3	0.32	0.86	−0.24	−0.3	0.09	0.13	−0.08
4	0.02	0.23	0.09	−0.3	0.57	−0.09	0.18
5	−1.80	0.01	0.09	−1.9	−0.09	1.86	−0.13
6	−0.39	−1.26	0.57	0.3	−0.96	3.32	2.66
7	2.84	−0.27	−0.09	3.2	−0.48	2.02	4.74
8	2.63	1.99	−0.96	1.6	0.21	1.08	2.89
9	0.66	1.84	−0.48	−0.7	−0.90	3.14	1.54
10	3.67	0.46	0.21	3.0	−1.29	2.78	4.49
11	5.97	2.57	−0.90	4.3	−0.33	0	3.97
12	3.99	4.18	−1.29	1.1			

$$S(0.3, 0.7) = \sum_{t=-4}^{12} [a_t]^2 = 89.2$$

TABLE 7.5   A second iteration of the calculation commenced in Table 7.4

t	$[a_t]$	$0.7[a_{t-1}]$	$-0.3[w_{t-1}]$	$[w_t]$	$-0.3[w_{t+1}]$	$0.7[e_{t+1}]$	$[e_t]$
0				−1.04	−0.60	1.64	0
1				2.0	−0.24	0.58	2.34
2				0.8	0.09	−0.06	0.83
3				−0.3	0.09	0.12	−0.09
4				−0.3	0.57	−0.10	0.17
5				−1.9	−0.09	1.85	−0.14
6				0.3	−0.96	3.30	2.64
7				3.2	−0.48	2.00	4.72
8				1.6	0.21	1.05	2.86
9				−0.7	−0.90	3.09	1.49
10				3.0	−1.29	2.71	4.42
11				4.3	−0.33	−0.10	3.87
12	3.99			1.1	0.74	−1.98	−0.14
13	0	2.79	−0.33	−2.46	0.22	−0.60	−2.84
14	0	0	0.74	−0.74	0.07	−0.18	−0.85
15	0	0	0.22	−0.22	0.02	−0.06	−0.26
16	0	0	0.07	−0.07	0.01	−0.02	−0.08
17	0	0	0.02	−0.02	0.00	−0.01	−0.03
18	0	0	0.01	−0.01	0.00		−0.01

$$S(0.3, 0.7) = \sum_{t=1}^{18} [e_t]^2 = 89.3$$

$[w_{-1}]$, ... In general, for the second cycle of calculations, we use the relations

$$[a_{n+i}|\phi, \theta, w] = 0 \qquad i = 1, 2, \ldots$$
$$[e_{n+i}|\phi, \theta, w] = 0 \qquad i > Q'$$

(7.1.16)

where Q' is chosen so that $[w_{n+i}]$ is negligible for $i > Q'$. Table 7.5 shows how the second cycle eventually leads to $[w_0] = -1.04$, which is the value obtained in the first cycle. Thus a further calculation for the a's would yield the same results as before.

Further, as follows from Appendix A7.4, $S(\phi, \theta|w)$ can be equally well computed from the sum of squares of the $[e_t]$'s. Using this fact, we find from Table 7.5 that

$$S(0.3, 0.7) = \sum_{t=1}^{18} [e_t]^2 = 89.3$$

which agrees very closely with the value 89.2 found from the a_t's.

We saw that, in the fitting of a process of order $(0, 1, 1)$ to the IBM Series B, the *conditional* sums of squares provided a very close approximation to the unconditional value. That this is not always so can be seen from the present example.

We have previously mentioned, in Section 7.1.3, two conditional sums of squares which might be used as approximations to the unconditional values. These were obtained
(1) by starting off the recursion at the first available observation, setting all unknown a's or e's to zero and all the w's equal to their unconditional expectations;
(2) by starting off the recursion at the pth observation using only observed values for the w's and zeroes for unknown a's or e's.
Two such conditional sums of squares and the two unconditional sums of squares
(3) obtained from the $[a_t]$'s
(4) obtained from the $[e_t]$'s
are compared below:
(1) Unknown w's and e's set equal to zero

$$\sum_{t=1}^{12} (e_t|0.3, 0.7, w_{13} = 0, e_{13} = 0, w)^2 = 101.0$$

(2) Unknown e's set equal to zero

$$\sum_{t=2}^{12} (e_t|0.3, 0.7, w_{12} = 1.1, e_{12} = 0, w)^2 = 82.4$$

(3) Unconditional calculation, first iteration

$$\sum_{t=-4}^{12} [a_t|0.3, 0.7, w]^2 = 89.2$$

(4) Unconditional calculation, second iteration

$$\sum_{t=1}^{18} [e_t|0.3, 0.7, \mathbf{w}]^2 = 89.3$$

The sum of squares in (1) is a very poor approximation to (3) and (4), although the discrepancy, which is over 10 % in this series of 12 values, would be diluted if the series were longer. This is because the transient introduced by the choice of starting value has essentially died out after twelve values. On the other hand, when allowance is made for the fact that the conditional sum of squares in (2) is based on only 11 rather than 12 squared quantities, it does seem to represent a more satisfactory approximation than (1). For reasons already given, if a conditional approximation is to be used, it should normally be of the form (2) rather than (1). However, as is discussed further in Chapter 9, for seasonal series, the conditional approximation becomes less satisfactory and the unconditional sum of squares should ordinarily be computed.

7.1.6 Graphical study of the sum of squares function

The sum of squares function $S(\theta)$ for the IBM data given in Table 7.2 is plotted in Figure 7.1. The overall minimum sum of squares is at about $\theta = -0.09, (\lambda = 1.09)$, which is the least squares estimate and (on the assumption of Normality) a close approximation to the *maximum likelihood estimate* of the parameter θ.

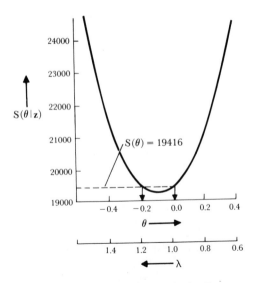

FIG. 7.1   Plot of $S(\theta)$ for Series B

   The graphical study of the sum of squares functions is readily extended to two parameters by evaluating the sum of squares over a suitable grid of parameter values and plotting contours. As we have discussed in Section 7.1.4, on the assumption of Normality, the contours are very nearly likelihood contours. For most practical purposes, rough contours drawn in by eye on the computer output grid are adequate. Figure 7.2 shows a grid of $S(\lambda_0, \lambda_1)$ values for Series B fitted with the IMA $(0, 2, 2)$ process

$$\nabla^2 z_t = (1 - \theta_1 B - \theta_2 B^2)a_t$$
$$= \{1 - (2 - \lambda_0 - \lambda_1)B - (\lambda_0 - 1)B^2\}a_t \qquad (7.1.17)$$

or in integrated form

$$z_t = \lambda_0 Sa_{t-1} + \lambda_1 S^2 a_{t-1} + a_t$$

The minimum sum of squares in Figure 7.2 is at about $\hat{\lambda}_0 = 1.09$, $\hat{\lambda}_1 = 0.0$. The plot thus confirms that the preferred model in this case is an IMA $(0, 1, 1)$ process. The device illustrated here, of fitting a model somewhat more elaborate than that expected to be needed, can provide a useful confirmation of the original identification. The elaboration of the model should be made, of course, in the direction "feared" to be necessary.

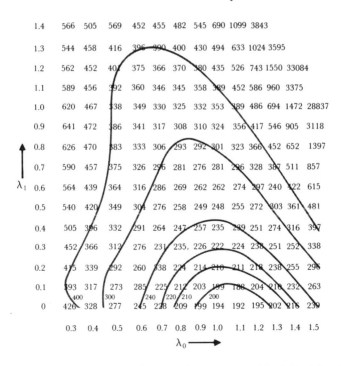

FIG. 7.2   Values of $S(\lambda_0, \lambda_1) \times 10^{-2}$ for Series B on a grid of (λ_0, λ_1) values and approximate contours

Plotting contours.   Devices for automatic plotting from computer output are becoming increasingly common. Using these, the plotting of contours from grid values is readily arranged. If it is desired to prepare precise contour plots by hand, a series of graphs of $S(\lambda_0, \lambda_1)$ against λ_0 may be made for various values of λ_1. Horizontal lines are then drawn at chosen values of S for which contours are required and the points of intersection transferred to the grid. If necessary, more detail may be obtained by repeating the process with S plotted against λ_1 for various fixed values of λ_0. The process is illustrated in Table 7.6 and in Figures 7.3 and 7.4 for an IMA (0, 2, 2) process fitted to the data of Series C. Table 7.6 shows a sum of squares grid for λ_0

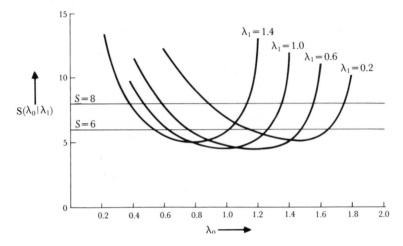

Fig. 7.3   Graphs of $S(\lambda_0, \lambda_1)$ versus λ_0 for fixed values of λ_1, used in the preparation of a contour plot for Series C

Table 7.6   Sum of squares grid for Series C on the tentative assumption that the model is of order (0, 2, 2). Values are shown within the invertible region

λ_1	0.0	0.2	0.4	0.6	0.8	λ_0 1.0	1.2	1.4	1.6	1.8	2.0
2.0		11.5	7.55	6.60	9.00						
1.8		11.1	7.08	5.87	6.51	14.0					
1.6		12.3	6.92	5.52	5.40	7.45					
1.4		13.8	7.24	5.41	4.89	5.56	12.1				
1.2		15.9	7.94	5.56	4.72	4.75	6.55				
1.0		20.3	9.02	5.99	4.79	4.44	5.03	11.0			
0.8		19.7	10.1	6.61	5.09	4.44	4.47	6.05			
0.6		23.3	11.4	7.48	5.65	4.72	4.40	4.87	10.3		
0.4		26.8	13.6	8.97	6.67	5.39	4.73	4.66	6.00		
0.2		41.3	19.7	12.5	8.91	6.90	5.74	5.18	5.40	10.2	
0.0	160	55.5	28.6	17.7	12.2		9.19	7.42	6.58	7.30	

and λ_1, while Figure 7.3 shows plots of $S(\lambda_0, \lambda_1)$ against λ_0 for various fixed values of λ_1. The intersections with the line $S(\lambda_0, \lambda_1) = 6$ are read off and used to plot the contour in Figure 7.4.

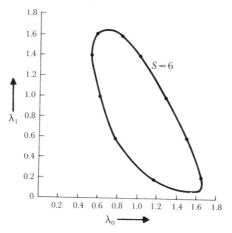

FIG. 7.4   Delineating the contour $S = 6$ for Series C

Three parameters.   When we wish to study the joint estimation situation for three parameters, two dimensional contour diagrams for a number of values of the third parameter can be drawn. For illustration, part of such a series of diagrams is shown in Figure 7.5 for Series A, C, and D. In each case the "elaborated" model

$$\nabla^2 z_t = (1 - \theta_1 B - \theta_2 B^2 - \theta_3 B^3) a_t$$

$$= \{1 - (2 - \lambda_{-1} - \lambda_0 - \lambda_1)B - (\lambda_0 + 2\lambda_{-1} - 1)B^2 + \lambda_{-1} B^3\} a_t$$

or

$$z_t = \lambda_{-1} a_{t-1} + \lambda_0 S a_{t-1} + \lambda_1 S^2 a_{t-1} + a_t$$

has been fitted, leading to the conclusion that the best fitting models of this type* are as shown in Table 7.7.

TABLE 7.7   IMA models fitted to Series A, C, and D

Series	$\hat{\lambda}_{-1}$	$\hat{\lambda}_0$	$\hat{\lambda}_1$	Fitted series
A	0	0.3	0	$z_t = 0.3 S a_{t-1} + a_t$
C	0	1.1	0.8	$z_t = 1.1 S a_{t-1} + 0.8 S^2 a_{t-1} + a_t$
D	0	0.9	0	$z_t = 0.9 S a_{t-1} + a_t$

* We show later in Section 7.2.5 that slightly better fits are obtained in some cases with closely related models containing autoregressive terms.

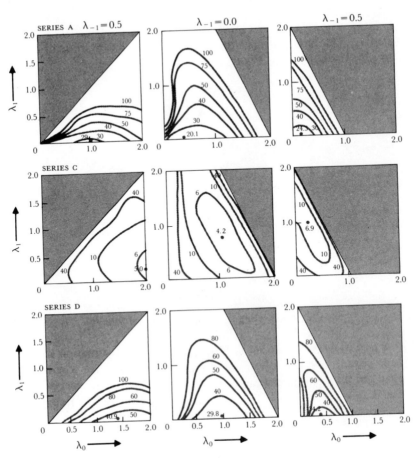

FIG. 7.5   Sums of squares contours for Series A, C, and D (shaded lines indicate boundaries of the invertibility regions)

The inclusion of additional parameters (particularly λ_{-1}) in this fitting process is not strictly necessary, but we have included them to illustrate the effect of overfitting and to show how closely our identification seems to be confirmed for these series.

7.1.7 Description of "well behaved" estimation situations—confidence regions

The likelihood function is not, of course, plotted merely to indicate maximum likelihood values. It is the whole course of this function which contains the totality of information coming from the data. In some fields of study, situations can occur where the likelihood function has two or more peaks (see, for instance the example in [60]) and also where the likelihood function contains sharp ridges and spikes. All of these situations have logical interpretations. In each case, the likelihood function is trying to tell us something which we ought to know. Thus, the existence of two peaks of approximately equal heights implies that there are two sets of values of the parameters which might explain the data. The existence of obliquely oriented ridges means that a value of one parameter, considerably different from its maximum likelihood value, could explain the data if accompanied by a value of the other which deviated appropriately. Characteristics of this kind determine what may be called the *estimation situation*. To understand the estimation situation, we must examine the likelihood function both analytically and graphically.

Need for care in interpreting the likelihood function.   Care is needed in interpreting the likelihood function. For example, results discussed later, which assume that the log likelihood is approximately quadratic near its maximum, will clearly not apply to the three-parameter estimation situations depicted in Figure 7.5. However, these examples are exceptional because here we are deliberately *overfitting* the model. If the simpler model is justified we should *expect* to find the likelihood function contours truncated near its maximum, by a boundary in the higher dimensional parameter space. However, quadratic approximations *could* be used if the simpler *identified* model rather than the overparameterized model was fitted.

Special care is needed when the maximum of the likelihood function may be on or near a boundary. Consider the situation shown in Figure 7.6 and suppose we know a priori that a parameter $\beta > \beta_0$. The maximum likelihood within the permissible range of β is at B, where $\beta = \beta_0$, not at A or at C.

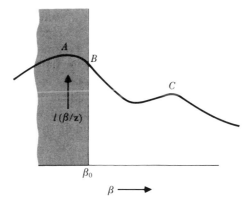

FIG. 7.6   A hypothetical likelihood function with a constraint $\beta > \beta_0$

It will be noticed that the first derivative of the likelihood is in this case *nonzero* at the maximum likelihood value and that the quadratic approximation is certainly not an adequate representation of the likelihood.

The treatment afforded the likelihood method has, in the past, often left much to be desired and ineptness in the practitioner has sometimes been mistaken for deficiency in the method. The treatment has often consisted of
(i) differentiating the log likelihood and setting first derivatives to zero to obtain the maximum likelihood (ML) estimates;
(ii) deriving approximate variances and covariances of these estimates from the second derivatives of the log likelihood or from the expected values of the second derivatives.

Mechanical application of the above can, of course, produce nonsensical answers. This is so, first, because of the elementary fact that setting derivatives to zero does not necessarily produce maxima, and second, because the information which the likelihood function contains is only fully expressed by the ML estimates and by the second derivatives of the log likelihood, if the quadratic approximation is adequate over the region of interest. To know whether this is so for a new estimation problem, a careful analytical and graphical investigation is usually required.

When a class of estimation problems (such as those arising from the estimation of parameters in ARMA models) is initially being investigated, it is important to plot the likelihood function rather extensively. After the behavior of a particular class of models is well understood, and knowledge of the situation indicates that it is safe to do so, we may take certain short cuts which we now consider. These results are described in greater detail in the Appendices A7.4 and A7.5. We begin by considering expressions for the variances and covariances of maximum likelihood estimates, appropriate when the log likelihood is approximately quadratic and the sample size is moderately large.

In what follows, it is convenient to define a vector $\boldsymbol{\beta}$ whose $k = p + q$ elements are the autoregressive and moving average parameters $\boldsymbol{\phi}$ and $\boldsymbol{\theta}$. Thus the complete set of $p + q + 1 = k + 1$ parameters of the ARMA process may be written as $\boldsymbol{\phi}, \boldsymbol{\theta}, \sigma_a$; or as $\boldsymbol{\beta}, \sigma_a$; or simply as $\boldsymbol{\xi}$.

Variances and covariances of ML estimates. For the appropriately parameterized ARMA model, it will often happen that over the relevant* region of the parameter space, the log likelihood will be approximately quadratic in the elements of $\boldsymbol{\beta}$ (that is of $\boldsymbol{\phi}$ and $\boldsymbol{\theta}$), so that

$$l(\boldsymbol{\xi}) = l(\boldsymbol{\beta}, \sigma_a) \simeq l(\hat{\boldsymbol{\beta}}, \sigma_a) + \frac{1}{2} \sum_{i=1}^{k} \sum_{j=1}^{k} l_{ij}(\beta_i - \hat{\beta}_i)(\beta_j - \hat{\beta}_j) \qquad (7.1.18)$$

where, to the approximation considered, the derivatives

$$l_{ij} = \frac{\partial^2 l(\boldsymbol{\beta}, \sigma_a)}{\partial \beta_i \partial \beta_j} \qquad (7.1.19)$$

are constant.

* Say over a 95 % confidence region.

For large n, the influence of the term $f(\phi, \theta)$, or equivalently, $f(\beta)$ in (7.1.5) can be ignored in most cases. Hence, $l(\beta, \sigma_a)$ will be essentially quadratic in β, if $S(\beta)$ is. Alternatively, $l(\beta, \sigma_a)$ will be essentially quadratic in β if the conditional expectations $[a_t|\beta, \mathbf{w}]$ in (7.1.6) are approximately locally linear in the elements of β.

For moderate and large samples, when the local quadratic approximation (7.1.18) is adequate, useful approximations to the variances and covariances of the estimates and approximate confidence regions may be obtained.

The information matrix for the parameters β. The $(k \times k)$ matrix $-E[l_{ij}] = \mathbf{I}(\beta)$ is referred to [56], [87] as the *information matrix* for the parameters β, where the expectation is taken over the distribution of $\mathbf{w}$. For a given value of σ_a, the *variance-covariance matrix* $\mathbf{V}(\hat{\beta})$ for the ML estimates $\hat{\beta}$ is, for large samples, given by the inverse of this information matrix, that is

$$\mathbf{V}(\hat{\beta}) \simeq \{-E[l_{ij}]\}^{-1} \qquad (7.1.20)$$

For example, if $k = 2$, the large sample variance-covariance matrix is

$$\mathbf{V}(\hat{\beta}) = \begin{bmatrix} V(\hat{\beta}_1) & \mathrm{cov}\,(\hat{\beta}_1, \hat{\beta}_2) \\ \mathrm{cov}\,(\hat{\beta}_1, \hat{\beta}_2) & V(\hat{\beta}_2) \end{bmatrix} \simeq - \begin{bmatrix} E[l_{11}] & E[l_{12}] \\ E[l_{12}] & E[l_{22}] \end{bmatrix}^{-1}$$

Now, using (7.1.5)

$$l_{ij} \simeq -\frac{S_{ij}}{2\sigma_a^2} \qquad (7.1.21)$$

where

$$S_{ij} = \frac{\partial^2 S(\beta|\mathbf{w})}{\partial\beta_i\partial\beta_j}$$

Furthermore, if for large samples, we approximate the expected values of l_{ij} or of S_{ij} by the values actually observed, then, using (7.1.20),

$$\mathbf{V}(\hat{\beta}) \simeq \{-E[l_{ij}]\}^{-1} \simeq 2\sigma_a^2\{E[S_{ij}]\}^{-1} \simeq 2\sigma_a^2\{S_{ij}\}^{-1} \qquad (7.1.22)$$

Thus, for $k = 2$,

$$\mathbf{V}(\hat{\beta}) \simeq 2\sigma_a^2 \begin{bmatrix} \dfrac{\partial^2 S(\beta)}{\partial\beta_1^2} & \dfrac{\partial^2 S(\beta)}{\partial\beta_1\partial\beta_2} \\[2ex] \dfrac{\partial^2 S(\beta)}{\partial\beta_1\partial\beta_2} & \dfrac{\partial^2 S(\beta)}{\partial\beta_2^2} \end{bmatrix}^{-1}$$

If $S(\beta)$ were exactly quadratic in β over the relevant region of the parameter space, then, all the derivatives S_{ij} would be constant over this region. In practice, the S_{ij} will vary somewhat, and we shall usually suppose these derivatives to be determined at or near the point $\hat{\boldsymbol{\beta}}$. Now it is shown in the Appendices A7.4 and A7.5 that an estimate* of σ_a^2 is provided by

$$\hat{\sigma}_a^2 = S(\hat{\boldsymbol{\beta}})/n \tag{7.1.23}$$

and that for large samples, $\hat{\sigma}_a^2$ and $\hat{\boldsymbol{\beta}}$ are uncorrelated. Finally, the elements of (7.1.22) may be estimated from

$$\text{cov}\,(\hat{\beta}_i, \hat{\beta}_j) \simeq 2\hat{\sigma}_a^2 S^{ij} \tag{7.1.24}$$

where the matrix $\{S^{ij}\}$ is given by

$$\{S^{ij}\} = \{S_{ij}\}^{-1}$$

and the expression (7.1.24) is understood to define the variance $V(\hat{\beta}_i)$ when $j = i$.

Approximate confidence regions for the parameters. In particular, these results allow us to obtain the approximate variances of our estimates. By taking the square root of these variances, we obtain approximate standard deviations, which are usually called the *standard errors* of the estimates. The standard error of an estimate $\hat{\beta}_i$ is denoted by SE $[\hat{\beta}_i]$. When we have to consider several parameters simultaneously, we need some means of judging the precision of the estimates *jointly*. One means of doing this is to determine a *confidence region*. It may be shown (see for example [61]) that a $1 - \varepsilon$ confidence region has the property that, if repeated samples of size n are imagined to be drawn from the same population and a confidence region constructed from each such sample, then a proportion $1 - \varepsilon$ of these regions will include the true parameter point.

If, for given $\sigma_a, l(\boldsymbol{\beta}, \sigma_a)$ is approximately quadratic in $\boldsymbol{\beta}$ in the neighborhood of $\hat{\boldsymbol{\beta}}$, then, using (7.1.20), (see also Appendix A7.1), an approximate $1 - \varepsilon$ confidence region will be defined by

$$-\sum_{i,j} E[l_{ij}](\beta_i - \hat{\beta}_i)(\beta_j - \hat{\beta}_j) < \chi_\varepsilon^2(k) \tag{7.1.25}$$

where $\chi_\varepsilon^2(k)$ is the significance point exceeded by a proportion ε of the χ^2 distribution, having k degrees of freedom.

Alternatively, using the approximation (7.1.22) and substituting the estimate of (7.1.23) for σ_a^2, the approximate confidence region is given by†

$$\sum_{i,j} S_{ij}(\beta_i - \hat{\beta}_i)(\beta_j - \hat{\beta}_j) < 2\hat{\sigma}_a^2 \chi_\varepsilon^2(k) \tag{7.1.26}$$

---

* Arguments can be advanced for using the divisor $n - k = n - p - q$ rather than n in (7.1.23), but for moderate sample sizes, this modification makes little difference.
† A somewhat closer approximation based on the F distribution, which takes account of the sampling distribution of $\hat{\sigma}^2$, may be employed. For moderate sample sizes this refinement makes little practical difference.

However, for a quadratic $S(\boldsymbol{\beta})$ surface

$$S(\boldsymbol{\beta}) - S(\hat{\boldsymbol{\beta}}) = \tfrac{1}{2} \sum_{i,j} S_{ij}(\beta_i - \hat{\beta}_i)(\beta_j - \hat{\beta}_j) \tag{7.1.27}$$

Thus, using (7.1.23) and (7.1.26), we finally obtain the result that the approximate $1 - \varepsilon$ confidence region is bounded by the contour on the sum of squares surface, for which

$$S(\boldsymbol{\beta}) = S(\hat{\boldsymbol{\beta}})\left\{1 + \frac{\chi^2_\varepsilon(k)}{n}\right\} \tag{7.1.28}$$

Examples of the calculation of approximate confidence intervals and regions.

(1) SERIES B: For Series B, values of $S(\lambda)$ and of its differences are shown in Table 7.8. The second difference of $S(\lambda)$ is not constant and thus $S(\lambda)$ is not

TABLE 7.8 $S(\lambda)$ and its first and second differences for various values of λ for Series B

$\lambda = 1 - \theta$	$S(\lambda)$	$\nabla(S)$	$\nabla^2(S)$
1.5	23,929	2,334	961
1.4	21,595	1,373	634
1.3	20,222	739	476
1.2	19,483	263	406
1.1	19,220	-143	390
1.0	19,363	-533	422
0.9	19,896	-955	509
0.8	20,851	$-1,464$	692
0.7	22,315	$-2,156$	1067
0.6	24,471	$-3,223$	
0.5	27,694		

strictly quadratic. However, in the range from $\lambda = 0.85$ to $\lambda = 1.35$, $\nabla^2(S)$ does not change greatly, so that (7.1.28) can be expected to provide a reasonably close approximation. With a minimum value $S(\hat{\lambda}) = 19{,}216$, the critical value $S(\lambda)$, defining an approximate 95 % confidence interval, is then given by

$$S(\lambda) = 19{,}216\left\{1 + \frac{3.84}{368}\right\} = 19{,}416$$

Reading off the values of λ corresponding to $S(\lambda) = 19{,}416$ in Figure 7.1, we obtain an approximate confidence interval $0.98 < \lambda < 1.19$.

Alternatively, we can employ (7.1.26). Using the second difference at $\lambda = 1.1$, given in Table 7.8, to approximate the derivative, we obtain

$$S_{11} = \frac{\partial^2 S}{\partial \lambda^2} \simeq \frac{390}{(0.1)^2}$$

Also, using (7.1.23), $\hat{\sigma}_a^2 = 19{,}216/368 = 52.2$. Thus, the 95% confidence interval, defined by (7.1.26), is

$$\frac{390}{(0.1)^2}(\lambda - 1.09)^2 < 2 \times 52.2 \times 3.84$$

that is

$$|\lambda - 1.09| < 0.10$$

Thus, the interval is $0.99 < \lambda < 1.19$, which agrees closely with the previous calculation.

In this example, where there is only a single parameter λ, the use of (7.1.25) and (7.1.26) is equivalent to using an interval

$$\hat{\lambda} \pm u_{\varepsilon/2}\hat{\sigma}(\hat{\lambda})$$

where $u_{\varepsilon/2}$ is the deviate which excludes a proportion $\varepsilon/2$ in the upper tail of the Normal distribution. An approximate standard error $\hat{\sigma}(\hat{\lambda}) = \sqrt{2\hat{\sigma}_a^2 S_{11}^{-1}}$ is obtained from (7.1.24). In the present example,

$$V(\hat{\lambda}) = 2\hat{\sigma}_a^2 S_{11}^{-1} = \frac{2 \times 52.2 \times 0.1^2}{390} = 0.00268$$

and the approximate standard error of $\hat{\lambda}$ is

$$\hat{\sigma}(\hat{\lambda}) = \sqrt{\mathrm{var}\,(\hat{\lambda})} = 0.052$$

Thus the approximate confidence interval is

$$\hat{\lambda} \pm 1.96\hat{\sigma}(\hat{\lambda}) = 1.09 \pm 0.10$$

as before.

Finally, we show later in Section 7.2 that it is possible to evaluate (7.1.20) analytically, for large samples from a MA (1) process, yielding

$$V(\hat{\lambda}) \simeq \frac{\lambda(2 - \lambda)}{n}$$

For the present example, substituting $\hat{\lambda} = 1.09$ for λ, we find

$$V(\hat{\lambda}) \simeq 0.00269$$

which agrees closely with the previous estimate and so yields the same standard error of 0.052, and the same confidence interval.

(2) SERIES C: In the identification of Series C, one model which was entertained was a (0, 2, 2) process. To illustrate the application of (7.1.28) for more than one parameter, Figure 7.7 shows an approximate 95% confidence region (shaded) for λ_0 and λ_1 of Series C. For this example $S(\hat{\lambda}) = 4.20$,

$n = 224$, and $\chi^2_{0.05}(2) = 5.99$, so that the approximate 95% confidence region is bounded by the contour for which

$$S(\lambda_0, \lambda_1) = 4.20\left\{1 + \frac{5.99}{224}\right\} = 4.31$$

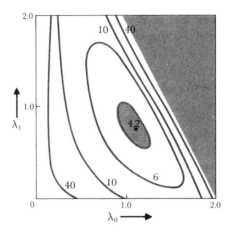

FIG. 7.7   Sum of squares contours with shaded approximate 95% confidence region for Series C, assuming a model of order $(0, 2, 2)$

7.2 NONLINEAR ESTIMATION

7.2.1 *General method of approach*

The plotting of the sum of squares function is of particular importance in the study of new estimation problems, because it ensures that any peculiarities in the estimation situation are shown up. When we are satisfied that anomalies are unlikely, other methods may be used.

We have seen that for most situations, the maximum likelihood estimates are closely approximated by the least squares estimates which make

$$S(\boldsymbol{\phi}, \boldsymbol{\theta}) = \sum_{t=-\infty}^{n} [a_t|\boldsymbol{\phi}, \boldsymbol{\theta}, \mathbf{w}]^2 = \sum_{t=-\infty}^{n} [a_t]^2$$

a minimum, and in practice, the infinite sum can be replaced by a manageable finite sum $\sum_{t=1-Q}^{n} [a_t]^2$.

In general, considerable simplification occurs in the minimization with respect to $\boldsymbol{\beta}$, of a sum of squares $\sum_{t=1}^{n} \{f_t(\boldsymbol{\beta})\}^2$, if each $f_t(\boldsymbol{\beta})$, $(t = 1, 2, \ldots, n)$ is a *linear* function of the parameters $\boldsymbol{\beta}$. We now show that the linearity status of $[a_t]$ is somewhat different in relation to the autoregressive parameters ϕ and to the moving average parameters θ.

For the purely autoregressive process, $[a_t] = \phi(B)[\tilde{w}_t]$ and

$$\frac{\partial[a_t]}{\partial\phi_i} = -[\tilde{w}_{t-i}] + \phi(B)\frac{\partial[\tilde{w}_t]}{\partial\phi_i}$$

Now for $u > 0$, $[\tilde{w}_u] = \tilde{w}_u$ and $\partial[\tilde{w}_u]/\partial\phi_i = 0$, while for $u \leqslant 0$, $[\tilde{w}_u]$ and $\partial[\tilde{w}_u]/\partial\phi_i$ are both functions of ϕ. Thus, except for the effect of "starting values", $[a_t]$ is linear in the ϕ's. By contrast, for the pure moving average process

$$[a_t] = \theta^{-1}(B)[\tilde{w}_t] \qquad \frac{\partial[a_t]}{\partial\theta_j} = \theta^{-2}(B)[\tilde{w}_{t-j}] + \theta^{-1}(B)\frac{\partial[\tilde{w}_t]}{\partial\theta_j}$$

so that the $[a_t]$'s are always nonlinear functions of the parameters.

We shall see in Section 7.3 that special simplifications occur in obtaining least squares and maximum likelihood estimates for the autoregressive process. We show in the present section how, by the iterative application of linear least squares, estimates may be obtained for any ARMA model.

Linearization of the model. In what follows, we continue to use β as a general symbol for the $k = p + q$ parameters (ϕ, θ). We need, then, to minimize

$$\sum_{t=1-Q}^{n} [a_t|\tilde{w}, \beta]^2 = \sum_{t=1-Q}^{n} [a_t]^2$$

Expanding $[a_t]$ in a Taylor series about its value corresponding to some guessed set of parameter values $\beta_0' = (\beta_{1,0}, \beta_{2,0}, \ldots, \beta_{k,0})$, we have approximately

$$[a_t] = [a_{t,0}] - \sum_{i=1}^{k} (\beta_i - \beta_{i,0})x_{i,t} \qquad\qquad (7.2.1)$$

where

$$[a_{t,0}] = [a_t|\mathbf{w}, \beta_0]$$

and

$$x_{i,t} = -\frac{\partial[a_t]}{\partial\beta_i}\bigg|_{\beta=\beta_0}$$

Now, if $\mathbf{X}$ is the $(n + Q) \times k$ matrix $\{x_{i,t}\}$, the $n + Q$ equations (7.2.1) may be expressed as

$$[\mathbf{a}_0] = \mathbf{X}(\beta - \beta_0) + [\mathbf{a}]$$

where $[\mathbf{a}_0]$ and $[\mathbf{a}]$ are column vectors with $n + Q$ elements.

The adjustments $\boldsymbol{\beta} - \boldsymbol{\beta}_0$, which minimize $S(\boldsymbol{\beta}) = S(\boldsymbol{\phi}, \boldsymbol{\theta}) = [\mathbf{a}]'[\mathbf{a}]$, may now be obtained by linear least squares, that is by "regressing" the $[a_0]$'s onto the x's. Because the $[a_t]$'s will not be exactly linear in the parameters $\boldsymbol{\beta}$, a single adjustment will not immediately produce least squares values. Instead, the adjusted values are substituted as new guesses and the process repeated until convergence occurs. Convergence is faster if reasonably good guesses, such as may be obtained at the identification stage, are used initially. If sufficiently bad initial guesses are used, the process may not converge at all.

7.2.2 Numerical estimates of the derivatives

The derivatives $x_{i,t}$ may be obtained directly, as we illustrate later. However, for machine computation, a general nonlinear least squares routine has been found very satisfactory, in which the derivatives are obtained numerically. This is done by perturbing the parameters "one at a time." Thus, for a given model, the values $[a_t|\mathbf{w}, \beta_{1,0}, \beta_{2,0}, \ldots, \beta_{k,0}]$ for $t = 1 - Q, \ldots, n$ are calculated recursively, using whatever preliminary "back forecasts" may be needed. The calculation is then repeated for $[a_t|\mathbf{w}, \beta_{1,0} + \delta_1, \beta_{2,0}, \ldots, \beta_{k,0}]$, then for $[a_t|\mathbf{w}, \beta_{1,0}, \beta_{2,0} + \delta_2, \ldots, \beta_{k,0}]$, and so on. The negative of the required derivative is then given to sufficient accuracy using

$$x_{i,t} = \{[a_t|\mathbf{w}, \beta_{1,0}, \ldots, \beta_{i,0}, \ldots, \beta_{k,0}] - [a_t|\mathbf{w}, \beta_{1,0}, \ldots, \beta_{i,0} + \delta_i, \ldots, \beta_{k,0}]\}/\delta_i$$
(7.2.2)

The above numerical method for obtaining derivatives has the advantage of universal applicability and requires us to program the calculation of the $[a_t]$'s only, and not their derivatives. General nonlinear estimation routines, which essentially require only input instructions on how to compute the $[a_t]$'s, are now becoming generally available [62]. In some versions it is necessary to choose the δ's in advance. In others, the program itself carries through a preliminary iteration to find suitable δ's. Some programs include special features to avoid overshoot and to speed up convergence [63].

Provided the least square solution is not on or near a constraining boundary, the value of $\mathbf{X} = \mathbf{X}_{\hat{\boldsymbol{\beta}}}$ from the final iteration, may be used to compute approximate variances, covariances, and confidence intervals. Thus, $(\mathbf{X}'_\beta \mathbf{X}_\beta)^{-1}\sigma_a^2$ will approximate the variance-covariance matrix of the $\hat{\beta}$'s, and σ_a^2 will be estimated by $S(\hat{\boldsymbol{\beta}})/n$.

Application to a (0, 1, 1) process. As a simple illustration, consider the fitting of Series A to a (0, 1, 1) process

$$w_t = (1 - \theta B)a_t$$

with $\mu = E[w_t] = 0$. The beginning of a calculation is shown in Table 7.9 for the guessed value $\theta_0 = 0.5$. The back-forecasted values for $[a_0]$ were actually obtained by setting $[e_7] = 0$, and using the back recursion $[e_t] = [w_t] + \theta[e_{t+1}]$. Greater accuracy would be achieved by beginning the back recursion further on in the series. Values of $[a_t]$ obtained by successive use of $[a_t] = \theta[a_{t-1}] + [w_t]$, for $\theta = 0.50$ and for $\theta = 0.51$, are shown in the fourth and fifth columns, together with values x_t, for the negative of the derivative, obtained using (7.2.2). To obtain a first adjustment for θ, we compute

$$\theta - \theta_0 = \frac{\sum\limits_{t=0}^{n} [a_{t,0}]x_t}{\sum\limits_{t=0}^{n} x_t^2}$$

TABLE 7.9   Illustration of numerical calculation of derivatives for data from Series A

t	z_t	w_t	$[a_{t,0}] = [a_t|0.50]$	$[a_t|0.51]$	$x_t = 100\{[a_t|0.50] - [a_t|0.51]\}$
0	17.0		0.2453	0.2496	−0.43
1	16.6	−0.40	−0.2773	−0.2727	−0.46
2	16.3	−0.30	−0.4387	−0.4391	0.04
3	16.1	−0.20	−0.4193	−0.4239	0.46
4	17.1	1.00	0.7903	0.7838	0.65
5	16.9	−0.20	0.1952	0.1997	−0.45
6	16.8	−0.10	−0.0024	0.0019	−0.43
7	17.4	0.60	0.5988	0.6010	−0.22
8	17.1	−0.30	−0.0006	0.0065	−0.71
9	17.0	−0.10	−0.1003	−0.0967	−0.36
10	16.7	−0.30	−0.3502	−0.3493	−0.09

In this example, using the whole series of 197 observations, convergence was obtained after four iterations. The course of the calculation was as follows:

Iteration	θ
0	0.50
1	0.63
2	0.68
3	0.69
4	0.70
5	0.70

In general, values ϕ and θ which minimize $S(\phi, \theta)$, can normally be found by this method to any degree of accuracy required. The method is especially attractive because, by its use, we do not need to program derivatives, and other than in the calculation of the $[a_t]$'s, no special provision need be made for end effects.

Program 3, which is described in the collection of Computer Programs in Part V of this volume, evaluates the derivatives numerically and incorporates the back forecasting feature. It can be used to calculate the least squares estimates for the parameters of any (p, d, q) process.

We now show that it is also possible to obtain derivatives directly, but additional recursive calculations are needed.

7.2.3 Direct evaluation of the derivatives

To illustrate the method, it will be sufficient to consider an ARMA $(1, 1)$ process, which can be written in either of the forms

$$e_t = w_t - \phi w_{t+1} + \theta e_{t+1}$$

$$a_t = w_t - \phi w_{t-1} + \theta a_{t-1}$$

We have seen in Section 7.1.4 how the two versions of the model may be used in alternation; one providing initial values with which to start off a recursion with the other. We assume that a first computation has already been made yielding values of $[e_t]$, of $[a_t]$, and of $[w_0], [w_{-1}], \ldots, [w_{1-Q}]$, as in Section 7.1.5, and that $[w_{-Q}], [w_{-Q-1}], \ldots$ and hence $[a_{-Q}], [a_{-Q-1}], \ldots$ are negligible. We now show that a similar dual calculation may be used in calculating derivatives.

Using the notation $a_t^{(\phi)}$ to denote the partial derivative $\partial[a_t]/\partial\phi$, we obtain

$$\begin{cases} e_t^{(\phi)} = w_t^{(\phi)} - \phi w_{t+1}^{(\phi)} + \theta e_{t+1}^{(\phi)} - [w_{t+1}] \\ a_t^{(\phi)} = w_t^{(\phi)} - \phi w_{t-1}^{(\phi)} + \theta a_{t-1}^{(\phi)} - [w_{t-1}] \end{cases} \tag{7.2.3} \label{7.2.4}$$

$$\begin{cases} e_t^{(\theta)} = w_t^{(\theta)} - \phi w_{t+1}^{(\theta)} + \theta e_{t+1}^{(\theta)} + [e_{t+1}] \\ a_t^{(\theta)} = w_t^{(\theta)} - \phi w_{t-1}^{(\theta)} + \theta a_{t-1}^{(\theta)} + [a_{t-1}] \end{cases} \tag{7.2.5}$$

(7.2.4), (7.2.6)

Now

$$\begin{aligned} [w_t] &= w_t \\ w_t^{(\phi)} &= w_t^{(\theta)} = 0 \end{aligned} \right\} \quad t = 1, 2, \ldots, n \tag{7.2.7}$$

and

$$[e_{-j}] = 0 \quad j = 0, 1, \ldots, n \tag{7.2.8}$$

Consider equations (7.2.3) and (7.2.4). By setting $e_{n+1}^{(\phi)} = 0$ in (7.2.3), we can commence a back recursion, which using (7.2.7) and (7.2.8) eventually allows us to compute $w_{-j}^{(\phi)}$ for $j = 0, 1, \ldots, Q - 1$. Since $a_{-Q}^{(\phi)}, a_{-Q-1}^{(\phi)}, \ldots$ can be taken to be zero, we can now use (7.2.4) to compute recursively the required derivatives $a_t^{(\phi)}$. In a similar way, (7.2.5) and (7.2.6) can be used to calculate the derivatives $a_t^{(\theta)}$.

To illustrate, consider again the calculation of $x_t = -\partial[a_t]/\partial\theta$ for the first part of Series A, performed wholly numerically in Table 7.9. Table 7.10 shows the corresponding calculations using

$$e_t^{(\theta)} = w_t^{(\theta)} + \theta e_{t+1}^{(\theta)} + [e_{t+1}]$$

$$-x_t = a_t^{(\theta)} = w_t^{(\theta)} + \theta a_{t-1}^{(\theta)} + [a_{t-1}]$$

TABLE 7.10   Illustration of recursive calculation of derivatives for data from Series A

t	$[a_{t-1}]$	$\theta a_{t-1}^{(\theta)}$	$x_t = -a_t^{(\theta)}$	$e_t^{(\theta)}$	$\theta e_{t+1}^{(\theta)}$	$[e_{t+1}]$
0			−0.43	$(-w_0^{(\theta)} = -0.43)$	0.06	−0.49
1	0.25	0.22	−0.47	0.12	0.30	−0.18
2	−0.28	0.24	0.04	0.60	0.36	0.24
3	−0.44	−0.02	0.46	0.73	−0.15	0.88
4	−0.42	−0.23	0.65	−0.30	−0.05	−0.25
5	0.79	−0.33	−0.46	−0.10	0	−0.10
6	0.20	0.23	−0.43	0		
7	0.00	0.22	−0.22			
8	0.60	0.11	−0.71			
9	0.00	0.35	−0.35			
10	−0.10	−0.18	−0.08			

The values of $[a_t]$ and $[e_t]$, which have already been computed, are first entered, and in the illustration the calculation of $e_t^{(\theta)}$ is commenced by setting $e_6^{(\theta)} = 0$. It will be seen that the values for x_t agree very closely with those set out in Table 7.9, obtained by the purely numerical procedure.

7.2.4 A general least squares algorithm for the conditional model

An approximation we have sometimes used with long series is to set starting values for the a's, and hence for the x's, to their unconditional expectations of zero and then to proceed directly with the forward recursions. Thus, for the previous example we could employ the equations

$$a_t = \theta a_{t-1} + w_t \qquad x_t = -a_t^{(\theta)} = \theta x_{t-1} - a_{t-1}$$

The effect is to introduce a transient into both the a_t and the x_t series, the latter being slower to die out since the x's depend on the a's. As one illustration, the values for the a's and the x's, obtained when this method was used in the present example, were calculated. It was found that, although not agreeing initially, the a_t's were in two-decimal agreement from $t = 4$ onward, and the x_t's from $t = 8$ onward. In some instances, where there is an abundance of data (say 200 or more observations), the effect of the approximation can be nullified at the expense of some loss of information, by discarding, say, the first 20 calculated values.

If we adopt the approximation, an interesting general algorithm for this conditional model results. The general model may be written

$$a_t = \theta^{-1}(B)\phi(B)\tilde{w}_t$$

where $w_t = \nabla^d z_t$, $\tilde{w}_t = w_t - \mu$ and

$$\theta(B) = 1 - \theta_1 B - \cdots - \theta_i B^i - \cdots - \theta_q B^q$$

$$\phi(B) = 1 - \phi_1 B - \cdots - \phi_j B^j - \cdots - \phi_p B^p$$

If the first guesses for the parameters $\boldsymbol{\beta} = (\boldsymbol{\phi}, \boldsymbol{\theta})$ are $\boldsymbol{\beta}_0 = (\boldsymbol{\phi}_0, \boldsymbol{\theta}_0)$, then

$$a_{t,0} = \theta_0^{-1}(B)\phi_0(B)\tilde{w}_t$$

and

$$-\left.\frac{\partial a_t}{\partial \phi_j}\right|_{\boldsymbol{\beta}_0} = u_{j,t} = u_{t-j} \qquad -\left.\frac{\partial a_t}{\partial \theta_i}\right|_{\boldsymbol{\beta}_0} = x_{i,t} = x_{t-i}$$

where

$$u_t = \theta_0^{-1}(B)\tilde{w}_t \qquad = \phi_0^{-1}(B)a_{t,0} \tag{7.2.9}$$

$$x_t = -\theta_0^{-2}(B)\phi_0(B)\tilde{w}_t = -\theta_0^{-1}(B)a_{t,0} \tag{7.2.10}$$

The a's, x's, and u's may be calculated recursively, with starting values for a's, x's, and u's set equal to zero, as follows

$$a_{t,0} = \tilde{w}_t - \phi_{1,0}\tilde{w}_{t-1} - \cdots - \phi_{p,0}\tilde{w}_{t-p} + \theta_{1,0} a_{t-1,0} + \cdots + \theta_{q,0} a_{t-q,0} \tag{7.2.11}$$

$$u_t = \theta_{1,0} u_{t-1} + \cdots + \theta_{q,0} u_{t-q} + \tilde{w}_t \tag{7.2.12}$$

$$= \phi_{1,0} u_{t-1} + \cdots + \phi_{p,0} u_{t-p} + a_{t,0} \tag{7.2.13}$$

$$x_t = \theta_{1,0} x_{t-1} + \cdots + \theta_{q,0} x_{t-q} - a_{t,0} \tag{7.2.14}$$

Corresponding to (7.2.1), the approximate linear regression equation becomes

$$a_{t,0} = \sum_{j=1}^{p} (\phi_j - \phi_{j,0})u_{t-j} + \sum_{i=1}^{q} (\theta_i - \theta_{i,0})x_{t-i} + a_t \tag{7.2.15}$$

The adjustments are then the regression coefficients of $a_{t,0}$ on the u_{t-j} and on x_{t-i}. By adding the adjustments to the first guesses $(\boldsymbol{\phi}_0, \boldsymbol{\theta}_0)$, a set of "second guesses" are formed and these now take the place of $(\boldsymbol{\phi}_0, \boldsymbol{\theta}_0)$ in a second iteration, in which new values of $a_{t,0}$, x_t, and u_t are computed until convergence eventually occurs.

An alternative form for the algorithm. The approximate linear expansion (7.2.15) can be written in the form

$$a_{t,0} = \sum_{j=1}^{p} (\phi_j - \phi_{j,0})B^j\phi_0^{-1}(B)a_{t,0} - \sum_{i=1}^{q} (\theta_i - \theta_{i,0})B^i\theta_0^{-1}(B)a_{t,0} + a_t$$

$$= -\{\phi(B) - \phi_0(B)\}\phi_0^{-1}(B)a_{t,0} + \{\theta(B) - \theta_0(B)\}\theta_0^{-1}(B)a_{t,0} + a_t$$

that is,

$$a_{t,0} = -\phi(B)\{\phi_0^{-1}(B)a_{t,0}\} + \theta(B)\{\theta_0^{-1}(B)a_{t,0}\} + a_t \qquad (7.2.16)$$

which presents the algorithm in an interesting form.

Application to an IMA $(0, 2, 2)$ *process.* To illustrate the calculation with the conditional approximation, consider the estimation of least squares values $\hat{\theta}_1, \hat{\theta}_2$ for Series C, using the model of order $(0, 2, 2)$

$$w_t = (1 - \theta_1 B - \theta_2 B^2)a_t$$

with

$$w_t = \nabla^2 z_t$$

$$a_{t,0} = w_t + \theta_{1,0}a_{t-1,0} + \theta_{2,0}a_{t-2,0}$$

$$x_t = -a_{t,0} + \theta_{1,0}x_{t-1} + \theta_{2,0}x_{t-2}$$

The calculations for initial values $\theta_{1,0} = 0.1$ and $\theta_{2,0} = 0.1$ are set out in Table 7.11.

TABLE 7.11   Nonlinear estimation of θ_1 and θ_2 for an IMA $(0, 2, 2)$ process

t	z_t	∇z_t	$\nabla^2 z_t = w_t$	$a_{t,0}$	x_{t-1}	x_{t-2}
1	26.6			*0*	*0*	*0*
2	27.0	0.4		*0*	*0*	*0*
3	27.1	0.1	−0.3	−0.300	*0*	*0*
4	27.2	0.1	0.0	−0.030	0.300	*0*
5	27.3	0.1	0.0	−0.033	0.060	0.300
6	26.9	−0.4	−0.5	−0.533	0.069	0.060
7	26.4	−0.5	−0.1	−0.156	0.546	0.069
8	26.0	−0.4	0.1	0.039	0.218	0.546
9	25.8	−0.2	0.2	0.189	0.038	0.218

The first adjustments to $\theta_{1,0}$ and $\theta_{2,0}$ are then found by "regressing" $a_{t,0}$ on x_{t-1} and x_{t-2}, and the process is repeated until convergence occurs. The iteration proceeded as shown in Table 7.12, using starting values $\theta_{1,0} = 0.1$, $\theta_{2,0} = 0.1$.

7.2.5 Summary of models fitted to Series A–F

In Table 7.13 we summarize the models fitted by the iterative least squares procedure of Sections 7.2.1 and 7.2.2 to Series A–F. The models fitted were identified in Chapter 6 and have been summarized in Table 6.4. We see from Table 7.13 that, in the case of Series A, C, and D, two possible models were identified and subsequently fitted. For Series A and D the alternative models involve the use of a stationary autoregressive operator $(1 - \phi B)$, instead of a

TABLE 7.12   Convergence of iterates of θ_1 and θ_2

Iteration	θ_1	θ_2
0	0.1000	0.1000
1	0.1247	0.1055
2	0.1266	0.1126
3	0.1286	0.1141
4	0.1290	0.1149
5	0.1292	0.1151
6	0.1293	0.1152
7	0.1293	0.1153
8	0.1293	0.1153

nonstationary operator $(1 - B)$. Examination of Table 7.13 shows that in both cases the autoregressive model results in a slightly smaller residual variance, although, as has already been pointed out, the models are very similar. Even though a slightly better fit is possible with a stationary model,

TABLE 7.13   Summary of models fitted to Series A–F (the values $(\pm)$ under each estimate denote the standard errors of those estimates)

Series	No. of observations	Fitted models	Residual* variance
A	197	$z_t - 0.92z_{t-1} = 1.45 + a_t - 0.58a_{t-1}$ $(\pm 0.04)\qquad\qquad\quad(\pm 0.08)$	0.097
		$\nabla z_t = a_t - 0.70a_{t-1}$ (± 0.05)	0.101
B	369	$\nabla z_t = a_t + 0.09a_{t-1}$ (± 0.05)	52.2
C	226	$\nabla z_t - 0.82\nabla z_{t-1} = a_t$ (± 0.04)	0.018
		$\nabla^2 z_t = a_t - 0.13a_{t-1} - 0.12a_{t-2}$ $(\pm 0.07)\quad\;(\pm 0.07)$	0.019
D	310	$z_t - 0.87z_{t-1} = 1.17 + a_t$ (± 0.03)	0.090
		$\nabla z_t = a_t - 0.06a_{t-1}$ (± 0.06)	0.096
E	100	$z_t = 14.35 + 1.42z_{t-1} - 0.73z_{t-2} + a_t$ $(\pm 0.07)\quad(\pm 0.07)$	228
		$z_t = 11.31 + 1.57z_{t-1} - 1.02z_{t-2} + 0.21z_{t-3} + a_t$ $(\pm 0.10)\quad(\pm 0.15)\quad(\pm 0.10)$	218
F	70	$z_t = 58.87 - 0.34z_{t-1} + 0.19z_{t-2} + a_t$ $(\pm 0.12)\quad(\pm 0.12)$	113

* Obtained from $S(\hat{\boldsymbol{\phi}}, \hat{\boldsymbol{\theta}})/n$.

the IMA $(0, 1, 1)$ model might be preferable in these cases on the grounds that, unlike the stationary model, it does not assume that the series has a fixed mean. This is especially important in predicting future values of the series. For if the level does change, a model with $d > 0$ will continue to track it, whereas a model for which $d = 0$ will be tied to a mean level which may have become out of date.

The limits under the coefficients in Table 7.13 represent the standard errors of the estimates obtained from the covariance matrix $(X'_{\hat{\beta}}X_{\hat{\beta}})^{-1}\hat{\sigma}_a^2$, as described in Section 7.2.1. Note that the estimate $\hat{\phi}_3$ in the AR (3) process, fitted to the sunspot Series E, is 2.1 times its standard error, indicating that a marginally better fit is obtained by the third-order autoregressive process, as compared with the second-order autoregressive process. This is in agreement with a conclusion reached by Moran [54].

7.2.6 Large sample information matrices and covariance estimates*

Denote by $[U:X]$, the $n \times (p + q)$ matrix of u's and x's defined in (7.2.13) and (7.2.14), when the elements of β_0 are the *true* values of the parameters, for a sample size n sufficiently large for end effects to be ignored. Then the information matrix for (ϕ, θ) for the mixed ARMA model is

$$I(\phi, \theta) = E\begin{bmatrix} U'U & \vdots & U'X \\ \cdots & + & \cdots \\ X'U & \vdots & X'X \end{bmatrix}\sigma_a^{-2} \qquad (7.2.17)$$

that is

$$= n\sigma_a^{-2}\begin{bmatrix} \gamma_{uu}(0) & \gamma_{uu}(1) & \cdots \gamma_{uu}(p-1) & \gamma_{ux}(0) & \gamma_{ux}(-1) & \cdots \gamma_{ux}(1-q) \\ \gamma_{uu}(1) & \gamma_{uu}(0) & \cdots \gamma_{uu}(p-2) & \gamma_{ux}(1) & \gamma_{ux}(0) & \cdots \gamma_{ux}(2-q) \\ \vdots & \vdots & \vdots & \vdots & \vdots & \vdots \\ \gamma_{uu}(p-1)\gamma_{uu}(p-2)\cdots\gamma_{uu}(0) & \gamma_{ux}(p-1)\gamma_{ux}(p-2)\cdots\gamma_{ux}(p-q) \\ \gamma_{ux}(0) & \gamma_{ux}(1) & \cdots \gamma_{ux}(p-1) & \gamma_{xx}(0) & \gamma_{xx}(1) & \cdots \gamma_{xx}(q-1) \\ \gamma_{ux}(-1) & \gamma_{ux}(0) & \cdots \gamma_{ux}(p-2) & \gamma_{xx}(1) & \gamma_{xx}(0) & \cdots \gamma_{xx}(q-2) \\ \vdots & \vdots & \vdots & \vdots & \vdots & \vdots \\ \gamma_{ux}(1-q)\gamma_{ux}(2-q)\cdots\gamma_{ux}(p-q) & \gamma_{xx}(q-1)\gamma_{xx}(q-2)\cdots\gamma_{xx}(0) \end{bmatrix} \qquad (7.2.18)$$

where $\gamma_{uu}(k)$ and $\gamma_{xx}(k)$ are the autocovariances for the u's and the x's and $\gamma_{ux}(k)$ are the cross covariances defined by

$$\gamma_{ux}(k) = \gamma_{xu}(-k) = E[u_t x_{t+k}] = E[x_t u_{t-k}]$$

The large sample covariance matrix for the maximum likelihood estimates

*The line in the margin indicates that material may be omitted at first reading.

may be obtained using

$$V(\hat{\phi}, \hat{\theta}) \simeq I^{-1}(\phi, \theta)$$

Estimates of $I(\phi, \theta)$ and hence of $V(\hat{\phi}, \hat{\theta})$ may be obtained by evaluating the u's and x's with $\beta_0 = \hat{\beta}$ and omitting the expectation sign in (7.2.17), or substituting standard sample estimates of the autocorrelations and cross correlations in (7.2.18). Theoretical large sample results can be obtained by noticing that, with the elements of β_0 equal to the true values of the parameters, equations (7.2.13) and (7.2.14) imply that the derived series u_t and x_t follow *autoregressive* processes defined by

$$\phi(B)u_t = a_t \qquad \theta(B)x_t = -a_t$$

It follows that the autocovariances which appear in (7.2.18) are those for pure autoregressive processes, and the cross covariances are the negative of those between two such processes generated by the same a's.

We illustrate the use of this result with a few examples.

Covariance matrix of parameter estimates for AR (p) and MA (q) processes. Let $\Gamma_p(\phi)$ be the $p \times p$ autocovariance matrix of p successive observations from an AR (p) process with parameters $\phi' = (\phi_1, \phi_2, \ldots, \phi_p)$. Then using (7.2.18), the $p \times p$ covariance matrix of the estimates $\hat{\phi}$ is given by

$$V(\hat{\phi}) \simeq n^{-1}\sigma_a^2 \Gamma_p^{-1}(\phi) \qquad (7.2.19)$$

Let $\Gamma_q(\theta)$ be the $q \times q$ autocovariance matrix of q successive observations from an AR (q) process with parameters $\theta' = (\theta_1, \theta_2, \ldots, \theta_q)$. Then, using (7.2.18), the $q \times q$ covariance matrix of the estimates $\hat{\theta}$ is

$$V(\hat{\theta}) \simeq n^{-1}\sigma_a^2 \Gamma_q^{-1}(\theta) \qquad (7.2.20)$$

It is occasionally useful to parameterize an ARMA process in terms of the zeroes of $\phi(B)$ and $\theta(B)$. In this case a particularly simple form is obtained for the covariance matrix of the parameter estimates.

Covariances for the zeroes of an ARMA process. Consider the ARMA (p, q) process parameterized in terms of its zeroes (assumed to be real), so that

$$\prod_{i=1}^{p} (1 - G_i B)\tilde{w}_t = \prod_{j=1}^{q} (1 - H_j B)a_t$$

or

$$a_t = \prod_{i=1}^{p} (1 - G_i B) \prod_{j=1}^{q} (1 - H_j B)^{-1} \tilde{w}_t$$

The derivatives of the a's are then such that

$$u_{i,t} = -\frac{\partial a_t}{\partial G_i} = (1 - G_i B)^{-1} a_{t-1}$$

$$x_{j,t} = -\frac{\partial a_t}{\partial H_j} = -(1 - H_j B)^{-1} a_{t-1}$$

Hence, using (7.2.18), for large samples, the information matrix for the roots is such that $n^{-1} I(G, H) =$

$$\begin{bmatrix}
(1-G_1^2)^{-1} & (1-G_1G_2)^{-1} & \cdots & (1-G_1G_p)^{-1} & -(1-G_1H_1)^{-1} \ldots -(1-G_1H_q)^{-1} \\
\vdots & \vdots & & \vdots & \vdots & \vdots \\
(1-G_1G_p)^{-1} & (1-G_2G_p)^{-1} & \cdots & (1-G_p^2)^{-1} & -(1-G_pH_1)^{-1} \cdots -(1-G_pH_q)^{-1} \\
\hline
-(1-G_1H_1)^{-1} & -(1-G_2H_1)^{-1} \cdots & -(1-G_pH_1)^{-1} & (1-H_1^2)^{-1} & \cdots & (1-H_1H_q)^{-1} \\
\vdots & \vdots & & \vdots & \vdots & \vdots \\
-(1-G_1H_q)^{-1} & -(1-G_2H_q)^{-1} \cdots & -(1-G_pH_q)^{-1} & (1-H_1H_q)^{-1} & \cdots & (1-H_q^2)^{-1}
\end{bmatrix}$$

$$(7.2.21)$$

AR (2). In particular therefore, for a second-order autoregressive process

$$(1 - G_1 B)(1 - G_2 B)\tilde{w}_t = a_t$$

$$V(\hat{G}_1, \hat{G}_2) \simeq n^{-1} \begin{bmatrix} (1-G_1^2)^{-1} & (1-G_1G_2)^{-1} \\ (1-G_1G_2)^{-1} & (1-G_2^2)^{-1} \end{bmatrix}^{-1}$$

$$= \frac{1}{n} \frac{1 - G_1 G_2}{(G_1 - G_2)^2} \begin{bmatrix} (1-G_1^2)(1-G_1G_2) & -(1-G_1^2)(1-G_2^2) \\ -(1-G_1^2)(1-G_2^2) & (1-G_2^2)(1-G_1G_2) \end{bmatrix} \quad (7.2.22)$$

Exactly parallel results will be obtained for a second-order moving average process.

ARMA (1, 1). Similarly, for the ARMA (1, 1) process

$$(1 - \phi B)\tilde{w}_t = (1 - \theta B)a_t$$

on setting $\phi = G_1$ and $\theta = H_1$ in (7.2.21), we obtain

$$V(\hat{\phi}, \hat{\theta}) \simeq n^{-1} \begin{bmatrix} (1-\phi^2)^{-1} & -(1-\phi\theta)^{-1} \\ -(1-\phi\theta)^{-1} & (1-\theta^2)^{-1} \end{bmatrix}^{-1}$$

$$= \frac{1}{n} \frac{1 - \phi\theta}{(\phi - \theta)^2} \begin{bmatrix} (1-\phi^2)(1-\phi\theta) & (1-\phi^2)(1-\theta^2) \\ (1-\phi^2)(1-\theta^2) & (1-\theta^2)(1-\phi\theta) \end{bmatrix} \quad (7.2.23)$$

The results for these two processes illustrate a duality property discussed in Appendix A7.5.

7.3 SOME ESTIMATION RESULTS FOR SPECIFIC MODELS

In Appendices A7.4 and A7.5, some estimation results for special cases are derived. These, and results obtained earlier in this chapter, are summarized here for reference.

7.3.1 Autoregressive processes

It is possible to obtain estimates of the parameters of a purely autoregressive process by solving certain *linear* equations. We show in Appendix A7.5:
(1) How exact least squares estimates may be obtained by solving a linear system of equations (see also Section (7.4.3)).
(2) How, by slight modification of the coefficients in these equations, a close approximation to the exact maximum likelihood equations may be obtained.
(3) How estimates which are approximations to the least squares estimates and to the maximum likelihood estimates may be obtained using the estimated autocorrelations as coefficients in the linear "Yule–Walker" equations.

The estimates obtained in (1) are, of course, identical with those given by direct minimization of $S(\boldsymbol{\phi})$, as described in general terms in Section 7.2. The estimates (3) are the well known approximations due to Yule and Walker. They are useful as first estimates for use at the identification stage, but can differ appreciably from the estimates (1) and (2) in some situations. To illustrate, we now compare the estimates (3), obtained from the Yule–Walker equations, with the least squares estimates (1), which have been summarized in Table 7.13.

Yule–Walker estimates.   The Yule–Walker estimates (6.3.6) are

$$\hat{\boldsymbol{\phi}} = \mathbf{R}^{-1}\mathbf{r}$$

where

$$\mathbf{R} = \begin{bmatrix} 1 & r_1 & \cdots & r_{p-1} \\ r_1 & 1 & \cdots & r_{p-2} \\ \cdot & \cdot & & \cdot \\ \cdot & \cdot & & \cdot \\ r_{p-1} & r_{p-2} & \cdots & 1 \end{bmatrix} \qquad \mathbf{r} = \begin{bmatrix} r_1 \\ r_2 \\ \cdot \\ \cdot \\ r_p \end{bmatrix} \qquad (7.3.1)$$

In particular, the estimates for first- and second-order autoregressive processes are

$$\text{AR}(1): \ \hat{\phi}_1 = r_1$$

$$\text{AR}(2): \ \hat{\phi}_1 = \frac{r_1(1 - r_2)}{1 - r_1^2} \qquad \hat{\phi}_2 = \frac{r_2 - r_1^2}{1 - r_1^2} \qquad (7.3.2)$$

It is shown in Appendix A7.5 that an approximation to $S(\hat{\boldsymbol{\phi}})$ is provided by

$$S(\hat{\boldsymbol{\phi}}) = \sum_{t=1}^{n} \tilde{w}_t^2 \{1 - \mathbf{r}'\hat{\boldsymbol{\phi}}\} \tag{7.3.3}$$

whence

$$\hat{\sigma}_a^2 = S(\hat{\boldsymbol{\phi}})/n = c_0(1 - \mathbf{r}'\hat{\boldsymbol{\phi}}) \tag{7.3.4}$$

where c_0 is the variance of the w's. A parallel expression relates σ_a^2 and γ_0, the theoretical variance of the w's, see (3.2.8), namely

$$\sigma_a^2 = \gamma_0(1 - \boldsymbol{\rho}'\boldsymbol{\phi})$$

where the elements of $\boldsymbol{\rho}$ and of $\boldsymbol{\phi}$ are the theoretical values. Thus, from (7.2.19), the covariance matrix for the estimates $\hat{\boldsymbol{\phi}}$ is

$$\mathbf{V}(\hat{\boldsymbol{\phi}}) \simeq n^{-1}\sigma_a^2\boldsymbol{\Gamma}^{-1} = n^{-1}(1 - \boldsymbol{\rho}'\boldsymbol{\phi})\mathbf{P}^{-1} \tag{7.3.5}$$

where $\boldsymbol{\Gamma}$ and $\mathbf{P}$ are the autocovariance and autocorrelation matrices of p successive values of the AR (p) process, as defined in (2.1.7).

In particular, for first- and second-order autoregressive processes, we find

$$\text{AR (1): } V(\hat{\phi}) \simeq n^{-1}(1 - \phi^2) \tag{7.3.6}$$

$$\text{AR (2): } \mathbf{V}(\hat{\phi}_1, \hat{\phi}_2) \simeq n^{-1}\begin{bmatrix} 1 - \phi_2^2 & -\phi_1(1 + \phi_2) \\ -\phi_1(1 + \phi_2) & 1 - \phi_2^2 \end{bmatrix} \tag{7.3.7}$$

Estimates of the variances and covariances are obtained by substituting estimates for the parameters in (7.3.5). Thus

$$\hat{\mathbf{V}}(\hat{\boldsymbol{\phi}}) = n^{-1}(1 - \mathbf{r}'\hat{\boldsymbol{\phi}})\mathbf{R}^{-1} \tag{7.3.8}$$

Examples. For Series C, D, E and F, which have been identified as possibly autoregressive (or in the case of Series C, autoregressive in the first difference), the lag 1 and lag 2 autocorrelations are shown below.

Series	Tentative identification	Degree of differencing	Relevant estimated autocorrelations	n
C	(1, 1, 0)	1	$r_1 = 0.805$	225
D	(1, 0, 0)	0	$r_1 = 0.861$	310
E	(2, 0, 0)	0	$r_1 = 0.806, r_2 = 0.428$	100
F	(2, 0, 0)	0	$r_1 = -0.390, r_2 = 0.304$	70

Using (7.3.2) we obtain the Yule–Walker estimates shown in Table 7.14, together with their standard errors calculated from (7.3.6) and (7.3.7).

Using (7.3.7), the correlation coefficient between the estimates of the two parameters of a second-order autoregressive process is given by

$$\rho(\hat{\phi}_1, \hat{\phi}_2) \simeq \frac{\text{cov}(\hat{\phi}_1, \hat{\phi}_2)}{\sqrt{V(\hat{\phi}_1)V(\hat{\phi}_2)}} = \frac{-\phi_1}{1 - \phi_2} = -\rho_1 \simeq -r_1$$

TABLE 7.14.   Yule–Walker estimates for Series C–F

Series	Estimates with standard errors		Correlation coefficients between estimates
C	$\hat{\phi}_1 =$ 0.81	± 0.04	
D	$\hat{\phi}_1 =$ 0.86	± 0.03	
E	$\begin{cases} \hat{\phi}_1 = 1.32 \\ \hat{\phi}_2 = -0.63 \end{cases}$	$\begin{array}{c} \pm 0.08 \\ \pm 0.08 \end{array}$	-0.81
F	$\begin{cases} \hat{\phi}_1 = -0.32 \\ \hat{\phi}_2 = 0.18 \end{cases}$	$\begin{array}{c} \pm 0.12 \\ \pm 0.12 \end{array}$	0.39

We notice that in the case of Series E, there is a high negative correlation between the estimates. This indicates that the confidence region for $\hat{\phi}_1$ and $\hat{\phi}_2$ will be attenuated along a NW to SE diagonal. This implies that the estimates are rather unstable and explains the relatively larger discrepancy between the least squares estimates of Table 7.13 and the Yule–Walker estimates of Table 7.14 for this particular series.

7.3.2 Moving average processes

Maximum likelihood estimates $\hat{\boldsymbol{\theta}}$ for moving average processes may, in simple cases, be obtained graphically, as illustrated in Section 7.1.6, or more generally, by the iterative calculation described in Section 7.2.1.

From (7.2.20), it follows that for moderate and large samples, the covariance matrix for the estimates of the parameters of a qth order moving average process is of the same form as the corresponding matrix for an autoregressive process of the same order.

Thus, for first- and second-order moving average processes, we find, corresponding to (7.3.6) and (7.3.7)

$$\text{MA}(1): \ V(\hat{\theta}) \simeq n^{-1}(1 - \theta^2) \tag{7.3.9}$$

$$\text{MA}(2): \ \mathbf{V}(\hat{\theta}_1, \hat{\theta}_2) \simeq n^{-1} \begin{bmatrix} 1 - \theta_2^2 & -\theta_1(1 + \theta_2) \\ -\theta_1(1 + \theta_2) & 1 - \theta_2^2 \end{bmatrix} \tag{7.3.10}$$

7.3.3 Mixed processes

Maximum likelihood estimates $(\hat{\boldsymbol{\phi}}, \hat{\boldsymbol{\theta}})$ for mixed processes, as for moving average processes, may be obtained graphically in simple cases, and more generally, by iterative calculation. For moderate and large samples, the

covariance matrix may be obtained by evaluating and inverting the information matrix (7.2.18). In the important special case of the ARMA (1, 1) process

$$(1 - \phi B)\tilde{w}_t = (1 - \theta B)a_t$$

we obtain, as in (7.2.23),

$$\mathbf{V}(\hat{\phi}, \hat{\theta}) \simeq n^{-1} \frac{1 - \phi\theta}{(\phi - \theta)^2} \begin{bmatrix} (1 - \phi^2)(1 - \phi\theta) & (1 - \phi^2)(1 - \theta^2) \\ (1 - \phi^2)(1 - \theta^2) & (1 - \theta^2)(1 - \phi\theta) \end{bmatrix} \qquad (7.3.11)$$

It will be noted that when $\phi = \theta$, the variances of $\hat{\phi}$ and $\hat{\theta}$ are infinite.
This is to be expected, for in this case the factor $(1 - \phi B) = (1 - \theta B)$ cancels
on both sides of the model, which becomes

$$\tilde{w}_t = a_t$$

This is a particular case of *parameter redundancy* which we discuss more
fully in Section 7.3.5.

7.3.4 Separation of linear and nonlinear components in estimation

It is occasionally of interest to make an analysis in which the estimation of
the parameters of the mixed model is separated into its basic linear and
nonlinear parts. Consider the general mixed model, which we write as

$$a_t = \phi(B)\theta^{-1}(B)\tilde{w}_t$$

or

$$a_t = \phi(B)(\varepsilon_t|\boldsymbol{\theta}) \qquad (7.3.12)$$

where

$$(\varepsilon_t|\boldsymbol{\theta}) = \theta^{-1}(B)\tilde{w}_t$$

that is

$$\tilde{w}_t = \theta(B)(\varepsilon_t|\boldsymbol{\theta}) \qquad (7.3.13)$$

For any given set of θ's, the ε's may be calculated recursively from (7.3.13),
which may be written

$$\varepsilon_t = \tilde{w}_t + \theta_1\varepsilon_{t-1} + \theta_2\varepsilon_{t-2} + \cdots + \theta_q\varepsilon_{t-q}$$

The recursion may be started by setting unknown ε's equal to zero. Having
calculated the ε's, the conditional estimates $\hat{\boldsymbol{\phi}}_\theta$ may be readily obtained.
These are the estimated autoregressive parameters in the linear model
(7.3.12), which may be written

$$a_t = \varepsilon_t - \phi_1\varepsilon_{t-1} - \phi_2\varepsilon_{t-2} - \cdots - \phi_p\varepsilon_{t-p} \qquad (7.3.14)$$

As we have explained in Section 7.3.1, the least squares estimates of the autoregressive parameters may be found by direct solution of a simple set of linear equations and approximated by the use of the Yule–Walker equations. In simple cases, we can examine the behavior of $S(\hat{\phi}_{\theta}, \theta)$ as well as obtain the absolute minimum sum of squares, by computing it on a grid of θ values and plotting contours.

An example using series C. According to one tentative identification for Series C, given in Table 6.4, it was possibly generated by a model of order $(1, 1, 0)$

$$(1 - \phi B)w_t = a_t$$

with $w_t = \nabla z_t$ and $E[w_t] = 0$. It was decided to examine the estimation situation for this data in relation to the somewhat more elaborate model

$$(1 - \phi B)w_t = (1 - \theta_1 B - \theta_2 B^2)a_t$$

Following the argument given above, the process may be thought of as resulting from a combination of the nonlinear model

$$\varepsilon_t = w_t + \theta_1 \varepsilon_{t-1} + \theta_2 \varepsilon_{t-2}$$

and the linear model

$$a_t = \varepsilon_t - \phi \varepsilon_{t-1}$$

For each choice of the nonlinear parameters $\theta = (\theta_1, \theta_2)$ within the invertibility region, a set of ε's was calculated recursively. Using the Yule–Walker approximation, an estimate $\hat{\phi}_{\theta} = r_1(\varepsilon)$ could now be obtained together with

$$S(\hat{\phi}_{\theta}, \theta) \simeq \sum_{t=1}^{n} \varepsilon_t^2 \{1 - r_1^2(\varepsilon)\}$$

This sum of squares was plotted for a grid of values of θ_1 and θ_2 and its contours are shown in Figure 7.8. We see that a minimum close to $\theta_1 = \theta_2 = 0$

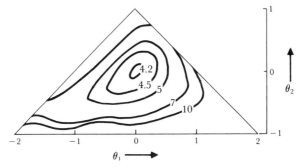

FIG. 7.8   Contours of $S(\hat{\phi}_{\theta}, \theta)$ plotted over the admissible parameter space for the θ's.

is indicated, at which point $r_1(\varepsilon) = 0.805$. Thus, within the whole class of models of order $(1, 1, 2)$, the simple $(1, 1, 0)$ model,

$$(1 - 0.8B)\nabla z_t = a_t$$

is confirmed as providing an adequate representation.

7.3.5 Parameter redundancy

The model

$$\phi(B)\tilde{w}_t = \theta(B)a_t$$

is identical with the model

$$(1 - \alpha B)\phi(B)\tilde{w}_t = (1 - \alpha B)\theta(B)a_t$$

in which both autoregressive and moving average operators are multiplied by the same factor $(1 - \alpha B)$. Serious difficulties in the estimation procedure will arise if a model is fitted which contains a redundant factor. Therefore, some care is needed in avoiding the situation where redundant or near-redundant factors occur. It is to be noted that the existence of redundancy is not necessarily obvious. For example, one can see the common factor in the ARMA $(2, 1)$ model

$$(1 - 1.3B + 0.4B^2)\tilde{w}_t = (1 - 0.5B)a_t$$

only after factoring the left hand side to obtain

$$(1 - 0.5B)(1 - 0.8B)\tilde{w}_t = (1 - 0.5B)a_t$$

that is,

$$(1 - 0.8B)\tilde{w}_t = a_t$$

In practice, it is not just exact cancellation which causes difficulties, but also near-cancellation. For example, suppose the true model was

$$(1 - 0.4B)(1 - 0.8B)\tilde{w}_t = (1 - 0.5B)a_t \qquad (7.3.15)$$

If an attempt was made to fit this model, extreme instability in the parameter estimates could be expected, because of near-cancellation of the factors $(1 - 0.4B)$ and $(1 - 0.5B)$, on the left and right sides. In this situation, combinations of parameter values yielding similar $[a]$'s and so similar likelihoods can be found, and a change of parameter value on the left can be nearly compensated by a suitable change on the right. The sum of squares contour surfaces in the three dimensional parameter space, will thus approach obliquely oriented cylinders and a line of "near least squares" solutions, rather than a clearly defined point minimum will be found.

From a slightly different viewpoint, we can write the model (7.3.15) in terms of an infinite autoregressive operator. Making the necessary expansion, we find

$$(1 - 0.700B - 0.030B^2 - 0.015B^3 - 0.008B^4 - \cdots)\tilde{w}_t = a_t$$

Thus, very nearly, the model is

$$(1 - 0.7B)\tilde{w}_t = a_t \tag{7.3.16}$$

The instability of the estimates, obtained by attempting to fit an ARMA (2, 1) model, would occur because we would be trying to fit three parameters in a situation that could almost be represented by one.

Preliminary identification as a means of avoiding parameter redundancy. A principal reason for going through the identification procedure prior to fitting the model, is to avoid difficulties arising from parameter redundancy, or, to be more positive, to achieve *parsimony* in parameterization. Thus, in the example just considered, for a time series of only a few hundred observations, the estimated autocorrelation function from data generated by (7.3.15) would be indistinguishable from that from data generated by the simple autoregressive process (7.3.16).

Thus we should be led to the fitting of the AR (1) process, which would normally be entirely adequate. Only with time series of several thousand observations would the need for a more elaborate model be detected and in these same circumstances, enough information would be available to obtain reasonably good estimates of the additional parameters.

Redundancy in the ARMA (1, 1) process. The simplest process where the possibility occurs of direct cancellation of factors is the ARMA (1, 1) process

$$(1 - \phi B)\tilde{w}_t = (1 - \theta B)a_t$$

In particular, if $\phi = \theta$, then whatever common value they have,

$$\tilde{w}_t = a_t$$

and the model implies that $\tilde{w}$ is generated by a white noise process. The data then cannot supply information about the common parameter, and, using (7.3.11), $\hat{\phi}$ and $\hat{\theta}$ have infinite variances. Furthermore, whatever the values of ϕ and θ, $S(\phi, \theta)$ must be constant on the line $\phi = \theta$, as it is for example in Figure 7.9 which shows a sum of squares plot for the data of Series A. For this data the least squares values $\hat{\phi} = 0.92$, $\hat{\theta} = 0.58$ correspond to a point which is not particularly close to the line $\phi = \theta$, and no difficulties occur in the estimation of these parameters.

In practice, if the identification technique we have recommended is adopted, these difficulties will be avoided. An ARMA (1, 1) process in which ϕ is very nearly equal to θ, will normally be identified as white noise,

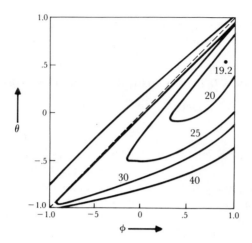

Fɪɢ. 7.9   Sums of squares plot for Series A

or if the difference is appreciable, as an AR (1) or MA (1) process with a single small coefficient.

In summary:

(1) We should avoid mixed processes containing near common factors and we should be alert to the difficulties that can result.
(2) We will automatically avoid such processes if we use identification and estimation procedures intelligently.

7.4 ESTIMATION USING BAYES' THEOREM

7.4.1 *Bayes' theorem*

In this section we again employ the symbol ξ to represent a general vector of parameters. Bayes' theorem tells us that if $p(\xi)$ is the probability distribution for ξ prior to the collection of the data, then $p(\xi|\mathbf{z})$, the distribution of ξ posterior to the data $\mathbf{z}$, is obtained by combining the prior distribution $p(\xi)$ and the likelihood $L(\xi|\mathbf{z})$ in the following way:

$$p(\xi|\mathbf{z}) = \frac{p(\xi)L(\xi|\mathbf{z})}{\int p(\xi)L(\xi|\mathbf{z})d\xi} \qquad (7.4.1)$$

The denominator merely ensures that $p(\xi|\mathbf{z})$ integrates to one. The important part of the expression is the numerator, from which we see that the posterior distribution is proportional to the prior distribution multiplied by the likelihood. Savage [64] has shown that prior and posterior probabilities can be interpreted as subjective probabilities. In particular, often before the data is available we have very little knowledge about ξ and we would be prepared to agree that, over the relevant region, it would have appeared a priori just as likely that it had one value as another. In this case $p(\xi)$ could be taken as *locally* uniform, and hence $p(\xi|\mathbf{z})$ would be proportional to the likelihood.

It should be noted that for this argument to hold, it is not necessary a priori for ξ to be uniform over its entire range (which for some parameters could be infinite). By requiring that it be *locally uniform* we mean that it be approximately uniform in the region in which the likelihood is appreciable and that it does not take an overwhelmingly large value outside that region.

Thus, if ξ was the weight of a chair, we could certainly say a priori that it weighed more than an ounce and less than a ton. It is also likely that when we obtained an observation z by weighing the chair on a weighing machine, which had an error standard deviation σ, then we could honestly say that we would have been equally happy with a priori values in the range $z \pm 3\sigma$. The exception would be if the weighing machine said that an apparently heavy chair weighed, say, 10 ounces. In this case the likelihood and the prior would be incompatible and we should not, of course, use Bayes' theorem to combine them, but would check the weighing machine and, if this turned out to be accurate, inspect the chair more closely.

There is, of course, some arbitrariness in this idea. Suppose we assumed the prior distribution of ξ to be locally uniform. Then this implies that the distribution of any linear function of ξ is also locally uniform. However the prior distribution of some nonlinear transformation, $\alpha = \alpha(\xi)$, (such as $\alpha = \log \xi$) could *not* be exactly locally uniform. This arbitrariness will usually have very little effect if we are able to obtain fairly precise estimates of ξ. We will then be considering ξ only over a small range, and over such a range the transformation from ξ to, say, $\log \xi$ would often be very nearly linear.

Jeffreys [65] has argued that it is best to choose the metric $\alpha(\xi)$ so that Fisher's measure of information $I_\alpha = -E[\partial^2 l/\partial \alpha^2]$ is independent of the value of α, and hence of ξ. This is equivalent to choosing $\alpha(\xi)$ so that the limiting variance of its maximum likelihood estimate is independent of ξ and is achieved by choosing the prior distribution of ξ to be proportional to $\sqrt{I_\xi}$.

Jeffreys justified this choice of prior on the basis of its invariance to the parameterization employed. Specifically, with this choice, the posterior distributions for $\alpha(\xi)$ and for ξ, where $\alpha(\xi)$ and ξ are connected by a $1:1$ transformation, are such that $p(\xi|\mathbf{z}) = p(\alpha|\mathbf{z})d\alpha/d\xi$. The same result may be obtained [66] by the following argument. If for large samples, the expected likelihood function for $\alpha(\xi)$ approaches a Normal curve, then the mean and variance of the curve summarize the information to be expected from the data. Suppose, now, a transformation $\alpha(\xi)$ can be found in which the approximating Normal curve has nearly constant variance whatever the true values of the parameter. Then, in this parameterization, the *only* information in prospect from the data is conveyed by the *location* of the expected likelihood function. To say we know essentially nothing a priori, relative to this prospective observational information, is to say that we regard different *locations* of α as equally likely a priori. Equivalently we say that α should be taken as locally uniform.

The generalization of Jeffreys' rule to deal with several parameters, is that the joint prior distribution of parameters ξ be taken proportional to

$$|I_\xi|^{1/2} = \left| -E\left\{\frac{\partial^2 l}{\partial \xi_i \partial \xi_j}\right\}\right|^{1/2} \tag{7.4.2}$$

It has been urged [67] that the likelihood itself is best considered and plotted in that metric α for which I_α is independent of α. If this is done, it will be noted that the likelihood function and the posterior density function with uniform prior are proportional.

7.4.2 Bayesian estimation of parameters

We now consider the estimation of the parameters in an ARIMA model from a Bayesian point of view.

It is shown in Appendix A7.4 that, the exact likelihood of a time series $\mathbf{z}$ of length $N = n + d$ from an ARIMA (p, d, q) process is of the form

$$L(\boldsymbol{\phi}, \boldsymbol{\theta}|\mathbf{z}) = \sigma_a^{-n} f(\boldsymbol{\phi}, \boldsymbol{\theta}) \exp\left\{-\frac{S(\boldsymbol{\phi}, \boldsymbol{\theta})}{2\sigma_a^2}\right\} \tag{7.4.3}$$

where

$$S(\boldsymbol{\phi}, \boldsymbol{\theta}) = \sum_{t=-\infty}^{n} [a_t|\mathbf{z}, \boldsymbol{\phi}, \boldsymbol{\theta}]^2 = \sum_{t=-\infty}^{n} [a_t|\mathbf{w}, \boldsymbol{\phi}, \boldsymbol{\theta}]^2 \tag{7.4.4}$$

If we have no prior information about σ_a, $\boldsymbol{\phi}$, or $\boldsymbol{\theta}$, and since information about σ_a would supply no information about $\boldsymbol{\phi}$ and $\boldsymbol{\theta}$, it is sensible, following Jeffreys, to employ a prior distribution for $\boldsymbol{\phi}$, $\boldsymbol{\theta}$, and σ_a of the form

$$p(\boldsymbol{\phi}, \boldsymbol{\theta}, \sigma_a) \propto |\mathbf{I}(\boldsymbol{\phi}, \boldsymbol{\theta})|^{1/2} \sigma_a^{-1}$$

It follows that the posterior distribution is

$$p(\boldsymbol{\phi}, \boldsymbol{\theta}, \sigma_a|\mathbf{z}) \propto \sigma_a^{-(n+1)} |\mathbf{I}(\boldsymbol{\phi}, \boldsymbol{\theta})|^{1/2} f(\boldsymbol{\phi}, \boldsymbol{\theta}) \exp\left\{-\frac{S(\boldsymbol{\phi}, \boldsymbol{\theta})}{2\sigma_a^2}\right\} \tag{7.4.5}$$

If we now integrate (7.4.5) from zero to infinity with respect to σ_a, we obtain the exact joint posterior distribution of the parameters $\boldsymbol{\phi}$ and $\boldsymbol{\theta}$ as

$$p(\boldsymbol{\phi}, \boldsymbol{\theta}|\mathbf{z}) \propto |\mathbf{I}(\boldsymbol{\phi}, \boldsymbol{\theta})|^{1/2} f(\boldsymbol{\phi}, \boldsymbol{\theta}) \{S(\boldsymbol{\phi}, \boldsymbol{\theta})\}^{-n/2} \tag{7.4.6}$$

7.4.3 Autoregressive processes

If z_t follows a process of order $(p, d, 0)$, then $w_t = \nabla^d z_t$ follows a pure auto-regressive process of order p. It is shown in Appendix A7.5 that, for such a process, the factors $|\mathbf{I}(\boldsymbol{\phi})|^{1/2}$ and $f(\boldsymbol{\phi})$, which in any case are dominated by the term in $S(\boldsymbol{\phi})$, essentially cancel. This yields the remarkably simple result that, given the assumptions, the parameters $\boldsymbol{\phi}$ of the AR (p) process in w have the posterior distribution

$$p(\boldsymbol{\phi}|\mathbf{z}) \propto \{S(\boldsymbol{\phi})\}^{-n/2} \tag{7.4.7}$$

On this argument then, the sum of squares contours which are approximate likelihood contours are, when nothing is known a priori, also contours of posterior probability.

The joint distribution of the autoregressive parameters. It is shown in Appendix A7.5 that for the pure AR process, the least squares estimates of the ϕ's which minimize $S(\boldsymbol{\phi}) = \boldsymbol{\phi}_u' \mathbf{D} \boldsymbol{\phi}_u$ are given by

$$\hat{\boldsymbol{\phi}} = \mathbf{D}_p^{-1} \mathbf{d} \tag{7.4.8}$$

where

$$\boldsymbol{\phi}_u = \begin{bmatrix} 1 \\ \hline \boldsymbol{\phi} \end{bmatrix} \quad \mathbf{d} = \begin{bmatrix} D_{12} \\ D_{13} \\ \vdots \\ D_{1,p+1} \end{bmatrix} \quad \mathbf{D}_p = \begin{bmatrix} D_{22} & D_{23} & \cdots & D_{2,p+1} \\ D_{23} & D_{33} & \cdots & D_{3,p+1} \\ \vdots & \vdots & \cdots & \vdots \\ D_{2,p+1} & D_{3,p+1} & \cdots & D_{p+1,p+1} \end{bmatrix}$$

$$\mathbf{D} = \begin{bmatrix} D_{11} & -\mathbf{d}' \\ \hline -\mathbf{d} & \mathbf{D}_p \end{bmatrix} \tag{7.4.9}$$

and

$$D_{ij} = D_{ji} = \tilde{w}_i \tilde{w}_j + \tilde{w}_{i+1} \tilde{w}_{j+1} + \cdots + \tilde{w}_{n+1-j} \tilde{w}_{n+1-i} \tag{7.4.10}$$

It follows that

$$S(\boldsymbol{\phi}) = v s_a^2 + (\boldsymbol{\phi} - \hat{\boldsymbol{\phi}})' \mathbf{D}_p (\boldsymbol{\phi} - \hat{\boldsymbol{\phi}}) \tag{7.4.11}$$

where

$$s_a^2 = S(\hat{\boldsymbol{\phi}})/v \qquad v = n - p \tag{7.4.12}$$

and

$$S(\hat{\boldsymbol{\phi}}) = \hat{\boldsymbol{\phi}}_u' \mathbf{D} \hat{\boldsymbol{\phi}}_u = D_{11} - \hat{\boldsymbol{\phi}}' \mathbf{D}_p \hat{\boldsymbol{\phi}} = D_{11} - \mathbf{d}' \mathbf{D}_p^{-1} \mathbf{d} \tag{7.4.13}$$

Thus we can write

$$p(\boldsymbol{\phi}|\mathbf{z}) \propto \left\{ 1 + \frac{(\boldsymbol{\phi} - \hat{\boldsymbol{\phi}})' \mathbf{D}_p (\boldsymbol{\phi} - \hat{\boldsymbol{\phi}})}{v s_a^2} \right\}^{-n/2} \tag{7.4.14}$$

Equivalently,

$$p(\boldsymbol{\phi}|\mathbf{z}) \propto \left\{ 1 + \frac{\frac{1}{2}\sum_i \sum_j S_{ij}(\phi_i - \hat{\phi}_i)(\phi_j - \hat{\phi}_j)}{vs_a^2} \right\}^{-n/2} \tag{7.4.15}$$

where

$$S_{ij} = \partial^2 S(\boldsymbol{\phi})/\partial\phi_i \partial\phi_j = 2D_{i+1,j+1}$$

It follows that, a posteriori, the parameters of an autoregressive process have a multiple t-distribution (A7.1.13), with $v = n - p$ degrees of freedom.

In particular, for the special case $p = 1$, $(\phi - \hat{\phi})/s_{\hat{\phi}}$ is distributed *exactly* in a Student t-distribution with $n - 1$ degrees of freedom where, using the general results given above, $\hat{\phi}$ and $s_{\hat{\phi}}$ are given by

$$\hat{\phi} = D_{12}/D_{22} \qquad s_{\hat{\phi}} = \left[\frac{1}{(n-1)} \frac{D_{11}}{D_{22}} \left\{ 1 - \frac{D_{12}^2}{D_{11}D_{22}} \right\} \right]^{1/2} \tag{7.4.16}$$

The quantity $s_{\hat{\phi}}$, for large samples, tends to $\{(1 - \phi^2)/n\}^{1/2}$ and in the sampling theory framework is identical with the large sample "standard error" for $\hat{\phi}$. However, it is to be remembered, when using this and similar expressions within the Bayesian framework, that it is the parameters (ϕ in this case) which are random variables. Quantities such as $\hat{\phi}$ and $s_{\hat{\phi}}$, which are functions of data that have already occurred, are regarded as fixed.

Normal approximation. For samples of size $n > 50$, in which we are usually interested, the Normal approximation to the t-distribution is adequate. Thus, very nearly, $\boldsymbol{\phi}$ has a joint p-variate Normal distribution $N\{\hat{\boldsymbol{\phi}}, \mathbf{D}_p^{-1}s_a^2\}$ having mean $\hat{\boldsymbol{\phi}}$ and variance-covariance matrix $\mathbf{D}_p^{-1}s_a^2$.

Bayesian regions of highest probability density. In summarizing what the posterior distribution has to tell us about the probability of various $\boldsymbol{\phi}$ values, it is useful to indicate a region of *highest probability density*, called for short [104] an HPD region. A Bayesian $1 - \varepsilon$ HPD region has the properties:

(1) Any parameter point inside the region has higher probability density than any point outside.
(2) The total posterior probability mass within the region is $1 - \varepsilon$.

Since $\boldsymbol{\phi}$ has a multiple t-distribution, it follows, using the result (A7.1.4) that,

$$\text{Prob}\,\{(\boldsymbol{\phi} - \hat{\boldsymbol{\phi}})'\mathbf{D}_p(\boldsymbol{\phi} - \hat{\boldsymbol{\phi}}) < ps_a^2 F_\varepsilon(p, v)\} = 1 - \varepsilon \tag{7.4.17}$$

defines the *exact* $1 - \varepsilon$ HPD region for $\boldsymbol{\phi}$. Now, for $v = n - p > 100$,

$$pF_\varepsilon(p, v) \simeq \chi_\varepsilon^2(p)$$

Also,

$$(\boldsymbol{\phi} - \hat{\boldsymbol{\phi}})'\mathbf{D}_p(\boldsymbol{\phi} - \hat{\boldsymbol{\phi}}) = \frac{1}{2}\sum_{i,j} S_{ij}(\phi_i - \hat{\phi}_i)(\phi_j - \hat{\phi}_j)$$

Thus, approximately, the HPD region defined in (7.4.17) is such that

$$\sum_{i,j} S_{ij}(\phi_i - \hat{\phi}_i)(\phi_j - \hat{\phi}_j) < 2s_a^2 \chi_\varepsilon^2(p) \qquad (7.4.18)$$

which, if we set $\hat{\sigma}_a^2 = s_a^2$, is identical with the confidence region defined by (7.1.26).

Although these approximate regions are identical, it will be remembered that their interpretation is different. From a sampling theory viewpoint, we say that if a confidence region is computed according to (7.1.26), then for each of a set of repeated samples, a proportion $1 - \varepsilon$ of these regions will include the true parameter point. From the Bayesian viewpoint, we are concerned only with the single sample z which has actually been observed. Assuming the relevance of the non-informative prior distribution which we have taken, the HPD region includes that proportion $1 - \varepsilon$ of the resulting probability distribution of ϕ, given z, which has the highest density. In other words, the probability that the value of ϕ, which gave rise to the data z, lies in the HPD region is $1 - \varepsilon$.

Using (7.4.11), (7.4.12) and (7.4.18), for large samples, the approximate Bayesian HPD region is bounded by a contour for which

$$S(\phi) = S(\hat{\phi}) \left\{ 1 + \frac{\chi_\varepsilon^2(p)}{n} \right\} \qquad (7.4.19)$$

which corresponds exactly with the confidence region defined by (7.1.28).

7.4.4 Moving average processes

If z follows an integrated moving average process of order $(0, d, q)$, then $w = \nabla^d z$ follows a pure moving average process of order q. It is shown in Appendix A7.5, that in this case the factors $|\mathbf{I}(\theta)|^{1/2}$ and $f(\theta)$ in (7.4.6), which in any case are dominated by $S(\theta)$, cancel for large samples. Thus, corresponding to (7.4.7), we find that the parameters θ, of the MA (q) process in w, have the posterior distribution

$$p(\theta|z) \propto \{S(\theta)\}^{-n/2} \qquad (7.4.20)$$

Again the sum of squares contours are, for moderate samples, essentially *exact* contours of posterior density. However, because $[a_t]$ is not a linear function of the θ's, $S(\theta)$ will not be exactly quadratic in θ, though for large samples it will often be nearly so within the relevant ranges. In that case, we have approximately

$$S(\theta) = vs_a^2 + \tfrac{1}{2} \sum_{i,j} S_{ij}(\theta_i - \hat{\theta}_i)(\theta_j - \hat{\theta}_j)$$

where $vs_a^2 = S(\hat{\theta})$ and $v = n - q$. It follows, after substituting for $S(\theta)$ in (7.4.20) and using the exponential approximation, that:

(1) For large samples, $\boldsymbol{\theta}$ is *approximately* distributed in a multi-Normal distribution $N\{\hat{\boldsymbol{\theta}}, 2\{S_{ij}\}^{-1}s_a^2\}$.

(2) An approximate HPD region is defined by (7.4.18) or (7.4.19), with q replacing p and $\boldsymbol{\theta}$ replacing $\boldsymbol{\phi}$.

Example: Posterior distribution of $\lambda = 1 - \theta$ for an IMA $(0, 1, 1)$ process.

To illustrate, Table 7.15 shows the calculation of the approximate posterior density distribution $p(\lambda|z)$ from the data of Series B. Column (2) of the table

TABLE 7.15   Calculation of approximate posterior density $p(\lambda|\mathbf{z})$ for Series B

λ	$10^{57} \times [S(\lambda)]^{-184.5}$	$p(\lambda|\mathbf{z})$
1.300	4	0.001
1.275	33	0.006
1.250	212	0.036
1.225	1007	0.171
1.200	3597	0.609
1.175	9956	1.685
1.150	21159	3.582
1.125	34762	5.884
1.100	44473	7.528
1.075	44988	7.615
1.050	35835	6.066
1.025	22563	3.819
1.000	11277	1.908
0.975	4540	0.769
0.950	1457	0.247
0.925	372	0.063
0.900	76	0.012
0.900	76	0.012
0.875	12	0.002
0.850	2	0.000
Total	236,325	40.000

shows the calculated ordinates $\{S(\lambda)\}^{-n/2} \times 10^{57}$, at intervals in λ, of $h = 0.025$. Their sum Σ is 236,325. By dividing the ordinates by $h\Sigma$, we produce a posterior density function, such that the area under the curve is, to a sufficient approximation, equal to 1.* The distribution is plotted in Figure 7.10. It is seen to be approximately Normal with its mode at $\hat{\lambda} = 1.09$ and having a standard deviation of about 0.05. A 95% Bayesian HPD interval covers essentially the same range, $0.98 < \lambda < 1.19$, as did the confidence interval.

* The approximate numerical integration to achieve a "standardized" curve of unit area, which could be carried out more exactly if desired, makes it possible to read off the actual probability densities. For almost all practical purposes this refinement is not needed. It is sufficient to compute and plot on some convenient scale $\{S(\lambda)\}^{-n/2}$, which is *proportional* to the actual density.

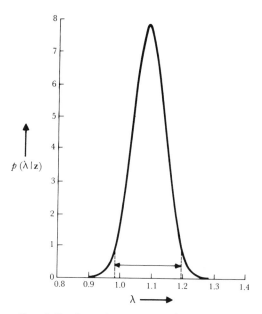

Fig. 7.10   Posterior density $p(\lambda|\mathbf{z})$ for Series B

7.4.5 Mixed processes

If z follows an ARIMA process of order (p, d, q), then $w = \nabla^d z$ follows an ARMA (p, q) process $\phi(B)\tilde{w}_t = \theta(B)a_t$. We show in Appendix A7.5 that, for such a process the factors $|\mathbf{I}(\boldsymbol{\phi}, \boldsymbol{\theta})|^{1/2}$ and $f(\boldsymbol{\phi}, \boldsymbol{\theta})$ in (7.4.5) do not exactly cancel. Instead we show that

$$|\mathbf{I}(\boldsymbol{\phi}, \boldsymbol{\theta})|^{1/2} f(\boldsymbol{\phi}, \boldsymbol{\theta}) = J(\boldsymbol{\phi}^*|\boldsymbol{\phi}, \boldsymbol{\theta}) \qquad (7.4.21)$$

In (7.4.21) the ϕ^*'s are the $p + q$ parameters obtained by multiplying the autoregressive and moving average operators

$$(1 - \phi_1^* B - \phi_2^* B^2 - \cdots - \phi_{p+q}^* B^{p+q}) = (1 - \phi_1 B - \cdots - \phi_p B^p)$$
$$\times (1 - \theta_1 B - \cdots - \theta_q B^q)$$

and J is the Jacobian of the transformation from $\boldsymbol{\phi}^*$ to $(\boldsymbol{\phi}, \boldsymbol{\theta})$, that is

$$p(\boldsymbol{\phi}, \boldsymbol{\theta}|\mathbf{z}) \propto J(\boldsymbol{\phi}^*|\boldsymbol{\phi}, \boldsymbol{\theta}) \{S(\boldsymbol{\phi}, \boldsymbol{\theta})\}^{-n/2} \qquad (7.4.22)$$

In particular, for the ARMA $(1, 1)$ process, $\phi_1^* = \phi + \theta$, $\phi_2^* = -\phi\theta$, $J = |\phi - \theta|$ and

$$p(\phi, \theta|\mathbf{z}) \propto |\phi - \theta|\{S(\phi, \theta)\}^{-n/2} \qquad (7.4.23)$$

In this case we see that the Jacobian will dominate in a region close to the line $\phi = \theta$, and will produce zero density on the line. This is sensible because the sum of squares $S(\phi, \theta)$ will take the finite value $\sum_{t=-1}^{n} \tilde{w}_t^2$ for any $\phi = \theta$, and corresponds to our entertaining the possibility that $\tilde{w}_t$ is white noise. However, in our derivation, we have not constrained the range of the parameters. The possibility that $\phi = \theta$ is thus associated with unlimited ranges for the (equal) parameters. The effect of limiting the parameter space by, for example, introducing the requirements for stationarity and invertibility $(-1 < \phi < 1, -1 < \theta < 1)$ would be to produce a small positive value for the density, but this refinement seems scarcely worthwhile.

The Bayesian analysis reinforces the point made earlier in Section 7.3.5, that estimation difficulties will be encountered with the mixed model, and in particular with iterative solutions, when there is near redundancy in the parameters. We have already seen that the use of preliminary identification will usually ensure that these situations are avoided.

APPENDIX A7.1   REVIEW OF NORMAL DISTRIBUTION THEORY

A7.1.1 *Partitioning of a positive definite quadratic form*

Consider the positive definite quadratic form $Q_p = \mathbf{x}'\boldsymbol{\Sigma}^{-1}\mathbf{x}$. Suppose the $p \times 1$ vector $\mathbf{x}$ is partitioned after the p_1th element, so that $\mathbf{x}' = (\mathbf{x}_1' \ \vdots \ \mathbf{x}_2') = (x_1, x_2, \ldots, x_{p_1}. \ \vdots \ x_{p_1+1}, \ldots, x_p)$, and suppose that the $p \times p$ matrix $\boldsymbol{\Sigma}$ is also partitioned after the p_1th row and column so that

$$\boldsymbol{\Sigma} = \begin{bmatrix} \boldsymbol{\Sigma}_{11} & \vdots & \boldsymbol{\Sigma}_{12} \\ \cdots & \vdots & \cdots \\ \boldsymbol{\Sigma}_{12}' & \vdots & \boldsymbol{\Sigma}_{22} \end{bmatrix}$$

Then, since

$$\mathbf{x}'\boldsymbol{\Sigma}^{-1}\mathbf{x} = (\mathbf{x}_1' \ \vdots \ \mathbf{x}_2')$$

$$\times \begin{bmatrix} \mathbf{I} & -\boldsymbol{\Sigma}_{11}^{-1}\boldsymbol{\Sigma}_{12} \\ \mathbf{0} & \mathbf{I} \end{bmatrix} \begin{bmatrix} \boldsymbol{\Sigma}_{11}^{-1} & \mathbf{0} \\ \mathbf{0} & (\boldsymbol{\Sigma}_{22} - \boldsymbol{\Sigma}_{12}'\boldsymbol{\Sigma}_{11}^{-1}\boldsymbol{\Sigma}_{12})^{-1} \end{bmatrix} \begin{bmatrix} \mathbf{I} & \mathbf{0} \\ -\boldsymbol{\Sigma}_{12}'\boldsymbol{\Sigma}_{11}^{-1} & \mathbf{I} \end{bmatrix} \begin{pmatrix} \mathbf{x}_1 \\ \mathbf{x}_2 \end{pmatrix}$$

$Q_p = \mathbf{x}'\boldsymbol{\Sigma}^{-1}\mathbf{x}$ can always be written as a sum of two quadratic forms Q_{p_1} and Q_{p_2}, containing p_1 and p_2 elements, respectively, where

$$\left. \begin{aligned} Q_p &= Q_{p_1} + Q_{p_2} \\ Q_{p_1} &= \mathbf{x}_1'\boldsymbol{\Sigma}_{11}^{-1}\mathbf{x}_1 \\ Q_{p_2} &= (\mathbf{x}_2 - \boldsymbol{\Sigma}_{12}'\boldsymbol{\Sigma}_{11}^{-1}\mathbf{x}_1)'(\boldsymbol{\Sigma}_{22} - \boldsymbol{\Sigma}_{12}'\boldsymbol{\Sigma}_{11}^{-1}\boldsymbol{\Sigma}_{12})^{-1}(\mathbf{x}_2 - \boldsymbol{\Sigma}_{12}'\boldsymbol{\Sigma}_{11}^{-1}\mathbf{x}_1) \end{aligned} \right\} \quad \text{(A7.1.1)}$$

We may also write for the determinant of $\boldsymbol{\Sigma}$

$$|\boldsymbol{\Sigma}| = |\boldsymbol{\Sigma}_{11}||\boldsymbol{\Sigma}_{22} - \boldsymbol{\Sigma}_{12}'\boldsymbol{\Sigma}_{11}^{-1}\boldsymbol{\Sigma}_{12}| \quad \text{(A7.1.2)}$$

A7.1.2 Two useful integrals

Let $\mathbf{z}'\mathbf{C}\mathbf{z}$ be a positive definite quadratic form in $\mathbf{z}$, which has q elements, so that $\mathbf{z}' = (z_1, z_2, \ldots, z_q)$, where $-\infty < z_i < \infty$, $i = 1, 2, \ldots, q$, and let a, b, and m be positive real numbers. Then it may be shown that

$$\int_R \left\{ a + \frac{\mathbf{z}'\mathbf{C}\mathbf{z}}{b} \right\}^{-(m+q)/2} d\mathbf{z} = \frac{(b\pi)^{q/2}\Gamma(m/2)}{a^{m/2}|\mathbf{C}|^{1/2}\Gamma\{(m+q)/2\}} \qquad \text{(A7.1.3)}$$

where the q-fold integral extends over the whole $\mathbf{z}$ space R and

$$\frac{\displaystyle\int_{\mathbf{z}'\mathbf{C}\mathbf{z} > qF_0} \left\{ 1 + \frac{\mathbf{z}'\mathbf{C}\mathbf{z}}{m} \right\}^{-(m+q)/2} d\mathbf{z}}{\displaystyle\int_R \left\{ 1 + \frac{\mathbf{z}'\mathbf{C}\mathbf{z}}{m} \right\}^{-(m+q)/2} d\mathbf{z}} = \int_{F_0}^{\infty} p(F|q, m)\, dF \qquad \text{(A7.1.4)}$$

where the function $p(F|q, m)$ is called the F distribution with q and m degrees of freedom and is defined by

$$p(F|q, m) = \frac{(q/m)^{q/2}\Gamma\{(m+q)/2\}}{\Gamma(q/2)\Gamma(m/2)} F^{(q-2)/2} \left(1 + \frac{q}{m}F \right)^{-(m+q)/2} \qquad \text{(A7.1.5)}$$

If m tends to infinity, then

$$\left\{ 1 + \frac{\mathbf{z}'\mathbf{C}\mathbf{z}}{m} \right\}^{-(m+q)/2} \qquad \text{tends to} \qquad e^{-(\mathbf{z}'\mathbf{C}\mathbf{z})/2}$$

and writing $qF = \chi^2$, we obtain from (A7.1.4) that

$$\frac{\displaystyle\int_{\mathbf{z}'\mathbf{C}\mathbf{z} > \chi_0^2} e^{-(\mathbf{z}'\mathbf{C}\mathbf{z})/2}\, d\mathbf{z}}{\displaystyle\int_R e^{-(\mathbf{z}'\mathbf{C}\mathbf{z})/2}\, d\mathbf{z}} = \int_{\chi_0^2}^{\infty} p(\chi^2|q)\, d\chi^2 \qquad \text{(A7.1.6)}$$

where the function $p(\chi^2|q)$ is called the χ^2 distribution with q degrees of freedom, and is defined by

$$p(\chi^2|q) = \frac{1}{2^{q/2}\Gamma(q/2)}(\chi^2)^{(q-2)/2}\, e^{-\chi^2/2} \qquad \text{(A7.1.7)}$$

Here and elsewhere $p(x)$ is used as a *general* notation to denote a probability density function for a random variable x.

A7.1.3 The Normal Distribution

The random variable x is said to be Normally distributed with mean μ and standard deviation σ, or to have a distribution $N(\mu, \sigma^2)$, if its probability density is

$$p(x) = (2\pi)^{-1/2}(\sigma^2)^{-1/2}\, e^{-(x-\mu)^2/2\sigma^2}. \qquad \text{(A7.1.8)}$$

Thus, the unit Normal deviate $u = (x - \mu)/\sigma$ has a distribution $N(0, 1)$. Table E at the end of the book shows ordinates $p(u = u_\varepsilon)$ and values u_ε such that $\Pr\{u > u_\varepsilon\} = \varepsilon$ for chosen values of ε.

Multi-Normal distribution. The vector of random variables $\mathbf{x}' = (x_1, x_2, \ldots, x_p)$ is said to have a joint p-variate Normal distribution $N\{\boldsymbol{\mu}, \boldsymbol{\Sigma}\}$ if its probability density function is

$$p(\mathbf{x}) = (2\pi)^{-p/2} |\boldsymbol{\Sigma}|^{-1/2} \, e^{-(\mathbf{x}-\boldsymbol{\mu})'\boldsymbol{\Sigma}^{-1}(\mathbf{x}-\boldsymbol{\mu})/2} \qquad (A7.1.9)$$

The probability density contours are ellipsoids defined by $(\mathbf{x} - \boldsymbol{\mu})'\boldsymbol{\Sigma}^{-1}(\mathbf{x} - \boldsymbol{\mu}) = \text{const.}$ For illustration, the elliptical contours for a bivariate Normal distribution are shown in Figure A7.1.

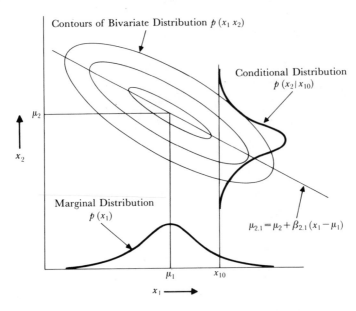

FIG. A7.1   Contours of a bivariate Normal distribution showing the marginal distribution $p(x_1)$ and the conditional distribution $p(x_2|x_{10})$ at $x_1 = x_{10}$

At the point $\mathbf{x} = \boldsymbol{\mu}$, the multivariate distribution has its maximum density

$$\max p(\mathbf{x}) = p(\boldsymbol{\mu}) = (2\pi)^{-p/2} |\boldsymbol{\Sigma}|^{-1/2}$$

The χ^2 distribution as the probability mass outside a density contour of the multivariate Normal. For the p-variate Normal distribution, (A7.1.9), the probability mass outside the density contour defined by

$$(\mathbf{x} - \boldsymbol{\mu})'\boldsymbol{\Sigma}^{-1}(\mathbf{x} - \boldsymbol{\mu}) = \chi_0^2$$

is given by the χ^2 integral with p degrees of freedom

$$\int_{\chi_0^2}^{\infty} p(\chi^2|p)\,d\chi^2$$

where the χ^2 density function is defined as in (A7.1.7). Table F at the end of the book shows values of $\chi_\varepsilon^2(p)$, such that $\Pr\{\chi^2 > \chi_\varepsilon^2(p)\} = \varepsilon$ for chosen values of ε.

Marginal and conditional distributions for the multivariate Normal distribution. Suppose the vector of $p = p_1 + p_2$ random variables is partitioned after the first p_1 elements, so that

$$\mathbf{x}' = (\mathbf{x}_1' : \mathbf{x}_2') = (x_1, x_2, \dots, x_{p_1} : x_{p_1+1}, \dots, x_{p_1+p_2})$$

and that the variance-covariance matrix is

$$\boldsymbol{\Sigma} = \begin{bmatrix} \boldsymbol{\Sigma}_{11} : \boldsymbol{\Sigma}_{12} \\ \dots : \dots \\ \boldsymbol{\Sigma}_{12}' : \boldsymbol{\Sigma}_{22} \end{bmatrix}$$

Then using (A7.1.1) and (A7.1.2), we can write the multivariate Normal distribution for the $p = p_1 + p_2$ variates as the *marginal* distribution of $\mathbf{x}_1$ multiplied by the *conditional* distribution of $\mathbf{x}_2$ given $\mathbf{x}_1$, that is,

$$p(\mathbf{x}) = p(\mathbf{x}_1, \mathbf{x}_2) = p(\mathbf{x}_1)p(\mathbf{x}_2|\mathbf{x}_1)$$
$$= (2\pi)^{-p_1/2}|\boldsymbol{\Sigma}_{11}|^{-1/2}\exp\{-(\mathbf{x}_1 - \boldsymbol{\mu}_1)'\boldsymbol{\Sigma}_{11}^{-1}(\mathbf{x}_1 - \boldsymbol{\mu}_1)/2\}$$
$$\times (2\pi)^{-p_2/2}|\boldsymbol{\Sigma}_{22.11}|^{-1/2}\exp\{-(\mathbf{x}_2 - \boldsymbol{\mu}_{2.1})'\boldsymbol{\Sigma}_{22.11}^{-1}(\mathbf{x}_2 - \boldsymbol{\mu}_{2.1})/2\}$$

$$\text{(A7.1.10)}$$

where

$$\boldsymbol{\Sigma}_{22.11} = \boldsymbol{\Sigma}_{22} - \boldsymbol{\Sigma}_{12}'\boldsymbol{\Sigma}_{11}^{-1}\boldsymbol{\Sigma}_{12} \qquad \text{(A7.1.11)}$$

and $\boldsymbol{\mu}_{2.1} = \boldsymbol{\mu}_2 + \boldsymbol{\beta}_{2.1}(\mathbf{x}_1 - \boldsymbol{\mu}_1) = E(\mathbf{x}_2|\mathbf{x}_1)$ define regression hyperplanes in $p_1 + p_2$ dimensional space, tracing the loci of the means of the p_2 elements of $\mathbf{x}_2$ as the p_1 elements of $\mathbf{x}_1$ vary. The $p_2 \times p_1$ matrix of regression coefficients is given by $\boldsymbol{\beta}_{2.1} = \boldsymbol{\Sigma}_{12}'\boldsymbol{\Sigma}_{11}^{-1}$.

Both marginal and conditional distributions for the multivariate Normal are therefore multivariate Normal distributions. It is seen that for the *multivariate Normal distribution*, the conditional distribution $p(\mathbf{x}_2|\mathbf{x}_1)$ is, *except for location*, identical whatever the value of $\mathbf{x}_1$.

Univariate marginals. In particular, the marginal density for a single element x_i ($i = 1, 2, \dots, p$) is $N(\mu_i, \sigma_i^2)$, a univariate Normal with mean equal to the ith element of $\boldsymbol{\mu}$ and variance equal to the ith diagonal element of $\boldsymbol{\Sigma}$.

Bivariate Normal. For illustration, the marginal and conditional distributions for a bivariate Normal are shown in Figure A7.1. In this case, the marginal distribution of x_1 is $N(\mu_1, \sigma_1^2)$, while the conditional distribution of x_2 given x_1 is

$$N\left\{\mu_2 + \rho\frac{\sigma_2}{\sigma_1}(x_1 - \mu_1), \sigma_2^2(1 - \rho^2)\right\}$$

where $\rho = (\sigma_1/\sigma_2)\beta_{2.1}$ is the correlation coefficient between x_1 and x_2.

A7.1.4 *Student's t distribution*

The random variable x is said to be distributed in a scaled t distribution $t(\mu, s^2, v)$, with mean μ and scale parameter s and with v degrees of freedom if

$$p(x) = (2\pi)^{-1/2}(s^2)^{-1/2}\left(\frac{v}{2}\right)^{-1/2}\Gamma\left(\frac{v+1}{2}\right)\Gamma^{-1}\left(\frac{v}{2}\right)\left\{1 + \frac{(x-\mu)^2}{vs^2}\right\}^{-(v+1)/2}$$

$$(A7.1.12)$$

Thus the standardized t deviate $t = (x - \mu)/s$ has distribution $t(0, 1, v)$. Table G at the end of the book shows values t_ε such that $\Pr(t > t_\varepsilon) = \varepsilon$ for chosen values of ε.

Approach to Normal distribution. For large v, the product

$$\left(\frac{v}{2}\right)^{-1/2}\Gamma\left(\frac{v+1}{2}\right)\Gamma^{-1}\left(\frac{v}{2}\right)$$

tends to unity, while the right-hand bracket in (A7.1.12) tends to $e^{-(1/2s^2)(x-\mu)^2}$. Thus, if for large v we write $s^2 = \sigma^2$, the t distribution tends to the Normal distribution (A7.1.8).

Multiple t distribution. Let $\boldsymbol{\mu}'$ be a $p \times 1$ vector $(\mu_1, \mu_2, \ldots, \mu_p)$ and S a $p \times p$ positive definite matrix. Then the vector random variable $\mathbf{x}$ is said [68], [69] to have a scaled t distribution $t(\boldsymbol{\mu}, \mathbf{S}, v)$ with means $\boldsymbol{\mu}$, scaling matrix $\mathbf{S}$, and v degrees of freedom if

$$p(\mathbf{x}) = (2\pi)^{-p/2}|\mathbf{S}|^{-1/2}\left(\frac{v}{2}\right)^{-p/2}\Gamma\left(\frac{v+p}{2}\right)\Gamma^{-1}\left(\frac{v}{2}\right)$$

$$\times \left\{1 + \frac{(\mathbf{x} - \boldsymbol{\mu})'\mathbf{S}^{-1}(\mathbf{x} - \boldsymbol{\mu})}{v}\right\}^{-(v+p)/2} \qquad (A7.1.13)$$

The probability contours of the multiple t distribution are ellipsoids defined by $(\mathbf{x} - \boldsymbol{\mu})'\mathbf{S}^{-1}(\mathbf{x} - \boldsymbol{\mu}) = \text{const.}$

Approach to the multi-Normal form. For large v, the product

$$\left(\frac{v}{2}\right)^{-p/2}\Gamma\left(\frac{v+p}{2}\right)\Gamma^{-1}\left(\frac{v}{2}\right)$$

tends to unity; also the right-hand bracket in (A7.1.13) tends to $e^{-(\mathbf{x}-\mathbf{\mu})'\mathbf{S}^{-1}(\mathbf{x}-\mathbf{\mu})/2}$. Thus, if for large v we write $\mathbf{S} = \mathbf{\Sigma}$, the multiple t tends to the multivariate Normal distribution (A7.1.9).

The F distribution as the probability mass outside a density contour of the multiple t distribution. Using (A7.1.4), the probability mass outside the density contour of the p-variate t distribution $t(\mathbf{\mu}, \mathbf{S}, v)$, defined by

$$(\mathbf{x} - \mathbf{\mu})'\mathbf{S}^{-1}(\mathbf{x} - \mathbf{\mu}) = pF_0$$

is given by the F integral with p and v degrees of freedom

$$\int_{F_0}^{\infty} p(F|p, v)\, dF$$

where the density function for F is defined by (A7.1.5). For large v, $\Pr\{F > F_\varepsilon(p, v)\} = \Pr\{\chi^2 > \chi_\varepsilon^2(p)\}$ with $pF = \chi^2$. Hence, as would be expected, the mass outside a density contour of the multiple t is, for large v, equal to the mass outside the corresponding density contour of the multivariate Normal distribution, which the multiple t distribution approaches.

Marginal t distribution. If the $p = p_1 + p_2$ dimensional vector $\mathbf{x}$, distributed as in equation (A7.1.13), is partitioned after the p_1th element so that $\mathbf{x}' = (\mathbf{x}_1'\!:\!\mathbf{x}_2')$, then with $\mathbf{S}$ similarly partitioned, we obtain

$$\mathbf{S} = \begin{bmatrix} \mathbf{S}_{11} & \vdots & \mathbf{S}_{12} \\ \cdots & \vdots & \cdots \\ \mathbf{S}_{12}' & \vdots & \mathbf{S}_{22} \end{bmatrix}$$

Writing

$$\mathbf{S}_{22.11} = \mathbf{S}_{22} - \mathbf{S}_{12}'\mathbf{S}_{11}^{-1}\mathbf{S}_{12}$$

$$\mathbf{\mu}_{2.1} = \mathbf{\mu}_2 + \mathbf{\beta}_{2.1}(\mathbf{x}_1 - \mathbf{\mu}_1), \quad \mathbf{\beta}_{2.1} = \mathbf{S}_{12}'\mathbf{S}_{11}^{-1}$$

then

$$p(\mathbf{x}_1, \mathbf{x}_2) = (2\pi)^{-(p_1 + p_2)/2} |\mathbf{S}_{11}|^{-1/2} |\mathbf{S}_{22.11}|^{-1/2} \left(\frac{v}{2}\right)^{-(p_1 + p_2)/2}$$

$$\times\ \Gamma\!\left(\frac{v + p_1 + p_2}{2}\right) \Gamma^{-1}\!\left(\frac{v}{2}\right)$$

$$\times \left\{ 1 + \frac{(\mathbf{x}_1 - \mathbf{\mu}_1)'\mathbf{S}_{11}^{-1}(\mathbf{x}_1 - \mathbf{\mu}_1)}{v} + \frac{(\mathbf{x}_2 - \mathbf{\mu}_{2.1})'\mathbf{S}_{22.11}^{-1}(\mathbf{x}_2 - \mathbf{\mu}_{2.1})}{v} \right\}^{-(v + p_1 + p_2)/2}$$

Now, using the preliminary result (A7.1.3), with

$$a = 1 + \frac{(\mathbf{x}_1 - \mathbf{\mu}_1)'\mathbf{S}_{11}^{-1}(\mathbf{x}_1 - \mathbf{\mu}_1)}{v}$$

$b = v$, $m = v + p_1$, $q = p_2$, $\mathbf{C} = \mathbf{S}_{22.11}^{-1}$, we obtain

$$p(\mathbf{x}_1) = (2\pi)^{-p_1/2}|\mathbf{S}_{11}|^{-1/2}\left(\frac{v}{2}\right)^{-p_1/2}\Gamma\left(\frac{v + p_1}{2}\right)\Gamma^{-1}\left(\frac{v}{2}\right)$$

$$\times \left\{1 + \frac{(\mathbf{x}_1 - \boldsymbol{\mu}_1)'\mathbf{S}_{11}^{-1}(\mathbf{x}_1 - \boldsymbol{\mu}_1)}{v}\right\}^{-(v + p_1)/2} \qquad (A7.1.14)$$

Thus, if a p dimensional vector $\mathbf{x}$ has the multiple t distribution of equation (A7.1.13), the marginal distribution of any p_1 variables $\mathbf{x}_1$ is the p_1-variate t distribution $t(\boldsymbol{\mu}_1, \mathbf{S}_{11}, v)$.

Univariate Marginals. In particular, the marginal distribution for a single element $x_i (i = 1, 2, \ldots, p)$ is $t(\mu_i, s_{ii}, v)$, a univariate t with mean equal to the ith element of $\boldsymbol{\mu}$ and scaling factor equal to the positive square root of the ith diagonal element of $\mathbf{S}$, and having v degrees of freedom.

Conditional t distributions. The conditional distribution $p(\mathbf{x}_2|\mathbf{x}_1)$ can be obtained from the ratio of the joint distribution $p(\mathbf{x}_1, \mathbf{x}_2)$ and the marginal distribution $p(\mathbf{x}_1)$; thus

$$p(\mathbf{x}_2|\mathbf{x}_{10}) = \frac{p(\mathbf{x}_{10}, \mathbf{x}_2)}{p(\mathbf{x}_{10})}$$

Making the division, we have

$$p(\mathbf{x}_2|\mathbf{x}_{10}) = \text{const}\left\{1 + \frac{(\mathbf{x}_2 - \boldsymbol{\mu}_{2.1})'\{c(\mathbf{x}_{10})\mathbf{S}_{22.11}^{-1}\}(\mathbf{x}_2 - \boldsymbol{\mu}_{2.1})}{v + p_1}\right\}^{-(v + p_1 + p_2)/2}$$

$$(A7.1.15)$$

where $c(\mathbf{x}_{10})$ (which is a fixed constant for *given* $\mathbf{x}_{10}$) is given by

$$c(\mathbf{x}_{10}) = \frac{v + p_1}{v + (\mathbf{x}_{10} - \boldsymbol{\mu}_1)'\mathbf{S}_{11}^{-1}(\mathbf{x}_{10} - \boldsymbol{\mu}_1)} \qquad (A7.1.16)$$

Thus the distribution of $\mathbf{x}_2$, for given $\mathbf{x}_1 = \mathbf{x}_{10}$, is the multiple t distribution

$$t\{\boldsymbol{\mu}_2 + \boldsymbol{\beta}_{2.1}(\mathbf{x}_{10} - \boldsymbol{\mu}_1), \quad c^{-1}(\mathbf{x}_{10})\mathbf{S}_{22.11}, \quad v + p_1\}$$

where $\boldsymbol{\beta}_{2.1} = \mathbf{S}_{12}'\mathbf{S}_{11}^{-1}$.

Bivariate t distribution. As before, some insight into the general multivariate situation can be gained by studying the bivariate distribution $t(\boldsymbol{\mu}, \mathbf{S}, v)$. A diagrammatic representation, parallel to Figure A7.1, is at first sight similar to the bivariate Normal. However, the marginal distributions are univariate t distributions having v degrees of freedom while the conditional distributions are t distributions with $v + 1$ degrees of freedom. Furthermore, the scale factor for the conditional distribution $p(x_2|x_{10})$, for example, would depend on x_{10}. This is to be contrasted with the conditional distribution for the Normal case which has the same variance irrespective of x_{10}.

APPENDIX A7.2   A REVIEW OF LINEAR LEAST SQUARES THEORY

A7.2.1 The normal equations

The model is assumed to be

$$w_i = \beta_1 x_{i1} + \beta_2 x_{i2} + \cdots + \beta_k x_{ik} + e_i \tag{A7.2.1}$$

where the w_i $(i = 1, 2, \ldots, n)$ are observations obtained from an experiment in which the independent variables $x_{i1}, x_{i2}, \ldots, x_{ik}$ take on known *fixed* values, the β_i are unknown parameters to be estimated from the data, and the e_i are uncorrelated errors having zero means and the same variance σ^2.

The relations (A7.2.1) may be assembled into a matrix relation

$$\begin{bmatrix} w_1 \\ w_2 \\ \cdot \\ \cdot \\ \cdot \\ w_n \end{bmatrix} = \begin{bmatrix} x_{11} & x_{12} & \cdots & x_{1k} \\ x_{21} & x_{22} & \cdots & x_{2k} \\ \cdot & \cdot & & \cdot \\ \cdot & \cdot & & \cdot \\ \cdot & \cdot & & \cdot \\ x_{n1} & x_{n2} & \cdots & x_{nk} \end{bmatrix} \begin{bmatrix} \beta_1 \\ \beta_2 \\ \cdot \\ \cdot \\ \cdot \\ \beta_k \end{bmatrix} + \begin{bmatrix} e_1 \\ e_2 \\ \cdot \\ \cdot \\ \cdot \\ e_n \end{bmatrix}$$

or

$$\mathbf{w} = \mathbf{X}\boldsymbol{\beta} + \mathbf{e} \tag{A7.2.2}$$

where $\mathbf{X}$ is assumed to be of full rank k. Gauss's Theorem of Least Squares may be stated [70] in the following form: the estimates $\hat{\boldsymbol{\beta}}' = (\hat{\beta}_1, \hat{\beta}_2, \ldots, \hat{\beta}_k)$ of the parameters $\boldsymbol{\beta}$, which are linear in the observations and which minimize the mean square error of any linear function $\lambda_1 \beta_1 + \lambda_2 \beta_2 + \cdots + \lambda_k \beta_k$ of the parameters, are obtained by minimizing the sum of squares.

$$S(\boldsymbol{\beta}) = \mathbf{e}'\mathbf{e} = (\mathbf{w} - \mathbf{X}\boldsymbol{\beta})'(\mathbf{w} - \mathbf{X}\boldsymbol{\beta}) \tag{A7.2.3}$$

To establish the minimum of $S(\boldsymbol{\beta})$, we note that the vector $\mathbf{w} - \mathbf{X}\boldsymbol{\beta}$ may be decomposed into two vectors $\mathbf{w} - \mathbf{X}\hat{\boldsymbol{\beta}}$ and $\mathbf{X}(\boldsymbol{\beta} - \hat{\boldsymbol{\beta}})$ according to

$$\mathbf{w} - \mathbf{X}\boldsymbol{\beta} = \mathbf{w} - \mathbf{X}\hat{\boldsymbol{\beta}} - \mathbf{X}(\boldsymbol{\beta} - \hat{\boldsymbol{\beta}}) \tag{A7.2.4}$$

Hence, provided we choose

$$(\mathbf{X}'\mathbf{X})\hat{\boldsymbol{\beta}} = \mathbf{X}'\mathbf{w} \tag{A7.2.5}$$

it follows that

$$S(\boldsymbol{\beta}) = S(\hat{\boldsymbol{\beta}}) + (\boldsymbol{\beta} - \hat{\boldsymbol{\beta}})'\mathbf{X}'\mathbf{X}(\boldsymbol{\beta} - \hat{\boldsymbol{\beta}}) \tag{A7.2.6}$$

and the vectors $\mathbf{w} - \mathbf{X}\hat{\boldsymbol{\beta}}$ and $\mathbf{X}(\boldsymbol{\beta} - \hat{\boldsymbol{\beta}})$ are orthogonal. Since the second term on the right of (A7.2.6) is a positive definite quadratic form, it follows that the minimum is attained when $\boldsymbol{\beta} = \hat{\boldsymbol{\beta}}$, where $\hat{\boldsymbol{\beta}}$ is given by the *normal equations* (A7.2.5).

A7.2.2 *Estimation of residual variance*

Using (A7.2.3) and (A7.2.5), the sum of squares at the minimum is

$$S(\hat{\boldsymbol{\beta}}) = \mathbf{w'w} - \hat{\boldsymbol{\beta}}'\mathbf{X'X}\hat{\boldsymbol{\beta}} \tag{A7.2.7}$$

Furthermore, if we define

$$s^2 = \frac{S(\hat{\boldsymbol{\beta}})}{n - k} \tag{A7.2.8}$$

it may be shown [71] that $E[s^2] = \sigma^2$, and hence s^2 provides an unbiased estimate of σ^2.

A7.2.3 *Covariance matrix of estimates*

This is defined by

$$\begin{aligned} \mathbf{V}(\hat{\boldsymbol{\beta}}) &= \text{cov}\,[\hat{\boldsymbol{\beta}}, \hat{\boldsymbol{\beta}}'] \\ &= \text{cov}\,[(\mathbf{X'X})^{-1}\mathbf{X'w}, \mathbf{w'X(X'X)}^{-1}] \\ &= (\mathbf{X'X})^{-1}\sigma^2 \end{aligned} \tag{A7.2.9}$$

since $\text{cov}\,(\mathbf{w}, \mathbf{w'}) = \mathbf{I}\sigma^2$.

A7.2.4 *Confidence regions*

Assuming normality [71], the quadratic forms $S(\hat{\boldsymbol{\beta}})$ and $(\boldsymbol{\beta} - \hat{\boldsymbol{\beta}})'\mathbf{X'X}(\boldsymbol{\beta} - \hat{\boldsymbol{\beta}})$ in (A7.2.6) are independently distributed as chi-squared random variables with $n - k$ and k degrees of freedom, respectively. Hence

$$\frac{(\boldsymbol{\beta} - \hat{\boldsymbol{\beta}})'\mathbf{X'X}(\boldsymbol{\beta} - \hat{\boldsymbol{\beta}})}{S(\hat{\boldsymbol{\beta}})} \frac{n - k}{k}$$

is distributed as $F(k, n - k)$. Using (A7.2.8), it follows that

$$(\boldsymbol{\beta} - \hat{\boldsymbol{\beta}})'(\mathbf{X'X})(\boldsymbol{\beta} - \hat{\boldsymbol{\beta}}) \leqslant ks^2 F_\varepsilon(k, n - k) \tag{A7.2.10}$$

defines a $1 - \varepsilon$ confidence region for $\boldsymbol{\beta}$.

A7.2.5 *Correlated errors*

Suppose that the errors have a *known* covariance matrix $\mathbf{V}$, where $\mathbf{V}^{-1} = \mathbf{PP'}/\sigma^2$. Then (A7.2.2) may be rewritten

$$\mathbf{P'w} = \mathbf{P'X}\boldsymbol{\beta} + \mathbf{P'e}$$

or

$$\mathbf{w^*} = \mathbf{X^*}\boldsymbol{\beta} + \mathbf{e^*} \tag{A7.2.11}$$

The covariance matrix of $e^* = P'e$ is

$$\text{cov}\,[P'e, e'P] = P'VP = I\sigma^2$$

Hence, we may apply ordinary least squares theory with $V = I\sigma^2$ to the transformed model (A7.2.11), in which w is replaced by $w^* = P'w$ and X by $X^* = P'X$.

APPENDIX A7.3 EXAMPLES OF THE EFFECT OF PARAMETER ESTIMATION ERRORS ON PROBABILITY LIMITS FOR FORECASTS

The variances and probability limits for the forecasts given in Section 5.2.4 are based on the assumption that the parameters (ϕ, θ) in the ARIMA model are known exactly. In practice, it is necessary to replace these by their estimates $(\hat{\phi}, \hat{\theta})$. To gain some insight into the effect of estimation errors on the variance of the forecast errors, we consider the special cases of the nonstationary IMA $(0, 1, 1)$ and the stationary first-order autoregressive processes. It is shown that, for these processes and for parameter estimates based on series of moderate length, the effect of such estimation errors is small.

IMA $(0, 1, 1)$ processes. Writing the model $\nabla z_t = a_t - \theta a_{t-1}$ for $t + l$, $t + l - 1, \ldots, t + 1$, and summing, we obtain

$$z_{t+l} - z_t = a_{t+l} + (1 - \theta)(a_{t+l-1} + \cdots + a_{t+1}) - \theta a_t$$

Denote by $\hat{z}_t(l|\theta)$ the lead l forecast when the parameter θ is known exactly. On taking conditional expectations at time t, for $l = 1, 2, \ldots$, we obtain

$$\hat{z}_t(1|\theta) = z_t - \theta a_t$$

$$\hat{z}_t(l|\theta) = \hat{z}_t(1|\theta) \qquad l \geqslant 2$$

Hence the lead l forecast error is

$$e_t(l|\theta) = z_{t+l} - \hat{z}_t(l|\theta)$$

$$= a_{t+l} + (1 - \theta)(a_{t+l-1} + \cdots + a_{t+1})$$

and the variance of the forecast error at lead time l is

$$V(l) = E_t[e_t^2(l|\theta)] = \sigma_a^2\{1 + (l - 1)\lambda^2\} \tag{A7.3.1}$$

where $\lambda = 1 - \theta$.

However, if θ is replaced by its estimate $\hat{\theta}$, obtained from a time series consisting of n values of $w_t = \nabla z_t$, then

$$\hat{z}_t(1|\hat{\theta}) = z_t - \hat{\theta} \hat{a}_t$$

$$\hat{z}_t(l|\hat{\theta}) = \hat{z}_t(1|\hat{\theta}) \qquad l \geqslant 2$$

where $\hat{a}_t = z_t - \hat{z}_{t-1}(1|\hat{\theta})$. Hence the lead l forecast error using $\hat{\theta}$ is

$$e_t(l|\hat{\theta}) = z_{t+l} - \hat{z}_t(l|\hat{\theta})$$
$$= z_{t+l} - z_t + \hat{\theta}\hat{a}_t$$
$$= e_t(l|\theta) - (\theta a_t - \hat{\theta}\hat{a}_t) \qquad (A7.3.2)$$

Since $\nabla z_t = (1 - \theta B)a_t = (1 - \hat{\theta}B)\hat{a}_t$, it follows that

$$\hat{a}_t = \left(\frac{1 - \theta B}{1 - \hat{\theta}B}\right)a_t$$

and on eliminating $\hat{a}_t$ from (A7.3.2), we obtain

$$e_t(l|\hat{\theta}) = e_t(l|\theta) - \frac{(\theta - \hat{\theta})}{1 - \hat{\theta}B}a_t$$

Now

$$\frac{\theta - \hat{\theta}}{1 - \hat{\theta}B}a_t = \frac{\theta - \hat{\theta}}{1 - \theta B}\left\{1 + \frac{(\theta - \hat{\theta})B}{1 - \theta B}\right\}^{-1}a_t$$

$$\simeq \frac{\theta - \hat{\theta}}{1 - \theta B}\left\{1 - \frac{(\theta - \hat{\theta})B}{1 - \theta B}\right\}a_t$$

$$= (\theta - \hat{\theta})(a_t + \theta a_{t-1} + \theta^2 a_{t-2} + \cdots)$$
$$- (\theta - \hat{\theta})^2(a_{t-1} + 2\theta a_{t-2} + 3\theta^2 a_{t-3} + \cdots)$$

$$(A7.3.3)$$

On the assumption that the forecast and the estimate $\hat{\theta}$ are based on essentially nonoverlapping data, $\hat{\theta}$ and $a_t, a_{t-1}, \ldots$ are independent. Also, $\hat{\theta}$ will be approximately Normally distributed about θ with variance $(1 - \theta^2)/n$, for moderate sized samples. On these assumptions the variance of the expression in (A7.3.3) may be shown to be

$$\frac{\sigma_a^2}{n}\left\{1 + \frac{3}{n}\frac{1 + \theta^2}{1 - \theta^2}\right\}$$

Thus, provided $|\theta|$ is not close to unity,

$$\text{var}\,[e_t(l|\hat{\theta})] \simeq \sigma_a^2\{1 + (l - 1)\lambda^2\} + \frac{\sigma_a^2}{n} \qquad (A7.3.4)$$

Clearly, the proportional change in the variance will be greatest for $l = 1$, when the exact forecast error reduces to σ_a^2. In this case, for parameter estimates based on a series of moderate length, the probability limits will be increased by a factor $(n + 1)/n$.

First-order autoregressive processes. Writing the model $\tilde{z}_t = \phi \tilde{z}_{t-1} + a_t$ at time $t + l$ and taking conditional expectations at time t, the lead l forecast, given the true value of the parameter, is

$$\hat{\tilde{z}}_t(l|\phi) = \phi \hat{\tilde{z}}_t(l - 1|\phi) = \phi^l \tilde{z}_t$$

Similarly,

$$\hat{\tilde{z}}_t(l|\hat{\phi}) = \hat{\phi} \hat{\tilde{z}}_t(l - 1|\hat{\phi}) = \hat{\phi}^l \tilde{z}_t$$

and hence

$$e_t(l|\hat{\phi}) = e_t(l|\phi) + (\phi^l - \hat{\phi}^l)\tilde{z}_t$$

It follows that

$$E_t[e_t^2(l|\hat{\phi})] = E_t[e_t^2(l|\phi)] + \tilde{z}_t^2 \, E_t[(\phi^l - \hat{\phi}^l)^2]$$

so that *on the average*

$$\text{var}\,[e_t(l|\hat{\phi})] \simeq \sigma_a^2 \frac{(1 - \phi^{2l})}{(1 - \phi^2)} + \sigma_a^2 \frac{E[(\phi^l - \hat{\phi}^l)^2]}{1 - \phi^2} \qquad (A7.3.5)$$

using (5.4.16). When $l = 1$,

$$\text{var}\,[e_t(1|\hat{\phi})] \simeq \sigma_a^2 + \left(\frac{\sigma_a^2}{1 - \phi^2}\right)\left(\frac{1 - \phi^2}{n}\right)$$

$$= \sigma_a^2\left(1 + \frac{1}{n}\right) \qquad (A7.3.6)$$

For $l > 1$, we have

$$\phi^l - \hat{\phi}^l = \phi^l - \{\phi - (\phi - \hat{\phi})\}^l = \phi^l - \phi^l\left\{1 - \frac{\phi - \hat{\phi}}{\phi}\right\}^l \simeq l\phi^{l-1}(\phi - \hat{\phi})$$

Thus, on the average,

$$\text{var}\,[e_t(l|\hat{\phi})] \simeq \text{var}\,[e_t(l|\phi)] + \frac{l^2\phi^{2(l-1)}}{n}\sigma_a^2$$

and the discrepancy is again of order n^{-1}.

APPENDIX A7.4 THE EXACT LIKELIHOOD FUNCTION FOR A MOVING AVERAGE PROCESS

To obtain the required likelihood function, we have to derive the probability density function for a series $\mathbf{w}_n' = (w_1, w_2, \ldots, w_n)$, assumed to be generated by an invertible moving average model of order q

$$\tilde{w}_t = a_t - \theta_1 a_{t-1} - \theta_2 a_{t-2} - \cdots - \theta_q a_{t-q} \qquad (A7.4.1)$$

where $\tilde{w}_t = w_t - \mu$, $\mu = E[w_t]$. Under the assumption that the a's and hence the $\tilde{w}$'s are Normally distributed, the joint density may be written

$$p(\mathbf{w}_n|\mathbf{\theta}, \sigma_a, \mu) = (2\pi\sigma_a^2)^{-n/2}|\mathbf{M}_n^{(0,q)}|^{1/2} \exp\left\{\frac{-\tilde{\mathbf{w}}_n'\mathbf{M}_n^{(0,q)}\tilde{\mathbf{w}}_n}{2\sigma_a^2}\right\} \qquad \text{(A7.4.2)}$$

with $(\mathbf{M}_n^{(p,q)-1}\sigma_a^2$ the $n \times n$ covariance matrix of the w's for an ARMA (p,q) process. When $w_t = \nabla^d z_t, d > 0$, it can sometimes be assumed that $\mu = 0$, in which case $\tilde{w}_t = w_t$. When this is not assumed and $E[w_t]$ is unknown, then μ may simply be regarded as an additional parameter to be estimated and included in the vector $\mathbf{\theta}$. We now consider a convenient way of evaluating $\tilde{\mathbf{w}}_n\mathbf{M}_n^{(0,q)}\tilde{\mathbf{w}}_n$, and for simplicity, we suppose that $\mu = 0$, so that $w_t = \tilde{w}_t$.

Using the model (A7.4.1), we can write down the $n + q$ equations

$$a_{1-q} = a_{1-q}$$
$$\vdots$$
$$a_{-1} = a_{-1}$$
$$a_0 = a_0$$
$$a_1 = w_1 + \theta_1 a_0 \quad + \theta_2 a_{-1} + \cdots + \theta_q a_{1-q}$$
$$a_2 = w_2 + \theta_1 a_1 \quad + \theta_2 a_0 \quad + \cdots + \theta_q a_{2-q}$$
$$\vdots$$
$$a_n = w_n + \theta_1 a_{n-1} + \theta_2 a_{n-2} + \cdots + \theta_q a_{n-q}$$

After substituting the expression for a_1 in that for a_2, the expressions for a_1 and a_2 in that for a_3, etc., we can write the $n + q$ dimensional vector $\mathbf{a}' = (a_{1-q}, a_{2-q}, \ldots, a_n)$ in terms of the n dimensional vector $\mathbf{w}_n = (w_1, w_2, \ldots, w_n)$ and the q dimensional vector of preliminary values $\mathbf{a}_*' = (a_{1-q}, a_{2-q}, \ldots, a_0)$. Thus

$$\mathbf{a} = \mathbf{L}\mathbf{w}_n + \mathbf{X}\mathbf{a}_*$$

where $\mathbf{L}$ is an $(n + q) \times n$ matrix and $\mathbf{X}$ is an $(n + q) \times q$ matrix whose elements are functions of the elements of $\mathbf{\theta}$, temporarily regarded as fixed.

Now the joint distribution of the $n + q$ values which are the elements of $\mathbf{a}$ is

$$p(\mathbf{a}|\sigma_a) = (2\pi\sigma_a^2)^{-(n+q)/2} \exp\left\{-\frac{\mathbf{a}'\mathbf{a}}{2\sigma_a^2}\right\}$$

Noting that the transformation has unit Jacobian, the joint distribution of $\mathbf{w}_n$ and $\mathbf{a}_*$ is

$$p(\mathbf{w}_n, \mathbf{a}_*|\mathbf{\theta}, \sigma_a) = (2\pi\sigma_a^2)^{-(n+q)/2} \exp\left\{-\frac{S(\mathbf{\theta}, \mathbf{a}_*)}{2\sigma_a^2}\right\}$$

where

$$S(\theta, \mathbf{a}_*) = (\mathbf{L}\mathbf{w}_n + \mathbf{X}\mathbf{a}_*)'(\mathbf{L}\mathbf{w}_n + \mathbf{X}\mathbf{a}_*) \tag{A7.4.3}$$

Now, let $\hat{\mathbf{a}}_*$ be the vector of values which minimize $S(\theta, \mathbf{a}_*)$. Then, using the result (A7.2.6)

$$S(\theta, \mathbf{a}_*) = S(\theta) + (\mathbf{a}_* - \hat{\mathbf{a}}_*)'\mathbf{X}'\mathbf{X}(\mathbf{a}_* - \hat{\mathbf{a}}_*)$$

where

$$S(\theta) = (\mathbf{L}\mathbf{w}_n + \mathbf{X}\hat{\mathbf{a}}_*)'(\mathbf{L}\mathbf{w}_n + \mathbf{X}\hat{\mathbf{a}}_*) \tag{A7.4.4}$$

is a function of the observations $\mathbf{w}_n$, but not of the preliminary values $\mathbf{a}_*$. Thus

$$p(\mathbf{w}_n, \mathbf{a}_*|\theta, \sigma_a)$$

$$= (2\pi\sigma_a^2)^{-(n+q)/2} \exp\left[-\frac{1}{2\sigma_a^2}\left\{ S(\theta) + (\mathbf{a}_* - \hat{\mathbf{a}}_*)'\mathbf{X}'\mathbf{X}(\mathbf{a}_* - \hat{\mathbf{a}}_*)\right\}\right]$$

However, since

$$p(\mathbf{w}_n, \mathbf{a}_*|\theta, \sigma_a) = p(\mathbf{w}_n|\theta, \sigma_a)p(\mathbf{a}_*|\mathbf{w}_n, \theta, \sigma_a)$$

it follows that

$$p(\mathbf{a}_*|\mathbf{w}_n, \theta, \sigma_a) = (2\pi\sigma_a^2)^{-q/2}|\mathbf{X}'\mathbf{X}|^{1/2} \exp\left\{ -\frac{(\mathbf{a}_* - \hat{\mathbf{a}}_*)'\mathbf{X}'\mathbf{X}(\mathbf{a}_* - \hat{\mathbf{a}}_*)}{2\sigma_a^2}\right\}$$

$$\tag{A7.4.5}$$

$$p(\mathbf{w}_n|\theta, \sigma_a) = (2\pi\sigma_a^2)^{-n/2}|\mathbf{X}'\mathbf{X}|^{-1/2} \exp -\frac{S(\theta)}{2\sigma_a^2} \tag{A7.4.6}$$

We can now deduce the following:

(1) From (A7.4.5), we see that $\hat{\mathbf{a}}_*$ is the conditional expectation of $\mathbf{a}_*$ given $\mathbf{w}_n$ and θ. Thus, using the notation introduced in Section 7.1.4,

$$\hat{\mathbf{a}}_* = [\mathbf{a}_*|\mathbf{w}_n, \theta] = [\mathbf{a}_*]$$

whence $[\mathbf{a}] = \mathbf{L}\mathbf{w}_n + \mathbf{X}[\mathbf{a}_*]$, and using (A7.4.4)

$$S(\theta) = \sum_{t=1-q}^{n} [a_t]^2 \tag{A7.4.7}$$

Now, although $\hat{\mathbf{a}}_*$ can be obtained by direct use of least squares, in practice, it is much more easily computed by using the fact that $\hat{\mathbf{a}}_* = [\mathbf{a}_*]$ and obtaining $[\mathbf{a}_*]$ by the "back-forecasting" technique of Sections 7.1.4 and 7.1.5.

(2) By comparing (A7.4.6) and (A7.4.2), we have

$$|\mathbf{X}'\mathbf{X}|^{-1} = |\mathbf{M}_n^{(0,q)}|$$

and

$$S(\boldsymbol{\theta}) = \mathbf{w}_n' \mathbf{M}_n^{(0,q)} \mathbf{w}_n$$

(3) To compute

$$S(\boldsymbol{\theta}) = \sum_{t=1-q}^{n} [a_t]^2$$

the quantities $[a_t] = [a_t|\mathbf{w}_n, \boldsymbol{\theta}]$ may be obtained by using the estimates $[\mathbf{a}_*]' = ([a_{1-q}], [a_{2-q}], \ldots, [a_0])$ obtained by back-forecasting for preliminary values, and computing the elements $[a_1], [a_2], \ldots, [a_n]$ recursively from

$$[a_t] = [w_t] + \theta_1[a_{t-1}] + \theta_2[a_{t-2}] + \cdots + \theta_q[a_{t-q}]$$

where $[w_t] = w_t$, $(t = 1, 2, \ldots, n)$.

(4) Finally, using (A7.4.6) and (A7.4.7), the unconditional likelihood is given exactly by

$$L(\boldsymbol{\theta}, \sigma_a | \mathbf{w}_n) = (\sigma_a^2)^{-n/2} |\mathbf{X}'\mathbf{X}|^{-1/2} \exp\left\{-\frac{1}{2\sigma_a^2} \sum_{t=1-q}^{n} [a_t]^2\right\} \qquad (A7.4.8)$$

For example, if $q = 1$, it is readily shown that $\mathbf{X}$ consists of an $(n+1)$ dimensional column vector whose elements are $1, \theta, \theta^2, \ldots, \theta^n$. Thus $\mathbf{X}'\mathbf{X} = \{1 - \theta^{2(n+1)}\}/(1 - \theta^2)$. For $n > 50$, the term $\theta^{2(n+1)}$ would ordinarily be negligible, and we should have the essentially exact result

$$L(\boldsymbol{\theta}, \sigma_a | \mathbf{w}_n) = (\sigma_a^2)^{-n/2} (1 - \theta^2)^{1/2} \exp\left\{-\frac{1}{2\sigma_a^2} \sum_{t=0}^{n} [a_t]^2\right\} \qquad (A7.4.9)$$

Extension to autoregressive and mixed processes. The method outlined above may be readily extended to provide the unconditional likelihood for the general mixed model

$$\phi(B)\tilde{w}_t = \theta(B)a_t \qquad (A7.4.10)$$

which, with $w_t = \nabla^d z_t$, defines the general ARIMA process. We first notice that this model may be written as an infinite moving average

$$\tilde{w}_t = (1 + \psi_1 B + \psi_2 B^2 + \cdots)a_t \qquad (A7.4.11)$$

Now, for processes of common interest, the weights $\psi_1, \psi_2, \ldots$ die out rather quickly, so that we can approximate (A7.4.11) to any desired accuracy by a *finite* moving average of some order Q

$$\tilde{w}_t = (1 + \psi_1 B + \psi_2 B^2 + \cdots + \psi_Q B^Q)a_t$$

A suitable value of Q can be chosen using the fact that, for a finite moving average of order Q, observations $Q + 1$ intervals apart are uncorrelated, and hence forecasts $[w]$ for lead times greater than Q will be zero. In the course of the calculations described in Section 7.1.5, back-forecasts are obtained for times $0, -1, -2, \ldots, -Q + 1$, and Q is chosen as the point beyond which the forecasted values $[\tilde{w}]$ have become essentially zero. Thus, in practice, we do not need to actually compute the ψ's, since the recursive calculations are made directly in terms of the general model (A7.4.10). An example of the calculation is given in Section 7.1.5.

Therefore, in general, the likelihood associated with a series $\mathbf{z}$ of $n + d$ values generated by any ARIMA process, is given by

$$L(\boldsymbol{\phi}, \boldsymbol{\theta}, \sigma_a | \mathbf{z}) = (2\pi\sigma_a^2)^{-n/2} |\mathbf{M}_n^{(p,q)}|^{1/2} \exp\left\{ -\frac{S(\boldsymbol{\phi}, \boldsymbol{\theta})}{2\sigma_a^2} \right\} \quad (\text{A7.4.12})$$

where

$$S(\boldsymbol{\phi}, \boldsymbol{\theta}) = \sum_{t=-\infty}^{n} [a_t | \mathbf{z}, \boldsymbol{\phi}, \boldsymbol{\theta}]^2 \quad (\text{A7.4.13})$$

and in practice the values $[a_t | \mathbf{z}, \boldsymbol{\phi}, \boldsymbol{\theta}] = [a_t | \mathbf{w}_n, \boldsymbol{\phi}, \boldsymbol{\theta}]$ can be computed recursively with the summation proceeding from some point $t = 1 - Q$, beyond which the $[a_t]$'s are negligible.

To illustrate, consider the first-order autoregressive process in w_t

$$w_t - \phi w_{t-1} = a_t \quad (\text{A7.4.14})$$

where w_t might be the dth difference $\nabla^d z_t$ of the actual observations and a series $\mathbf{z}$ of length $n + d$ observations is available. To compute the likelihood (A7.4.12), we require

$$S(\phi) = \sum_{t=-\infty}^{n} [a_t | \mathbf{w}_n, \phi]^2 = \sum_{t=-\infty}^{1} [a_t | \mathbf{w}_n, \phi]^2 + \sum_{t=2}^{n} (w_t - \phi w_{t-1})^2$$

The required conditional expectations (back forecasts) are

$$[w_j | \mathbf{w}_n, \phi] = \phi^{1-j} w_1 \qquad j = 0, -1, -2, \ldots$$

whence

$$[a_t | \mathbf{w}_n, \phi] = \phi^{-t+1}(1 - \phi^2) w_1 \qquad t = 1, 0, -1, \ldots$$

Hence

$$S(\phi) = (1 - \phi^2) w_1^2 + \sum_{t=2}^{n} (w_t - \phi w_{t-1})^2 \quad (\text{A7.4.15})$$

a result which may be obtained more directly by methods discussed in Appendix A7.5.

APPENDIX A7.5 THE EXACT LIKELIHOOD FUNCTION FOR AN AUTOREGRESSIVE PROCESS

We now suppose that a given series $\mathbf{w}_n' = (w_1, w_2, \ldots, w_n)$ is generated by the pth-order stationary autoregressive model

$$w_t - \phi_1 w_{t-1} - \phi_2 w_{t-2} - \cdots - \phi_p w_{t-p} = a_t$$

where, temporarily the w's are assumed to have mean $\mu = 0$, but, as before, the argument can be extended to the case where $\mu \neq 0$. Assuming Normality for the a's and hence for the w's, the joint probability density function of the w's is,

$$p(\mathbf{w}_n | \boldsymbol{\phi}, \sigma_a) = (2\pi\sigma_a^2)^{-n/2} |\mathbf{M}_n^{(p,0)}|^{1/2} \exp -\left\{ \frac{\mathbf{w}_n' \mathbf{M}_n^{(p,0)} \mathbf{w}_n}{2\sigma_a^2} \right\} \quad (A7.5.1)$$

and because of the reversible character of the general process, the $n \times n$ matrix $\mathbf{M}_n^{(p,0)}$ is symmetric about *both* of its principal diagonals. We shall say that such a matrix is *doubly* symmetric. Now

$$p(\mathbf{w}_n | \boldsymbol{\phi}, \sigma_a) = p(w_{p+1}, w_{p+2}, \ldots, w_n | \mathbf{w}_p, \boldsymbol{\phi}, \sigma_a) p(\mathbf{w}_p | \boldsymbol{\phi}, \sigma_a)$$

where $\mathbf{w}_p' = (w_1, w_2, \ldots, w_p)$. The first factor on the right may be obtained by making use of the distribution

$$p(a_{p+1}, \ldots, a_n) = (2\pi\sigma_a^2)^{-(n-p)/2} \exp -\left\{ \frac{1}{2\sigma_a^2} \sum_{t=p+1}^{n} a_t^2 \right\} \quad (A7.5.2)$$

For fixed $\mathbf{w}_p$, $(a_{p+1}, \ldots, a_n)$ and $(w_{p+1}, \ldots, w_n)$ are related by the transformation

$$a_{p+1} = w_{p+1} - \phi_1 w_p - \cdots - \phi_p w_1$$
$$\vdots$$
$$a_n = w_n - \phi_1 w_{n-1} - \cdots - \phi_p w_{n-p}$$

which has unit Jacobian. Thus we obtain

$$p(w_{p+1}, \ldots, w_n | \mathbf{w}_p, \boldsymbol{\phi}, \sigma_a)$$
$$= (2\pi\sigma_a^2)^{-(n-p)/2} \exp \left\{ -\frac{1}{2\sigma_a^2} \sum_{t=p+1}^{n} (w_t - \phi_1 w_{t-1} - \cdots - \phi_p w_{t-p})^2 \right\}$$

Also

$$p(\mathbf{w}_p | \boldsymbol{\phi}, \sigma_a) = (2\pi\sigma_a^2)^{-p/2} |\mathbf{M}_p^{(p,0)}|^{1/2} \exp \left\{ -\frac{1}{2\sigma_a^2} \mathbf{w}_p' \mathbf{M}_p^{(p,0)} \mathbf{w}_p \right\}$$

Thus

$$p(\mathbf{w}_n | \boldsymbol{\phi}, \sigma_a) = (2\pi\sigma_a^2)^{-n/2} |\mathbf{M}_p^{(p,0)}|^{1/2} \exp \left\{ \frac{-S(\boldsymbol{\phi})}{2\sigma_a^2} \right\} \quad (A7.5.3)$$

where

$$S(\boldsymbol{\phi}) = \sum_{i=1}^{p} \sum_{j=1}^{p} m_{ij}^{(p)} w_i w_j + \sum_{t=p+1}^{n} (w_t - \phi_1 w_{t-1} - \cdots - \phi_p w_{t-p})^2 \qquad (A7.5.4)$$

Also,

$$\mathbf{M}_p^{(p,0)} = \{m_{ij}^{(p)}\} = \{\gamma_{|i-j|}\}^{-1}\sigma_a^2 = \begin{bmatrix} \gamma_0 & \gamma_1 & \cdots & \gamma_{p-1} \\ \gamma_1 & \gamma_0 & \cdots & \gamma_{p-2} \\ \vdots & \vdots & & \vdots \\ \gamma_{p-1} & \gamma_{p-2} & \cdots & \gamma_0 \end{bmatrix}^{-1} \sigma_a^2 \qquad (A7.5.5)$$

where $\gamma_0, \gamma_1, \ldots, \gamma_{p-1}$ are the theoretical autocovariances of the process and $|\mathbf{M}_p^{(p,0)}| = |\mathbf{M}_n^{(p,0)}|$.

Now let $n = p + 1$, so that

$$\mathbf{w}_{p+1}'\mathbf{M}_{p+1}^{(p,0)}\mathbf{w}_{p+1} = \sum_{i=1}^{p} \sum_{j=1}^{p} m_{ij}^{(p)} w_i w_j$$

$$+ (w_{p+1} - \phi_1 w_p - \phi_2 w_{p-1} - \cdots - \phi_p w_1)^2$$

Then

$$\mathbf{M}_{p+1}^{(p)} =$$

$$\begin{bmatrix} & & & \vdots & 0 \\ & \mathbf{M}_p^{(p)} & & \vdots & 0 \\ & & & \vdots & \vdots \\ \hline 0 & 0 & \cdots & \vdots & 0 \end{bmatrix} + \begin{bmatrix} \phi_p^2 & \phi_p\phi_{p-1} & \cdots & \vdots & -\phi_p \\ \phi_p\phi_{p-1} & \phi_{p-1}^2 & \cdots & \vdots & -\phi_{p-1} \\ \vdots & \vdots & & \vdots & \vdots \\ & & & \vdots & -\phi_1 \\ \hline -\phi_p & -\phi_{p-1} & \cdots & \vdots & 1 \end{bmatrix}$$

and the elements of $\mathbf{M}_p^{(p)} = \mathbf{M}_p^{(p,0)}$ can now be deduced from the consideration that both $\mathbf{M}_p^{(p)}$ and $\mathbf{M}_{p+1}^{(p)}$ are doubly symmetric. Thus, for example,

$$\mathbf{M}_2^{(1)} = \begin{bmatrix} m_{11}^{(1)} + \phi_1^2 & -\phi_1 \\ -\phi_1 & 1 \end{bmatrix} = \begin{bmatrix} 1 & -\phi_1 \\ -\phi_1 & m_{11}^{(1)} + \phi_1^2 \end{bmatrix}$$

and after equating elements in the two matrices, we have

$$M_1^{(1)} = m_{11}^{(1)} = 1 - \phi_1^2$$

Proceeding in this way, we find for processes of orders 1 and 2

$$M_1^{(1)} = 1 - \phi_1^2 \qquad |\mathbf{M}_1^{(1)}| = 1 - \phi_1^2$$

$$\mathbf{M}_2^{(2)} = \begin{bmatrix} 1 - \phi_2^2 & -\phi_1(1 + \phi_2) \\ -\phi_1(1 + \phi_2) & 1 - \phi_2^2 \end{bmatrix}$$

$$|\mathbf{M}_2^{(2)}| = (1 + \phi_2)^2\{(1 - \phi_2)^2 - \phi_1^2\}$$

For example, when $p = 1$,

$p(\mathbf{w}_n|\phi, \sigma_a)$

$$= (2\pi\sigma_a^2)^{-n/2}(1 - \phi^2)^{1/2} \exp\left[-\frac{1}{2\sigma_a^2}\left\{(1 - \phi^2)w_1^2 + \sum_{t=2}^{n}(w_t - \phi w_{t-1})^2\right\}\right]$$

which checks with the result obtained in (A7.4.15). The above process of generation must lead to matrices $M_p^{(p)}$, whose elements are *quadratic* in the ϕ's.

Thus, it is clear from (A7.5.4) that, not only is $S(\phi) = \mathbf{w}_n'\mathbf{M}_n^{(p)}\mathbf{w}_n$ a quadratic form in the w's, but it is also quadratic in the parameters ϕ. Writing $\phi_u' = (1, \phi_1, \phi_2, \ldots, \phi_p)$, it is clearly true that, for some $(p + 1) \times (p + 1)$ matrix $\mathbf{D}$ whose elements are quadratic functions of the w's,

$$\mathbf{w}_n'\mathbf{M}_n^{(p)}\mathbf{w}_n = \phi_u'\mathbf{D}\phi_u$$

Now write

$$\mathbf{D} = \begin{bmatrix} D_{11} & -D_{12} & -D_{13} & \cdots & -D_{1,p+1} \\ -D_{12} & D_{22} & D_{23} & \cdots & D_{2,p+1} \\ \vdots & \vdots & \vdots & & \vdots \\ -D_{1,p+1} & D_{2,p+1} & D_{3,p+1} & \cdots & D_{p+1,p+1} \end{bmatrix} \qquad (A7.5.6)$$

Inspection of (A7.5.4) shows that the elements D_{ij} are "symmetric" sums of squares and lagged products, defined by

$$D_{ij} = D_{ji} = w_i w_j + w_{i+1} w_{j+1} + \cdots + w_{n+1-j} w_{n+1-i} \qquad (A7.5.7)$$

where the sum D_{ij} contains $n - (i - 1) - (j - 1)$ terms.

Finally, we can write the *exact* probability density, and hence the exact likelihood, as

$$p(\mathbf{w}_n|\phi, \sigma_a) = L(\phi, \sigma_a|\mathbf{w}_n) = (2\pi\sigma_a^2)^{-n/2}|\mathbf{M}_p^{(p)}|^{1/2} \exp\left\{\frac{-S(\phi)}{2\sigma_a^2}\right\} \qquad (A7.5.8)$$

where

$$S(\phi) = \mathbf{w}_p'\mathbf{M}_p^{(p)}\mathbf{w}_p + \sum_{t=p+1}^{n}(w_t - \phi_1 w_{t-1} - \cdots - \phi_p w_{t-p})^2 = \phi_u'\mathbf{D}\phi_u$$

$$(A7.5.9)$$

and the log likelihood is

$$l(\phi, \sigma_a|\mathbf{w}_n) = -\frac{n}{2}\ln\sigma_a^2 + \frac{1}{2}\ln|\mathbf{M}_p^{(p)}| - \frac{S(\phi)}{2\sigma_a^2} \qquad (A7.5.10)$$

Maximum likelihood estimates. Differentiating with respect to σ_a and each of the ϕ's in (A7.5.10), we obtain

$$\frac{\partial l}{\partial \sigma_a} = -\frac{n}{\sigma_a} + \frac{S(\phi)}{\sigma_a^3} \tag{A7.5.11}$$

$$\frac{\partial l}{\partial \phi_j} = M_j + \sigma_a^{-2}\{D_{1,j+1} - \phi_1 D_{2,j+1} - \cdots - \phi_p D_{p+1,j+1}\} \quad j = 1, 2, \ldots, p \tag{A7.5.12}$$

where

$$M_j = \frac{\partial\{\frac{1}{2}\ln|\mathbf{M}_p^{(p)}|\}}{\partial \phi_j}$$

Whence maximum likelihood estimates may be obtained by equating these expressions to zero and solving the resultant equations.

We have at once from (A7.5.11)

$$\hat{\sigma}_a^2 = \frac{S(\hat{\phi})}{n} \tag{A7.5.13}$$

Estimates of ϕ. A difficulty occurs in dealing with the equations (A7.5.12) since, in general, the quantities $M_j(j = 1, 2, \ldots, p)$ are complicated functions of the ϕ's. We consider briefly three alternative approximations.

(1) Least squares estimates:

Whereas the expected value of $S(\phi)$ is proportional to n, the value of $|\mathbf{M}_p^{(p)}|$ is independent of n and for moderate or large samples, (A7.5.8) is dominated by the term in $S(\phi)$ and the term in $|\mathbf{M}_p^{(p)}|$ is, by comparison, small.

If we ignore the influence of this term, then

$$l(\phi, \sigma_a|\mathbf{w}_n) \simeq -\frac{n}{2}\ln \sigma_a^2 - \frac{S(\phi)}{2\sigma_a^2} \tag{A7.5.14}$$

and the estimates $\hat{\phi}$ of ϕ obtained by maximization of (A7.5.14) are the least squares estimates obtained by minimizing $S(\phi)$. Now, from (A7.5.9), $S(\phi) = \phi_u' \mathbf{D} \phi_u$, where $\mathbf{D}$ is a $(p + 1) \times (p + 1)$ matrix of symmetric sums of squares and products, defined in (A7.5.7). Thus, on differentiating, the minimizing values are

$$\left.\begin{array}{l} D_{12} = \hat{\phi}_1 D_{22} + \hat{\phi}_2 D_{23} + \cdots + \hat{\phi}_p D_{2,p+1} \\ D_{13} = \hat{\phi}_1 D_{23} + \hat{\phi}_2 D_{33} + \cdots + \hat{\phi}_p D_{3,p+1} \\ D_{1,p+1} = \hat{\phi}_1 D_{2,p+1} + \hat{\phi}_2 D_{3,p+1} + \cdots + \hat{\phi}_p D_{p+1,p+1} \end{array}\right\} \tag{A7.5.15}$$

which, in an obvious matrix notation, can be written

$$\mathbf{d} = \mathbf{D}_p \hat{\phi}$$

so that

$$\hat{\boldsymbol{\phi}} = \mathbf{D}_p^{-1}\mathbf{d}$$

These least squares estimates maximize the posterior density (7.4.15).

(2) Approximate maximum likelihood estimates:

We now recall an earlier result (3.2.3) which may be written

$$\gamma_j - \phi_1\gamma_{j-1} - \phi_2\gamma_{j-2} - \cdots - \phi_p\gamma_{j-p} = 0 \qquad j > 0 \quad \text{(A7.5.16)}$$

Also, on taking expectations in (A7.5.12) and using the fact that $E[\partial l/\partial \phi_j] = 0$, we obtain

$$M_j\sigma_a^2 + (n-j)\gamma_j - (n-j-1)\phi_1\gamma_{j-1} - (n-j-2)\phi_2\gamma_{j-2} - \cdots$$

$$-(n-j-p)\phi_p\gamma_{j-p} = 0 \tag{A7.5.17}$$

After multiplying (A7.5.16) by n and subtracting the result from (A7.5.17), we obtain

$$M_j\sigma_a^2 = j\gamma_j - (j+1)\phi_1\gamma_{j-1} - \cdots - (j+p)\phi_p\gamma_{j-p}$$

Therefore, on using $D_{i+1,j+1}/(n-j-i)$ as an estimate of $\gamma_{|j-i|}$, a natural estimate of $M_j\sigma_a^2$ is

$$j\frac{D_{1,j+1}}{n-j} - (j+1)\phi_1\frac{D_{2,j+1}}{n-j-1} - \cdots - (j+p)\phi_p\frac{D_{p+1,j+1}}{n-j-p}$$

Substituting this estimate in (A7.5.12) yields

$$\frac{\partial l}{\partial \phi_j} \simeq n\sigma_a^{-2}\left\{\frac{D_{1,j+1}}{n-j} - \phi_1\frac{D_{2,j+1}}{n-j-1} - \cdots - \phi_p\frac{D_{p+1,j+1}}{n-j-p}\right\}$$

$$(j = 1, 2, \ldots, p) \quad \text{(A7.5.18)}$$

leading to a set of linear equations of the form (A7.5.15), but now with

$$D_{ij}^* = nD_{ij}/\{n - (i-1) - (j-1)\}$$

replacing D_{ij}.

(3) Yule–Walker estimates:

Finally, if n is moderate or large, as an approximation, we may replace the symmetric sums of squares and products in (A7.5.15) by n times the appropriate autocovariance estimate. For example, D_{ij}, where $|i - j| = k$, would be replaced by $nc_k = \sum_{t=1}^{n-k} \tilde{w}_t\tilde{w}_{t+k}$. On dividing throughout in the equations that result, by nc_0, we obtain the following relations expressed in terms of the estimated autocorrelations $r_k = c_k/c_0$:

$$r_1 = \hat{\phi}_1 \quad\quad + \hat{\phi}_2 r_1 \quad + \quad \ldots \quad + \hat{\phi}_p r_{p-1}$$
$$r_2 = \hat{\phi}_1 r_1 \quad + \hat{\phi}_2 \quad\quad + \quad \ldots \quad + \hat{\phi}_p r_{p-2}$$
$$\vdots \quad\quad \vdots \quad\quad \vdots \quad\quad\quad\quad \vdots$$
$$r_p = \hat{\phi}_1 r_{p-1} + \hat{\phi}_2 r_{p-2} + \quad \ldots \quad + \hat{\phi}_p$$

These are the well known Yule–Walker equations.

In the matrix notation (7.3.1) they can be written $\mathbf{r} = \mathbf{R}\hat{\boldsymbol{\phi}}$, so that

$$\hat{\boldsymbol{\phi}} = \mathbf{R}^{-1}\mathbf{r} \tag{A7.5.19}$$

which corresponds to the equations (3.2.7), with $\mathbf{r}$ substituted for $\boldsymbol{\rho}_p$ and $\mathbf{R}$ for $\mathbf{P}_p$.

To illustrate the differences among the three estimates, take the case $p = 1$. Then $M_1 \sigma_a^2 = -\gamma_1$ and, corresponding to (A7.5.12), the exact maximum likelihood estimate of ϕ is the solution of

$$-\gamma_1 + \sum_{t=2}^{n} w_t w_{t-1} - \phi \sum_{t=2}^{n-1} w_t^2 = 0$$

Approximation (1) corresponds to ignoring the term γ_1 altogether, yielding

$$\hat{\phi} = \frac{\sum_{t=2}^{n} w_t w_{t-1}}{\sum_{t=2}^{n-1} w_t^2} = \frac{D_{12}}{D_{22}}$$

(2) corresponds to substituting the estimate $\sum_{t=2}^{n} w_t w_{t-1}/(n-1)$ for γ_1, yielding

$$\hat{\phi} = \frac{\sum_{t=2}^{n} w_t w_{t-1}/(n-1)}{\sum_{t=2}^{n-1} w_t^2/(n-2)} = \frac{n-2}{n-1}\frac{D_{12}}{D_{22}}$$

(3) replaces numerator and denominator by standard autocovariance estimates (2.1.10), yielding

$$\hat{\phi} = \frac{\sum_{t=2}^{n} w_t w_{t-1}}{\sum_{t=1}^{n} w_t^2} = \frac{c_1}{c_0} = r_1 = \frac{D_{12}}{D_{11}}$$

Usually, as in this example, for moderate and large samples, the differences between the estimates given by the various approximations will be small. We normally use the least squares estimates given by (1). These estimates can of course be computed directly from (A7.5.15). However, assuming an electronic computer to be available, it is scarcely worthwhile to treat autoregressive processes separately, and we have found it simplest, even when the fitted process is autoregressive, to employ the general iterative algorithm described in Section 7.2.1, which computes least squares estimates for any ARMA process.

Estimate of σ_a^2. Using approximation (3) with (A7.5.9) and (A7.5.13),

$$\hat{\sigma}_a^2 = \frac{S(\hat{\boldsymbol{\phi}})}{n} = c_0(1 : \hat{\boldsymbol{\phi}}')\begin{bmatrix} -\dfrac{1}{-\mathbf{r}} & \dfrac{-\mathbf{r}'}{\mathbf{R}} \end{bmatrix}\left(-\dfrac{1}{\hat{\boldsymbol{\phi}}}-\right)$$

On multiplying out the right-hand side and recalling that $\mathbf{r} - \mathbf{R}\hat{\boldsymbol{\phi}} = 0$, we find

$$\hat{\sigma}_a^2 = c_0(1 - \mathbf{r}'\hat{\boldsymbol{\phi}}) = c_0(1 - \mathbf{r}'\mathbf{R}^{-1}\mathbf{r}) = c_0(1 - \hat{\boldsymbol{\phi}}'\mathbf{R}\hat{\boldsymbol{\phi}}) \quad \text{(A7.5.20}a)$$

It is readily shown that σ_a^2 can be similarly written in terms of the *theoretical* correlations

$$\sigma_a^2 = \gamma_0(1 - \boldsymbol{\rho}'\boldsymbol{\phi}) = \gamma_0(1 - \boldsymbol{\rho}'\mathbf{P}_p^{-1}\boldsymbol{\rho}) = \gamma_0(1 - \boldsymbol{\phi}'\mathbf{P}_p\boldsymbol{\phi}) \quad \text{(A7.5.20}b)$$

agreeing with the result (3.2.8).

Parallel expressions for $\hat{\sigma}_a^2$ may be obtained for approximations (1) and (2).

The information matrix. Differentiating for a second time in (A7.5.11) and (A7.5.18), we obtain

$$-\frac{\partial^2 l}{\partial \sigma_a^2} = -\frac{n}{\sigma_a^2} + \frac{3S(\boldsymbol{\phi})}{\sigma_a^4} \quad \text{(A7.5.21}a)$$

$$\frac{\partial^2 l}{\partial \sigma_a \partial \phi_j} \simeq -\sigma_a^{-1}\frac{\partial l}{\partial \phi_j} \quad \text{(A7.5.21}b)$$

$$-\frac{\partial^2 l}{\partial \phi_i \partial \phi_j} \simeq \frac{n}{\sigma_a^2}\left\{\frac{D_{i+1,j+1}}{n-i-j}\right\} \quad \text{(A7.5.21}c)$$

Now, since

$$E\left[\frac{\partial l}{\partial \phi_j}\right] = 0$$

it follows that, for moderate or large samples,

$$E\left[-\frac{\partial^2 l}{\partial \sigma_a \partial \phi_j}\right] \simeq 0$$

and

$$|\mathbf{I}(\boldsymbol{\phi}, \sigma_a)| \simeq |\mathbf{I}(\boldsymbol{\phi})|I(\sigma_a)$$

where

$$I(\sigma_a) = E\left[-\frac{\partial^2 l}{\partial \sigma_a^2}\right] = \frac{2n}{\sigma_a^2}$$

Now, using (A7.5.21c)

$$\mathbf{I}(\boldsymbol{\phi}) = -E\left[\frac{\partial^2 l}{\partial \phi_i \partial \phi_j}\right] \simeq \frac{n}{\sigma_a^2}\boldsymbol{\Gamma}_p = \frac{n\gamma_0}{\sigma_a^2}\mathbf{P}_p = n(\mathbf{M}_p^{(p)})^{-1} \quad \text{(A7.5.22)}$$

Hence

$$\left| \mathbf{I}(\boldsymbol{\phi}, \sigma_a) \right| \simeq \frac{2n^{p+1}}{\sigma_a^2} \left| \mathbf{M}_p^{(p,0)} \right|^{-1}$$

Variances and covariances of estimates of autoregressive parameters. Now, in circumstances fully discussed in [87], the inverse of the information matrix supplies the asymptotic variance–covariance matrix of the maximum likelihood (ML) estimates. Moreover if the log-likelihood is approximately quadratic and the maximum is not close to a boundary, even if the sample size is only moderate, the elements of this matrix will normally provide adequate approximations to the variances and covariances of the estimates.

Thus, using (A7.5.22) and (A7.5.20b),

$$\left.\begin{aligned} \mathbf{V}(\hat{\boldsymbol{\phi}}) &= \mathbf{I}^{-1}(\hat{\boldsymbol{\phi}}) \simeq n^{-1}\mathbf{M}_p^{(p)} = n^{-1}\sigma_a^2\boldsymbol{\Gamma}_p^{-1} \\ &= n^{-1}(1 - \boldsymbol{\rho}'\mathbf{P}_p^{-1}\boldsymbol{\rho})\mathbf{P}_p^{-1} \\ &= n^{-1}(1 - \boldsymbol{\phi}'\mathbf{P}_p\boldsymbol{\phi})\mathbf{P}_p^{-1} = n^{-1}(1 - \boldsymbol{\rho}'\boldsymbol{\phi})\mathbf{P}_p^{-1} \end{aligned}\right\} \qquad \text{(A7.5.23)}$$

In particular, for autoregressive processes of first and second order

$$V(\hat{\phi}) \simeq n^{-1}(1 - \phi^2)$$

$$\mathbf{V}(\hat{\phi}_1, \hat{\phi}_2) \simeq n^{-1} \begin{bmatrix} 1 - \phi_2^2 & -\phi_1(1 + \phi_2) \\ -\phi_1(1 + \phi_2) & 1 - \phi_2^2 \end{bmatrix} \qquad \text{(A7.5.24)}$$

Estimates of the variances and covariances may be obtained by substituting estimates for the parameters in (A7.5.24). For example, we may substitute r's for ρ's and $\hat{\boldsymbol{\phi}}$ for $\boldsymbol{\phi}$ in (A7.5.23) to obtain

$$\hat{\mathbf{V}}(\hat{\boldsymbol{\phi}}) = n^{-1}(1 - \mathbf{r}'\hat{\boldsymbol{\phi}})\mathbf{R}^{-1} \qquad \text{(A7.5.25)}$$

Extension of the autoregressive results to more general processes. There is an interesting relationship which allows us to extend the estimation results obtained for autoregressive models to more general models. Suppose observations are generated by the ARMA model of order (p, q)

$$\phi(B)w_t = \theta(B)a_t \qquad \text{(A7.5.26)}$$

where $w_t = \nabla^d z_t$. We can write the model in the form

$$\prod_{i=1}^{p} (1 - G_i B)w_t = \prod_{j=1}^{q} (1 - H_j B)a_t \qquad \text{(A7.5.27)}$$

Now, let $H_1^0, H_2^0, \ldots, H_q^0$ be values of the moving average parameters differing slightly from $H_1, H_2, \ldots, H_q$, and suppose we make a transformation

$$\prod_{j=1}^{q} (1 - H_j^0 B)^2 x_t = w_t \qquad \text{(A7.5.28)}$$

By approximating unknown preliminary values by zeroes, we can use this expression to generate a corresponding series $\mathbf{x}$ for each time series $\mathbf{w}$ of n observations, generated by the model (A7.5.27). Denote the transformation by $\mathbf{T}^{-1}\mathbf{x} = \mathbf{w}$, so that $\mathbf{x} = \mathbf{Tw}$ and since $\mathbf{T}^{-1}$ is a triangular matrix with diagonal elements all equal to unity, $|\mathbf{T}| = 1$. Now write $H_1 = H_1^0 + \delta_1$, $H_2 = H_2^0 + \delta_2, \ldots, H_q = H_q^0 + \delta_q$. Then the likelihood function associated with parameter values $(\mathbf{G}, \mathbf{H})$, that is with values $(\mathbf{G}, \mathbf{H}^0 + \boldsymbol{\delta})$, is

$$L\{\mathbf{G}, \mathbf{H}^0 + \boldsymbol{\delta}, \sigma_a^2 | \mathbf{w}, (p, q)\}$$

This choice of parameters implies that the w_t are generated from the ARMA (p, q) model

$$\prod_{i=1}^{p} (1 - G_i B)w_t = \prod_{j=1}^{q} \{1 - (H_j^0 + \delta_j)B\}a_t$$

and hence that the x_t are generated by the model

$$\prod_{i=1}^{p} (1 - G_i B) \prod_{j=1}^{q} (1 - H_j^0 B)^2 x_t = \prod_{j=1}^{q} \{1 - (H_j^0 + \delta_j)B\}a_t$$

or, for small values of δ, very nearly by the AR $(p + q)$ model

$$\prod_{i=1}^{p} (1 - G_i B) \prod_{j=1}^{q} \{1 - (H_j^0 - \delta_j)B\}x_t = a_t$$

The corresponding likelihood will be written as

$$L\{\mathbf{G}, \mathbf{H}^0 - \boldsymbol{\delta}, \sigma_a^2 | \mathbf{x}, (p + q, 0)\}$$

Now, for moderate or large samples, the end approximation arising in transforming sample values, will have little effect and also the likelihood will be small for large values of δ, so that only small deviations will be of importance. Thus it follows that

$$L\{\mathbf{G}, \mathbf{H}^0 + \boldsymbol{\delta}, \sigma_a^2 | \mathbf{w}, (p, q)\} \simeq L\{\mathbf{G}, \mathbf{H}^0 - \boldsymbol{\delta}, \sigma_a^2 | \mathbf{x}, (p + q, 0)\} \qquad \text{(A7.5.29)}$$

These results follow immediately:

(1) Suppose the information matrix for parameters $(\mathbf{G}, \mathbf{H})$ associated with the ARMA (p, q) model

$$\prod_{i=1}^{p} (1 - G_i B)w_t = \prod_{j=1}^{q} (1 - H_j B)a_t$$

is $\mathbf{I}\{\mathbf{G}, \mathbf{H} | (p, q)\}$. Suppose, correspondingly, that the information matrix for the parameters $(\mathbf{G}, \mathbf{H})$ in the pure AR $(p + q)$ model

$$\prod_{i=1}^{p} (1 - G_i B) \prod_{j=1}^{q} (1 - H_j B)w_t = a_t$$

is

$$\mathbf{I}\{\mathbf{G}, \mathbf{H}|(p + q, 0)\} = \begin{bmatrix} \mathbf{I}_{GG} & \vdots & \mathbf{I}_{GH} \\ \hline \mathbf{I}'_{GH} & \vdots & \mathbf{I}_{HH} \end{bmatrix}$$

where the matrix is partitioned after the pth row and column. Then, for moderate and large samples

$$\mathbf{I}\{\mathbf{G}, \mathbf{H}|(p, q)\} \simeq \mathbf{I}\{\mathbf{G}, -\mathbf{H}|(p + q, 0)\} = \begin{bmatrix} \mathbf{I}_{GG} & \vdots & -\mathbf{I}_{GH} \\ \hline -\mathbf{I}'_{GH} & \vdots & \mathbf{I}_{HH} \end{bmatrix} \qquad (A7.5.30)$$

(2) It follows, to the degree of approximation considered, that the determinants of the information matrices for the ARMA (p, q) process and the AR $(p + q)$ process are identical.

(3) Since, for moderate and large samples, the reciprocal of the information matrix provides a close approximation to the covariance matrix $\mathbf{V}(\mathbf{G}, \mathbf{H})$ of the parameters, we have correspondingly

$$\mathbf{V}\{\mathbf{G}, \mathbf{H}|(p, q)\} \simeq \mathbf{V}\{\mathbf{G}, -\mathbf{H}|(p + q, 0)\} \qquad (A7.5.31)$$

(4) Now, suppose we write

$$\prod_{i=1}^{p} (1 - G_i B) \prod_{j=1}^{q} (1 - H_j B) = 1 - \phi_1^* B - \phi_2^* B^2 - \cdots - \phi_{p+q}^* B^{p+q}$$

Then, using (A7.5.22),

$$|\mathbf{I}(\boldsymbol{\phi}^*)| \simeq n^{p+q} |\mathbf{M}_{p+q}^{(p+q,0)}|^{-1} = n^{p+q} |\mathbf{M}_n^{(p,q)}|^{-1} \qquad (A7.5.32)$$

Also

$$|\mathbf{I}(\mathbf{G}, \mathbf{H})| = |\mathbf{I}(\boldsymbol{\phi}^*)| J^2(\boldsymbol{\phi}^*|\mathbf{G}, \mathbf{H}) \qquad (A7.5.33)$$

where

$$J(\boldsymbol{\phi}^*|\mathbf{G}, \mathbf{H}) = \left| \frac{\partial \boldsymbol{\phi}^*}{\partial(\mathbf{G}, \mathbf{H})} \right| \qquad (A7.5.34)$$

is the Jacobian of the transformation.

Moving average processes—variances and covariances of ML estimates. It follows immediately that, for large samples, the covariance matrix for the ML estimates from a pure moving average is precisely the same as that for a pure autoregressive process of the same order. Thus, for moving average processes of orders 1 and 2, we have corresponding to (A7.5.24)

$$V(\hat{\theta}) \simeq n^{-1}(1 - \theta^2) \quad V(\hat{\theta}_1, \hat{\theta}_2) \simeq n^{-1} \begin{bmatrix} 1 - \theta_2^2 & -\theta_1(1 + \theta_2) \\ -\theta_1(1 + \theta_2) & 1 - \theta_2^2 \end{bmatrix}$$

Mixed moving average-autoregressive processes—variances and covariances of ML estimates. To illustrate the use of (A7.5.31), consider the ARMA $(1, 1)$ process

$$(1 - \phi B)w_t = (1 - \theta B)a_t$$

which we relate to the AR (2) process

$$(1 - \phi B)(1 - \theta B)x_t = a_t \tag{A7.5.35}$$

The variance-covariance matrix for $\hat{\phi}$ and $\hat{\theta}$ in the autoregressive model (A7.5.35) is, using (7.2.22),

$$\mathbf{V}\{\hat{\phi}, \hat{\theta}|(2, 0)\} \simeq n^{-1}\frac{1 - \phi\theta}{(\phi - \theta)^2}\begin{bmatrix} (1 - \phi^2)(1 - \phi\theta) & -(1 - \phi^2)(1 - \theta^2) \\ -(1 - \phi^2)(1 - \theta^2) & (1 - \theta^2)(1 - \phi\theta) \end{bmatrix}$$

Hence, the corresponding variances and covariances for the ARMA $(1, 1)$ process are

$$\mathbf{V}\{\hat{\phi}, \hat{\theta}|(1, 1)\} \simeq n^{-1}\frac{1 - \phi\theta}{(\phi - \theta)^2}\begin{bmatrix} (1 - \phi^2)(1 - \phi\theta) & (1 - \phi^2)(1 - \theta^2) \\ (1 - \phi^2)(1 - \theta^2) & (1 - \theta^2)(1 - \phi\theta) \end{bmatrix}$$

APPENDIX A7.6 SPECIAL NOTE ON ESTIMATION OF MOVING AVERAGE PARAMETERS

If the least squares iteration is allowed to stray outside the invertibility region, parameter values can readily be found that apparently provide sums of squares smaller than the true minimum. However, these do not provide appropriate estimates and are quite meaningless. To illustrate, suppose a series has been generated by the first-order moving average model $w_t = (1 - \theta B)a_t$ with $-1 < \theta < 1$. Then the series could equally well have been generated by the corresponding backward process $w_t = (1 - \theta F)e_t$ with $\sigma_e^2 = \sigma_a^2$. Now the latter process can also be written as $w_t = (1 - \theta^{-1}B)\alpha_t$ where now θ^{-1} is *outside* the invertibility region. However, in this representation $\sigma_\alpha^2 = \sigma_a^2\theta^2$ and is itself a function of θ. Therefore, a valid estimate of θ^{-1} will not be provided by minimizing $\sum \alpha_t^2 = \theta^2 \sum a_t^2$. Indeed this has its minimum at $\theta^{-1} = \infty$.

The difficulty may be avoided

(a) by using as starting values rough preliminary estimates within the invertibility region obtained at the identification stage,
(b) by checking that all moving average estimates, obtained after convergence has apparently occurred, lie within the invertibility region.

It is also possible to write least squares programs such that estimates are constrained to lie within the invertibility region.

8

Model Diagnostic Checking

The model having been identified and the parameters estimated, *diagnostic checks* are then applied to the fitted model. One useful method of checking a model is to *overfit*, that is, to estimate the parameters in a model somewhat more general than that which we believe to be true. This method assumes that we can guess the direction in which the model is likely to be inadequate. Therefore, it is necessary to supplement this approach by less specific checks applied to the residuals from the fitted model. These allow the data themselves to suggest modifications to the model. We shall describe two such checks which employ (1) the autocorrelation function of the residuals and (2) the cumulative periodogram of the residuals.

8.1 CHECKING THE STOCHASTIC MODEL

8.1.1 General philosophy

Suppose that, using a particular time series, the model has been identified and the parameters estimated using the methods described in Chapters 6 and 7. The question remains of deciding whether this model is adequate. If there should be evidence of serious inadequacy, we shall need to know how the model should be modified in the next iterative cycle. What we are doing is only partially described by the words, "testing goodness of fit." We need to discover *in what way* a model is inadequate, so as to suggest appropriate modification. To illustrate, by reference to familiar procedures outside time series analysis, the scrutiny of residuals for the Analysis of Variance, described by Anscombe and Tukey [72], [73], and the criticism of factorial experiments, leading to Normal plotting and other methods due to Daniel [74], would be called diagnostic checks.

No model form ever represents the truth absolutely. It follows that, given sufficient data, statistical tests can discredit models which could nevertheless be entirely adequate for the purpose at hand. Alternatively, tests can fail to indicate serious departures from assumptions because these tests are insensitive to the types of discrepancies that occur. The best policy is to devise the most sensitive statistical procedures possible but be prepared, for sufficient reason, to employ models which exhibit slight lack of fit. Know the facts as clearly as they can be known—then use judgement.

Clearly, diagnostic checks must be such that they *place the model in jeopardy*. That is to say, they must be sensitive to discrepancies which are likely to happen. No system of diagnostic checks can ever be comprehensive, since it is always possible that characteristics in the data of an unexpected kind could be overlooked. However, if diagnostic checks, which have been thoughtfully devised, are applied to a model fitted to a reasonably large body of data and fail to show serious discrepancies, then we shall rightly feel more comfortable about using that model.

8.1.2 Overfitting

One technique which can be used for diagnostic checking is *overfitting*. Having identified what is believed to be a correct model, we actually fit a more elaborate one. This puts the identified model in jeopardy, because the more elaborate model contains additional parameters covering feared directions of discrepancy. Careful thought should be given to the question of how the model should be augmented. In particular, in accordance with the discussion on model redundancy in Section 7.3.5, it would be foolish to add factors *simultaneously* to both sides of the ARMA model. If the analysis fails to show that the additions are needed, we shall not, of course, have proved that our model is correct. A model is only capable of being "proved" in the biblical sense of being put to the test. As was recommended by Saint Paul in his first epistle to the Thessalonians, what we can do is to "Prove all things; hold fast to that which is good."

An example of overfitting. As an example, we consider again some IBM stock price data. For this analysis, data was employed which is listed as Series B', in the collection of time series at the end of this volume. This series consists of IBM stock prices for the period* June 29, 1959–June 30, 1960. The (0, 1, 1) model

$$\nabla z_t = (1 - \theta B)a_t$$

with $\hat{\lambda}_0 = 1 - \hat{\theta} = 0.90$, was identified and fitted to the 255 available observations.

The (0, 1, 1) model can equally well be expressed in the integrated form (4.3.3) as

$$z_t = \lambda_0 S a_{t-1} + a_t$$

---

*The IBM stock data previously considered, and referred to as Series B, covers a different period, namely, May 17, 1961–November 2, 1962.

The extended model which was considered in the overfitting procedure was the $(0, 3, 3)$ process

$$\nabla^3 z_t = (1 - \theta_1 B - \theta_2 B^2 - \theta_3 B^3)a_t$$

or in integrated form

$$z_t = \lambda_0 S a_{t-1} + \lambda_1 S^2 a_{t-1} + \lambda_2 S^3 a_{t-1} + a_t$$

The immediate motivation for extending the model in this particular way, was to test a suggestion made by Brown [2] that the series should be forecasted by an adaptive *quadratic* forecast function. Now, it was shown in Chapter 5, that an IMA $(0, q, q)$ process has for its optimal forecasting function, an adaptive polynomial of degree $q - 1$. Thus, for the above extended $(0, 3, 3)$ model the optimal, lead $- l$, forecast function is the quadratic polynomial in l

$$\hat{z}_t(l) = b_0^{(t)} + b_1^{(t)}l + b_2^{(t)}l^2$$

where the coefficients $b_0^{(t)}$, $b_1^{(t)}$ and $b_2^{(t)}$ are adjusted as each new piece of data becomes available.

However, the model we have identified is an IMA $(0, 1, 1)$ process, yielding a forecast function

$$\hat{z}_t(l) = b_0^{(t)} \qquad\qquad (8.1.1)$$

This is "a polynomial in l" of degree zero. Hence, the model implies that the forecast $\hat{z}_t(l)$ is independent of l; that is, the forecast at any particular time t is the same for one step ahead, two steps ahead, and so on. In other words, the series contains information only on the future *level* of the series, and nothing about slope or curvature. At first sight this is somewhat surprising because, using hindsight, quite definite linear and curvilinear trends appear to be present in the series. Therefore, it is worthwhile to check whether nonzero values of λ_1 and λ_2, which would produce predictable trends, actually occur. Sum of squares grids for $S(\lambda_1, \lambda_2|\lambda_0)$ are shown for $\lambda_0 = 0.7$, 0.9, and 1.1, in Figure 8.1, from which it can be seen that the minimum is close to $\hat{\lambda}_0 = 0.9$, $\hat{\lambda}_1 = 0$, and $\hat{\lambda}_2 = 0$. It is also clear that values of $\lambda_1 > 0$ and $\lambda_2 > 0$ lead to higher sums of squares and, therefore, departures from the identified IMA $(0, 1, 1)$ model in these directions are counter-indicated. This implies, in particular, that a quadratic forecast function would give worse instead of better forecasts, than those obtained from (8.1.1), as was indeed shown to be the case in Section A5.3.3.

8.2 DIAGNOSTIC CHECKS APPLIED TO RESIDUALS

The method of overfitting, by extending the model in a particular direction, assumes that we know what kind of discrepancies are to be feared. Procedures less dependent upon such knowledge are based on the analysis of *residuals*.

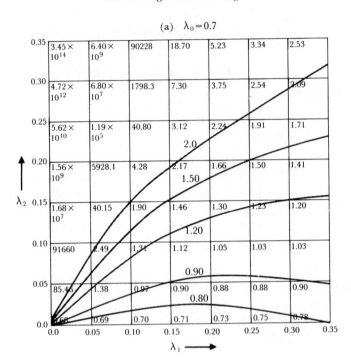

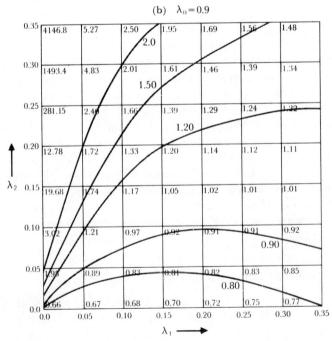

Fig 8-1a,b

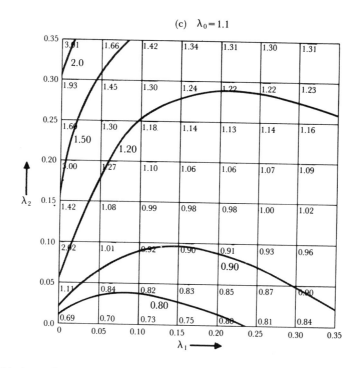

FIG. 8.1(a, b, c)   Sum of squares grids and contours for Series B' for extended model
of order (0, 3, 3)

It cannot be too strongly emphasized that *visual inspection of a plot of the
residuals themselves* is an indispensable first step in the checking process.

8.2.1 Autocorrelation check

Suppose a model

$$\phi(B)\tilde{w}_t = \theta(B)a_t$$

with $w_t = \nabla^d z_t$, has been fitted and ML estimates $(\hat{\phi}, \hat{\theta})$ obtained for the
parameters. Then we shall refer to the quantities

$$\hat{a}_t = \hat{\theta}^{-1}(B)\hat{\phi}(B)\tilde{w}_t \qquad (8.2.1)$$

as the *residuals*. Now it is possible to show that, if the model is adequate,

$$\hat{a}_t = a_t + 0\!\left(\frac{1}{\sqrt{n}}\right)$$

As the series length increases, the $\hat{a}_t$'s become close to the white noise a_t's.
Therefore, one might expect that study of the $\hat{a}_t$'s could indicate the existence

and nature of model inadequacy. In particular, recognizable patterns in the estimated autocorrelation function of the $\hat{a}_t$'s could point to appropriate modifications in the model. This point is discussed further in Section 8.3.

Now suppose the form of the model were correct and that we *knew* the true parameter values ϕ and θ. Then, using (2.1.13) and a result of Anderson [52], the estimated autocorrelations $r_k(a)$, of the a's, would be uncorrelated and distributed approximately Normally about zero with variance n^{-1}, and hence, with a standard error of $n^{-1/2}$. We could use these facts to assess approximately the statistical significance of apparent departures of these autocorrelations from zero.

Now, in practice, we do not know the *true* parameter values. We have only the estimates $(\hat{\phi}, \hat{\theta})$ from which, using (8.2.1), we can calculate not the a's but the $\hat{a}$'s. The autocorrelations $r_k(\hat{a})$ of the $\hat{a}$'s can yield valuable evidence concerning lack of fit and the possible nature of model inadequacy. However, it was pointed out by Durbin [75] that it might be dangerous to assess the statistical significance of apparent discrepancies of these autocorrelations $r_k(\hat{a})$ from their theoretical zero values on the basis of a standard error $n^{-1/2}$, appropriate to the $r_k(a)$'s. He was able to show, for example, that for the AR (1) process with parameter ϕ, the variance of $r_1(\hat{a})$ is $\phi^2 n^{-1}$, which can be very substantially *less* than n^{-1}. The large sample variances and covariances for all the autocorrelations of the $\hat{a}$'s from any ARMA process were subsequently derived by Box and Pierce [77]. They showed that, while in all cases, a reduction in variance can occur for low lags and that at these low lags the $r_k(\hat{a})$'s can be highly correlated, these effects usually disappear rather quickly at high lags. Thus, the use of $n^{-1/2}$ as the standard error for $r_k(\hat{a})$ would underestimate the statistical significance of apparent departures from zero of the autocorrelations at low lags but could usually be employed for moderate or high lags.

For illustration, the large sample one standard error limits and two standard error limits of the $r_k(\hat{a})$'s, for two first-order autoregressive processes and two second-order autoregressive processes, are shown in Figure 8.2. These also supply the corresponding approximate standard errors for moving average processes with the same parameters (as indicated in Figure 8.2).

It may be concluded that, except at moderately high lags, $n^{-1/2}$ must be regarded as supplying an upper bound for the standard errors of the $r_k(\hat{a})$'s, rather than the standard errors themselves. If, for low lags, we use the standard error $n^{-1/2}$ for the $r_k(\hat{a})$'s, we may seriously *underestimate* the significance of apparent discrepancies.

8.2.2 A portmanteau lack of fit test

Rather than consider the $r_k(\hat{a})$'s individually, an indication is often needed of whether, say, the first 20 autocorrelations of the $\hat{a}$'s, *taken as a whole,* indicate inadequacy of the model. Suppose that we have the first K auto-correlations* $r_k(\hat{a})$ $(k = 1, 2, \ldots, K)$ from any ARIMA (p, d, q) process, then it is possible to show [77] that, if the fitted model is appropriate,

$$Q = n \sum_{k=1}^{K} r_k^2(\hat{a}) \tag{8.2.2}$$

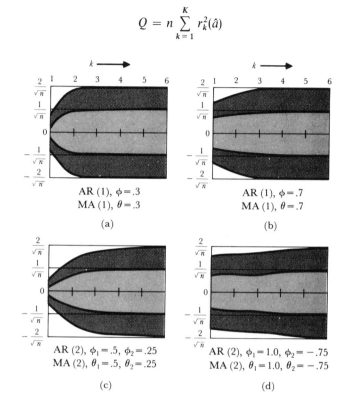

FIG. 8.2   Standard error limits for residual autocorrelations $r_k(\hat{a})$

is approximately distributed as $\chi^2(K - p - q)$, where $n = N - d$ is the number of w's used to fit the model. On the other hand, if the model is inappropriate, the average values of Q will be inflated. Therefore, an approximate, general, or "portmanteau" test of the hypothesis of model adequacy, designed to take account of the difficulties discussed above, may be made by referring an observed value of Q to a table of the percentage points of χ^2 (such as Table F at the end of this volume).

---

* It is assumed here that K is taken sufficiently large so that the weights ψ_j in the model, written in the form

$$\tilde{w}_t = \phi^{-1}(B)\theta(B)a_t = \psi(B)a_t$$

will be negligibly small after $j = K$.

The $(0, 2, 2)$ *model fitted to Series C.* To illustrate the portmanteau criterion (8.2.2), Table 8.1 shows the first 25 autocorrelations $r_k(\hat{a})$ of the residuals from the IMA $(0, 2, 2)$ process $\nabla^2 z_t = (1 - 0.13B - 0.12B^2)a_t$, which was one of the models fitted to Series C in Chapter 7.

TABLE 8.1   Autocorrelations $r_k(\hat{a})$ of residuals from the model $\nabla^2 z_t = (1 - 0.13B - 0.12B^2)a_t$ fitted to Series C

k	$r_k(\hat{a})$	k	$r_k(\hat{a})$	k	$r_k(\hat{a})$	k	$r_k(\hat{a})$	k	$r_k(\hat{a})$
1	0.020	6	−0.033	11	−0.129	16	−0.050	21	0.007
2	0.032	7	0.022	12	0.063	17	0.153	22	0.132
3	−0.125	8	−0.056	13	−0.084	18	−0.092	23	0.012
4	−0.078	9	−0.130	14	0.022	19	−0.005	24	−0.012
5	−0.011	10	0.093	15	−0.006	20	−0.015	25	−0.127

Since there are $n = 224$ w's, the approximate upper bound for the standard error of a single autocorrelation is $1/\sqrt{224} \approx 0.07$. Compared with this standard error bound, the values $r_3(\hat{a}) = -0.125$, $r_9(\hat{a}) = -0.130$, $r_{11}(\hat{a}) = -0.129$, $r_{17}(\hat{a}) = 0.153$, $r_{22}(\hat{a}) = 0.132$, and $r_{25}(\hat{a}) = -0.127$, are all rather large. Of course, occasional large deviations occur even in random series, but taking these results as a whole, there must certainly be a suspicion of lack of fit of the model.

To make a more formal assessment, we refer

$$Q = 224\{(0.020)^2 + (0.032)^2 + \cdots + (-0.127)^2\} = 33.7$$

to a χ^2 table with 23 degrees of freedom. The 10% and 5% points for χ^2, with 23 degrees of freedom, are 32.0 and 35.2, respectively. Therefore, there is some doubt as to the adequacy of this model.

The $(1, 1, 0)$ *model fitted to Series C.* The first 25 autocorrelations of the residuals from the model $(1 - 0.82B)\nabla z_t = a_t$, which we decided in Chapter 7 gave a preferable representation of Series C, are shown in Table 8.2. For this model, $Q = 225 \sum_{k=1}^{25} r_k^2(\hat{a}) = 28.9$. Comparison with the χ^2 table for 24 degrees of freedom shows that there is no ground here for questioning this model.

TABLE 8.2   Autocorrelations $r_k(\hat{a})$ of residuals from the model $(1 - 0.82B)\nabla z_t = a_t$ fitted to Series C

k	$r_k(\hat{a})$	k	$r_k(\hat{a})$	k	$r_k(\hat{a})$	k	$r_k(\hat{a})$	k	$r_k(\hat{a})$
1	−0.007	6	0.019	11	−0.098	16	−0.039	21	0.001
2	−0.002	7	0.073	12	0.074	17	0.165	22	0.129
3	−0.061	8	−0.030	13	−0.054	18	−0.083	23	0.014
4	−0.014	9	−0.097	14	0.034	19	−0.004	24	−0.017
5	0.047	10	0.133	15	0.002	20	−0.009	25	−0.129

Table 8.3 summarizes the values of the criterion Q, based on 25 residual autocorrelations for the models fitted to Series A–F in Table 7.13.

TABLE 8.3   Summary of results of portmanteau test applied to residuals of various models fitted to Series A–F

Series	$n = N - d$	Fitted model	Q	Degrees of freedom
A	197	$z_t - 0.92z_{t-1} = 1.45 + a_t - 0.58a_{t-1}$	26.5	23
	196	$\nabla z_t = a_t - 0.70a_{t-1}$	29.9	24
B	368	$\nabla z_t = a_t + 0.09a_{t-1}$	37.1	24
C	225	$\nabla z_t - 0.82\nabla z_{t-1} = a_t$	28.9	24
	224	$\nabla^2 z_t = a_t - 0.13a_{t-1} - 0.12a_{t-2}$	33.7	23
D	310	$z_t - 0.87z_{t-1} = 1.17 + a_t$	10.8	24
	309	$\nabla z_t = a_t - 0.06a_{t-1}$	18.0	24
E	100	$z_t - 1.42z_{t-1} + 0.73z_{t-2} = 14.35 + a_t$	23.7	23
	100	$z_t - 1.57z_{t-1} + 1.02z_{t-2} - 0.21z_{t-3} = 11.31 + a_t$	23.7	22
F	70	$z_t + 0.34z_{t-1} - 0.19z_{t-2} = 58.87 + a_t$	11.3	23

Inspection of Table 8.3 shows that only two suspiciously large values of Q occur. One is the value 33.7 obtained after fitting a $(0, 2, 2)$ model to Series C, which we have discussed already. The other is the value $Q = 37.1$ obtained after fitting a $(0, 1, 1)$ model to Series B. This suggests some model inadequacy, since the 5% and 2.5% points for χ^2 with 24 degrees of freedom are 36.4 and 39.3, respectively. We consider the possible nature of this inadequacy in the next section.

8.2.3 Model inadequacy arising from changes in parameter values

One interesting form of model inadequacy, which may be imagined, occurs when the *form* of model remains the same but the parameters change over a prolonged period of time. Evidence exists which might explain in these terms, the possible inadequacy of the $(0, 1, 1)$ model fitted to the IBM data.

Table 8.4 shows the results obtained by fitting $(0, 1, 1)$ processes separately to the first and second halves of Series B, as well as to the complete series.

Denoting the estimates obtained from the two halves by $\hat{\lambda}^{(1)}$ and $\hat{\lambda}^{(2)}$, we find that the standard error of $\hat{\lambda}^{(1)} - \hat{\lambda}^{(2)}$ is $\sqrt{(0.070)^2 + (0.074)^2} = 0.102$. Since the difference $\hat{\lambda}^{(1)} - \hat{\lambda}^{(2)} = 0.26$ is 2.6 times its standard error, it is likely that a real change in λ has occurred. Inspection of the Q values suggests that the $(0, 1, 1)$ model, with parameters appropriately modified for different time periods, might explain the series more exactly.

TABLE 8.4   Comparison of IMA $(0, 1, 1)$ models fitted to first and second halves of
Series B

	n	$\hat{\theta}$	$\hat{\lambda} = 1 - \hat{\theta}$	$\hat{\sigma}(\hat{\lambda}) = \left\{ \dfrac{\hat{\lambda}(1 - \hat{\lambda})}{n} \right\}^{1/2}$	Q	Degrees of freedom
First half	184	−0.29	1.29	±0.070	22.6	24
Second half	183	−0.03	1.03	±0.074	33.9	24
Complete series	368	−0.09	1.09	±0.052	37.1	24

8.2.4 Cumulative periodogram check

In some situations, particularly in the fitting of seasonal time series, which
are discussed in Chapter 9, it may be feared that we have not adequately
taken into account the *periodic* characteristics of the series. Therefore, we
are on the lookout for periodicities in the residuals. The autocorrelation
function will not be a sensitive indicator of such departures from randomness, because periodic effects will typically dilute themselves among several
autocorrelations. The periodogram, on the other hand, is specifically
designed for the detection of periodic patterns in a background of white
noise.

The periodogram of a time series $a_t, t = 1, 2, \ldots, n$, as defined in Section
2.2.1 is

$$I(f_i) = \frac{2}{n} \left[\left(\sum_{t=1}^{n} a_t \cos 2\pi f_i t \right)^2 + \left(\sum_{t=1}^{n} a_t \sin 2\pi f_i t \right)^2 \right] \qquad (8.2.3)$$

where $f_i = i/n$ is the frequency. Thus, it is a device for correlating the a_t's
with sine and cosine waves of different frequencies. A pattern with given
frequency f_i in the residuals is reinforced when correlated with a sine or cosine
wave at that same frequency, and so produces a large value of $I(f_i)$.

The cumulative periodogram.   It has been shown by Bartlett [78] (see
also [27]) that the *cumulative periodogram* provides an effective means for
the detection of periodic nonrandomness.

The power spectrum $p(f)$ for white noise has a constant value $2\sigma_a^2$ over
the frequency domain 0–0.5 cycles. Consequently, the cumulative spectrum
for white noise

$$P(f) = \int_0^f p(g) \, dg \qquad (8.2.4)$$

plotted against f is a straight line running from $(0, 0)$ to $(0.5, \sigma_a^2)$, that is, $P(f)/\sigma_a^2$ is a straight line running from $(0, 0)$ to $(0.5, 1)$.

As mentioned in Section 2.2.3, $I(f)$ provides an estimate of the power spectrum at frequency f. In fact, for white noise, $E[I(f)] = 2\sigma_a^2$, and hence the estimate is unbiased. It follows that $1/n \sum_{i=1}^{j} I(f_i)$ provides an unbiased estimate of the integrated spectrum $P(f_j)$, and

$$C(f_j) = \frac{\sum_{i=1}^{j} I(f_i)}{ns^2} \tag{8.2.5}$$

an estimate of $P(f_j)/\sigma_a^2$, where s^2 is an estimate of σ_a^2. We shall refer to $C(f_j)$ as the *normalized cumulative periodogram*.

Now if the model were adequate and the parameters known *exactly*, then the a's could be computed from the data and would yield a white noise series. For a white noise series, the plot of $C(f_j)$ against f_j would be scattered about a straight line joining the points $(0, 0)$ and $(0.5, 1)$. On the other hand, model inadequacies would produce nonrandom a's, whose cumulative periodogram could show systematic deviations from this line. In particular, periodicities in the a's would tend to produce a series of neighboring values of $I(f_j)$ which were large. These large ordinates would reinforce each other in $C(f_j)$ and form a bump on the expected straight line.

In practice, we do not know the exact values of the parameters, but only their estimated values. We do not have the a's, but only the estimated residual $\hat{a}$'s. However, for large samples, the periodogram for the $\hat{a}$'s will have similar properties to that for the a's. Thus careful inspection of the periodogram of the $\hat{a}$'s can provide a useful additional diagnostic check, particularly for indicating periodicities inadequately taken account of.

Example: Series C. We have seen that Series C is well fitted by the model of order $(1, 1, 0)$

$$(1 - 0.82B)\nabla z_t = a_t$$

and somewhat less well by the IMA $(0, 2, 2)$ model

$$\nabla^2 z_t = (1 - 0.13B - 0.12B^2)a_t$$

which is almost equivalent to it. We illustrate the cumulative periodogram test by showing what happens when we analyze the residual a's after fitting to the series the inadequate IMA $(0, 1, 1)$ model

$$\nabla z_t = (1 - \theta B)a_t$$

When the model is thus restricted, the least squares estimate of θ is found to be -0.65. The normalized cumulative periodogram plot of the residuals from this model is shown in Figure 8.3(a). Study of the figure shows immediately that there are marked departures from linearity in the cumulative

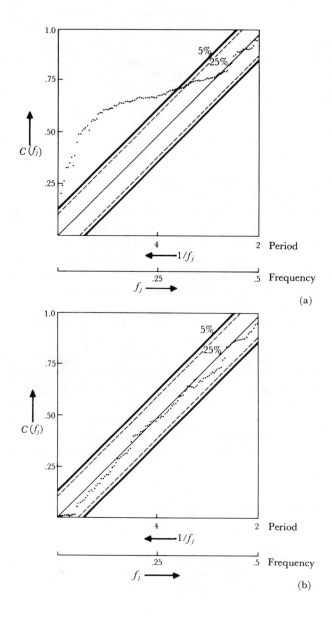

Fig. 8.3 Series C: Cumulative periodograms of residuals from best fitting processes (a) of order (0, 1, 1) and (b) of order (0, 2, 2)

periodogram. These departures are very pronounced at low frequencies and are what might be expected if, for example, there were insufficient differencing. Figure 8.3(b) shows the corresponding plot for the best fitting IMA (0, 2, 2) process. The points of the cumulative periodogram now cluster more closely about the expected line, although, as we have seen in Table 8.3, other evidence points to the inadequacy of this model.

It is wise to indicate on the diagram the period as well as the frequency. This makes for easy identification of the bumps which occur when residuals contain periodicities. For example, in monthly sales data, bumps near periods 12, 24, 36, etc., might indicate that seasonal effects were inadequately accounted for.

The probability relationship between the cumulative periodogram and the integrated spectrum is precisely the same as that between the empirical cumulative frequency function and the cumulative distribution function. For this reason we can assess deviations of the periodogram from that expected if the $\hat{a}$'s were white noise, by use of the Kolmogorov–Smirnov test [79]. Using this test, we can place limit lines about the theoretical line. The limit lines are such that if the $\hat{a}_t$ series were white noise, then the cumulative periodogram would deviate from the straight line sufficiently to cross these limits only with the stated probability. Now, because the $\hat{a}$'s are fitted values and not the true a's, we know that even when the model is correct they will not precisely follow a white noise process. Thus, as a test for model inadequacy, application of the Kolmogorov–Smirnov limits will indicate only approximate probabilities. However, it is worthwhile to show these limits on the cumulative periodogram to provide a rough guide as to what deviations to regard with scepticism and what to take more note of.

The limit lines are such that, for a truly random series, they would be crossed a proportion ε of the time. They are drawn at distances $\pm K_\varepsilon/\sqrt{q}$ above and below the theoretical line, where $q = (n - 2)/2$ for n even and $(n - 1)/2$ for n odd. Approximate values for K_ε are given in Table 8.5.

TABLE 8.5   Coefficients for calculating approximate probability limits for cumulative periodogram test

ε	0.01	0.05	0.10	0.25
K_ε	1.63	1.36	1.22	1.02

For Series C, $q = (224–2)/2 = 111$, and the 5% limit lines inserted on Figure 8.3 deviate from the theoretical line by amounts $\pm 1.36/\sqrt{111} = \pm 0.13$. Similarly, the 25% limit lines deviate by $\pm 1.02/\sqrt{111} = \pm 0.10$.

Conclusion. Each of the checking procedures mentioned above has essential advantages and disadvantages. Checks based on the study of the estimated autocorrelation function and the cumulative periodogram, although they can point out *unsuspected* peculiarities of the series, may not be particularly sensitive. Tests for specific departures by overfitting are more sensitive, but may fail to warn of trouble other than that specifically anticipated.

8.3 USE OF RESIDUALS TO MODIFY THE MODEL

8.3.1 Nature of the correlations in the residuals when an incorrect model is used

When the autocorrelation function of the residuals from some fitted model has indicated model inadequacy, it becomes necessary to consider in what way the model ought to be modified. In Section 8.3.2 we show how the auto-correlations of the residuals can be used to suggest such modifications. By way of introduction, we consider the effect of fitting an incorrect model on the autocorrelation function of the residuals.

Suppose the correct model is

$$\phi(B)\tilde{w}_t = \theta(B)a_t$$

but that an incorrect model

$$\phi_0(B)\tilde{w}_t = \theta_0(B)b_t$$

is used. Then the residuals b_t, in the incorrect model, will be correlated and since

$$b_t = \theta_0^{-1}(B)\theta(B)\phi_0(B)\phi^{-1}(B)a_t \qquad (8.3.1)$$

the autocovariance generating function of the b's will be

$$\sigma_a^2\{\theta_0^{-1}(B)\theta_0^{-1}(F)\theta(B)\theta(F)\phi_0(B)\phi_0(F)\phi^{-1}(B)\phi^{-1}(F)\} \qquad (8.3.2)$$

For example, suppose in an IMA $(0, 1, 1)$ process that, instead of the correct value θ, we use some other value θ_0. Then the residuals b_t would follow the mixed process of order $(1, 0, 1)$

$$(1 - \theta_0 B)b_t = (1 - \theta B)a_t$$

and using (3.4.8)

$$\rho_1 = \frac{(1 - \theta\theta_0)(\theta_0 - \theta)}{1 + \theta^2 - 2\theta\theta_0}$$

$$\rho_j = \rho_1\theta_0^{j-1} \qquad j = 2, 3, \ldots$$

For example, suppose that in the IMA $(0, 1, 1)$ process

$$\nabla z_t = (1 - \theta B)a_t$$

we took $\theta_0 = 0.8$ when the correct value was $\theta = 0$. Then

$$\theta_0 = 0.8 \qquad \theta = 0.0$$

$$\rho_1 = 0.8 \qquad \rho_j = 0.8^j$$

Thus the b's would be highly correlated and would follow the autoregressive process

$$(1 - 0.8B)b_t = a_t$$

8.3.2 Use of residuals to modify the model

Suppose that the residuals b_t from the model

$$\phi_0(B)\nabla^{d_0}z_t = \theta_0(B)b_t \tag{8.3.3}$$

appear to be nonrandom. Using the autocorrelation function of b_t, the methods of Chapter 6 may now be applied to identify a model

$$\bar{\phi}(B)\nabla^{\bar{d}}b_t = \bar{\theta}(B)a_t \tag{8.3.4}$$

for the b_t series. On eliminating b_t between (8.3.3) and (8.3.4), we arrive at a new model

$$\phi_0(B)\bar{\phi}(B)\nabla^{d_0}\nabla^{\bar{d}}z_t = \theta_0(B)\bar{\theta}(B)a_t \tag{8.3.5}$$

which can now be fitted and diagnostically checked.

For example, suppose that a series had been wrongly identified as an IMA $(0, 1, 1)$ process and fitted to give the model

$$\nabla z_t = (1 + 0.6B)b_t \tag{8.3.6}$$

Suppose also that a model

$$\nabla b_t = (1 - 0.8B)a_t \tag{8.3.7}$$

was identified for this residual series. Then on eliminating b_t between (8.3.6) and (8.3.7), we would obtain

$$\nabla^2 z_t = (1 - 0.2B - 0.48B^2)a_t$$

which would suggest that an IMA $(0, 2, 2)$ process should now be entertained.

9

Seasonal Models

In Chapters 4–8 we have considered the properties of a class of linear stochastic models, which are of value in representing stationary and non-stationary time series, and we have seen how these models may be used for forecasting. We then considered the practical problems of identification, fitting, and diagnostic checking which arise when relating these models to actual data. The present chapter applies these methods to analyzing and forecasting seasonal series and also provides an opportunity to show how the ideas of the previous chapters fit together.

9.1 PARSIMONIOUS MODELS FOR SEASONAL TIME SERIES

Figure 9.1 shows the totals of international airline passengers for 1952, 53, and 54. It is part of a longer series (twelve years of data) quoted by Brown [2] and listed in the collection of time series given at the end of the volume as Series G. The series shows a marked seasonal pattern since travel is at its highest in the late summer months, while a secondary peak occurs in the spring. Many other series, particularly sales data, show similar seasonal characteristics.

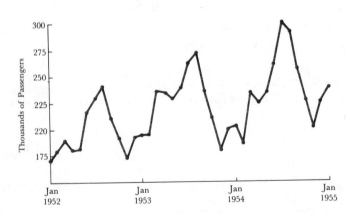

FIG. 9.1 Totals of international airline passengers in thousands (part of Series G)

In general, we say that a series exhibits periodic behavior with period s, when similarities in the series occur after s basic time intervals. In the example above, the basic time interval is one month and the period is $s = 12$ months. However, examples occur when s can take on other values. For example, $s = 4$ for quarterly data showing seasonal effects within years. It sometimes happens that there is more than one periodicity. Thus, because bills tend to be paid monthly, we would expect weekly business done by a bank to show a periodicity of about 4 within months, while monthly business shows a periodicity of 12.

9.1.1 Fitting versus forecasting

One of the deficiencies in the analysis of time series in the past has been the confusion between *fitting* a series and *forecasting* it. For example, suppose that a time series has shown a tendency to increase over a particular period and also to follow a seasonal pattern. A common method of analysis is to decompose the series arbitrarily into three components—a "trend," a "seasonal component," and a "random component." The trend might be fitted by a polynomial and the seasonal component by a Fourier series. A forecast was then made by projecting these fitted functions.

Such methods can give extremely misleading results. For example, we have already seen that the behavior of IBM stock prices (Series B) is closely approximated by the random walk model

$$z_t = \sum_{j=-\infty}^{t} a_j \tag{9.1.1}$$

which implies that $\hat{z}_t(l) = z_t$. In other words, the best forecast of future values of the stock is very nearly today's price. Now, it is true that short lengths of Series B do look as if they might be fitted by quadratic curves. This simply reflects the fact that a sum of random deviates can sometimes have this appearance. However, there is no basis for the use of a quadratic forecast function, which produces very poor forecasts. Of course, genuine systematic effects which can be explained physically should be taken into account by the inclusion of a suitable deterministic component in the model. For example, if it is known that heat is being steadily added to a system, then it would be sensible to explain the resulting increase in temperature by means of a suitable deterministic function of time, in addition to the stochastic component.

9.1.2 Seasonal models involving adaptive sines and cosines

The general linear model

$$\tilde{z}_t = \sum_{j=1}^{\infty} \pi_j \tilde{z}_{t-j} + a_t = \sum_{j=1}^{\infty} \psi_j a_{t-j} + a_t \tag{9.1.2}$$

with suitable values for the coefficients π_j and ψ_j, is entirely adequate to describe many seasonal time series. The problem is to choose a suitable system of *parsimonious parameterization* for such models. As we have said before, this is not a mathematical problem, but is a question of finding out how the world tends to behave. To do this one can only proceed by trying out ideas on actual time series and developing those concepts which seem fruitful.

We have seen that, for nonseasonal series, it is usually possible to obtain a useful and parsimonious representation in the form

$$\varphi(B)\tilde{z}_t = \theta(B)a_t \tag{9.1.3}$$

Moreover, the generalized autoregressive operator $\varphi(B)$ determines the eventual forecast function, which is the solution of the difference equation

$$\varphi(B)\hat{z}_t(l) = 0$$

where B is understood to operate on l. In representing seasonal behavior, we shall want the forecast function to trace out a periodic pattern. Our first thought might be that $\varphi(B)$ should produce a forecast function consisting of a mixture of sines and cosines, and possibly mixed with polynomial terms, to allow for changes in the level of the series and changes in the seasonal pattern. Such a forecast function could arise perfectly naturally within the structure of the general model (9.1.3). For example, with monthly data, a forecast function which is a sine wave with a twelve-month period, adaptive in phase and amplitude, will satisfy the difference equation

$$(1 - \sqrt{3}B + B^2)\hat{z}_t(l) = 0$$

where B is understood to operate on l. However, it is not true that periodic behavior is necessarily represented *economically* by mixtures of sines and cosines. Many sine-cosine components would, for example, be needed to represent sales data affected by Christmas, Easter, and other seasonal buying. To take an extreme case, sales of fireworks in Britain are largely confined to the weeks immediately prior to November 5, when the abortive attempt of Guy Fawkes to blow up the Houses of Parliament is celebrated. An attempt to represent the "single spike" of fireworks sales data directly by sines and cosines might be unprofitable. It is clear that a more careful consideration of the problem is needed.

Now, in our previous analysis, we have not necessarily estimated *all* the components of $\varphi(B)$. Where differencing d times was needed to induce stationarity, we have written $\varphi(B) = \phi(B)(1 - B)^d$, which is equivalent to setting d roots of the equation $\varphi(B) = 0$, equal to unity. When such a representation proved adequate, we could proceed with the simpler analysis of $w_t = \nabla^d z_t$. Thus we have used $\nabla = 1 - B$ as a simplifying operator. In other problems, different types of simplifying operators might be appropriate.

For example, the consumption of fuel oil for heat is highly dependent on ambient temperature which, because the earth rotates around the sun, is known to follow approximately a sine wave with period 12 months. In analyzing sales of fuel oil, it might then be sensible to introduce $1 - \sqrt{3}B + B^2$ as a simplifying operator, constituting one of the contributing components of the generalized autoregressive operator $\varphi(B)$. If such a representation proved useful, we could then proceed with the simpler analysis of $w_t = (1 - \sqrt{3}B + B^2)z_t$. This operator, it may be noted, is of the homogeneous nonstationary variety, having zeroes $e^{\pm(i2\pi/12)}$ on the unit circle.

9.1.3 The general multiplicative seasonal model

The simplifying operator $1 - B^s$. The fundamental fact about seasonal time series with period s, is that observations which are s intervals apart are similar. Therefore, one might expect that the operation $B^s z_t = z_{t-s}$ would play a particularly important role in the analysis of seasonal series, and furthermore, since nonstationarity is to be expected in the series z_t, $z_{t-s}, z_{t-2s}, \ldots$, the simplifying operation $\nabla_s z_t = (1 - B^s)z_t = z_t - z_{t-s}$ might be useful. This stable nonstationary operator $1 - B^s$ has s zeroes $e^{i(2\pi k/s)}$ ($k = 0, 1, \ldots, s - 1$) evenly spaced on the unit circle. Furthermore, the eventual forecast function satisfies $(1 - B^s)\hat{z}_t(l) = 0$ and so may (but need not) be represented by a full complement of sines and cosines

$$\hat{z}_t(l) = b_0^{(t)} + \sum_{j=1}^{[s/2]} \left\{ b_{1j}^{(t)} \cos\left(\frac{2\pi jl}{s}\right) + b_{2j}^{(t)} \sin\left(\frac{2\pi jl}{s}\right) \right\}$$

where the b's are adaptive coefficients, and where $[s/2] = \frac{1}{2}s$ if s is even and $[s/2] = \frac{1}{2}(s - 1)$ if s is odd.

The multiplicative model. When we have a series exhibiting seasonal behavior with known periodicity s, it is of value to set down the data in the form of a table containing s columns, such as Table 9.1, which shows the logarithms of the airline data. For seasonal data special care is needed in selecting an appropriate transformation. In this example (see Section 9.3.5) data analysis supports the use of the logarithm.

The arrangement of Table 9.1 emphasizes the fact that, in periodic data, there are not one but two time intervals of importance. For this example, these intervals correspond to months and to years. Specifically, we expect relationships to occur (a) between observations for successive months in a particular year and (b) between the observations for the same month in successive years. The situation is somewhat like that in a two-way analysis of variance model, where similarities can be expected between observations in the same column and between observations in the same row.

TABLE 9.1 Natural logarithms of monthly passenger totals (measured in thousands) in international air travel (Series G)

	Jan.	Feb.	Mar.	Apr.	May	June	July	Aug.	Sept.	Oct.	Nov.	Dec.
1949	4.718	4.771	4.883	4.860	4.796	4.905	4.997	4.997	4.913	4.779	4.644	4.771
1950	4.745	4.836	4.949	4.905	4.828	5.004	5.136	5.136	5.063	4.890	4.736	4.942
1951	4.977	5.011	5.182	5.094	5.147	5.182	5.293	5.293	5.215	5.088	4.984	5.112
1952	5.142	5.193	5.263	5.199	5.209	5.384	5.438	5.489	5.342	5.252	5.147	5.268
1953	5.278	5.278	5.464	5.460	5.434	5.493	5.576	5.606	5.468	5.352	5.193	5.303
1954	5.318	5.236	5.460	5.425	5.455	5.576	5.710	5.680	5.557	5.434	5.313	5.434
1955	5.489	5.451	5.587	5.595	5.598	5.753	5.897	5.849	5.743	5.613	5.468	5.628
1956	5.649	5.624	5.759	5.746	5.762	5.924	6.023	6.004	5.872	5.724	5.602	5.724
1957	5.753	5.707	5.875	5.852	5.872	6.045	6.146	6.146	6.001	5.849	5.720	5.817
1958	5.829	5.762	5.892	5.852	5.894	6.075	6.196	6.225	6.001	5.883	5.737	5.820
1959	5.886	5.835	6.006	5.981	6.040	6.157	6.306	6.326	6.138	6.009	5.892	6.004
1960	6.033	5.969	6.038	6.133	6.157	6.282	6.433	6.407	6.230	6.133	5.966	6.068

Referring to the airline data of Table 9.1, the seasonal effect implies that an observation for a particular month, say April, is related to the observations for previous Aprils. Suppose the tth observation z_t is for the month of April. We might be able to link this observation z_t to observations in previous Aprils by a model of the form

$$\Phi(B^s)\nabla_s^D z_t = \Theta(B^s)\alpha_t \qquad (9.1.4)$$

where $s = 12$, $\nabla_s = 1 - B^s$ and $\Phi(B^s)$, $\Theta(B^s)$ are polynomials in B^s of degrees P and Q, respectively, and satisfying stationarity and invertibility conditions. Similarly, a model

$$\Phi(B^s)\nabla_s^D z_{t-1} = \Theta(B^s)\alpha_{t-1} \qquad (9.1.5)$$

might be used to link the current behavior for March with previous March observations, and so on, for each of the twelve months. Moreover, it would usually be reasonable to assume that the parameters Φ and Θ contained in these monthly models would be approximately the same for each month.

Now the error components $\alpha_t, \alpha_{t-1}, \ldots,$ in these models would not in general be uncorrelated. For example, the total of airline passengers in April, 1960, while related to previous April totals, would also be related to totals in March of 1960, February of 1960, and January of 1960, etc. Thus we would expect that α_t in (9.1.4) would be related to α_{t-1} in (9.1.5) and to α_{t-2}, etc. Therefore, to take care of such relationships, we introduce a second model

$$\phi(B)\nabla^d \alpha_t = \theta(B)a_t \qquad (9.1.6)$$

where now a_t is a white noise process, and $\phi(B)$ and $\theta(B)$ are polynomials in B of degrees p and q, respectively, and satisfying stationarity and invertibility conditions, and $\nabla = \nabla_1 = 1 - B$.

Substituting (9.1.6) in (9.1.4), we finally obtain a general *multiplicative* model

$$\phi_p(B)\Phi_P(B^s)\nabla^d\nabla_s^D z_t = \theta_q(B)\Theta_Q(B^s)a_t \qquad (9.1.7)$$

where for this particular example, $s = 12$. Also, in (9.1.7) the subscripts p, P, q, Q have been added to remind the reader of the orders of the various operators. The resulting multiplicative process will be said to be *of order* $(p, d, q) \times (P, D, Q)_s$. A similar argument can be used to obtain models with three or more periodic components to take care of multiple seasonalities.

9.2 REPRESENTATION OF THE AIRLINE DATA BY A MULTIPLICATIVE $(0, 1, 1) \times (0, 1, 1)_{12}$ MODEL

In the remainder of this chapter, we consider the basic forms for seasonal models of the kind just introduced and their potential for forecasting. We also consider the problems of identification, estimation and diagnostic checking which arise in relating such models to data. No new principles are needed to do this, but merely an application of the procedures and ideas we have already discussed in detail in Chapters 6–8. We proceed in this Section 9.2 by discussing a particular example in considerable detail. In Section 9.3 we discuss those aspects of the general seasonal model which call for special mention.

The detailed illustration in this section will consist of the relating of a model of order $(0, 1, 1) \times (0, 1, 1)_{12}$ to the airline data of Series G. We consider in Section 9.2.1 the model itself; in Section 9.2.2 its forecasting; in Section 9.2.3 its identification; in Section 9.2.4 its fitting; and finally in Section 9.2.5 its diagnostic checking.

9.2.1 The multiplicative $(0, 1, 1) \times (0, 1, 1)_{12}$ model

We have already seen that a simple and widely applicable stochastic model for the analysis of nonstationary time series, which contains no seasonal component, is the IMA $(0, 1, 1)$ process. Suppose, following the argument of Section 9.1.3 that we employed such a model

$$\nabla_{12}z_t = (1 - \Theta B^{12})\alpha_t$$

for linking z's one year apart. Suppose further that we employed a similar model

$$\nabla\alpha_t = (1 - \theta B)a_t$$

for linking α's one month apart, where in general θ and Θ will have different values. Then, on combining these expressions, we would obtain the seasonal multiplicative model

$$\nabla\nabla_{12}z_t = (1 - \theta B)(1 - \Theta B^{12})a_t \qquad (9.2.1)$$

of order $(0, 1, 1) \times (0, 1, 1)_{12}$. The model written explicitly is

$$z_t - z_{t-1} - z_{t-12} + z_{t-13} = a_t - \theta a_{t-1} - \Theta a_{t-12} + \theta\Theta a_{t-13} \qquad (9.2.2)$$

The invertibility region for this model, required by the condition that the roots of $(1 - \theta B)(1 - \Theta B^{12}) = 0$ lie outside the unit circle, is defined by the inequalities

$$-1 < \theta < 1 \qquad -1 < \Theta < 1$$

Note that the moving average operator $(1 - \theta B)(1 - \Theta B^{12}) = 1 - \theta B - \Theta B^{12} + \theta\Theta B^{13}$, on the right of (9.2.1), is of order $q + sQ = 1 + 12(1) = 13$.

We shall show in Sections 9.2.3, 9.2.4, and 9.2.5 that the logged airline data is well fitted by a model of the form (9.2.1), where to a sufficient approximation, $\hat{\theta} = 0.4$, $\hat{\Theta} = 0.6$, and $\hat{\sigma}_a^2 = 1.34 \times 10^{-3}$. It is convenient as a preliminary to consider how, using this model and with these parameter values inserted, future values of the series may be forecast.

9.2.2 Forecasting

In Chapter 4 we saw that there are three basically different ways of considering the general model, each giving rise in Chapter 5 to a different way of viewing the forecast. We consider now these three approaches for the forecasting of the seasonal model (9.2.1).

Difference equation approach. Forecasts are best *computed* directly from the difference equation itself. Thus, since

$$z_{t+l} = z_{t+l-1} + z_{t+l-12} - z_{t+l-13} + a_{t+l} - \theta a_{t+l-1}$$
$$- \Theta a_{t+l-12} + \theta\Theta a_{t+l-13} \qquad (9.2.3)$$

after setting $\theta = 0.4$, $\Theta = 0.6$, the minimum mean square error forecast at lead time l and origin t is given immediately by

$$\hat{z}_t(l) = [z_{t+l-1} + z_{t+l-12} - z_{t+l-13} + a_{t+l} - 0.4a_{t+l-1}$$
$$- 0.6a_{t+l-12} + 0.24a_{t+l-13}] \qquad (9.2.4)$$

As in chapter 5, we refer to

$$[z_{t+l}] = E[z_{t+l}|\theta, \Theta, z_t, z_{t-1},\ldots]$$

as the conditional expectation of z_{t+l} taken at origin t. In this expression the parameters are supposed exactly known and knowledge of the series z_t, $z_{t-1},\ldots$ is supposed to extend into the remote past.

Practical application depends upon the facts that

(a) invertible models fitted to actual data usually yield forecasts which depend appreciably only on recent values of the series

(b) the forecasts are insensitive to small changes in parameter values such as are introduced by estimation errors.

Now

$$[z_{t+j}] = \begin{cases} z_{t+j} & j \leqslant 0 \\ \hat{z}_t(j) & j > 0 \end{cases} \qquad (9.2.5)$$

$$[a_{t+j}] = \begin{cases} a_{t+j} & j \leqslant 0 \\ 0 & j > 0 \end{cases} \qquad (9.2.6)$$

Thus, to obtain the forecasts, as in Chapter 5, we simply replace unknown z's by forecasts, and unknown a's by zeroes. The known a's are, of course, the one-step ahead forecast errors already computed, that is, $a_t = z_t - \hat{z}_{t-1}(1)$.

For example, to obtain the three months ahead forecast, we have

$$z_{t+3} = z_{t+2} + z_{t-9} - z_{t-10} + a_{t+3} - 0.4a_{t+2} - 0.6a_{t-9} + 0.24a_{t-10}$$

Taking conditional expectations at the origin t,

$$\hat{z}_t(3) = \hat{z}_t(2) + z_{t-9} - z_{t-10} - 0.6a_{t-9} + 0.24a_{t-10}$$

that is

$$\hat{z}_t(3) = \hat{z}_t(2) + z_{t-9} - z_{t-10} - 0.6\{z_{t-9} - \hat{z}_{t-10}(1)\} \\ + 0.24\{z_{t-10} - \hat{z}_{t-11}(1)\}$$

Hence

$$\hat{z}_t(3) = \hat{z}_t(2) + 0.4z_{t-9} - 0.76z_{t-10} + 0.6\hat{z}_{t-10}(1) - 0.24\hat{z}_{t-11}(1) \qquad (9.2.7)$$

which expresses the forecast in terms of previous z's and previous forecasts of z's. Although separate expressions for each lead time may readily be written down, computation of the forecasts is best carried out by using the single expression (9.2.4) directly; the elements of its right-hand side being defined by (9.2.5) and (9.2.6).

Figure 9.2 shows the forecasts for lead times up to 36 months, all made at the arbitrarily selected origin, July 1957. We see that the simple model, containing only two parameters, faithfully reproduces the seasonal pattern and supplies excellent forecasts. It is to be remembered, of course, that like all predictions obtained from the general linear stochastic model, the forecast function is adaptive. When changes occur in the seasonal pattern, these will be appropriately projected into the forecast. It will be noticed that when the one month ahead forecast is too high, there is a tendency for all future forecasts from the point to be high. This is to be expected because, as has been noted in Appendix A5.1, forecast errors from the same origin,

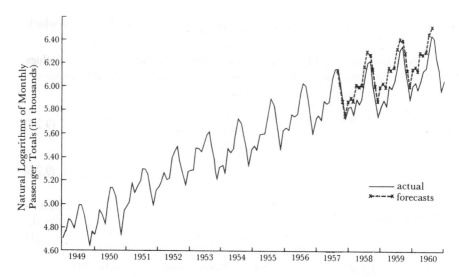

FIG 9.2   Series G, with forecasts for $1, 2, 3, \ldots, 36$ months ahead, all made from an arbitrarily selected origin, July 1957

but for different lead times, are highly correlated. Of course, a forecast for a long lead time, such as 36 months, may necessarily contain a fairly large error. However, in practice, an initially remote forecast will be continually updated and, as the lead shortens, greater accuracy will be possible.

The preceding forecasting procedure is robust to moderate changes in the values of the parameters. Thus, if we used $\theta = 0.5$ and $\Theta = 0.5$, instead of the values $\theta = 0.4$ and $\Theta = 0.6$, the forecasts would not be greatly affected. This is true even for forecasts made several steps ahead, e.g. 12 months. The approximate effect on the one step ahead forecasts of modifying the values of the parameters can be seen by studying the sum of squares surface. Thus, we know that the approximate confidence region for the k parameters $\boldsymbol{\beta}$ is bounded, in general, by the contour $S(\boldsymbol{\beta}) = S(\hat{\boldsymbol{\beta}})\{1 + \chi_\varepsilon^2(k)/n\}$, which includes the true parameter point with probability $1 - \varepsilon$. Therefore, we know that, had the *true* parameter values been employed, with this same probability the mean square of the one step ahead forecast errors could not have been increased by a factor greater than $1 + \chi_\varepsilon^2(k)/n$.

The forecast function, its updating and the forecast error variance.   As we have said in Chapter 5, in practice, the difference equation procedure is by far the simplest and most convenient way for actually *computing* forecasts and updating them. However, the difference equation itself does not reveal very much about the *nature* of the forecasts so computed, and about their updating. It is to throw light on these aspects (and not to provide alternative computational procedures) that we now consider the forecasts from other points of view.

The forecast function.   Using (5.1.12),

$$z_{t+l} = \hat{z}_t(l) + e_t(l) \tag{9.2.8}$$

where

$$e_t(l) = a_{t+l} + \psi_1 a_{t+l-1} + \cdots + \psi_{l-1} a_{t+1}$$

Now, the moving average operator on the right of (9.2.1) is of order 13. Hence, from (5.3.2) and for $l > 13$, the forecasts satisfy the difference equation

$$(1 - B)(1 - B^{12})\hat{z}_t(l) = 0 \qquad l > 13 \tag{9.2.9}$$

where in this equation B operates on l.

We now write $l = (r, m)$, $r = 0, 1, 2, \ldots$ and $m = 1, 2, \ldots, 12$, to represent a lead time of r years and m months, so that for example, $l = 15 = (1, 3)$. Then, the forecast function, which is the solution of (9.2.9), with starting conditions given by the first 13 forecasts, is of the form

$$\hat{z}_t(l) = \hat{z}_t(r, m) = b_{0,m}^{(t)} + r b_1^{(t)} \qquad l > 0 \tag{9.2.10}$$

This forecast function contains 13 adjustable coefficients $b_{0,1}^{(t)}, b_{0,2}^{(t)}, \ldots, b_{0,12}^{(t)}, b_1^{(t)}$. These represent 12 monthly contributions and one yearly contribution and are determined by the first 13 forecasts. The nature of this function is more clearly understood from Figure 9.3, which shows a forecast function of this kind, but with the period $s = 5$, so that there are six adjustable coefficients $b_{0,1}^{(t)}, b_{0,2}^{(t)}, \ldots, b_{0,5}^{(t)}, b_1^{(t)}$.

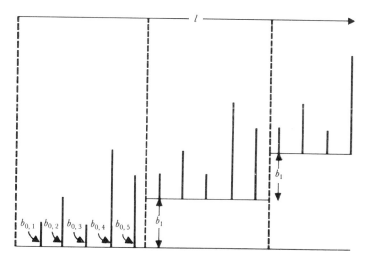

FIG. 9.3   A seasonal forecast function generated by the model $\nabla\nabla_s z_t = (1 - \theta B) \times (1 - \Theta B^s)a_t$, with $s = 5$

The ψ weights. To determine updating formulae and to obtain the variance of the forecast error $e_t(l)$, we need the ψ weights in the form $z_t = \sum_{j=0}^{\infty} \psi_j a_{t-j}$ of the model. We can write the moving average operator in (9.2.1) in the form

$$(1 - \theta B)(1 - \Theta B^{12}) = (\nabla + \lambda B)(\nabla_{12} + \Lambda B^{12})$$

where $\lambda = 1 - \theta$, $\Lambda = 1 - \Theta$, $\nabla_{12} = 1 - B^{12}$. Hence, the model (9.2.1) may be written

$$\nabla \nabla_{12} z_t = (\nabla + \lambda B)(\nabla_{12} + \Lambda B^{12}) a_t$$

On summing, we obtain

$$z_t = \lambda S a_{t-1} + \Lambda S_{12} a_{t-12} + \lambda \Lambda S S_{12} a_{t-13} + a_t$$

where $S_{12} = 1 + B^{12} + B^{24} + \dots$ Thus the ψ weights for this process are

$$\psi_0 = 1$$

$$\psi_1 = \psi_2 = \dots = \psi_{11} = \lambda \qquad\qquad \psi_{12} = \lambda + \Lambda$$

$$\psi_{13} = \psi_{14} = \dots = \psi_{23} = \lambda(1 + \Lambda) \qquad \psi_{24} = \lambda(1 + \Lambda) + \Lambda$$

$$\psi_{25} = \psi_{26} = \dots = \psi_{35} = \lambda(1 + 2\Lambda) \qquad \psi_{36} = \lambda(1 + 2\Lambda) + \Lambda$$

and so on. Writing ψ_j as $\psi_{r,m}$, where $r = 0, 1, 2, \dots$ and $m = 1, 2, 3, \dots, 12$, refer respectively to years and months, we obtain

$$\psi_{r,m} = \lambda(1 + r\Lambda) + \delta\Lambda \qquad\qquad (9.2.11)$$

where

$$\delta = \begin{cases} 1, & \text{when } m = 12 \\ 0, & \text{when } m \neq 12 \end{cases}$$

Updating. The general updating formula (5.2.5) is

$$\hat{z}_{t+1}(l) = \hat{z}_t(l + 1) + \psi_l a_{t+1}$$

Thus, if $m \neq s = 12$,

$$b_{0,m}^{(t+1)} + r b_1^{(t+1)} = b_{0,m+1}^{(t)} + r b_1^{(t)} + (\lambda + r\lambda\Lambda) a_{t+1}$$

and on equating coefficients of r, the updating formulae are

$$\begin{cases} b_{0,m}^{(t+1)} = b_{0,m+1}^{(t)} + \lambda a_{t+1} \\ b_1^{(t+1)} = b_1^{(t)} + \lambda\Lambda a_{t+1} \end{cases} \qquad\qquad (9.2.12)$$

Alternatively, if $m = s = 12$,

$$b_{0,12}^{(t+1)} + r b_1^{(t+1)} = b_{0,1}^{(t)} + (r + 1) b_1^{(t)} + (\lambda + \Lambda + r\lambda\Lambda) a_{t+1}$$

and in this case,

$$\begin{cases} b_{0,12}^{(t+1)} = b_{0,1}^{(t)} + b_1^{(t)} + (\lambda + \Lambda)a_{t+1} \\ b_1^{(t+1)} = b_1^{(t)} + \lambda\Lambda a_{t+1} \end{cases} \tag{9.2.13}$$

In studying these relations, it should be remembered that $b_{0,m}^{(t+1)}$ will be the updated version of $b_{0,m+1}^{(t)}$. Thus, if the origin t was January of a particular year, then $b_{0,2}^{(t)}$ would be the coefficient for March. After a month had elapsed we should move the forecast origin to February, and the updated version for the March coefficient would now be $b_{0,1}^{(t+1)}$.

The forecast error variance. Knowledge of the ψ weights enables us to calculate the variance of the forecast errors at any lead time l, using the result (5.1.16), namely

$$V(l) = \{1 + \psi_1^2 + \cdots + \psi_{l-1}^2\}\sigma_a^2 \tag{9.2.14}$$

Setting $\lambda = 0.6$, $\Lambda = 0.4$, $\sigma_a^2 = 1.34 \times 10^{-3}$ in (9.2.11) and (9.2.14), the estimated standard deviations $\hat{\sigma}(l)$ of the log forecast errors of the airline data for lead times 1 to 36 are shown in Table 9.2.

TABLE 9.2   Estimated standard deviations of forecast errors for logarithms of airline series at various lead times

Forecast lead times	1	2	3	4	5	6	7	8	9	10	11	12
$\hat{\sigma}(l) \times 10^{-2}$		3.7	4.3	4.8	5.3	5.8	6.2	6.6	6.9	7.2	7.6	8.0	8.2

Forecast lead times	13	14	15	16	17	18	19	20	21	22	23	24
$\hat{\sigma}(l) \times 10^{-2}$		9.0	9.5	10.0	10.5	10.9	11.4	11.7	12.1	12.6	13.0	13.3	13.6

Forecast lead times	25	26	27	28	29	30	31	32	33	34	35	36
$\hat{\sigma}(l) \times 10^{-2}$		14.4	15.0	15.5	16.0	16.4	17.0	17.4	17.8	18.3	18.7	19.2	19.6

The forecasts as a weighted mean of previous observations. If we write the model in the form

$$z_t = \sum_{j=1}^{\infty} \pi_j z_{t-j} + a_t$$

the one step ahead forecast is

$$\hat{z}_t(1) = \sum_{j=1}^{\infty} \pi_j z_{t+1-j}$$

The π weights may be obtained by equating coefficients in

$$(1 - B)(1 - B^{12}) = (1 - \theta B)(1 - \Theta B^{12})(1 - \pi_1 B - \pi_2 B^2 - \cdots)$$

Thus

$$
\left.
\begin{aligned}
\pi_j &= \theta^{j-1}(1-\theta) \quad j = 1, 2, \ldots, 11 \\
\pi_{12} &= \theta^{11}(1-\theta) + (1-\Theta) \\
\pi_{13} &= \theta^{12}(1-\theta) - (1-\theta)(1-\Theta) \\
(1 &- \theta B - \Theta B^{12} + \theta\Theta B^{13})\pi_j = 0 \quad j \geqslant 14
\end{aligned}
\right\}
\qquad (9.2.15)
$$

These weights are plotted in Figure 9.4 for the parameter values $\theta = 0.4$ and $\Theta = 0.6$.

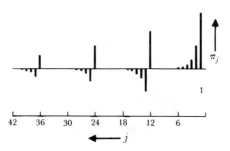

Fig. 9.4 π weights for $(0, 1, 1) \times (0, 1, 1)_{12}$ process fitted to Series G $(\theta = 0.4, \Theta = 0.6)$

The reason that the weight function takes the particular form shown in the figure may be understood as follows: The process (9.2.1) may be written

$$
a_{t+1} = \left\{ 1 - \frac{\lambda B}{1 - \theta B} \right\} \left\{ 1 - \frac{\Lambda B^{12}}{1 - \Theta B^{12}} \right\} z_{t+1} \qquad (9.2.16)
$$

We now use the notation $\mathrm{EWMA}_\lambda(z_t)$ to mean an exponentially weighted moving average, with parameter λ, of values $z_t, z_{t-1}, z_{t-2}, \ldots$, so that

$$
\mathrm{EWMA}_\lambda(z_t) = \frac{\lambda}{1 - \theta B} z_t = \lambda z_t + \lambda\theta z_{t-1} + \lambda\theta^2 z_{t-2} + \cdots
$$

Similarly, we use $\mathrm{EWMA}_\Lambda(z_t)$ to mean an exponentially weighted moving average, with parameter Λ, of values $z_t, z_{t-12}, z_{t-24}, \ldots$, so that

$$
\mathrm{EWMA}_\Lambda(z_t) = \frac{\Lambda}{1 - \Theta B^{12}} z_t = \Lambda z_t + \Lambda\Theta z_{t-12} + \Lambda\Theta^2 z_{t-24} + \cdots
$$

Substituting $\hat{z}_t(1) = z_{t+1} - a_{t+1}$ in (9.2.16), we obtain

$$
\hat{z}_t(1) = \mathrm{EWMA}_\lambda(z_t) + \mathrm{EWMA}_\Lambda(z_{t-11} - \mathrm{EWMA}_\lambda z_{t-12}) \quad (9.2.17)
$$

Thus, the forecast is an EWMA taken over previous months, modified by a second EWMA of discrepancies found between similar monthly EWMA's and actual performance in previous years.

For example, suppose we were attempting to predict December sales for a department store. These sales would include a heavy component from Christmas buying. The first term on the right of (9.2.17) would be an EWMA taken over previous months up to November. However, we know this will be an underestimate, so we correct it by taking a second EWMA over previous years of the *discrepancies* between actual December sales and the corresponding monthly EWMA's taken over previous months in those years.

The forecasts for lead times $l > 1$ can be generated from the π weights by substituting forecasts of shorter lead time for unknown values. Thus

$$\hat{z}_t(2) = \pi_1 \hat{z}_t(1) + \pi_2 z_t + \pi_3 z_{t-1} + \cdots$$

Alternatively, explicit values for the weights applied directly to z_t, z_{t-1}, $z_{t-2}, \ldots$ may be computed, for example, from (5.3.9) or from (A5.2.3).

9.2.3 Identification

The identification of the nonseasonal IMA $(0, 1, 1)$ process depends upon the fact that, after taking first differences, the autocorrelations for all lags beyond the first are zero. For the multiplicative $(0, 1, 1) \times (0, 1, 1)_{12}$ process (9.2.1), the only nonzero autocorrelations of $\nabla\nabla_{12} z_t$ are those at lags 1, 11, 12, and 13. In fact, the autocovariances are

$$\left.\begin{aligned}
\gamma_0 &= (1 + \theta^2)(1 + \Theta^2)\sigma_a^2 \\
\gamma_1 &= -\theta(1 + \Theta^2)\sigma_a^2 \\
\gamma_{11} &= \theta\Theta\sigma_a^2 \\
\gamma_{12} &= -\Theta(1 + \theta^2)\sigma_a^2 \\
\gamma_{13} &= \theta\Theta\sigma_a^2
\end{aligned}\right\} \tag{9.2.18}$$

Table 9.3 shows the estimated autocorrelations of the logged airline data for:
(a) the original logged series, z_t;
(b) the logged series differenced with respect to months only, ∇z_t;
(c) the logged series differenced with respect to years only, $\nabla_{12} z_t$;
(d) the logged series differenced with respect to months and years, $\nabla\nabla_{12} z_t$.

The autocorrelations for z are large and fail to die out at higher lags. While simple differencing reduces the correlations in general, a very heavy periodic component remains. This is evidenced particularly by very large correlations at lags 12, 24, 36 and 48. Simple differencing with respect to period twelve results in correlations which are first persistently positive and then persistently negative. By contrast, the differencing $\nabla\nabla_{12}$ markedly reduces correlation throughout.

TABLE 9.3   Estimated autocorrelations of various differences of the logged airline data

	Lags		Autocorrelations											
(a) z	1–12	0.95	0.90	0.85	0.81	0.78	0.76	0.74	0.73	0.73	0.74	0.76	0.76
	13–24	0.72	0.66	0.62	0.58	0.54	0.52	0.50	0.49	0.50	0.50	0.52	0.52
	25–36	0.48	0.44	0.40	0.36	0.34	0.31	0.30	0.29	0.30	0.30	0.31	0.32
	37–48	0.29	0.24	0.21	0.17	0.15	0.12	0.11	0.10	0.10	0.11	0.12	0.13
(b) ∇z	1–12	0.20	−0.12	−0.15	−0.32	−0.08	0.03	−0.11	−0.34	−0.12	−0.11	0.21	0.84
	13–24	0.22	−0.14	−0.12	−0.28	−0.05	0.01	−0.11	−0.34	−0.11	−0.08	0.20	0.74
	25–36	0.20	−0.12	−0.10	−0.21	−0.06	0.02	−0.12	−0.29	−0.13	−0.04	0.15	0.66
	37–48	0.19	−0.13	−0.06	−0.16	−0.06	0.01	−0.11	−0.28	−0.11	−0.03	0.12	0.59
(c) $\nabla_{12} z$	1–12	0.71	0.62	0.48	0.44	0.39	0.32	0.24	0.19	0.15	−0.01	−0.12	−0.24
	13–24	−0.14	−0.14	−0.10	−0.15	−0.10	−0.11	−0.14	−0.16	−0.11	−0.08	0.00	−0.05
	25–36	−0.10	−0.09	−0.13	−0.15	−0.19	−0.20	−0.19	−0.15	−0.22	−0.23	−0.27	−0.22
	37–48	−0.18	−0.16	−0.14	−0.10	−0.05	0.02	0.04	0.10	0.15	0.22	0.29	0.30
(d) $\nabla\nabla_{12} z$	1–12	−0.34	0.11	−0.20	0.02	0.06	0.03	−0.06	0.00	0.18	−0.08	0.06	−0.39
	13–24	0.15	−0.06	0.15	−0.14	0.07	0.02	−0.01	−0.12	0.04	−0.09	0.22	−0.02
	25–36	−0.10	0.05	−0.03	0.05	−0.02	−0.05	−0.05	0.20	−0.12	0.08	−0.15	−0.01
	37–48	0.05	0.03	−0.02	−0.03	−0.07	0.10	−0.09	0.03	−0.04	−0.04	0.11	−0.05

On the assumption that the model is of the form (9.2.1), the variances for the estimated higher lag autocorrelations are approximated by Bartlett's formula (2.1.13) which in this case becomes

$$\text{var}\,[r_k] \simeq \frac{1 + 2(\rho_1^2 + \rho_{11}^2 + \rho_{12}^2 + \rho_{13}^2)}{n} \qquad k > 13 \qquad (9.2.19)$$

Substituting estimated correlations for the ρ's and setting $n = 144 - 13 = 131$ in (9.2.19), where $n = 131$ is the number of differences $\nabla\nabla_{12}z_t$, we obtain a standard error $\hat{\sigma}(r) \simeq 0.11$.

In Table 9.4 observed frequencies of the 35 autocorrelations r_k, $k > 13$, are compared with those for a Normal distribution having mean zero and standard deviation 0.11. This rough check suggests that the model is worthy of further investigation.

TABLE 9.4 Comparison of observed and expected frequencies for autocorrelations of $\nabla\nabla_{12}z_t$ at lags greater than 13

	Expected from Normal distribution mean zero and std. dev. 0.11	Observed*
$0 < \vert r \vert < 0.11$	23.9	27.5
$0.11 < \vert r \vert < 0.22$	9.5	7.0
$0.22 < \vert r \vert$	1.6	0.5
	35.0	35.0

* Observations on the cell boundary are allocated half to each adjacent cell.

Preliminary estimates. As with the nonseasonal model, by equating observed correlations to their expected values, approximate values can be obtained for the parameters θ and Θ. On substituting the sample estimates $r_1 = -0.34$ and $r_{12} = -0.39$ in the expressions

$$\rho_1 = \frac{-\theta}{1 + \theta^2} \qquad \rho_{12} = \frac{-\Theta}{1 + \Theta^2}$$

we obtain rough estimates $\hat{\theta} \simeq 0.39$ and $\hat{\Theta} \simeq 0.48$. A table summarizing the behavior of the autocorrelation function for some specimen seasonal models, useful in identification and in obtaining preliminary estimates of the parameters, is given in Appendix A9.1.

9.2.4 Estimation

Contours of the sum of squares function $S(\theta, \Theta)$ for the airline data fitted to the model (9.2.1) are shown in Figure 9.5, together with the appropriate

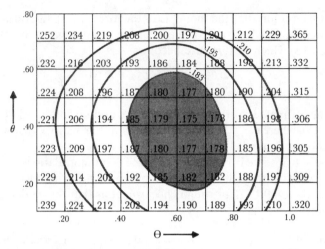

FIG. 9.5   Series G fitted by the model $\nabla\nabla_{12}z_t = (1 - \theta B)(1 - \Theta B^{12})a_t$: contours of $S(\theta, \Theta)$ with shaded 95% confidence region

95% confidence region. The least squares estimates are seen to be very nearly $\hat{\theta} = 0.4$ and $\hat{\Theta} = 0.6$. The grid of values for $S(\theta, \Theta)$ was computed using the technique described in Chapter 7. It was shown there that, given n observations $\mathbf{w}$ from a linear process defined by

$$\phi(B)w_t = \theta(B)a_t$$

or equivalently by

$$\phi(F)w_t = \theta(F)e_t$$

the quadratic form $\mathbf{w}'\mathbf{M}_n\mathbf{w}$, which appears in the exponent of the likelihood, can always be expressed as a sum of squares of the conditional expectation of a's or e's, that is

$$\mathbf{w}'\mathbf{M}_n\mathbf{w} = S(\phi, \theta) = \sum_{t=-\infty}^{n} [a_t]^2 = \sum_{t=1}^{\infty} [e_t]^2$$

where

$$[a_t] = [a_t|\mathbf{w}, \phi, \theta]$$
$$[e_t] = [e_t|\mathbf{w}, \phi, \theta]$$

Furthermore, it was shown that $S(\phi, \theta)$ plays a central role in the estimation of the parameters ϕ and θ both from a sampling theory and a likelihood or Bayesian point of view.

The computation for seasonal models follows precisely the same course as described in Section 7.1.5 for nonseasonal models. We illustrate by considering the computation of $S(\theta, \Theta)$ for the airline data in relation to the model

$$\nabla\nabla_{12} z_t = w_t = (1 - \theta B)(1 - \Theta B^{12}) a_t$$

A format for the computation of the $[a]$'s is shown in Table 9.5. If there are N observations of z, then in general, with a difference operator $\nabla^d \nabla_s^D$, we can compute $n = N - d - sD$ values of w. Therefore, it is convenient to use a numbering system so that the first observation in the z series has a subscript $1 - d - sD$. The first observation in the w series then has a subscript 1 and the last has subscript n. There are $N = 144$ observations in the airline series. Accordingly, in Table 9.5 these are designated as z_{-12}, $z_{-11}, \ldots, z_{131}$. The w's, obtained by differencing, then form the series $w_1, w_2, \ldots, w_n$, where $n = 131$. These values are set out in the center of the table.

TABLE 9.5   Airline data: Computation table for the $[a]$'s, and hence for $S(\theta, \Theta)$

z_t	t	$[a_t]$	$[w_t]$	$[e_t]$
		0	0	0
z_{-12}	-12	$[a_{-12}]$	$[w_{-12}]$	0
z_{-11}	-11	$[a_{-11}]$	$[w_{-11}]$	0
$\vdots$	$\vdots$	$\vdots$	$\vdots$	$\vdots$
z_0	0	$[a_0]$	$[w_0]$	0
z_1	1	$[a_1]$	w_1	$[e_1]$
z_2	2	$[a_2]$	w_2	$[e_2]$
$\vdots$	$\vdots$	$\vdots$	$\vdots$	$\vdots$
z_{131}	131	$[a_{131}]$	w_{131}	$[e_{131}]$

The fundamental formulae, on which backward and forward recursions are based, may be obtained, as before, by taking conditional expectations in the backward and forward forms of the model. In this instance they yield

$$[e_t] = [w_t] + \theta[e_{t+1}] + \Theta[e_{t+12}] - \theta\Theta[e_{t+13}] \qquad (9.2.20)$$

$$[a_t] = [w_t] + \theta[a_{t-1}] + \Theta[a_{t-12}] - \theta\Theta[a_{t-13}] \qquad (9.2.21)$$

In general, for seasonal models, we might have a stationary autoregressive operator $\phi(B)\Phi(B^s)$ of degree $(p + sP)$. If we wished the back computation of Section 7.1.5 to commence as far back in the series as possible, the recursion would begin with the calculation of an approximate value for $[e_{n-p-sP}]$, obtained by setting unknown $[e]$'s equal to zero. In the present example,

$p = P = 0$ and hence, using (9.2.20), we can begin with

$$[e_{131}] = w_{131} + \{\theta \times 0\} + \{\Theta \times 0\} - \{\theta\Theta \times 0\}$$

$$[e_{130}] = w_{130} + \{\theta \times [e_{131}]\} + \{\Theta \times 0\} - \{\theta\Theta \times 0\}$$

and so on until e_1 is obtained. Recalling that $[e_{-j}] = 0$ when $j \geqslant 0$, we can now use (9.2.20) to compute the back forecasts $[w_0], [w_{-1}], \ldots, [w_{-12}]$. Furthermore, values of $[w_{-j}]$ for $j > 12$ are all zero, and since each $[a_t]$ is a function of previously occurring $[w]$'s, it follows (and otherwise is obvious directly from the form of the model) that $[a_{-j}] = 0$, $j > 12$. Thus, (9.2.21) may now be used directly to compute the $[a]$'s, and hence to calculate $S(\theta, \Theta) = \sum_{t=-12}^{131} [a_t]^2$. In almost all cases of interest, the transients introduced by the approximation at the beginning of the back recursion will have negligible effect on the calculation of the preliminary $[w]$'s, so that $S(\theta, \Theta)$ computed in this way will be virtually exact. However, it is possible, as indicated in Section 7.1.5, to continue the "up and down" iteration. The next iteration would involve recomputing the $[e]$'s, starting off the iteration using forecasts $[w_{n+1}], [w_{n+2}], \ldots, [w_{n+13}]$ obtained from the $[a]$'s already calculated.

Iterative calculation of least squares estimates $\hat{\theta}, \hat{\Theta}$. As discussed in Section 7.2, while it is essential to plot sums of squares surfaces in a new situation, or whenever difficulties arise, an iterative linearization technique may be used in straightforward situations to supply the least squares estimates and their approximate standard errors. The procedure has been set out in Section 7.2.1, and no new difficulties arise in estimating the parameters of seasonal models.

For the present example, we can write approximately

$$a_{t,0} = (\theta - \theta_0)x_{1,t} + (\Theta - \Theta_0)x_{2,t} + a_t$$

where

$$x_{1,t} = -\left.\frac{\partial a_t}{\partial \theta}\right|_{\theta_0,\Theta_0} \qquad x_{2,t} = -\left.\frac{\partial a_t}{\partial \Theta}\right|_{\theta_0,\Theta_0}$$

and where θ_0 and Θ_0 are guessed values and $a_{t,0} = [a_t|\theta_0, \Theta_0]$. As explained and illustrated in Section 7.2.2, the derivatives are most easily computed numerically. Proceeding in this way, and using as starting values the preliminary estimates $\hat{\theta} = 0.39$, $\hat{\Theta} = 0.48$ obtained in Section 9.2.3 from the estimated autocorrelations, the iteration proceeded as in Table 9.6.

Thus, values of the parameters correct to two decimals, which is the most that would be needed in practice, are available in three iterations. The estimated variance of the residuals is $\hat{\sigma}_a^2 = 1.34 \times 10^{-3}$. From the inverse of the matrix of sums of squares and products of the x's on the last iteration, the standard errors of the estimates may now be calculated. The least squares estimates followed by their standard errors are then:

$$\hat{\theta} = 0.40 \pm 0.08$$

$$\hat{\Theta} = 0.61 \pm 0.07$$

agreeing closely with the values obtained from the sum of squares plot.

TABLE 9.6   Iterative estimation of θ and Θ for the logged airline data

Iteration	θ	Θ
Starting Values	0.390	0.480
1	0.404	0.640
2	0.395	0.612
3	0.396	0.614
4	0.396	0.614

The estimation procedure can be curtailed as follows. Given moderately good estimates for the parameters obtained from the autocorrelations, we compute back forecasts for the w's as before. These same starting values are then used for all iterations until the process converges. At this point one further iteration, using revised estimates of the back forecasts, will usually bring final convergence. However, with a fast computer, refinements of this sort are hardly worthwhile. Alternatively, proceeding as in Section 7.2.3, the derivatives may be obtained to any degree of required accuracy by recursive calculation.

Large sample variances and covariances for the estimates.   As in Section 7.2.6, large sample formulae for the variances and covariances of the parameter estimates may be obtained. In this case,

$$x_{1,t} \simeq -(1 - \theta B)^{-1} a_{t-1} = -\sum_{j=0}^{\infty} \theta^j B^j a_{t-1}$$

$$x_{2,t} \simeq -(1 - \Theta B^{12})^{-1} a_{t-12} = -\sum_{i=0}^{\infty} \Theta^i B^{12i} a_{t-12}$$

Therefore, for large samples, the information matrix is

$$\mathbf{I}(\theta, \Theta) = n \begin{bmatrix} (1 - \theta^2)^{-1} & \theta^{11}(1 - \theta^{12}\Theta)^{-1} \\ \theta^{11}(1 - \theta^{12}\Theta)^{-1} & (1 - \Theta^2)^{-1} \end{bmatrix}$$

Provided $|\theta|$ is not close to unity, the off-diagonal term is negligible and approximate values for the variances and covariances of $\hat{\theta}$ and $\hat{\Theta}$ are

$$V(\hat{\theta}) \simeq n^{-1}(1 - \theta^2) \qquad V(\hat{\Theta}) \simeq n^{-1}(1 - \Theta^2)$$

$$\text{cov}\,(\hat{\theta}, \hat{\Theta}) \simeq 0 \tag{9.2.22}$$

In the present example, substituting the values $\hat{\theta} = 0.40$, $\hat{\Theta} = 0.61$, and $n = 131$, we obtain

$$V(\hat{\theta}) \simeq 0.0064 \qquad V(\hat{\Theta}) \simeq 0.0048$$

and

$$\sigma(\hat{\theta}) \simeq 0.08 \qquad \sigma(\hat{\Theta}) \simeq 0.07$$

which, to this accuracy, are identical with the values obtained directly from the iteration. It is also interesting to note that the parameter estimates $\hat{\theta}$ and $\hat{\Theta}$, associated with months and with years respectively, are virtually uncorrelated.

9.2.5 Diagnostic checking

Before proceeding further, we check the adequacy of fit of the model by examining the residuals from the fitted process.

Autocorrelation check. The estimated autocorrelations of the residuals $\hat{a}_t = \nabla\nabla_{12}z_t + 0.40\hat{a}_{t-1} + 0.61\hat{a}_{t-12} - 0.24\hat{a}_{t-13}$ are shown in Table 9.7.

A number of individual correlations appear rather large compared with the upper bound 0.09 of their standard error, and the value $r_{23} = 0.22$, which is about 2.5 times this upper bound, is particularly discrepant. However, among 48 random deviates one would expect some large deviations.

An overall check is provided by the quantity $Q = n \sum_{k=1}^{48} r_k^2(\hat{a})$, which (see Section 8.2.2) is approximately distributed as χ^2 with 46 degrees of freedom, since two parameters have been fitted. The observed value of Q is $131 \times 0.2726 = 35.7$, and on the hypothesis of adequacy of the model, deviations greater than this would be expected in about 86% of cases. The check does not provide any evidence of inadequacy in the model.

Periodogram check. The cumulative periodogram (see Section 8.2.4) for the residuals is shown in Figure 9.6. The Kolmogrov–Smirnoff 5% and 25% probability limits which, as we have seen in Section 8.2.4, supply a very rough guide to the significance of apparent deviations, fail in this instance to indicate any significant departure from the assumed model.

TABLE 9.7   Estimated autocorrelations of the residuals from fitting the model $\nabla\nabla_{12}z_t = (1 - 0.40B)(1 - 0.61B^{12})a_t$ to logged airline data (Series G)

Lag k	Autocorrelations $r_k(\hat{a})$												Standard Error (upper bound)
1–12	0.02	0.02	−0.13	−0.14	0.05	0.06	−0.07	−0.04	0.10	−0.08	0.02	−0.01	0.09
13–24	0.03	0.04	0.05	−0.16	0.03	0.00	−0.11	−0.10	−0.03	−0.03	0.22	0.03	0.09
25–36	−0.02	0.06	−0.04	−0.06	−0.05	−0.08	−0.05	0.12	−0.13	0.00	−0.06	−0.02	0.09
37–48	0.11	0.07	−0.02	−0.05	−0.10	−0.02	−0.04	0.00	−0.08	0.03	0.04	0.06	0.09

$$\sum_{k=1}^{48} r_k^2(\hat{a}) = 0.2726$$

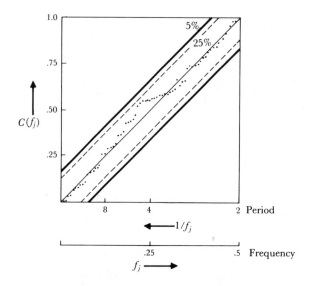

FIG. 9.6 Cumulative periodogram check on residuals from model $\nabla\nabla_{12}z_t = (1 - 0.40B)(1 - 0.61B^{12})a_t$ fitted to Series G

9.3 SOME ASPECTS OF MORE GENERAL SEASONAL MODELS

9.3.1 Multiplicative and non-multiplicative models [']

In previous sections we discussed methods of dealing with seasonal time series, and in particular, we examined an example of a multiplicative model. We have seen how this can provide a useful representation with remarkably few parameters. It now remains to study other seasonal models of this kind, and in so far as new considerations arise, the associated processes of identification, estimation, diagnostic checking and forecasting.

Suppose, in general, we have a seasonal effect associated with period s. Then the general class of multiplicative models may be typified in the manner shown in Table 9.8. In the multiplicative model it is supposed that the "between periods" development of the series is represented by some model

$$\Phi_P(B^s)\nabla_s^D z_{r,m} = \Theta_Q(B^s)\alpha_{r,m}$$

while "within periods" the α's are related by

$$\phi_p(B)\nabla^d \alpha_{r,m} = \theta_q(B)a_{r,m}$$

Obviously, we could change the order in which we considered the two types of models and in either case obtain the general multiplicative model

$$\phi_p(B)\Phi_P(B^s)\nabla^d\nabla_s^D z_{r,m} = \theta_q(B)\Theta_Q(B^s)a_{r,m} \qquad (9.3.1)$$

TABLE 9.8   Two way table for multiplicative seasonal model

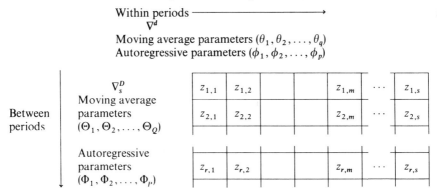

Within periods ————————————————→
∇^d
Moving average parameters $(\theta_1, \theta_2, \ldots, \theta_q)$
Autoregressive parameters $(\phi_1, \phi_2, \ldots, \phi_p)$

	∇^D_s Moving average parameters $(\Theta_1, \Theta_2, \ldots, \Theta_Q)$	$z_{1,1}$	$z_{1,2}$			$z_{1,m}$	$\cdots$	$z_{1,s}$
Between periods		$z_{2,1}$	$z_{2,2}$			$z_{2,m}$	$\cdots$	$z_{2,s}$
	Autoregressive parameters $(\Phi_1, \Phi_2, \ldots, \Phi_p)$	$z_{r,1}$	$z_{r,2}$			$z_{r,m}$	$\cdots$	$z_{r,s}$

where $a_{r,m}$ is a white noise process with zero mean. In practice, the usefulness of models such as (9.3.1) depends on how far it is possible to parameterize actual time series parsimoniously in these terms. In fact, this has been possible for a variety of seasonal time series coming from widely different sources [80].

It is not possible to obtain a completely adequate fit with multiplicative models for all series. One modification which is sometimes useful allows the mixed moving average operator to be non-multiplicative. By this is meant that we replace the operator $\theta_q(B)\Theta_Q(B^s)$ on the right-hand side of (9.3.1) by a more general moving average operator $\theta^*_{q^*}(B)$. Alternatively, or in addition, it may be necessary to replace the autoregressive operator $\phi_p(B)\Phi_P(B^s)$ on the left by a more general autoregressive operator $\phi^*_{p^*}(B)$. Some specimens of non-multiplicative models are given in Appendix A9.1. These are numbered 4, 4a, 5 and 5a.

In those cases where a non-multiplicative model is found necessary, experience suggests that the best fitting multiplicative model can provide a good starting point from which to construct a better non-multiplicative model. The situation is reminiscent of the problems encountered in analyzing two-way analysis of variance tables, where additivity of row and column constants may or may not be an adequate assumption, but may provide a good point of departure.

Our general strategy for relating multiplicative or non-multiplicative models to data is that which we have already discussed and illustrated in some detail in Section 9.2. Using the autocorrelation function for guidance:

(1) The series is differenced with respect to ∇ and ∇_s, so as to produce stationarity.

(2) By inspection of the autocorrelation function of the suitably differenced series, a tentative model is selected.

(3) From the values of appropriate autocorrelations of the differenced

series, preliminary estimates of the parameters are obtained. These can
be used as starting values in the search for the least squares estimates.
(4) After fitting, the diagnostic checking process applied to the residuals
either may lead to the acceptance of the tentative model or, alternatively,
may suggest ways in which it can be improved, leading to refitting and
repetition of the diagnostic checks.

9.3.2 Identification

A useful aid in model identification is the list in Appendix A9.1, giving the
covariance structure of $w_t = \nabla^d \nabla_s^D$ for a number of simple seasonal models.
This list makes no claim to be comprehensive. However, it is believed that it
does include some of the frequently encountered models, and the reader
should have no difficulty in discovering the characteristics of others that
seem representationally useful. It should be emphasized that rather simple
models (such as models 1 and 2 in Appendix A9.1) have provided adequate
representations for many seasonal series.

A fact of considerably utility in deriving the autocovariances of a multiplicative process is that, for such a process, the autocovariance generating
function (3.1.11) is the product of the generating functions of the components.
Thus in (9.3.1) if the component processes

$$\phi_p(B)\nabla^d z_t = \theta_q(B)\alpha_t \qquad \Phi_P(B^s)\nabla_s^D \alpha_t = \Theta_Q(B^s)a_t$$

have autocovariance generating function $\gamma(B)$ and $\Gamma(B^s)$, then the autocovariance generating function for z in (9.3.1) is

$$\gamma(B)\Gamma(B^s)$$

Another point to be remembered is that it may be useful to parametrize
more general models in terms of their departures from related multiplicative
forms in a manner now illustrated.

The three parameter non-multiplicative operator

$$1 - \theta_1 B - \theta_{12} B^{12} - \theta_{13} B^{13} \tag{9.3.2}$$

employed in models 4 and 5 may be written

$$(1 - \theta_1 B)(1 - \theta_{12} B^{12}) - \kappa B^{13}$$

where

$$\kappa = \theta_1 \theta_{12} - (-\theta_{13})$$

An estimate of κ which was large compared with its standard error would
indicate the need for a non-multiplicative model in which the value of θ_{13}
is not tied to the values of θ_1 and θ_{12}. On the other hand, if κ is small, then
on writing $\theta_1 = \theta$, $\theta_{12} = \Theta$, the model approximates the multiplicative
$(0, 1, 1) \times (0, 1, 1)_{12}$ model.

9.3.3 Estimation

No new problems arise in the estimation of the parameters of general seasonal models. The unconditional sum of squares is computed quite generally by the methods set out fully in Section 7.1.5 and illustrated further in Section 9.2.4. As always, contour plotting can illuminate difficult situations. In well behaved situations, iterative least squares with numerical determination of derivatives yield rapid convergence to the least squares estimates, together with approximate variances and covariances of the estimates. Recursive procedures can be derived in each case which allow direct calculation of derivatives, if desired.

Large sample variances and covariances of the estimates. The large sample information matrix $I(\phi, \theta, \Phi, \Theta)$ is given by evaluating $E[X'X]$, where as in Section 7.2.6, X is the $n \times (p + q + P + Q)$ matrix of derivatives with reversed signs. Thus for the general multiplicative model

$$a_t = \theta^{-1}(B)\Theta^{-1}(B^s)\phi(B)\Phi(B^s)w_t$$

where

$$w_t = \nabla^d\nabla_s^D z_t$$

the required derivatives are

$$\frac{\partial a_t}{\partial \theta_i} = \theta^{-1}(B)B^i a_t \qquad\qquad \frac{\partial a_t}{\partial \Theta_i} = \Theta^{-1}(B^s)B^{si} a_t$$

$$\frac{\partial a_t}{\partial \phi_j} = -\phi^{-1}(B)B^j a_t \qquad\qquad \frac{\partial a_t}{\partial \Phi_j} = -\Phi^{-1}(B^s)B^{sj} a_t$$

Approximate variances and covariances of the estimates are obtained as before, by inverting the matrix $I(\phi, \theta, \Phi, \Theta)$.

9.3.4 Eventual forecast functions for various seasonal models

We now consider the characteristics of the eventual forecast functions for a number of seasonal models. For a seasonal model with single periodicity s, the eventual forecast function at origin t for lead time l, is the solution of the difference equation

$$\phi(B)\Phi(B^s)\nabla^d\nabla_s^D \hat{z}_t(l) = 0$$

Table 9.9 shows this solution for various choices of the difference equation; also shown is the number of initial values on which the behavior of the forecast function depends.

TABLE 9.9 Eventual forecast functions for various autoregressive operators. Coefficients b are all adaptive and depend upon forecast origin t

Autoregressive Operator	Eventual Forecast Function $\hat{z}(r, m)$	Number of Initial Values on Which Forecast Function Depends
(1) $\quad 1 - \Phi B^s$	$\mu + (b_{0,m} - \mu)\Phi^r$	s
(2) $\quad 1 - B^s$	$b_{0,m}$	s
(3) $\ (1 - B)(1 - \Phi B^s)$	$b_0 + (b_{0,m} - b_0)\Phi^r + b_1 \left\{ \dfrac{1 - \Phi^r}{1 - \Phi} \right\}$	$s + 1$
(4) $\ (1 - B)(1 - B^s)$	$b_{0,m} + b_1 r$	$s + 1$
(5) $\ (1 - \phi B)(1 - B^s)$	$b_{0,m} + b_1 \phi^{m-1} \left\{ \dfrac{1 - \phi^{sr}}{1 - \phi^s} \right\}$	$s + 1$
(6) $\ (1 - B)(1 - B^s)^2$	$b_{0,m} + b_{1,m} r + \frac{1}{2} b_2 r(r - 1)$	$2s + 1$
(7) $\ (1 - B)^2(1 - B^s)$	$b_{0,m} + \{b_1 + (m - 1)b_2\}r$ $+ \frac{1}{2} b_2 sr(r - 1)$	$s + 2$

In Figure 9.7 the behavior of each forecast function is illustrated for $s = 4$. It will be convenient to regard the lead time $l = rs + m$ as referring to a forecast r years and m quarters ahead. In the diagram, an appropriate number of initial values (required to start the forecast off and indicated by bold dots) has been set arbitrarily and the course of the forecast function traced to the end of the fourth period. When the difference equation involves an autoregressive parameter, its value has been set equal to 0.5.

The constants $b_{0,m}$, b_1, etc., appearing in the solutions in Table 9.9, should strictly be indicated by $b_{0,m}^{(t)}$, $b_1^{(t)}$, etc., since each one depends on the origin t of the forecast and these constants are adaptively modified each time the origin changes. The superscript t has been omitted temporarily to simplify notation. The operator labelled (1) is stationary, containing a fixed mean μ. It is autoregressive in the seasonal pattern, which decays with each period, approaching closer and closer to the mean.

Operator (2) is nonstationary in the seasonal component. The forecasts for a particular quarter are linked from year to year by a polynomial of degree 0. Thus, the basic forecast of the seasonal component is exactly reproduced in forecasts of future years.

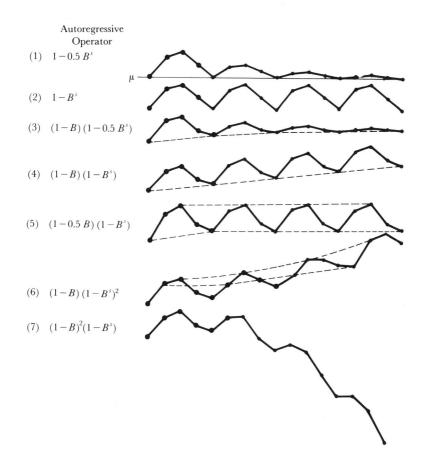

Autoregressive
Operator

(1) $1 - 0.5\,B^s$

(2) $1 - B^s$

(3) $(1 - B)\,(1 - 0.5\,B^s)$

(4) $(1 - B)\,(1 - B^s)$

(5) $(1 - 0.5\,B)\,(1 - B^s)$

(6) $(1 - B)\,(1 - B^s)^2$

(7) $(1 - B)^2(1 - B^s)$

FIG. 9.7   Behavior of the seasonal forecast function for various choices of the general seasonal autoregressive operator

Operator (3) is nonstationary with respect to the basic interval but stationary in the seasonal component. Figure 9.7 (3) shows the general level of the forecast approaching asymptotically the new level

$$b_0 + b_1/(1 - \Phi)$$

where, at the same time, the superimposed predictable component of the stationary seasonal effect dies out exponentially.

Operator (4) is the limiting case of the operator (3) as Φ approaches unity. The operator is nonstationary with respect to both the basic interval and the periodic component. The basic initial forecast pattern is reproduced, as is the incremental yearly increase. This is the type of forecast function given by the multiplicative $(0, 1, 1) \times (0, 1, 1)_{12}$ process fitted to the airline date.

Operator (5) is nonstationary in the seasonal pattern but stationary with respect to the basic interval. The pattern approaches exponentially an asymptotic basic pattern

$$\hat{z}_t(\infty, m) = b_{0,m} + b_1 \phi^{m-1}/(1 - \phi^s)$$

Operator (6) is nonstationary in both the basic interval and the seasonal component. An overall quadratic trend occurs over years, and a particular kind of modification occurs in the seasonal pattern. Individual quarters not only have their own level $b_{0,m}$ but also their own rate of change of level $b_{1,m}$. Therefore, when this kind of forecast function is appropriate, we can have a situation where, for example, as the lead time is increased, the difference in summer over spring sales can be forecast to increase from one year to the next, while at the same time the difference in autumn over summer sales can be forecast to decrease.

Operator (7) is again nonstationary in both basic interval and in the seasonal component, and there is again a quadratic tendency over years with the incremental changes in the forecasts from one quarter to the next changing linearly. However, in this case they are restricted to have a common *rate* of change.

9.3.5 Choice of Transformation

It is particularly true for seasonal models that the weighted averages of previous data values, which comprise the forecasts, may extend far back into the series. Care is therefore needed in choosing a transformation in terms of which a parsimonious linear model will closely apply over a sufficient stretch of the series. Simple graphical analysis can often suggest such a transformation. Thus an appropriate transformation may be suggested by determining in what metric the amplitude of the seasonal component is roughly independent of the level of the series. To illustrate how a data based transformation may be chosen more exactly denote the *untransformed* airline data by x, and let us assume that some power transformation ($z = x^\lambda$ for $\lambda \neq 0$, $z = \ln x$ for $\lambda = 0$) may be needed to make the model (9.2.1) appropriate. Then, as suggested in Section 4.1.3 the approach of Box and Cox [42] may be followed, and the maximum likelihood value obtained by fitting the model to $x^{(\lambda)} = (x^\lambda - 1)/\lambda \dot{x}^{\lambda-1}$ for various values of λ, and choosing λ which results in the smallest residual sum of squares S_λ. In this expression $\dot{x}$ is the geometric mean of the series, and it is easily shown that $x^{(0)} = \dot{x} \ln x$.

For the airline data we find

λ	-0.4	-0.3	-0.2	-0.1	0	0.1	0.2	0.3	0.4
S_λ	13825.5	12794.6	12046.0	11627.2	11458.1	11554.3	11784.3	12180.0	12633.2

The maximum likelihood value is thus close to $\lambda = 0$ confirming for this particular example the appropriateness of the logarithmic transformation.

APPENDIX A9.1   AUTOCOVARIANCES FOR SOME SEASONAL MODELS

Model	(Autocovariances of w_t)/σ_a^2	Special characteristics
$(1)\ w_t = (1 - \theta B)(1 - \Theta B^s)a_t$ $w_t = a_t - \theta a_{t-1} - \Theta a_{t-s} + \theta\Theta a_{t-s-1}$ $s \geqslant 3$	$\gamma_0 = (1 + \theta^2)(1 + \Theta^2)$ $\gamma_1 = -\theta(1 + \Theta^2)$ $\gamma_{s-1} = \theta\Theta$ $\gamma_s = -\Theta(1 + \theta^2)$ $\gamma_{s+1} = \gamma_{s-1}$ All other autocovariances are zero.	(a) $\gamma_{s-1} = \gamma_{s+1}$ (b) $\rho_{s-1} = \rho_{s+1} = \rho_1\rho_s$
$(2)\ (1 - \Phi B^s)w_t = (1 - \theta B)(1 - \Theta B^s)a_t$ $w_t - \Phi w_{t-s} = a_t - \theta a_{t-1} - \Theta a_{t-s}$ $\qquad\qquad + \theta\Theta a_{t-s-1}$ $s \geqslant 3$	$\gamma_0 = (1 + \theta^2)\left[1 + \dfrac{(\Theta - \Phi)^2}{1 - \Phi^2}\right]$ $\gamma_1 = -\theta\left[1 + \dfrac{(\Theta - \Phi)^2}{1 - \Phi^2}\right]$ $\gamma_{s-1} = \theta\left[\Theta - \Phi - \dfrac{\Phi(\Theta - \Phi)^2}{1 - \Phi^2}\right]$ $\gamma_s = -(1 + \theta^2)\left[\Theta - \Phi - \dfrac{\Phi(\Theta - \Phi)^2}{1 - \Phi^2}\right]$ $\gamma_{s+1} = \gamma_{s-1}$ $\gamma_j = \Phi\gamma_{j-s} \quad j \geqslant s + 2$ For $s \geqslant 4$, $\gamma_2, \gamma_3, \ldots, \gamma_{s-2}$ are all zero.	(a) $\gamma_{s-1} = \gamma_{s+1}$ (b) $\gamma_j = \Phi\gamma_{j-s} \qquad j \geqslant s + 2$

APPENDIX A9.1 (contd.)

Model	(Autocovariances of w_t)$/\sigma_a^2$	Special characteristics
(3) $w_t = (1 - \theta_1 B - \theta_2 B^2)$ $\times (1 - \Theta_1 B^s - \Theta_2 B^{2s})a_t$ $w_t = a_t - \theta_1 a_{t-1} - \theta_2 a_{t-2} - \Theta_1 a_{t-s}$ $+ \theta_1 \Theta_1 a_{t-s-1} + \theta_2 \Theta_1 a_{t-s-2}$ $- \Theta_2 a_{t-2s} + \theta_1 \Theta_2 a_{t-2s-1}$ $+ \theta_2 \Theta_2 a_{t-2s-2}$ $s \geqslant 5$	$\gamma_0 = (1 + \theta_1^2 + \theta_2^2)(1 + \Theta_1^2 + \Theta_2^2)$ $\gamma_1 = -\theta_1(1 - \theta_2)(1 + \Theta_1^2 + \Theta_2^2)$ $\gamma_2 = -\theta_2(1 + \Theta_1^2 + \Theta_2^2)$ $\gamma_{s-2} = \theta_2 \Theta_1(1 - \Theta_2)$ $\gamma_{s-1} = \theta_1 \Theta_1(1 - \theta_2)(1 - \Theta_2)$ $\gamma_s = -\Theta_1(1 + \theta_1^2 + \theta_2^2)(1 - \Theta_2)$ $\gamma_{s+1} = \gamma_{s-1}$ $\gamma_{s+2} = \gamma_{s-2}$ $\gamma_{2s-2} = \theta_2 \Theta_2$ $\gamma_{2s-1} = \theta_1 \Theta_2(1 - \theta_2)$ $\gamma_{2s} = -\Theta_2(1 + \theta_1^2 + \theta_2^2)$ $\gamma_{2s+1} = \gamma_{2s-1}$ $\gamma_{2s+2} = \gamma_{2s-2}$ All other autocovariances are zero.	(a) $\gamma_{s-2} = \gamma_{s+2}$ (b) $\gamma_{s-1} = \gamma_{s+1}$ (c) $\gamma_{2s-2} = \gamma_{2s+2}$ (d) $\gamma_{2s-1} = \gamma_{2s+1}$

APPENDIX A9.1 (contd.)

Model	(Autocovariances of w_t)/σ_a^2	Special characteristics
(3a) *Special Case of Model 3* $w_t = (1 - \theta_1 B - \theta_2 B^2)(1 - \Theta B^s)a_t$ $w_t = a_t - \theta_1 a_{t-1} - \theta_2 a_{t-2} - \Theta a_{t-s}$ $\quad + \theta_1 \Theta a_{t-s-1} + \theta_2 \Theta a_{t-s-2}$ $s \geqslant 5$	$\gamma_0 = (1 + \theta_1^2 + \theta_2^2)(1 + \Theta^2)$ $\gamma_1 = -\theta_1(1 - \theta_2)(1 + \Theta^2)$ $\gamma_2 = -\theta_2(1 + \Theta^2)$ $\gamma_{s-2} = \theta_2 \Theta$ $\gamma_{s-1} = \theta_1 \Theta(1 - \theta_2)$ $\gamma_s = -\Theta(1 + \theta_1^2 + \theta_2^2)$ $\gamma_{s+1} = \gamma_{s-1}$ $\gamma_{s+2} = \gamma_{s-2}$ All other autocovariances are zero.	(a) $\gamma_{s-2} = \gamma_{s+2}$ (b) $\gamma_{s-1} = \gamma_{s+1}$
(3b) *Special Case of Model 3* $w_t = (1 - \theta B)(1 - \Theta_1 B^s - \Theta_2 B^{2s})a_t$ $w_t = a_t - \theta a_{t-1} - \Theta_1 a_{t-s} + \theta \Theta_1 a_{t-s-1}$ $\quad - \Theta_2 a_{t-2s} + \theta \Theta_2 a_{t-2s-1}$ $s \geqslant 3$	$\gamma_0 = (1 + \theta^2)(1 + \Theta_1^2 + \Theta_2^2)$ $\gamma_1 = -\theta(1 + \Theta_1^2 + \Theta_2^2)$ $\gamma_{s-1} = \theta \Theta_1(1 - \Theta_2)$ $\gamma_s = -\Theta_1(1 + \theta^2)(1 - \Theta_2)$ $\gamma_{s+1} = \gamma_{s-1}$ $\gamma_{2s-1} = \theta \Theta_2$ $\gamma_{2s} = -\Theta_2(1 + \theta^2)$ $\gamma_{2s+1} = \gamma_{2s-1}$ All other autocovariances are zero.	(a) $\gamma_{s-1} = \gamma_{s+1}$ (b) $\gamma_{2s-1} = \gamma_{2s+1}$

APPENDIX A9.1 (contd.)

Model	(Autocovariances of w_t)/σ_a^2	Special characteristics
(4) $w_t = (1 - \theta_1 B - \theta_s B^s - \theta_{s+1}B^{s+1})a_t$ $w_t = a_t - \theta_1 a_{t-1} - \theta_s a_{t-s}$ $\quad - \theta_{s+1}a_{t-s-1}$ $s \geqslant 3$	$\gamma_0 = 1 + \theta_1^2 + \theta_s^2 + \theta_{s+1}^2$ $\gamma_1 = -\theta_1 + \theta_s\theta_{s+1}$ $\gamma_{s-1} = \theta_1\theta_s$ $\gamma_s = \theta_1\theta_{s+1} - \theta_s$ $\gamma_{s+1} = -\theta_{s+1}$ All other autocovariances are zero.	(a) In general, $\gamma_{s-1} \neq \gamma_{s+1}$ $\gamma_1\gamma_s \neq \gamma_{s+1}$
(4a) *Special Case of Model 4* $w_t = (1 - \theta_1 B - \theta_s B^s)a_t$ $w_t = a_t - \theta_1 a_{t-1} - \theta_s a_{t-s}$ $s \geqslant 3$	$\gamma_0 = 1 + \theta_1^2 + \theta_s^2$ $\gamma_1 = -\theta_1$ $\gamma_{s-1} = \theta_1\theta_s$ $\gamma_s = -\theta_s$ All other autocovariances are zero.	(a) Unlike model 4, $\gamma_{s+1} = 0$

APPENDIX A9.1 (contd.)

Model	(Autocovariances of w_t) σ_a^2	Special characteristics
(5) $(1 - \Phi B^s)w_t = (1 - \theta_1 B - \theta_s B^s$ $\qquad - \theta_{s+1}B^{s+1})a_t$ $w_t - \Phi w_{t-s} = a_t - \theta_1 a_{t-1} - \theta_s a_{t-s}$ $\qquad\qquad - \theta_{s+1}a_{t-s-1}$ $s \geq 3$	$\gamma_0 = 1 + \theta_1^2 + \dfrac{(\theta_s - \Phi)^2}{1 - \Phi^2} + \dfrac{(\theta_{s+1} + \theta_1\Phi)^2}{1 - \Phi^2}$ $\gamma_1 = -\theta_1 + \dfrac{(\theta_s - \Phi)(\theta_{s+1} + \theta_1\Phi)}{1 - \Phi^2}$ $\gamma_{s-1} = (\theta_s - \Phi)\left[\theta_1 + \Phi\dfrac{(\theta_{s+1} + \theta_1\Phi)}{1 - \Phi^2}\right]$ $\gamma_s = -(\theta_s - \Phi)\left[1 - \Phi\dfrac{(\theta_s - \Phi)}{1 - \Phi^2}\right]$ $\qquad + (\theta_{s+1} + \theta_1\Phi)\left[\theta_1 + \Phi\dfrac{(\theta_{s+1} + \theta_1\Phi)}{1 - \Phi^2}\right]$ $\gamma_{s+1} = -(\theta_{s+1} + \theta_1\Phi)\left[1 - \Phi\dfrac{(\theta_s - \Phi)}{1 - \Phi^2}\right]$ $\gamma_j = \Phi\gamma_{j-s}$ $j \geq s + 2$ For $s \geq 4$, $\gamma_2, \ldots, \gamma_{s-2}$ are all zero.	(a) $\gamma_{s-1} \neq \gamma_{s+1}$ (b) $\gamma_j = \Phi\gamma_{j-s}$ $j \geq s + 2$
(5a) *Special Case of Model 5* $(1 - \Phi B^s)w_t = (1 - \theta_1 B - \theta_s B^s)a_t$ $w_t - \Phi w_{t-s} = a_t - \theta_1 a_{t-1} - \theta_s a_{t-s}$ $\qquad\qquad s \geq 3$	$\gamma_0 = 1 + \dfrac{\theta_1^2 + (\theta_s - \Phi)^2}{1 - \Phi^2}$ $\gamma_1 = -\theta_1\left[1 - \dfrac{(\theta_s - \Phi)}{1 - \Phi^2}\right]$ $\gamma_{s-1} = \dfrac{\theta_1(\theta_s - \Phi)}{1 - \Phi^2}$ $\gamma_s = \dfrac{\Phi\theta_1^2 - (\theta_s - \Phi)(1 - \Phi\theta_s)}{1 - \Phi^2}$ $\gamma_j = \Phi\gamma_{j-s}$ $j \geq s + 1$ For $s \geq 4$, $\gamma_2, \ldots, \gamma_{s-2}$ are all zero.	(a) Unlike model 5, $\gamma_{s+1} = \Phi\gamma_1$

Part III

Transfer Function Model Building

Suppose X measures the level of an *input* to a system. For example, X might be the concentration of some constituent in the feed to a chemical process. Suppose that the level of X influences the level of a system *output Y*. For example, Y might be the yield of product from the chemical process. It will usually be the case that, because of the inertia of the system, a change in X from one level to another will have no immediate effect on the output but, instead, will produce delayed response with Y eventually coming to equilibrium at a new level. We refer to such a change as a *dynamic* response. A model which describes this dynamic response is called a *transfer function model*. We shall suppose that observations of input and output are made at equispaced intervals of time. The associated transfer function model will then be called a *discrete* transfer function model.

Models of this kind can describe not only the behavior of industrial processes but also that of economic and business systems. Transfer function model building is important because it is only when the dynamic characteristics of a system are understood that intelligent direction, manipulation and control of the system is possible.

Even under carefully controlled conditions, influences other than X will affect Y. We refer to the combined effect on Y of such influences as the *disturbance* or the *noise*. A model such as can be related to real data must take account not only of the dynamic relationship associating X and Y but also of the noise infecting the system. Such joint models are obtained by combining a deterministic transfer function model with a stochastic noise model.

In Chapter 10 following we introduce a class of linear transfer function models capable of representing many of the dynamic relationships commonly met in practice. Chapter 11 shows how, taking account of corrupting noise, they may be related to data. This relating of model to data is accomplished by processes of *identification*, *estimation* and *diagnostic checking*, which closely parallel those already described.

10

Transfer Function Models

In this chapter we introduce a class of discrete linear transfer function models. These models can be used to represent commonly occurring dynamic situations and are parsimonious in their use of parameters.

10.1 LINEAR TRANSFER FUNCTION MODELS

We suppose that pairs of observations (X_t, Y_t) are available at equispaced intervals of time of an input X and an output Y from some dynamic system, as illustrated in Figure 10.1. In some situations, both X and Y are essentially continuous but are observed only at discrete times. It then makes sense to consider not only what the data has to tell us about the model representing transfer from one discrete series to another, but also what the discrete model might be able to tell us about the corresponding continuous model.

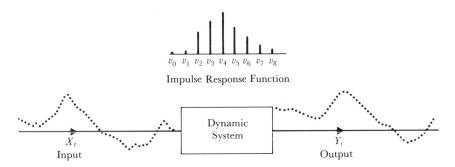

v_0 v_1 v_2 v_3 v_4 v_5 v_6 v_7 v_8

Impulse Response Function

X_t
Input

Dynamic
System

Y_t
Output

FIG. 10.1   Input to, and output from, a dynamic system

In other examples the discrete series are all that exist, and there is no background continuous process. Where we relate continuous and discrete systems, we shall use the basic sampling interval as the unit of time. That is to say, periods of time will be measured by the number of sampling intervals they occupy. Also a discrete observation X_t will be deemed to have occurred "at time t."

When we consider the value of a continuous variable, say Y at time t, we denote it by $Y(t)$. If t happens to be a time at which a discrete variable Y is observed, then its value is denoted by Y_t. When we wish to emphasize the dependence of a discrete output Y, not only on time but also on the level of the input X, we write $Y_t(X)$.

10.1.1 The discrete transfer function

With suitable inputs and outputs, which are left to the imagination of the reader, the dynamic system of Figure 10.1 might represent an industrial process, the economy of a country, or the behavior of a particular corporation or government department.

From time to time, we shall refer to the *steady state* level of the output obtained when the input is held at some fixed value. By this we mean the value $Y_\infty(X)$ at which the discrete output from a stable system *eventually* comes to equilibrium when the input is held at the fixed level X. Very often, over the range of interest, the relationship between $Y_\infty(X)$ and X will be approximately linear. Hence, if we use Y and X to denote *deviations* from convenient origins situated on the line, we can write the steady state relationship as

$$Y_\infty = gX \qquad (10.1.1)$$

where g is called the *steady state gain*, and it is understood that Y_∞ is a function of X.

Now, suppose the level of the input is being varied and that X_t and Y_t represent *deviations* at time t from equilibrium. Then it frequently happens that, to an adequate approximation, the inertia of the system can be represented by a *linear filter* of the form

$$
\begin{aligned}
Y_t &= v_0 X_t + v_1 X_{t-1} + v_2 X_{t-2} + \cdots \\
&= (v_0 + v_1 B + v_2 B^2 + \cdots) X_t \qquad (10.1.2) \\
&= v(B) X_t
\end{aligned}
$$

in which the output deviation at some time t is represented as a linear aggregate of input deviations at times $t, t-1, \ldots$. The operator $v(B)$ is called the *transfer function* of the filter.

The impulse response function. The weights $v_0, v_1, v_2, \ldots$ in (10.1.2) are called the *impulse response function* of the system. The impulse response function is shown in Figure 10.1 in the form of a bar chart. When there is no immediate response, one or more of the initial v's, say $v_0, v_1, \ldots, v_{b-1}$, will be equal to zero.

According to (10.1.2), the output deviation can be regarded as a linear aggregate of a series of superimposed impulse response functions scaled by

the deviations X_t. This is illustrated in Figure 10.2, which shows a hypothetical impulse response function and the transfer it induces from the
input to the output. In the situation illustrated, the input and output are
initially in equilibrium. The deviations that occur in the input at times
$t = 1$, $t = 2$, and $t = 3$ produce impulse response patterns of deviations
in the output, which add together to produce the overall output response.

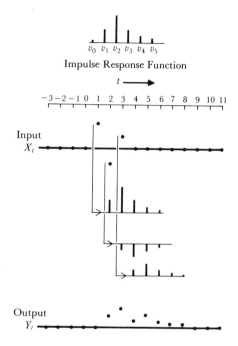

FIG. 10.2   Linear transfer from input X to output Y

Relation between the incremental changes.   Denote by

$$y_t = Y_t - Y_{t-1} = \nabla Y_t$$

and by

$$x_t = X_t - X_{t-1} = \nabla X_t$$

the *incremental changes* in Y and X. We often wish to relate such changes.
On differencing (10.1.2), we obtain

$$y_t = v(B)x_t$$

Thus, we see that the incremental changes y_t and x_t satisfy the same transfer
function model as do Y_t and X_t.

Stability. If the infinite series $v_0 + v_1B + v_2B^2 + \ldots$ converges for $|B| \leqslant 1$, then the system is said to be *stable*. We shall only be concerned here with stable systems and consequently impose this condition on the models we study. The stability condition implies that a finite incremental change in the input results in a finite incremental change in the output.

Now, suppose X is held indefinitely at the value $+1$. Then, according to (10.1.1), Y will adjust and maintain itself at the value g. On substituting in (10.1.2) the values $Y_t = g$, $1 = X_t = X_{t-1} = X_{t-2} = \ldots$, we obtain

$$\sum_{j=0}^{\infty} v_j = g \qquad (10.1.3)$$

Thus, for a stable system the sum of the impulse response weights converges and is equal to the steady state gain of the system.

Parsimony. It would often be unsatisfactory to parameterize the system in terms of the v's of (10.1.2). To be thus prodigal in the use of parameters could, at the estimation stage, lead to inaccurate and unstable estimation of the transfer function. Furthermore, it is usually inappropriate to estimate the weights v_j directly because for many real situations the v's would be functionally related, as we now see.

10.1.2 Continuous dynamic models represented by differential equations

A first order dynamic system. Consider Figure 10.3. Suppose that at time t, $X(t)$ is the *volume* of water in tank A and $Y_1(t)$ the volume of water in tank B, which is connected to A by a pipe. For the time being we ignore tank C,

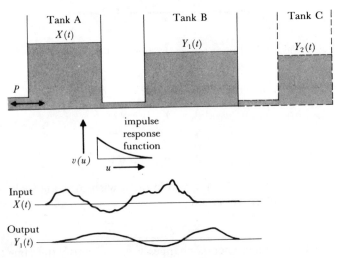

FIG. 10.3   Representation of a simple dynamic system

shown by dotted lines. Now suppose that water can be forced in or out of A through pipe P and that mechanical devices are available which make it possible to force the level and hence the volume X in A to follow any desired pattern *irrespective* of what happens in B.

Now if the volume X in the first tank is held at some *fixed* level, water will flow from one tank to the other until the levels are equal. If we now reset the volume X to some other value, again a flow between the tanks will occur until equilibrium is reached. The volume in B at equilibrium as a function of the fixed volume in A yields the steady state relationship

$$Y_{1\infty} = g_1 X \qquad (10.1.4)$$

In this case the steady state gain g_1 physically represents the ratio of the cross-sectional areas of the two tanks. If the levels are not in equilibrium at some time t, it is to be noted that the difference in the water level between the tanks is proportional to $g_1 X(t) - Y_1(t)$.

Suppose now that, by forcing liquid in and out of pipe P, the volume $X(t)$ is made to follow a pattern like that labelled "Input $X(t)$" in Figure 10.3. Then the volume $Y_1(t)$ in B will correspondingly change in some pattern such as that labelled on the figure as "Output $Y_1(t)$." In general the function $X(t)$ which is responsible for driving the system is called the *forcing function*.

To relate output and input we note that to a close approximation, the rate of flow through the pipe will be proportional to the difference in head. That is,

$$\frac{dY_1}{dt} = \frac{1}{T_1}\{g_1 X(t) - Y_1(t)\} \qquad (10.1.5)$$

where T_1 is a constant. The differential equation (10.1.5) may be rewritten in the form

$$(1 + T_1 D)Y_1(t) = g_1 X(t) \qquad (10.1.6)$$

where $D = d/dt$. The dynamic system so represented by a first order differential equation is often referred to as a first order dynamic system. The constant T_1 is called the *time constant* of the system. The same first order model can approximately represent the behavior of many simple systems. For example, $Y_1(t)$ might be the outlet temperature of water from a water heater, and $X(t)$ the flow rate of water into the heater.

It is possible to show, see for example [27], that the solution of a linear differential equation such as (10.1.6) can be written in the form

$$Y_1(t) = \int_0^\infty v(u)X(t - u)\, du \qquad (10.1.7)$$

where in general $v(u)$ is the (continuous) impulse response function. We see that $Y_1(t)$ is generated from $X(t)$ as a continuously weighted aggregate, just as Y_t is generated from X_t as a discretely weighted aggregate in (10.1.2). Furthermore, we see that the role of weight function played by $v(u)$ in the continuous case is precisely parallel to that played by v_j in the discrete situation. For the particular first-order system defined by (10.1.6),

$$v(u) = g_1 T_1^{-1} e^{-u/T_1}$$

Thus, the impulse response in this case undergoes simple exponential decay, as indicated in Figure 10.3.

In the continuous case, determination of the output for a completely arbitrary forcing function, such as shown in Figure 10.3, is normally accomplished by simulation on an analog computer, or by using numerical procedures on a digital machine. Solutions are available analytically only for special forcing functions. Suppose, for example, that with the hydraulic system empty, $X(t)$ was suddenly raised to a level $X(t) = 1$ and maintained at that value. Then we shall refer to the forcing function, which was at a steady level of zero and changed instantaneously to a steady level of unity, as a (unit) *step function*. The response of the system to such a function, called the *step response* of the system, is derived by solving the differential equation (10.1.6) with a unit step input, to obtain

$$Y_1(t) = g_1(1 - e^{-t/T_1}) \tag{10.1.8}$$

Thus, the level in tank B rises exponentially in the manner shown in Figure 10.4. Now, when $t = T_1$, $Y_1(t) = g_1(1 - e^{-1}) = 0.632 g_1$. Thus, the time constant T_1 is the time required after the initiation of a step input for the first-order system (10.1.6) to reach 63.2% of its final equilibrium level.

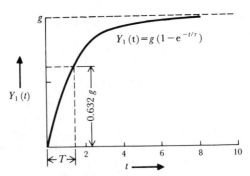

FIG. 10.4   Response of a first-order system to a unit step change

Sometimes there is an initial period of pure *delay* or *dead time* before the response to a given input change begins to take effect. For example, if there were a long length of pipe between A and B in Figure 10.3, a sudden change in level in A could not begin to take effect until liquid had flowed down the pipe. Suppose the delay thus introduced occupies τ units of time. Then the response of the delayed system would be represented by a differential equation like (10.1.6), but with $t - \tau$ replacing t on the right-hand side, so that

$$(1 + T_1 D)Y_1(t) = g_1 X(t - \tau) \qquad (10.1.9)$$

The corresponding impulse and step response functions for this system would be of precisely the same shape as for the undelayed system, but the functions would be translated along the horizontal axis a distance τ.

A second-order dynamic system. Consider Figure 10.3 once more. Imagine a three tank system in which a pipe leads from tank B to a third tank C, the volume of liquid in which is denoted by $Y_2(t)$. Let T_2 be the time constant for the additional system and g_2 its steady state gain. Then $Y_2(t)$ and $Y_1(t)$ are related by the differential equation

$$(1 + T_2 D)Y_2(t) = g_2 Y_1(t)$$

After substitution in (10.1.6), we obtain a *second-order* differential equation linking the output from the third tank and the input to the first

$$\{1 + (T_1 + T_2)D + T_1 T_2 D^2\}Y_2(t) = gX(t) \qquad (10.1.10)$$

where $g = g_1 g_2$. For such a system the impulse response function is a mixture of exponentials

$$v(u) = g(e^{-u/T_1} - e^{-u/T_2})/(T_1 - T_2) \qquad (10.1.11)$$

and the response to a unit step is given by

$$Y_2(t) = g\left\{1 - \frac{T_1 e^{-t/T_1} - T_2 e^{-t/T_2}}{T_1 - T_2}\right\} \qquad (10.1.12)$$

The continuous curve R in Figure 10.5 shows the response to a unit step for the system

$$(1 + 3D + 2D^2)Y_2(t) = 5X(t)$$

for which $T_1 = 1$, $T_2 = 2$, $g = 5$. Note that, unlike the first-order system, the second-order system has a step response which has zero slope initially.

A more general second-order system is defined by

$$(1 + \Xi_1 D + \Xi_2 D^2)Y(t) = gX(t) \qquad (10.1.13)$$

where

$$\Xi_1 = T_1 + T_2 \qquad \Xi_2 = T_1 T_2 \qquad (10.1.14)$$

and the constants T_1 and T_2 may be complex. If we write

$$T_1 = \frac{1}{\zeta}e^{i\lambda} \qquad T_2 = \frac{1}{\zeta}e^{-i\lambda} \qquad (10.1.15)$$

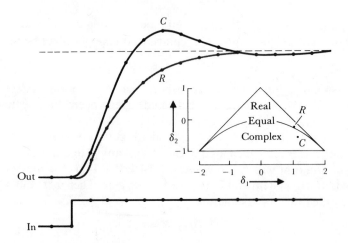

Fig. 10.5   Step responses of coincident, discrete and continuous second-order systems having characteristic equations with real roots (Curve R) and complex roots (Curve C)

then (10.1.13) becomes

$$\left(1 + \frac{2\cos\lambda}{\zeta}D + \frac{1}{\zeta^2}\right)Y(t) = gX(t) \tag{10.1.16}$$

The impulse response function (10.1.11) then reduces to

$$v(u) = g\frac{\zeta e^{-\zeta u\cos\lambda}\sin(\zeta u\sin\lambda)}{\sin\lambda} \tag{10.1.17}$$

and the response (10.1.12) to a unit step, to

$$Y(t) = g\left\{1 - \frac{e^{-\zeta t\cos\lambda}\sin(\zeta t\sin\lambda + \lambda)}{\sin\lambda}\right\} \tag{10.1.18}$$

The continuous curve C in Figure 10.5 shows the response to a unit step for the system

$$(1 + \sqrt{2}D + 2D^2)Y(t) = 5X(t)$$

for which $\lambda = 60°$ and $\zeta = \sqrt{2}/2$. It will be noticed that the response overshoots the value $g = 5$ and then comes to equilibrium as a damped sine wave. This behavior is typical of underdamped systems, as they are called. In general a second-order system is said to be *overdamped, critically damped* or *underdamped* depending on whether the constants T_1 and T_2 are real, real and equal, or complex. The overdamped system has a step response which is a mixture of exponentials, given by (10.1.12), and will always remain below the asymptote $Y(\infty) = g$. As with the first-order system, the response can be made subject to a period of dead time by replacing t on the right-hand side of (10.1.13) by $t - \tau$. Many quite complicated dynamic systems can be closely approximated by such second-order systems with delay.

More elaborate linear dynamic systems can be represented by allowing not only the level of the forcing function $X(t)$ but also its rate of change dX/dt and higher derivatives to influence the behavior of the system. Thus, a general model for representing (continuous) dynamic systems is the linear differential equation

$$(1 + \Xi_1 D + \cdots + \Xi_R D^R)Y(t) = g(1 + H_1 D + \cdots + H_S D^S)X(t - \tau)$$

$$(10.1.19)$$

10.2 DISCRETE DYNAMIC MODELS REPRESENTED BY DIFFERENCE EQUATIONS

10.2.1 The general form of the difference equation

Corresponding to the continuous representation (10.1.19), discrete dynamic systems are often parsimoniously represented by the general linear *difference* equation

$$(1 + \xi_1 \nabla + \cdots + \xi_r \nabla^r)Y_t = g(1 + \eta_1 \nabla + \cdots + \eta_s \nabla^s)X_{t-b} \quad (10.2.1)$$

which we refer to as a transfer function model of order (r, s). The difference equation (10.2.1) may also be written in terms of the backward shift operator $B = 1 - \nabla$ as

$$(1 - \delta_1 B - \cdots - \delta_r B^r)Y_t = (\omega_0 - \omega_1 B - \cdots - \omega_s B^s)X_{t-b} \quad (10.2.2)$$

or as

$$\delta(B)Y_t = \omega(B)X_{t-b}$$

Equivalently, writing $\Omega(B) = \omega(B)B^b$, the model becomes

$$\delta(B)Y_t = \Omega(B)X_t \qquad (10.2.3)$$

Comparing (10.2.3) with (10.1.2) we see that the transfer function for this model is

$$v(B) = \delta^{-1}(B)\Omega(B) \qquad (10.2.4)$$

Thus the transfer function is represented by the ratio of two polynomials in B.

Dynamics of ARIMA stochastic models   The ARIMA model

$$\varphi(B)z_t = \theta(B)a_t$$

used for the representation of a time series $\{z_t\}$ relates z_t and a_t by the linear filtering operation

$$z_t = \varphi^{-1}(B)\theta(B)a_t$$

where a_t is white noise. Thus the ARIMA model postulates that a time series can be usefully represented as an output from a dynamic system to which the input is white noise and for which the transfer function can be parsimoniously expressed as the ratio of two polynomials in B.

Stability of the discrete models. The requirement of stability for the discrete transfer function models exactly parallels that of stationarity for the ARMA stochastic models. In general, for stability we require that the roots of the characteristic equation

$$\delta(B) = 0$$

with B regarded as a variable, lie outside the unit circle. In particular, this implies that, for the first-order model, the parameter δ_1 satisfies

$$-1 < \delta_1 < 1$$

and for the second-order model (see for example Figure 10.5), the parameters δ_1, δ_2 satisfy

$$\delta_2 + \delta_1 < 1$$
$$\delta_2 - \delta_1 < 1$$
$$-1 < \delta_2 < 1$$

On writing (10.2.2) in full as

$$Y_t = \delta_1 Y_{t-1} + \cdots + \delta_r Y_{t-r} + \omega_0 X_{t-b} - \omega_1 X_{t-b-1} - \cdots - \omega_s X_{t-b-s}$$

we see that if X_t is held indefinitely at a value $+1$, Y_t will eventually reach the value

$$g = \frac{\omega_0 - \omega_1 - \cdots - \omega_s}{1 - \delta_1 - \cdots - \delta_r} \tag{10.2.5}$$

which expresses the steady state gain in terms of the parameters of the model.

10.2.2 Nature of the transfer function

If we employ a transfer function model defined by the difference equation (10.2.2), then substituting

$$Y_t = v(B)X_t \tag{10.2.6}$$

in (10.2.2), we obtain the identity

$$(1 - \delta_1 B - \delta_2 B^2 - \cdots - \delta_r B^r)(v_0 + v_1 B + v_2 B^2 + \cdots)$$

$$= (\omega_0 - \omega_1 B - \cdots - \omega_s B^s)B^b \tag{10.2.7}$$

On equating coefficients of B, we find

$$
\begin{aligned}
v_j &= 0 & j &< b \\
v_j &= \delta_1 v_{j-1} + \delta_2 v_{j-2} + \cdots + \delta_r v_{j-r} + \omega_0 & j &= b \\
v_j &= \delta_1 v_{j-1} + \delta_2 v_{j-2} + \cdots + \delta_r v_{j-r} - \omega_{j-b} & j &= b+1, b+2, \ldots, b+s \\
v_j &= \delta_1 v_{j-1} + \delta_2 v_{j-2} + \cdots + \delta_r v_{j-r} & j &> b+s
\end{aligned}
\right\}
$$

$$(10.2.8)$$

The weights $v_{b+s}, v_{b+s-1}, \ldots, v_{b+s-r+1}$ supply r starting values for the difference equation

$$\delta(B)v_j = 0 \qquad j > b+s$$

The solution

$$v_j = f(\delta, \omega, j)$$

of this difference equation applies to all values v_j for which $j \geqslant b + s - r + 1$.

Thus, in general, the impulse response weights v_j consist of
(i) b zero values $v_0, v_1, \ldots, v_{b-1}$;
(ii) a further $s - r + 1$ values $v_b, v_{b+1}, \ldots, v_{b+s-r}$, following no fixed pattern (no such values occur if $s < r$);
(iii) values v_j with $j \geqslant b + s - r + 1$ following the pattern dictated by the rth order difference equation which has r starting values v_{b+s}, $v_{b+s-1}, \ldots, v_{b+s-r+1}$. Starting values v_j for $j < b$ will, of course, be zero.

The step response.   We now write $V(B)$ for the generating function of the step response weights V_j. Thus

$$V(B) = V_0 + V_1 B + V_2 B^2 + \cdots = v_0 + (v_0 + v_1)B$$

$$+ (v_0 + v_1 + v_2)B^2 + \cdots \quad (10.2.9)$$

and

$$v(B) = (1 - B)V(B) \qquad (10.2.10)$$

Substitution of (10.2.10) in (10.2.7) yields the identity

$$(1 - \delta_1^* B - \delta_2^* B^2 - \cdots - \delta_{r+1}^* B^{r+1})(V_0 + V_1 B + V_2 B^2 + \cdots)$$

$$= (\omega_0 - \omega_1 B - \cdots - \omega_s B^s)B^b \qquad (10.2.11)$$

with

$$(1 - \delta_1^* B - \delta_2^* B^2 - \cdots - \delta_{r+1}^* B^{r+1}) = (1 - B)(1 - \delta_1 B - \cdots - \delta_r B^r)$$

$$(10.2.12)$$

The identity (10.2.11) for the step response weights V_j exactly parallels the identity (10.2.7) for the impulse response weights, except that the left-hand operator $\delta^*(B)$ is of order $r + 1$ instead of r.

Using the results of (10.2.8), it follows that the step response function is defined by

(i) b zero values $V_0, V_1, \ldots, V_{b-1}$;

(ii) a further $s - r$ values $V_b, V_{b+1}, V_{b+s-r-1}$ following no fixed pattern (no such values occur if $s < r + 1$);

(iii) values V_j, with $j \geq b + s - r$, which follow the pattern dictated by the $(r + 1)$th order difference equation $\delta^*(B)V_j = 0$ which has $r + 1$ starting values $V_{b+s}, V_{b+s-1}, \ldots, V_{b+s-r}$. Starting values V_j for $j < b$ will, of course, be zero.

10.2.3 First- and second-order discrete transfer function models

Details of transfer function models for all combinations of $r = 0, 1, 2$ and $s = 0, 1, 2$ are shown in Table 10.1. Specific examples of the models, with bar charts showing step response and impulse response, are given in Figure 10.6.

The equations at the end of Table 10.1 allow the parameters ξ, g, η of the V form of the model to be expressed in terms of the parameters δ, ω of the B form. These equations are given for the most general of the models considered, namely that for which $r = 2$ and $s = 2$. All the other models are special cases of this one, and the corresponding equations for these are obtained by setting appropriate parameters to zero. For example, if $r = 1$ and $s = 1, \xi_2 = \eta_2 = \delta_2 = \omega_2 = 0$, then

$$\delta_1 = \frac{\xi_1}{1 + \xi_1} \qquad \omega_0 = \frac{g(1 + \eta_1)}{1 + \xi_1} \qquad \omega_1 = \frac{g\eta_1}{1 + \xi_1}$$

In Figure 10.6 the starting values for the difference equations satisfied by the impulse and step responses, respectively, are indicated by circles on the bar charts.

Discussion of the tabled models. The models, whose properties are summarized in Table 10.1 and Figure 10.6, will repay careful study since they are useful in representing many commonly met dynamic systems. In all the models the operator B^b on the right ensures that the first nonzero term in the impulse response function is v_b. In the examples in Figure 10.6 the value of g is supposed equal to unity, and b is supposed equal to three.

Models with $r = 0$. With r and s both equal to zero, the impulse response consists of a single value $v_b = \omega_0 = g$. The output is proportional to the input, but is displaced by b time intervals. More generally, if we have an operator of order s on the right, then the instantaneous input will be delayed b intervals and will be spread over $s + 1$ values in proportion to $v_b = \omega_0$, $v_{b+1} = -\omega_1, \ldots, v_{b+s} = -\omega_s$. The step response is obtained by summing the impulse response and eventually satisfies the difference equation $(1 - B)V_j = 0$ with starting value $V_{b+s} = g = \omega_0 - \omega_1 - \cdots - \omega_s$.

r, s, b	∇ Form	B Form	Impulse Response v_j	Step Response $V_j = \sum_{i=0}^{j} v_i$
003	$Y_t = X_{t-3}$	$Y_t = B^3 X_t$	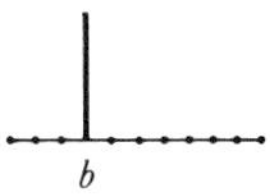 b	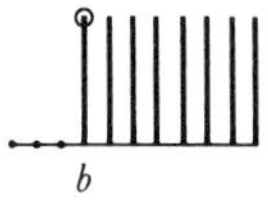 b
013	$Y_t = (1 - .5\nabla)\, X_{t-3}$	$Y_t = (.5 + .5B)\, B^3 X_t$	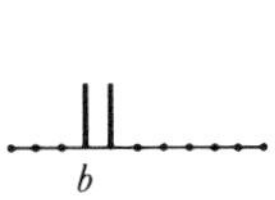 b	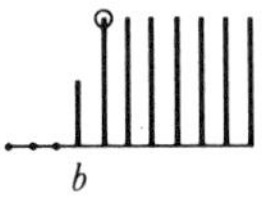 b
023	$Y_t = (1 - \nabla + .25\nabla^2)\, X_{t-3}$	$Y_t = (.25 + .50B + .25B^2)\, B^3 X_t$	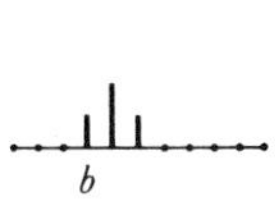 b	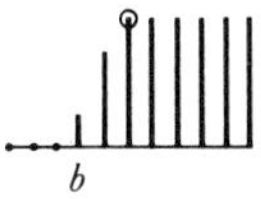 b
103	$(1 + \nabla)\, Y_t = X_{t-3}$	$(1 - .5B)\, Y_t = .5 B^3 X_t$	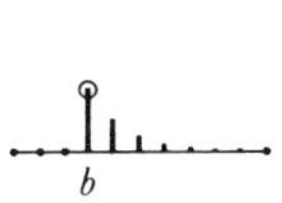 b	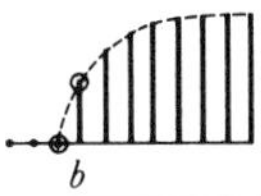 b
113	$(1 + \nabla)\, Y_t = (1 - .5\nabla)\, X_{t-3}$	$(1 - .5B)\, Y_t = (.25 + .25B)\, B^3 X_t$	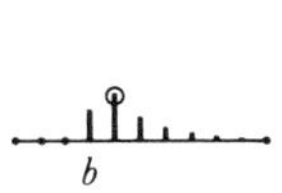 b	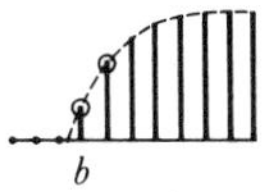 b
123	$(1 + \nabla)\, Y_t = (1 - \nabla + .25\nabla^2)\, X_{t-3}$	$(1 - .5B)\, Y_t = (.125 + .25B + .125B^2)\, B^3 X_t$	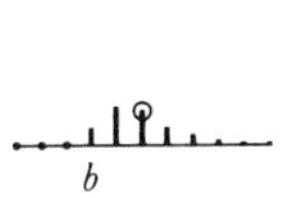 b	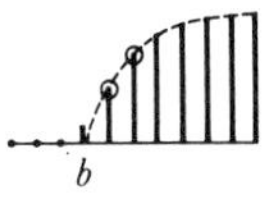 b
203	$(1 - .25\nabla + .5\nabla^2)\, Y_t = X_{t-3}$	$(1 - .6B + .4B^2)\, Y_t = .8 B^3 X_t$	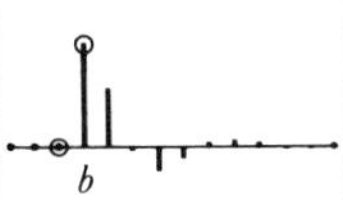 b	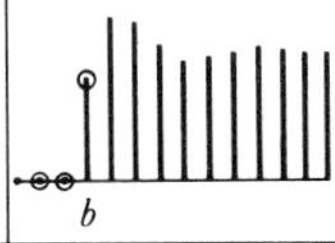 b
213	$(1 - .25\nabla + .5\nabla^2)\, Y_t = (1 - .5\nabla)\, X_{t-3}$	$(1 - .6B + .4B^2)\, Y_t = (.4 + .4B)\, B^3 X_t$	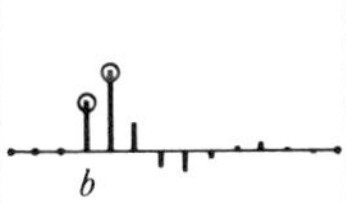 b	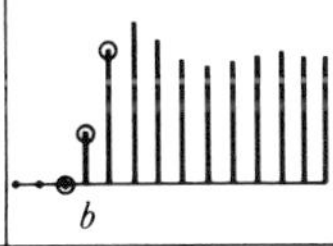 b
223	$(1 - .25\nabla + .5\nabla^2)\, Y_t = (1 - \nabla + .25\nabla^2)\, X_{t-3}$	$(1 - .6B + .4B^2)\, Y_t = (.2 + .4B + .2B^2)\, B^3 X_t$	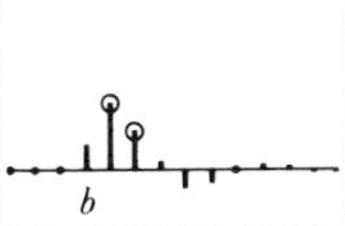 b	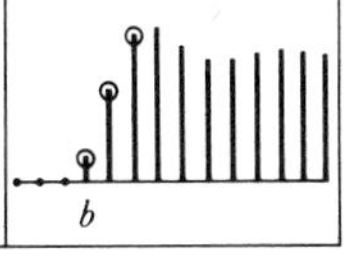 b

FIG. 10.6   Examples of impulse and step response functions with gain $g = 1$

TABLE 10.1   Impulse response functions for transfer function models of the form $\delta_r(B)Y_t = \omega_s(B)B^b X_t$

$r\,s\,b$	∇ form	B form	Impulse response v_j	
$00b$	$Y_t = gX_{t-b}$	$Y_t = \omega_0 B^b X_t$	0 ω_0 0	$j < b$ $j = b$ $j > b$
$01b$	$Y_t = g(1 + \eta_1\nabla)X_{t-b}$	$Y_t = (\omega_0 - \omega_1 B)B^b X_t$	0 ω_0 $-\omega_1$ 0	$j < b$ $j = b$ $j = b+1$ $j > b+1$
$02b$	$Y_t = g(1 + \eta_1\nabla + \eta_2\nabla^2)X_{t-b}$	$Y_t = (\omega_0 - \omega_1 B - \omega_2 B^2)B^b X_t$	0 ω_0 $-\omega_1$ $-\omega_2$ 0	$j < b$ $j = b$ $j = b+1$ $j = b+2$ $j > b+2$
$10b$	$(1 + \xi_1\nabla)Y_t = gX_{t-b}$	$(1 - \delta_1 B)Y_t = \omega_0 B^b X_t$	0 ω_0 $\delta_1 v_{j-1}$	$j < b$ $j = b$ $j > b$
$11b$	$(1 + \xi_1\nabla)Y_t = g(1 + \eta_1\nabla)X_{t-b}$	$(1 - \delta_1 B)Y_t = (\omega_0 - \omega_1 B)B^b X_t$	0 ω_0 $\delta_1\omega_0 - \omega_1$ $\delta_1 v_{j-1}$	$j < b$ $j = b$ $j = b+1$ $j > b+1$

TABLE 10.1   (contd.)

r s b	∇ form	B form	Impulse response v_j
12b	$(1 + \xi_1\nabla)Y_t = g(1 + \eta_1\nabla + \eta_2\nabla^2)X_{t-b}$	$(1 - \delta_1 B)Y_t = (\omega_0 - \omega_1 B - \omega_2 B^2)B^b X_t$	$\begin{array}{ll} 0 & j < b \\ \omega_0 & j = b \\ \delta_1\omega_0 - \omega_1 & j = b+1 \\ \delta_1^2\omega_0 - \delta_1\omega_1 - \omega_2 & j = b+2 \\ \delta_1 v_{j-1} & j > b+2 \end{array}$
20b	$(1 + \xi_1\nabla + \xi_2\nabla^2)Y_t = gX_{t-b}$	$(1 - \delta_1 B - \delta_2 B^2)Y_t = \omega_0 B^b X_t$	$\begin{array}{ll} 0 & j < b \\ \omega_0 & j = b \\ \delta_1 v_{j-1} + \delta_2 v_{j-2} & j > b \end{array}$
21b	$(1 + \xi_1\nabla + \xi_2\nabla^2)Y_t = g(1 + \eta_1\nabla)X_{t-b}$	$(1 - \delta_1 B - \delta_2 B^2)Y_t = (\omega_0 - \omega_1 B)B^b X_t$	$\begin{array}{ll} 0 & j < b \\ \omega_0 & j = b \\ \delta_1\omega_0 - \omega_1 & j = b+1 \\ \delta_1 v_{j-1} + \delta_2 v_{j-2} & j > b+1 \end{array}$
22b	$(1 + \xi_1\nabla + \xi_2\nabla^2)Y_t = g(1 + \eta_1\nabla + \eta_2\nabla^2)X_{t-b}$	$(1 - \delta_1 B - \delta_2 B^2)Y_t = (\omega_0 - \omega_1 B - \omega_2 B^2)B^b X_t$	$\begin{array}{ll} 0 & j < b \\ \omega_0 & j = b \\ \delta_1\omega_0 - \omega_1 & j = b+1 \\ (\delta_1^2 + \delta_2)\omega_0 & \\ \quad - \delta_1\omega_1 - \omega_2 & j = b+2 \\ \delta_1 v_{j-1} + \delta_2 v_{j-2} & j > b+2 \end{array}$

$$\xi_1 = \frac{\delta_1 + 2\delta_2}{1 - \delta_1 - \delta_2} \qquad \xi_2 = \frac{-\delta_2}{1 - \delta_1 - \delta_2}$$

$$g = \frac{\omega_0 - \omega_1 - \omega_2}{1 - \delta_1 - \delta_2}$$

$$\eta_1 = \frac{\omega_1 + 2\omega_2}{\omega_0 - \omega_1 - \omega_2}$$

$$\eta_2 = \frac{-\omega_2}{\omega_0 - \omega_1 - \omega_2}$$

$$\delta_1 = \frac{\xi_1 + 2\xi_2}{1 + \xi_1 + \xi_2} \qquad \delta_2 = \frac{-\xi_2}{1 + \xi_1 + \xi_2}$$

$$\omega_0 = \frac{g(1 + \eta_1 + \eta_2)}{1 + \xi_1 + \xi_2}$$

$$\omega_1 = \frac{g(\eta_1 + 2\eta_2)}{1 + \xi_1 + \xi_2}$$

$$\omega_2 = \frac{-g\eta_2}{1 + \xi_1 + \xi_2}$$

$$1 - \delta_1 - \delta_2 = (1 + \xi_1 + \xi_2)^{-1}$$

Models with $r = 1$.   With $s = 0$, the impulse response tails off exponentially (geometrically) from the initial starting value $v_b = \omega_0 = g/(1 + \xi_1) = g(1 - \delta_1)$. The step response increases exponentially till it attains the value $g = 1$. If the exponential step response is extrapolated backwards as indicated by the dotted line it cuts the time axis at time $b - 1$. This corresponds to the fact that $V_{b-1} = 0$ as well as $V_b = v_b$ are starting values for the appropriate difference equation $(1 - \delta B)(1 - B)V_j = 0$.

With $s = 1$, there is an initial value $v_b = \omega_0 = g(1 + \eta_1)/(1 + \xi_1)$ of the impulse response which does not follow a pattern. The exponential pattern induced by the difference equation $v_j = \delta_1 v_{j-1}$ associated with the left-hand operator begins with the starting value $v_{b+1} = (\delta_1 \omega_0 - \omega_1) = g(\xi_1 - \eta_1)/(1 + \xi_1)^2$.

The step response function follows an exponential curve, determined by the difference equation $(1 - \delta B)(1 - B)V_j = 0$, which approaches g asymptotically from the starting value $V_b = v_b$ and $V_{b+1} = v_b + v_{b+1}$. An exponential curve projected by the dotted line backwards through the points will, in general, cut the time axis at some intermediate point in the time interval. We show in Section 10.3 that certain discrete models, which approximate continuous first-order systems having *fractional* periods of delay, may in fact be represented by a first-order difference equation with an operator of order $s = 1$ on the right.

With $s = 2$, there are two values v_b and v_{b+1} for the impulse response which do not follow a pattern, followed by exponential fall off beginning with v_{b+2}. Correspondingly, there is a single preliminary value V_b in the step response which does not coincide with the exponential curve projected by the dotted line. This curve is, as before, determined by the difference equation $(1 - \delta B)(1 - B)V_j = 0$ but with starting values V_{b+1} and V_{b+2}.

Models with $r = 2$.   The flexibility of the model with $s = 0$ is limited because the first starting value of the impulse response is fixed to be zero. More useful models are obtained for $s = 1$ and $s = 2$. The use of these models in approximating continuous second-order systems is discussed in Section 10.3 and in Appendix A10.1.

The behavior of the dynamic weights v_j, which eventually satisfy

$$v_j - \delta_1 v_{j-1} - \delta_2 v_{j-2} = 0 \qquad j > b + s \qquad (10.2.13)$$

depends on the nature of the roots S_1^{-1} and S_2^{-1}, of the *characteristic equation*

$$(1 - \delta_1 B - \delta_2 B^2) = (1 - S_1 B)(1 - S_2 B) = 0$$

This dependence is shown in Table 10.2. As in the continuous case, the model may be overdamped, critically damped or underdamped, depending on the nature of the roots of the characteristic equation.

When the roots are complex, the solution of (10.2.13) will follow a damped sine wave, as in the examples of second-order systems in Figure 10.6. When the roots are real the solution will be the sum of two exponentials. As in the continuous case considered in Section 10.1.2, the system can then be thought of as equivalent to two discrete first-order systems arranged in series and having parameters S_1 and S_2.

TABLE 10.2   Dependence of nature of second order system on the roots of $(1 - \delta_1 B - \delta_2 B^2) = 0$

Roots (S_1^{-1}, S_2^{-1})	Condition	Damping
Real	$\delta_1^2 + 4\delta_2 > 0$	Overdamped
Real and equal	$\delta_1^2 + 4\delta_2 = 0$	Critically damped
Complex	$\delta_1^2 + 4\delta_2 < 0$	Underdamped

The weights V_j for the step response eventually satisfy a difference equation

$$(V_j - g) - \delta_1(V_{j-1} - g) - \delta_2(V_{j-2} - g) = 0$$

which is of the same form as (10.2.13). Thus the behavior of the step response V_j about its asymptotic value g parallels the behavior of the impulse response about the time axis. In the situation where there are complex roots the step response "overshoots" the value g and then oscillates about this value until it reaches equilibrium. When the roots are real and positive, the step response, which is the sum of two exponential terms, approaches its asymptote g without crossing it. However, if there are negative real roots, the step response may overshoot and oscillate as it settles down to its equilibrium value.

In Figure 10.5 the dots indicate two discrete step responses, labelled R and C, respectively, in relation to a discrete step input indicated by dots at the bottom of the figure. The difference equation models* corresponding to R and C are:

$$R: \qquad (1 - 0.97B + 0.22B^2)Y_t = 5(0.15 + 0.09B)X_{t-1}$$

$$C: \qquad (1 - 1.15B + 0.49B^2)Y_t = 5(0.19 + 0.15B)X_{t-1}$$

Also shown in Figure 10.5 is a diagram of the stability region with the parameter points (δ_1, δ_2) marked for each of the two models. Note that the system described by model R, which has real positive roots, has no overshoot while that for model C, which has complex roots, does have overshoot.

10.2.4 Recursive computation of output for any input

It would be extremely tedious if it were necessary to use the impulse response form (10.1.2) of the model to compute the output for a given input. For-

---

* The parameters in these models were in fact selected, in a manner to be discussed in Section 10.3.2, so that at the discrete points, the step responses exactly matched those of the continuous systems introduced in Section 10.1.2.

tunately this is not necessary. Instead we may employ the difference equation model directly. In this way it is a simple matter to compute the output recursively for any input whatsoever. For example, consider the model with $r = 1$, $s = 0$, $b = 1$ and with $\xi = 1$ and $g = 5$. Thus

$$(1 + \nabla)Y_t = 5X_{t-1}$$

or equivalently,

$$(1 - 0.5B)Y_t = 2.5X_{t-1} \qquad (10.2.14)$$

Table 10.3 shows the calculation of Y_t when the input X_t is (a) a unit pulse input; (b) a unit step input; (c) a "general" input. In all cases it is assumed that the output has the initial value $Y_0 = 0$. To perform the recursive calculation, the difference equation is written out with Y_t on the left. Thus

$$Y_t = 0.5Y_{t-1} + 2.5X_{t-1}$$

and, for example, in the case of the "general" input

$$Y_1 = (0.5) \times (0) + (2.5) \times (0) = 0$$

$$Y_2 = (0.5) \times (0) + (2.5) \times (1.5) = 3.75$$

and so on.

TABLE 10.3   Calculation of output from discrete first-order system for impulse, step and general input

	(a)		(b)		(c)	
t	X_t Impulse input	Y_t Output	X_t Step input	Y_t Output	X_t General input	Y_t Output
0	0	0	0	0	0	0
1	1	0	1	0	1.5	0
2	0	2.50	1	2.50	0.5	3.75
3	0	1.25	1	3.75	2.0	3.12
4	0	0.62	1	4.38	1.0	6.56
5	0	0.31	1	4.69	-2.5	5.78
6	0	0.16	1	4.84	0.5	-3.36

These inputs and outputs are plotted in Figure 10.7 (a), (b) and (c).

In general, we see that having written the transfer function model in the form

$$Y_t = \delta_1 Y_{t-1} + \cdots + \delta_r Y_{t-r} + \omega_0 X_{t-b} - \omega_1 X_{t-b-1} - \cdots - \omega_s X_{t-b-s}$$

it is an easy matter to compute the discrete output for any discrete input. To start off the recursion we need to know certain initial values. This need is not of course a shortcoming of the method of calculation but comes about because, with a transfer function model, the initial values of Y *will* depend on values of X which occurred before observation was begun. In practice, when the necessary initial values are not known, we can substitute mean values for unknown Y's and X's (zero's if these quantities are considered as deviations from their means). The early calculated values will then depend upon this choice of the starting values. However for a stable system, the effect of this choice will be negligible after a period sufficient for the impulse response to become negligible. If this period is p_0 time intervals, then an alternative procedure is to compute $Y_{p_0}, Y_{p_0+1}, \ldots$ directly from the impulse response until enough values are available to set the recursion going.

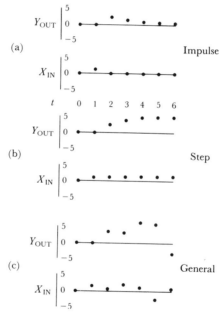

FIG. 10.7   Response of a first-order system to an impulse, a step and a "general" input

10.3 RELATION BETWEEN DISCRETE AND CONTINUOUS MODELS

The discrete dynamic model, defined by a linear difference equation, is of importance in its own right. It provides a sensible class of transfer functions and needs no other justification. In many examples no question will arise of attempting to relate the discrete model to a supposed underlying continuous one because no underlying continuous series properly exists. However, in

some cases, for example where instantaneous observations are taken periodically on a chemical reactor, the discrete record can be used to tell us
something about the continuous system. In particular, control engineers are
used to thinking in terms of the time constants and dead times of continuous
systems and may best understand the results of the discrete analysis when
so expressed.

As before, we denote a continuous output and input at time t by $Y(t)$ and
$X(t)$, respectively. Suppose the output and input are related by the linear
filtering operation

$$Y(t) = \int_0^\infty v(u)X(t - u)\, du$$

Suppose now that only discrete observations (X_t, Y_t), $(X_{t-1}, Y_{t-1}), \ldots$ of
output and input are available at equispaced intervals of time $t, t - 1, \ldots$
and that the discrete output and input are related by the discrete linear
filter

$$Y_t = \sum_{j=0}^\infty v_j X_{t-j}$$

Then, for certain special cases, and with appropriate assumptions, useful
relationships may be established between the discrete and continuous
models.

10.3.1 Response to a pulsed input

A special case, which is of importance in the design of the discrete control
schemes discussed in Part IV of this book, arises when the opportunity for
adjustment of the process occurs immediately after observation of the
output, so that the input variable is allowed to remain at the same level
between observations. The typical appearance of the resulting square wave,
or *pulsed input* as we shall call it, is shown in Figure 10.8. We denote the
fixed level at which the input is held during the period $t - 1 < \tau < t$ by
X_{t-1+}.

Consider a continuous linear system which has b whole periods of delay
plus a fractional period c of further delay. Thus, in terms of previous notation, $b + c = \tau$. Then we can represent the output from the system as

$$Y(t) = \int_0^\infty v(u)X(t - u)\, du$$

where the impulse response function $v(u)$ is zero for $u < b + c$. Now for a
pulsed input, as is seen from Figure 10.9, the output at time t will be given

exactly by

$$Y(t) = \left\{ \int_{b+c}^{b+1} v(u)\, du \right\} X_{t-b-1+} + \left\{ \int_{b+1}^{b+2} v(u)\, du \right\} X_{t-b-2+} + \cdots$$

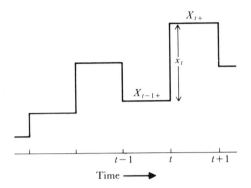

FIG. 10.8   An example of a pulsed input

Thus

$$Y(t) = Y_t = v_b X_{t-b-1+} + v_{b+1} X_{t-b-2+} + \cdots$$

Therefore, for a *pulsed input*, there exists a discrete linear filter which is such that, at times t, $t - 1$, $t - 2, \ldots$, the continuous output $Y(t)$ *exactly* equals the discrete output.

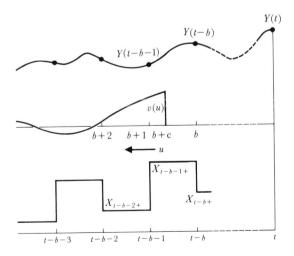

FIG. 10.9   Transfer to output from a pulsed input

Given a pulsed input, consider the output Y_t from a discrete model

$$\xi(\nabla)Y_t = \eta(\nabla)X_{t-b-1+} \tag{10.3.1}$$

of order (r, r) in relation to the continuous output from the Rth order model

$$(1 + \Xi_1 D + \Xi_2 D^2 + \cdots + \Xi_R D^R)Y(t) = X(t - b - c) \tag{10.3.2}$$

subject to the same input. It is shown in Appendix A10.1 that, for suitably chosen values of the parameters (Ξ, c), the outputs will coincide exactly if $R = r$. Furthermore, if $c = 0$, the output from the continuous model (10.3.2) will be identical at the discrete times with that of a discrete model (10.3.1) of order $(r, r - 1)$. We refer to the related continuous and discrete models as *discretely coincident* systems. If, then, a discrete model of the form (10.3.1) of order (r, r) has been obtained, then on the assumption *that the continuous model would be represented by the rth order differential equation (10.3.2)*, the parameters, and in particular the time constants for the discretely coincident continuous system, may be written explicitly in terms of the parameters of the discrete model.

The parameter relationships for a delayed second-order system have been derived in Appendix A10.1. From these the corresponding relationships for simpler systems may be obtained by setting appropriate constants equal to zero, as we shall now discuss.

10.3.2 Relationships for first- and second-order coincident systems

Undelayed first-order system—B form.   The continuous system satisfying

$$(1 + T_1 D)Y(t) = gX(t) \tag{10.3.3}$$

is, for a pulsed input, discretely coincident with the discrete system satisfying

$$(1 - \delta B)Y_t = \omega_0 X_{t-1+} \tag{10.3.4}$$

where

$$\delta = e^{-1/T} \quad T = (-\ln\delta)^{-1} \quad \omega_0 = g(1 - \delta) \tag{10.3.5}$$

∇ *form.*   Alternatively, the difference equation may be written

$$(1 + \xi\nabla)Y_t = gX_{t-1+} \tag{10.3.6}$$

where

$$\xi = \delta/(1 - \delta) \tag{10.3.7}$$

To illustrate, we reconsider the example of Section 10.2.4 for the "general" input. The output for this case is calculated in Table 10.3 (c) and plotted in Figure 10.7 (c). Suppose that, in fact, we had a continuous system

$$(1 + 1.44D)Y(t) = 5X(t)$$

Then this would be discretely coincident with the discrete model (10.2.14) actually considered, namely

$$(1 - 0.5B)Y_t = 2.5X_{t-1+}$$

If the input and output were continuous and the input were pulsed, then the actual course of the response would be that shown by the continuous lines in Figure 10.10. The output would in fact follow a series of exponential curves. Each dotted line shows the further course which the response would take if no further change in the input was made. The curves correspond exactly at the discrete sample points with the discrete output already calculated in Table 10.3 (c) and plotted in Figure 10.7 (c).

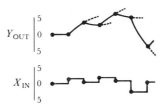

FIG. 10.10   Continuous response of the system $(1 + 1.44D)Y(t) = 5X(t)$ to a pulsed input

Delayed first-order system—B form.   The continuous system satisfying

$$(1 + TD)Y(t) = gX(t - b - c) \qquad (10.3.8)$$

is, for a pulsed input, discretely coincident with the discrete system satisfying

$$(1 - \delta B)Y_t = (\omega_0 - \omega_1 B)X_{t-b-1+} \qquad (10.3.9)$$

where

$$\delta = e^{-1/T} \qquad \omega_0 = g(1 - \delta^{1-c}) \qquad \omega_1 = g(\delta - \delta^{1-c}) \qquad (10.3.10)$$

∇ *form.*   Alternatively, the difference equation may be written

$$(1 + \xi\nabla)Y_t = g(1 + \eta\nabla)X_{t-b-1+} \qquad (10.3.11)$$

where

$$\xi = \delta/(1 - \delta) \qquad -\eta = \delta(\delta^{-c} - 1)/(1 - \delta) \qquad (10.3.12)$$

Now

$$(1 + \eta\nabla)X_{t-b-1+} = (1 + \eta)X_{t-b-1+} - \eta X_{t-b-2+} \qquad (10.3.13)$$

can be regarded as an interpolation at an increment $(-\eta)$ between X_{t-b-1+} and X_{t-b-2+}. Table 10.4 allows the corresponding parameters $(\xi, -\eta)$ and (T, c) of the discrete and continuous models to be determined for a range of alternatives.

TABLE 10.4   Values of $-\eta$ for various values of T and c for a first-order system with delay. Also shown are corresponding values of ξ and δ.

Value of δ	Value of ξ	Value of T	$c = 0.9$	$c = 0.7$	$c = 0.5$	$c = 0.3$	$c = 0.1$
0.9	9.00	9.49	0.90	0.69	0.49	0.29	0.10
0.8	4.00	4.48	0.89	0.68	0.47	0.28	0.09
0.7	2.33	2.80	0.88	0.66	0.46	0.26	0.09
0.6	1.50	1.95	0.88	0.64	0.44	0.25	0.08
0.5	1.00	1.44	0.87	0.62	0.41	0.23	0.07
0.4	0.67	1.09	0.85	0.60	0.39	0.21	0.06
0.3	0.43	0.83	0.84	0.57	0.35	0.19	0.05
0.2	0.25	0.62	0.82	0.52	0.31	0.15	0.04
0.1	0.11	0.43	0.77	0.45	0.24	0.11	0.03

Undelayed second-order system—B form.   The continuous system satisfying

$$(1 + T_1 D)(1 + T_2 D)Y(t) = gX(t) \tag{10.3.14}$$

is, for a pulsed input, discretely coincident with the system

$$(1 - \delta_1 B - \delta_2 B^2)Y_t = (\omega_0 - \omega_1 B)X_{t-1+} \tag{10.3.15}$$

or equivalently, with the system

$$(1 - S_1 B)(1 - S_2 B)Y_t = (\omega_0 - \omega_1 B)X_{t-1+} \tag{10.3.16}$$

where

$$\left. \begin{array}{l} S_1 = e^{-1/T_1} \quad S_2 = e^{-1/T_2} \\ \omega_0 = g(T_1 - T_2)^{-1}\{T_1(1 - S_1) - T_2(1 - S_2)\} \\ \omega_1 = g(T_1 - T_2)^{-1}\{T_1 S_2(1 - S_1) - T_2 S_1(1 - S_2)\} \end{array} \right\} \tag{10.3.17}$$

∇ *form.*   Alternatively, the difference equation may be written

$$(1 + \xi_1 \nabla + \xi_2 \nabla^2)Y_t = g(1 + \eta_1 \nabla)X_{t-1+} \tag{10.3.18}$$

where

$$-\eta_1 = (1 - S_1)^{-1}(1 - S_2)^{-1}(T_1 - T_2)^{-1}\{T_2 S_1(1 - S_2) - T_1 S_2(1 - S_1)\} \tag{10.3.19}$$

may be regarded as the increment of an interpolation between X_{t-1+} and X_{t-2+}. Values for ξ_1 and ξ_2 in terms of the δ's can be obtained directly using the results given in Table 10.1.

As a specific example, Figure 10.5 shows the step response for two discrete systems we have considered before, together with the corresponding continuous responses from the discretely coincident systems.

The pairs of models are

Curve C Continuous: $(1 + 1.41D + 2D^2)Y(t) = 5X(t)$

Discrete: $(1 - 1.15B + 0.49B^2)Y_t = 5(0.19 + 0.15B)X_{t-1+}$

Curve R Continuous: $(1 + 2D)(1 + D)Y(t) = 5X(t)$

Discrete: $(1 - 0.97B + 0.22B^2)Y_t = 5(0.15 + 0.09B)X_{t-1+}$

The continuous curves were drawn using (10.1.18) and (10.1.12), which give the continuous step responses for second-order systems having respectively complex and real roots.

The discrete representation of the response of a second-order continuous system with delay to a pulsed input is given in Appendix A10.1.

10.3.3 Approximating general continuous models by discrete models

Perhaps we should emphasize once more that the discrete transfer function models do not need to be justified in terms of, or related to, continuous systems. They are of importance in their own right in allowing a discrete output to be calculated from a discrete input. However, in some instances, such relationships are of interest.

For continuous systems the pulsed input arises of itself in control problems when the convenient way to operate is to make an observation on the output Y and then immediately to make any adjustment that may be needed on the input variable X. Thus the input variable stays at a fixed level between observations, and we have a pulsed input. The relationships established in the previous sections may then be applied immediately. In particular, these relationships indicate that, with the notation we have used, the *undelayed* discrete system is represented by

$$\xi(\nabla)Y_t = \eta(\nabla)X_{t-1+}$$

in which the subscript $t - 1+$ on X is one step behind the subscript t on Y.

Use of discrete models when continuous records are available. Even though we have a continuous record of input and output, it may be convenient to determine the dynamic characteristics of the system by discrete methods, as we shall describe in Chapter 11. Thus, if pairs of values are read off with a sufficiently short sampling interval, then very little is lost by replacing the continuous record by the discrete one.

One way in which the discrete results may then be used to approximate the continuous transfer function is to treat the input as though it were pulsed, that is to treat the input record as if the discrete input observed at time j extended from just after $j - \frac{1}{2}$ to $j + \frac{1}{2}$, as in Figure 10.11. Thus $X(t) = X_j$, $(j - \frac{1}{2} < t \leqslant j + \frac{1}{2})$. We can then relate the discrete result to that of the continuous record by using the pulsed input equations with X_t replacing X_{t+} and with $b + c - \frac{1}{2}$ replacing $b + c$, that is with one half a time period subtracted from the delay. The continuous record will normally be read at a sufficiently small sampling interval so that sudden changes do not occur between the sampled points. In this case the approximation will be very close.

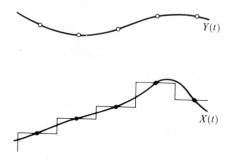

FIG. 10.11   Replacement of continuous input by pulsed input

10.3.4 *Transfer function models with added noise*

In practice, the output Y could not be expected to follow exactly the pattern determined by the transfer function model, even if that model were entirely adequate. Disturbances of various kinds other than X normally corrupt the system. A disturbance might originate at any point in the system, but it is often convenient to consider it in terms of its net effect on the output Y, as indicated in Figure 1.5. If we assume that the disturbance, or noise N_t, is independent of the level of X and is additive with respect to the influence of X, then we can write

$$Y_t = \delta^{-1}(B)\omega(B)X_{t-b} + N_t \qquad (10.3.20)$$

If the noise model can be represented by an ARIMA (p, d, q) process

$$N_t = \varphi^{-1}(B)\theta(B)a_t$$

where a_t is white noise, the model (10.3.20) can be written finally as

$$Y_t = \delta^{-1}(B)\omega(B)X_{t-b} + \varphi^{-1}(B)\theta(B)a_t \qquad (10.3.21)$$

In the next chapter we describe methods for identifying, fitting and checking combined transfer-function noise models of the form (10.3.21).

APPENDIX A10.1 CONTINUOUS MODELS WITH PULSED INPUTS

We showed in Section 10.3.1 (see also Figure 10.9) that, for a pulsed input, the output from any delayed continuous linear system

$$Y(t) = \int_0^\infty v(u)X(t-u)\,du$$

where $v(u) = 0$, $u < b + c$, is exactly given at the discrete times t, $t - 1$, $t - 2, \ldots$ by the discrete linear filter

$$Y_t = v(B)X_{t-1+}$$

where the weights $v_0, v_1, \ldots, v_{b-1}$ are zero and the weights $v_b, v_{b+1}, \ldots$ are given by

$$v_b = \int_{b+c}^{b+1} v(u)\,du \qquad\qquad (A10.1.1)$$

$$v_{b+j} = \int_{b+j}^{b+j+1} v(u)\,du \quad j \geqslant 1 \qquad (A10.1.2)$$

Now suppose the dynamics of the continuous system is represented by the Rth order linear differential equation

$$\Xi(D)Y(t) = gX(t - b - c) \qquad\qquad (A10.1.3)$$

which may be written in the form

$$\prod_{h=1}^{R}(1 + T_h D)Y(t) = gX(t - b - c)$$

where $T_1, T_2, \ldots, T_R$ may be real or complex. We now show that, for a pulsed input, the output from this continuous system is discretely coincident with that from a discrete difference equation model of order (r, r), or of order $(r, r - 1)$ if $c = 0$. Now $v(u)$ is zero for $u < b + c$ and for $u \geqslant b + c$ is in general nonzero and satisfies the differential equation

$$\prod_{h=1}^{R}(1 + T_h D)v(u - b - c) = 0 \qquad\qquad u \geqslant b + c$$

Thus

$$v(u) = 0 \qquad\qquad u < b + c$$

$$v(u) = \alpha_1 e^{-(u-b-c)/T_1} + \alpha_2 e^{-(u-b-c)/T_2} + \cdots + \alpha_R e^{-(u-b-c)/T_R} \qquad u \geqslant b + c$$

whence, using (A10.1.1) and (A10.1.2),

$$v_b = \sum_{h=1}^{R} \alpha_h T_h \{1 - e^{-(1-c)/T_h}\} \qquad\qquad (A10.1.4)$$

$$v_{b+j} = \sum_{h=1}^{R} \alpha_h T_h (1 - e^{-1/T_h}) e^{c/T_h} e^{-j/T_h} \qquad j \geqslant 1 \qquad \text{(A10.1.5)}$$

It will be noted that, in the particular case when $c = 0$, the weights v_{b+j} are given by (A10.1.2) for $j = 0$ as well as for $j > 0$.

Now consider the difference equation model of order (r, s)

$$\delta(B)Y_t = \omega(B)B^b X_{t-1+} \qquad \text{(A10.1.6)}$$

If we write

$$\Omega(B) = \omega(B)B^b$$

the discrete transfer function $v(B)$ for this model satisfies

$$\delta(B)v(B) = \Omega(B) \qquad \text{(A10.1.7)}$$

As we have observed in (10.2.8), by equating coefficients in (A10.1.7) we obtain b zero weights $v_0, v_1, \ldots, v_{b-1}$, and, if $s \geqslant r$, a further $s - r + 1$ values $v_b, v_{b+1}, \ldots, v_{b+s-r}$ which do not follow a pattern. The weights v_j eventually satisfy

$$\delta(B)v_j = 0 \qquad j > b + s \qquad \text{(A10.1.8)}$$

with $v_{b+s}, v_{b+s-1}, \ldots, v_{b+s-r+1}$ supplying the required r starting values. Now write

$$\delta(B) = \prod_{h=1}^{r} (1 - S_h B)$$

where $S_1^{-1}, S_2^{-1}, \ldots, S_r^{-1}$ are the roots of the equation $\delta(B) = 0$. Then the solution of (A10.1.8) is of the form

$$v_j = A_1(\omega)S_1^j + A_2(\omega)S_2^j + \cdots + A_r(\omega)S_r^j \qquad j > b + s - r$$

$$\text{(A10.1.9)}$$

where the coefficients $A_h(\omega)$ are suitably chosen so that the solutions of (A10.1.9) for $j = s - r + 1, s - r + 2, \ldots, s$ generate the starting values $v_{b+s-r+1}, \ldots, v_{b+s}$ and the notation $A_h(\omega)$ is used as a reminder that the A's are functions of $\omega_0, \omega_1, \ldots, \omega_s$. Thus, if we set $s = r$, for given parameters (ω, δ) in (A10.1.6), and hence for given parameters (ω, S), there will be a corresponding set of values $A_h(\omega)$ $(h = 1, 2, \ldots, r)$ which produce the appropriate r starting values $v_{b+1}, v_{b+2}, \ldots, v_{b+r}$. Furthermore, we know that $v_b = \omega_0$. Thus

$$v_b = \omega_0 \qquad \text{(A10.1.10)}$$

$$v_{b+j} = \sum_{h=1}^{r} A_h(\omega)S_h^j \qquad \text{(A10.1.11)}$$

and we can equate the values of the weights in (A10.1.4) and (A10.1.5), which come from the differential equation, to those in (A10.1.10) and (A10.1.11), which come from the difference equation. To do this we must set

$$R = r \qquad S_h = e^{-1/T_h}$$

and the remaining $r + 1$ equations

$$\omega_0 = \sum_{h=1}^{r} \alpha_h T_h (1 - S_h^{1-c})$$

$$A_h(\omega) = \alpha_h T_h (1 - S_h) S_h^{-c}$$

determine $c, \alpha_1, \alpha_2, \ldots, \alpha_r$ in terms of the S's and ω's.

When $c = 0$, we set $s = r - 1$, and for given parameters (ω, S) in the difference equation, there will then be a set of r values $A_h(\omega)$ which are functions of $\omega_0, \omega_1, \ldots, \omega_{r-1}$, which produce the r starting values $v_b, v_{b+1}, \ldots, v_{b+r-1}$ and which can be equated to the values given by (A10.1.5) for $j = 0, 1, \ldots, r - 1$. To do this we set

$$R = r \qquad S_h = e^{-1/T_h}$$

and the remaining r equations

$$A_h(\omega) = \alpha_h T_h (1 - S_h)$$

determine $\alpha_1, \alpha_2, \ldots, \alpha_r$ in terms of the S's and ω's.

It follows, in general, that for a pulsed input the output at times $t, t - 1, \ldots$ from the continuous rth order dynamic system defined by

$$\Xi(D) Y(t) = g X(t - b - c) \qquad (A10.1.12)$$

is identical to the output from a discrete model

$$\xi(\nabla) Y_t = g \eta(\nabla) X_{t-b-1+} \qquad (A10.1.13)$$

of order (r, r) with the parameters suitably chosen. Furthermore, if $c = 0$, the output from the continuous model (A10.1.12) is identical at the discrete times to that of a model (A10.1.13) of order $(r, r - 1)$.

We now derive the discrete model corresponding to the second-order system with delay, from which the results given in Section 10.3.2 may be obtained as special cases.

Second-order system with delay. Suppose the differential equation relating input and output for a continuous system is given by

$$(1 + T_1 D)(1 + T_2 D) Y(t) = g X(t - b - c) \qquad (A10.1.14)$$

Then the continuous impulse response function is

$$v(u) = g(T_1 - T_2)^{-1}(e^{-(u-b-c)/T_1} - e^{-(u-b-c)/T_2}) \qquad u > b + c$$

$$(A10.1.15)$$

For a pulsed input, the output at discrete times $t, t - 1, t - 2, \ldots$ will be related to the input by the difference equation

$$(1 + \xi_1 \nabla + \xi_2 \nabla^2) Y_t = g(1 + \eta_1 \nabla + \eta_2 \nabla^2) X_{t-b-1+} \quad (A10.1.16)$$

with suitably chosen values of the parameters. This difference equation can also be written

$$(1 - \delta_1 B - \delta_2 B^2) Y_t = (\omega_0 - \omega_1 B - \omega_2 B^2) X_{t-b-1+}$$

or

$$(1 - S_1 B)(1 - S_2 B) Y_t = (\omega_0 - \omega_1 B - \omega_2 B^2) X_{t-b-1+}$$

$$(A10.1.17)$$

Using (A10.1.1) and (A10.1.2) and writing

$$S_1 = e^{-1/T_1} \qquad S_2 = e^{-1/T_2}$$

we obtain

$$v_b = \int_{b+c}^{b+1} v(u)\, du = g(T_1 - T_2)^{-1} \{ T_1(1 - S_1^{1-c}) - T_2(1 - S_2^{1-c}) \}$$

$$v_{b+j} = \int_{b+j}^{b+j+1} v(u)\, du = g(T_1 - T_2)^{-1} \{ T_1 S_1^{-c}(1 - S_1) S_1^j - T_2 S_2^{-c}(1 - S_2) S_2^j \}$$

$$j \geqslant 1$$

Thus

$$(T_1 - T_2)v(B) = gB^b T_1 \{ 1 - S_1^{1-c} + S_1^{-c}(1 - S_1)(1 - S_1 B)^{-1} S_1 B \}$$
$$- gB^b T_2 \{ 1 - S_2^{1-c} + S_2^{-c}(1 - S_2)(1 - S_2 B)^{-1} S_2 B \}$$

But from (A10.1.17),

$$v(B) = \frac{B^b(\omega_0 - \omega_1 B - \omega_2 B^2)}{(1 - S_1 B)(1 - S_2 B)}$$

whence we obtain

$$\left.\begin{aligned}
\omega_0 &= g(T_1 - T_2)^{-1} \{ T_1(1 - S_1^{1-c}) - T_2(1 - S_2^{1-c}) \} \\
\omega_1 &= g(T_1 - T_2)^{-1} \{ (S_1 + S_2)(T_1 - T_2) + T_2 S_2^{1-c}(1 + S_1) \\
&\quad - T_1 S_1^{1-c}(1 + S_2) \} \\
\omega_2 &= g S_1 S_2 (T_1 - T_2)^{-1} \{ T_2(1 - S_2^{-c}) - T_1(1 - S_1^{-c}) \}
\end{aligned}\right\} \quad (A10.1.18)$$

and

$$\delta_1 = S_1 + S_2 = e^{-1/T_1} + e^{-1/T_2} \qquad \delta_2 = -S_1 S_2 = -e^{-(1/T_1)-(1/T_2)}$$

$$(A10.1.19)$$

Complex roots.   If T_1 and T_2 are complex, corresponding expressions are obtained by substituting

$$T_1 = \zeta^{-1} e^{i\lambda} \qquad T_2 = \zeta^{-1} e^{-i\lambda} \qquad (i^2 = -1)$$

yielding

$$\left.\begin{aligned}
\omega_0 &= g\left\{1 - \frac{e^{-\zeta(1-c)\cos\lambda}\sin\{\zeta(1-c)\sin\lambda + \lambda\}}{\sin\lambda}\right\} \\
\omega_2 &= g\delta_2\left\{1 - \frac{e^{\zeta c \cos\lambda}\sin(-\zeta c \sin\lambda + \lambda)}{\sin\lambda}\right\} \\
\omega_1 &= \omega_0 - \omega_2 - (1 - \delta_1 - \delta_2)g
\end{aligned}\right\} \qquad \text{(A10.1.20)}$$

where

$$\begin{aligned}
\delta_1 &= 2e^{-\zeta\cos\lambda}\cos(\zeta\sin\lambda) \\
\delta_2 &= -e^{-2\zeta\cos\lambda}
\end{aligned} \qquad \text{(A10.1.21)}$$

APPENDIX A10.2 NONLINEAR TRANSFER FUNCTIONS AND LINEARIZATION

The linearity (or additivity) of the transfer function models we have considered implies that the overall response to the sum of a number of individual inputs will be the sum of the individual responses to those inputs. Specifically, that if $Y_t^{(1)}$ is the response at time t to an input history $\{X_t^{(1)}\}$ and $Y_t^{(2)}$ is the response at time t to an input history $\{X_t^{(2)}\}$, then the response at time t to an input history $\{X_t^{(1)} + X_t^{(2)}\}$ would be $Y_t^{(1)} + Y_t^{(2)}$, and similarly for continuous inputs and outputs. In particular, if the input level is multiplied by some constant, the output level is multiplied by this same constant. In practice this assumption is probably never quite true, but it supplies a useful approximation for many practical situations.

Models for nonlinear systems may sometimes be obtained by allowing the parameters to depend upon the level of the input in some prescribed manner. For example, suppose that a system were being studied over a range where Y had a maximum η, and for any X the steady state relation could be approximated by the quadratic expression

$$Y_\infty = \eta - \tfrac{1}{2}k(\mu - X)^2$$

where Y and X are, as before, deviations from a convenient origin. Then

$$g(X) = \frac{dY_\infty}{dX} = k(\mu - X)$$

and the dynamic behavior of the system might then be capable of representation by the first-order difference equation (10.3.4) but with variable gain proportional to $k(\mu - X)$. Thus

$$Y_t = \delta Y_{t-1} + k(\mu - X_{t-1+})(1 - \delta)X_{t-1+} \qquad \text{(A10.2.1)}$$

Dynamics of a simple chemical reactor. It sometimes happens that we can make a theoretical analysis of a physical problem which will yield the appropriate form for the transfer function. In particular this allows us to see very specifically what is involved in the linearized approximation.

As an example, suppose that a pure chemical A is continuously fed through a stirred tank reactor, and in the presence of a catalyst a certain proportion of it is changed to a product B, with no change of overall volume; hence the material continuously leaving the reactor consists of a mixture of B and unchanged A.

Suppose that initially the system is in equilibrium and that with quantities measured in suitable units

(1) μ is the rate at which A is fed to the reactor (and consequently is also the rate at which the mixture of A and B leaves the reactor).
(2) η is the proportion of unchanged A at the outlet, so that $1 - \eta$ is the proportion of the product B at the outlet.
(3) V is the volume of the reactor.
(4) k is a constant determining the rate at which the product B is formed.

Suppose that the reaction is "first-order" with respect to A which means that the rate at which B is formed and A is used up is proportional to the amount of A present. Then the rate of formation of B is $kV\eta$, but the rate at which B is leaving the outlet is $\mu(1 - \eta)$, and since the system is in equilibrium,

$$\mu(1 - \eta) = kV\eta \qquad \text{(A10.2.2)}$$

Now, suppose the equilibrium of the system is disturbed, the rate of feed to the reactor at time t being $\mu + X(t)$ and the corresponding concentration of A in the outlet being $\eta + Y(t)$. Now the rate of chemical formation of B, which now equals $kV(\eta + Y(t))$, will in general no longer exactly balance the rate at which B is flowing out of the system, which now equals $[\mu + X(t)][(1 - \eta - Y(t)]$. The difference in these two quantities is the rate of increase in the amount of B within the reactor, which equals $-V[dY(t)/dt]$. Thus

$$-V\frac{dY(t)}{dt} = kV[\eta + Y(t)] - [\mu + X(t)][1 - \eta - Y(t)] \qquad \text{(A10.2.3)}$$

Using (A10.2.2) and rearranging, (A10.2.3) may be written

$$(kV + \mu + VD)Y(t) = X(t)[1 - \eta - Y(t)]$$

or

$$(1 + TD)Y(t) = g\left(1 - \frac{Y(t)}{1 - \eta}\right)X(t) \qquad \text{(A10.2.4)}$$

where

$$T = \frac{V}{kV + \mu} \qquad g = \frac{(1 - \eta)}{kV + \mu} \qquad \text{(A10.2.5)}$$

Now (A10.2.4) is a nonlinear differential equation, since it contains a term $X(t)$ multiplied by $Y(t)$. However, in some practical circumstances, it could be adequately approximated by a linear differential equation, as we now show.

Processes operate under a wide range of conditions, but certainly a not unusual situation might be one where $100(1 - \eta)$, the percentage conversion of feed A to product B, was say 80%, and $100Y(t)$, the percentage fluctuation that was of practical interest, was say 4%. In this case the factor $1 - Y(t)/(1 - \eta)$ would vary from 0.95 to 1.05 and, to a good approximation, could be replaced by unity. The nonlinear differential equation (A10.2.4) could then be replaced by the linear first-order differential equation

$$(1 + TD)Y(t) = gX(t)$$

where T and g are as defined in Section 10.1.2. If the system were observed at discrete intervals of time, this equation could be approximated by a linear difference equation.

Situations can obviously occur when nonlinearities are of importance. This is particularly true of optimization studies, where the range of variation for the variables may be large. A device which is sometimes useful when the linear assumption is not adequate is to represent the dynamics by a set of linear models applicable over different ranges of the input variables. However, for discrete systems it is often less clumsy to work directly with a nonlinear difference equation which can be "solved" recursively rather than analytically. For example, we might replace the nonlinear differential equation (A10.2.4) by the nonlinear difference equation

$$(1 + \xi_1 \nabla)Y_t = g(1 + \eta_{12}Y_{t-1})X_{t-1}$$

11

Identification, Fitting, and Checking of Transfer Function Models

In Chapter 10 a parsimonious class of discrete linear transfer function models was introduced

$$Y_t - \delta_1 Y_{t-1} - \cdots - \delta_r Y_{t-r} = \omega_0 X_{t-b} - \omega_1 X_{t-b-1} - \cdots - \omega_s X_{t-b-s}$$

or

$$Y_t = \delta^{-1}(B)\omega(B)X_{t-b}$$

In these models X_t and Y_t were deviations from equilibrium of the system input and output. In practice the system will be infected by disturbances, or noise, whose net effect is to corrupt the output predicted by the transfer function model by an amount N_t. The combined transfer function-noise model may then be written as

$$Y_t = \delta^{-1}(B)\omega(B)X_{t-b} + N_t$$

In this chapter, methods are described for identifying, fitting, and checking transfer function-noise models when simultaneous pairs of observations (X_1, Y_1), $(X_2, Y_2), \ldots, (X_N, Y_N)$ of the input and output are available at discrete equispaced times $1, 2, \ldots, N$.

Engineering methods for estimating transfer functions are usually based on the choice of special inputs to the system, for example, step and sine wave inputs [81] and "pulse" inputs [82]. These methods have been useful when the system is affected by small amounts of noise, but are less satisfactory otherwise. In the presence of appreciable noise, it is necessary to use statistical methods for estimating the transfer function. Two previous approaches which have been tried for this problem are direct estimation of the impulse response in the time domain, and direct estimation of the gain and phase characteristics in the frequency domain, as described, for example, in [6], [7], and [27]. These methods are often unsatisfactory because they involve the estimation of too many parameters. For example, to determine the gain and phase characteristics, it is necessary to estimate two parameters at each frequency. The approach adopted in this chapter is to estimate the

370

parameters in parsimonious difference equation models. Throughout most of the chapter we assume that the input X_t is itself a stochastic process. Models of the kind discussed are useful in representing and forecasting certain multiple time series.

11.1 THE CROSS CORRELATION FUNCTION

In the same way that the autocorrelation function was used to identify stochastic models, the data analysis tool employed for the identification of transfer function models is the *cross correlation function* between the input and output. In this section we describe the basic properties of the cross correlation function and in the next section show how it can be used to identify transfer function models.

11.1.1 Properties of the cross covariance and cross correlation functions

Bivariate stochastic processes. We have seen in Chapter 2 that, to analyze a statistical time series, it is useful to regard it as a realization of a hypothetical population of time series called a stochastic process.

Now, suppose that we wish to describe an input time series X_t and the corresponding output time series Y_t from some physical system. For example, Figure 11.1 shows continuous data representing the (coded) input gas feed rate and corresponding output CO_2 concentration, from a gas furnace.

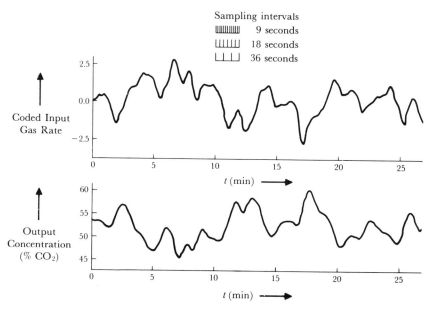

FIG. 11.1   Input gas rate and output CO_2 concentration from a gas furnace

Then we can regard this pair of time series as realizations of a hypothetical population of pairs of time series, called a *bivariate stochastic process* (X_t, Y_t). We shall assume that the data are read off at equispaced times yielding a pair of discrete time series, generated by a discrete bivariate process, and that values of the time series at times $t_0 + h, t_0 + 2h, \ldots, t_0 + Nh$ are denoted by $(X_1, Y_1), (X_2, Y_2), \ldots, (X_N, Y_N)$.

In this chapter, extensive illustrative use is made of the gas furnace data read at intervals of nine seconds (see Figure 11.1). The values (X_t, Y_t) so obtained are listed as Series J in the Collection of Time Series in Part V at the end of this volume.

The cross covariance and cross correlation functions. We have seen in Chapter 2 that a stationary Gaussian stochastic process can be described by its mean μ and autocovariance function γ_k, or equivalently by its mean μ, variance σ^2, and autocorrelation function ρ_k. Moreover, since $\gamma_k = \gamma_{-k}$ and $\rho_k = \rho_{-k}$, the autocovariance and autocorrelation functions need only be plotted for nonnegative values of the lag $k = 0, 1, 2, \ldots$.

In general, a bivariate stochastic process (X_t, Y_t) need not be stationary. However, as in Chapter 4, we assume that the appropriately differenced process (x_t, y_t), where $x_t = \nabla^d X_t$, $y_t = \nabla^d Y_t$, is stationary. The stationarity assumption implies in particular that the constituent processes x_t and y_t have constant means μ_x and μ_y and constant variances σ_x^2 and σ_y^2. If, in addition, it is assumed that the bivariate process is Gaussian, or Normal, then it is uniquely characterized by its means μ_x, μ_y and its covariance matrix. Figure 11.2 shows the different kinds of covariances that need to be considered.

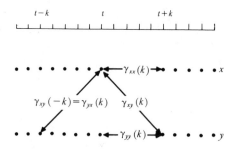

FIG. 11.2   Autocovariances and cross covariances of a bivariate stochastic process

The autocovariance coefficients of each constituent series at lag k are defined by the usual formula

$$\gamma_{xx}(k) = E[(x_t - \mu_x)(x_{t+k} - \mu_x)] = E[(x_t - \mu_x)(x_{t-k} - \mu_x)]$$

$$\gamma_{yy}(k) = E[(y_t - \mu_y)(y_{t+k} - \mu_y)] = E[(y_t - \mu_y)(y_{t-k} - \mu_y)]$$

where we now use the extended notation $\gamma_{xx}(k)$ and $\gamma_{yy}(k)$ for the auto-covariances of the x and y series. The only other covariances which can appear in the covariance matrix are the *cross covariance* coefficients between x and y at lag $+k$

$$\gamma_{xy}(k) = E[(x_t - \mu_x)(y_{t+k} - \mu_y)] \qquad k = 0, 1, 2, \ldots \qquad (11.1.1)$$

and the cross covariance coefficients between y and x at lag $+k$

$$\gamma_{yx}(k) = E[(y_t - \mu_y)(x_{t+k} - \mu_x)] \qquad k = 0, 1, 2, \ldots . \qquad (11.1.2)$$

Note that, in general, $\gamma_{xy}(k)$ will not be the same as $\gamma_{yx}(k)$. However, since

$$\gamma_{xy}(k) = E[(x_{t-k} - \mu_x)(y_t - \mu_y)] = E[(y_t - \mu_y)(x_{t-k} - \mu_x)] = \gamma_{yx}(-k)$$

we need only define one function $\gamma_{xy}(k)$ for $k = 0, \pm1, \pm2, \pm \ldots$. The function $\gamma_{xy}(k)$, defined for $k = 0, \pm1, \pm2, \pm \ldots$, is called the *cross covariance function* of the bivariate process. Similarly the dimensionless quantity

$$\rho_{xy}(k) = \frac{\gamma_{xy}(k)}{\sigma_x \sigma_y} \qquad k = 0, \pm1, \pm2, \pm \ldots \qquad (11.1.3)$$

is called the *cross correlation* coefficient at lag k, and the function $\rho_{xy}(k)$, defined for $k = 0, \pm1, \pm2, \pm \ldots$, the *cross correlation function* of the bivariate process.

Since $\rho_{xy}(k)$ is not in general equal to $\rho_{xy}(-k)$, the cross correlation function, in contrast to the autocorrelation function, is not symmetric about $k = 0$. In fact it will often happen that the cross correlation function is zero over some range $-\infty$ to i or i to $+\infty$. For example, consider the cross covariance function between a and z for the "delayed" first-order autoregressive process

$$(1 - \phi B)\tilde{z}_t = a_{t-b} \qquad -1 < \phi < 1 \quad b > 0$$

where a_t has zero mean. Then since

$$\tilde{z}_{t+k} = a_{t+k-b} + \phi a_{t+k-b-1} + \phi^2 a_{t+k-b-2} + \cdots$$

the cross covariance function between a and z is

$$\gamma_{az}(k) = E[a_t \tilde{z}_{t+k}] = \begin{cases} \phi^{k-b} \sigma_a^2 & k \geqslant b \\ 0 & k < b \end{cases}$$

Hence for the delayed autoregressive process, the cross correlation function is

$$\rho_{az}(k) = \begin{cases} \phi^{k-b} \dfrac{\sigma_a}{\sigma_z} = \phi^{k-b}(1 - \phi^2)^{1/2} & k \geqslant b \\ 0 & k < b \end{cases}$$

Figure 11.3 shows this cross correlation function when $b = 2$ and $\phi = 0.6$.

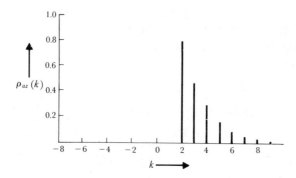

FIG. 11.3   Cross correlation function between a and z for delayed autoregressive process $\tilde{z}_t - 0.6\tilde{z}_{t-1} = a_{t-2}$

11.1.2 Estimation of the cross covariance and cross correlation functions

We assume that, after differencing, the original input and output time series d times, there are $n = N - d$ pairs of values $(x_1, y_1), (x_2, y_2), \ldots, (x_n, y_n)$ available for analysis. Then it is shown, for example in [27], that an estimate $c_{xy}(k)$ of the cross covariance coefficient at lag k is provided by

$$
c_{xy}(k) = \begin{cases} \dfrac{1}{n} \displaystyle\sum_{t=1}^{n-k} (x_t - \bar{x})(y_{t+k} - \bar{y}) & k = 0, 1, 2, \ldots \\[2ex] \dfrac{1}{n} \displaystyle\sum_{t=1}^{n+k} (y_t - \bar{y})(x_{t-k} - \bar{x}) & k = 0, -1, -2, \ldots \end{cases} \tag{11.1.4}
$$

where $\bar{x}$, $\bar{y}$ are the means of the x series and y series, respectively. Similarly, the estimate $r_{xy}(k)$ of the cross correlation coefficient $\rho_{xy}(k)$ at lag k may be obtained by substituting in (11.1.3) the estimates $c_{xy}(k)$ for $\gamma_{xy}(k)$, $s_x = \sqrt{c_{xx}(0)}$ for σ_x and $s_y = \sqrt{c_{yy}(0)}$ for σ_y, yielding

$$
r_{xy}(k) = \frac{c_{xy}(k)}{s_x s_y} \qquad k = 0, \pm 1, \pm 2, \pm \ldots . \tag{11.1.5}
$$

An example.   In practice we would need at least 50 pairs of observations to obtain a useful estimate of the cross correlation function. However, to illustrate the formulae (11.1.4) and (11.1.5), we compute an estimate of the cross correlation function at lags $+1$ and -1 for the following series of 5 pairs of observations

t	1	2	3	4	5
x_t	11	7	8	12	14
y_t	7	10	6	7	10

Now $\bar{x} = 10.4$, $\bar{y} = 8$, so that the deviations from the means are

t	1	2	3	4	5
$x_t - \bar{x}$	0.6	-3.4	-2.4	1.6	3.6
$y_t - \bar{y}$	-1.0	2.0	-2.0	-1.0	2.0

Hence

$$\sum_{t=1}^{4} (x_t - \bar{x})(y_{t+1} - \bar{y}) = (0.6)(2.0) + (-3.4)(-2.0) + (-2.4)(-1.0)$$
$$+ (1.6)(2.0)$$
$$= 13.60$$

and

$$c_{xy}(1) = 13.60/5 = 2.720$$

Using $s_x = 2.577$, $s_y = 1.673$, we obtain

$$r_{xy}(1) = \frac{c_{xy}(1)}{s_x s_y} = \frac{2.720}{(2.577)(1.673)} = 0.63$$

Similarly $\sum_{t=1}^{4} (y_t - \bar{y})(x_{t+1} - \bar{x}) = -8.20$. Hence, $c_{xy}(-1) = -1.640$
and

$$r_{xy}(-1) = \frac{-1.640}{(2.577)(1.673)} = -0.38$$

Figure 11.4 shows the estimated cross correlation function $r_{XY}(k)$ between the input and output for the discrete gas furnace data obtained by reading

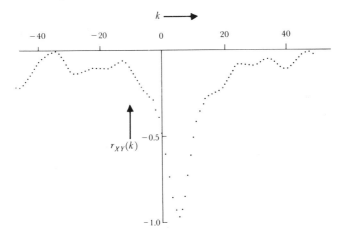

FIG. 11.4   Cross correlation function between input and output for coded gas furnace
data read at 9 second intervals

the continuous data of Figure 11.1 at intervals of 9 seconds. Note that the cross correlation function is not symmetrical about zero and has a well defined peak at $k = +5$, indicating that the output lags behind the input. The cross correlations are negative. This is to be expected since (see Figure 11.1) an *increase* in the coded input produces a *decrease* in the output.

11.1.3 Approximate standard errors of cross correlation estimates

A crude check as to whether certain values of the cross correlation function $\rho_{xy}(k)$ could be effectively zero may be made by comparing the corresponding cross correlation estimates with their approximate standard errors obtained from a formula by Bartlett [78]. He shows that the covariance between two cross correlation estimates $r_{xy}(k)$ and $r_{xy}(k + l)$ is, on the Normal assumption, and $k \geqslant 0$, given by

$$\text{cov}\,[r_{xy}(k), r_{xy}(k + l)]$$

$$\simeq (n - k)^{-1} \sum_{v=-\infty}^{+\infty} [\rho_{xx}(v)\rho_{yy}(v + l) + \rho_{xy}(-v)\rho_{xy}(v + 2k + l)$$

$$+ \rho_{xy}(k)\rho_{xy}(k + l)\{\rho_{xy}^2(v) + \tfrac{1}{2}\rho_{xx}^2(v) + \tfrac{1}{2}\rho_{yy}^2(v)\}$$

$$- \rho_{xy}(k)\{\rho_{xx}(v)\rho_{xy}(v + k + l) + \rho_{xy}(-v)\rho_{yy}(v + k + l)\}$$

$$- \rho_{xy}(k + l)\{\rho_{xx}(v)\rho_{xy}(v + k) + \rho_{xy}(-v)\rho_{yy}(v + k)\}] \qquad (11.1.6)$$

In particular, setting $l = 0$,

$$\text{var}\,[r_{xy}(k)] \simeq (n - k)^{-1} \sum_{v=-\infty}^{+\infty} [\rho_{xx}(v)\rho_{yy}(v) + \rho_{xy}(k + v)\rho_{xy}(k - v)$$

$$+ \rho_{xy}^2(k)\{\rho_{xy}^2(v) + \tfrac{1}{2}\rho_{xx}^2(v) + \tfrac{1}{2}\rho_{yy}^2(v)\}$$

$$- 2\rho_{xy}(k)\{\rho_{xx}(v)\rho_{xy}(v + k) + \rho_{xy}(-v)\rho_{yy}(v + k)\}] \qquad (11.1.7)$$

As noted by Bartlett, formulae which apply to important special cases are derivable from these general expressions. For example, if it is supposed that $x_t \equiv y_t$, it becomes appropriate to set

$$\rho_{xx}(v) = \rho_{yy}(v) = \rho_{xy}(v) = \rho_{xy}(-v)$$

On making this substitution in (11.1.6) and (11.1.7) one obtains an expression for the covariance between two autocorrelation estimates and more particularly the expression for the variance of an autocorrelation estimate given in (2.1.11).

It is often the case that two processes are appreciably cross correlated only over some rather narrow range of lags. Suppose it is postulated that

$\rho_{xy}(v)$ is nonzero *only* over some range $Q_1 \leqslant v \leqslant Q_2$. Then
(a) if neither k, $k + l$ nor $k + \frac{1}{2}l$ are included in this range, then all terms in
   (11.1.6) except the first are zero, and

$$\text{cov}\,[r_{xy}(k), r_{xy}(k + l)] \simeq (n - k)^{-1} \sum_{v=-\infty}^{\infty} \rho_{xx}(v)\rho_{yy}(v + l). \quad (11.1.8)$$

(b) if k is not included in this range, then in a similar way (11.1.7) reduces to

$$\text{var}\,[r_{xy}(k)] \simeq (n - k)^{-1} \sum_{v=-\infty}^{\infty} \rho_{xx}(v)\rho_{yy}(v) \quad (11.1.9)$$

In particular, on the hypothesis that the two processes have *no cross correlation*, it follows that the simple formulae (11.1.8) and (11.1.9) apply for *all* lags k and $k + l$.

Another special case of some interest occurs when two processes are *not cross correlated and one is white noise*. Suppose $y_t = a_t$ is generated by a white noise process but x_t is autocorrelated, then from (11.1.8)

$$\text{cov}\,[r_{xa}(k), r_{xa}(k + l)] \simeq (n - k)^{-1}\rho_{xx}(l) \quad (11.1.10)$$

$$\text{var}\,[r_{xa}(k)] \simeq (n - k)^{-1} \quad (11.1.11)$$

Whence it follows that

$$\rho[r_{xa}(k), r_{xa}(k + l)] \simeq \rho_{xx}(l). \quad (11.1.12)$$

Thus in this case the cross correlations have the *same* autocorrelation function as the process generating x_t. Thus even though a_t and x_t are *not* cross correlated, the cross correlation function can be expected to vary about zero with standard deviation $(n - k)^{-1/2}$ *in a systematic pattern* typical of the behaviour of the autocorrelation function $\rho_{xx}(l)$.

Finally if two processes are *both* white noise and are not cross correlated then the covariance between cross correlations will be zero.

11.2 IDENTIFICATION OF TRANSFER FUNCTION MODELS

We now show how to *identify* a combined transfer function-noise model

$$Y_t = \delta^{-1}(B)\omega(B)X_{t-b} + N_t$$

for a linear system corrupted by noise N_t at the output and assumed to be generated by an ARIMA process which is statistically independent* of the input X_t. Specifically, the objective at this stage is to obtain some idea of the orders r and s of the left-hand and right-hand operators in the transfer

---

* When the input is at our choice, we can guarantee that it is independent of N_t by *generating* X_t according to some random process.

function model and to derive initial guesses for the parameters δ, ω, and the delay parameter b. In addition we aim to make rough guesses of the parameters p, d, q of the ARIMA process describing the noise at the output and to obtain initial estimates of the parameters ϕ and θ in that model. The tentative transfer function and noise models so obtained can then be used as a starting point for more efficient estimation methods described in Section 11.3.

An outline of the identification procedure. Suppose that the transfer function model

$$Y_t = v(B)X_t + N_t \tag{11.2.1}$$

may be parsimoniously parametrized in the form

$$Y_t = \delta^{-1}(B)\omega(B)X_{t-b} + N_t \tag{11.2.2}$$

where $\delta(B) = 1 - \delta_1 B - \delta_2 B^2 - \cdots - \delta_r B^r$ and $\omega(B) = \omega_0 - \omega_1 B - \cdots - \omega_s B^s$. The identification procedure consists of

(1) deriving rough estimates $\hat{v}_j$ of the impulse response weights v_j in (11.2.1);

(2) using the estimates $\hat{v}_j$ so obtained to make guesses of the orders r and s of the right-hand and left-hand operators in (11.2.2), and of the delay parameter b;

(3) substituting the estimates $\hat{v}_j$ in the equations (10.2.8) with values of r, s, and b obtained from (2) to obtain initial estimates of the parameters δ and ω in (11.2.2).

Knowing the $\hat{v}_j$, values of b, r, and s may be guessed using the following facts established in Section 10.2.2. For a model of the form of (11.2.2) the impulse response weights v_j consist of

(i) b zero values $v_0, v_1, \ldots, v_{b-1}$;

(ii) a further $s - r + 1$ values $v_b, v_{b+1}, \ldots, v_{b+s-r}$ following no fixed pattern (no such values occur if $s < r$);

(iii) values v_j with $j \geqslant b + s - r + 1$ which follow the pattern dictated by an rth order difference equation which has r starting values $v_{b+s}, \ldots,$ $v_{b+s-r+1}$. Starting values v_j for $j < b$ will, of course, be zero.

Differencing of the input and output. The basic tool which is employed here in the identification process is the cross correlation function between input and output. When the processes are nonstationary it is assumed that stationarity can be induced by suitable differencing. Nonstationary behavior is suspected if the estimated auto- and cross-correlation functions of the (X_t, Y_t) series fail to damp out quickly. We assume that a degree of differencing* d necessary to induce stationarity has been achieved when the estimated auto- and cross-correlations $r_{xx}(k)$, $r_{yy}(k)$, and $r_{xy}(k)$ of $x_t = \nabla^d X_t$ and $y_t = \nabla^d Y_t$ damp out quickly. In practice, d is usually 0, 1, or 2.

---

*The procedures outlined can equally well be used when different degrees of differencing are employed for input and output.

Identification of the impulse response function without prewhitening.
Suppose that, after differencing d times, the model (11.2.1) can be written
in the form

$$y_t = v_0 x_t + v_1 x_{t-1} + v_2 x_{t-2} + \cdots + n_t \qquad (11.2.3)$$

where $y_t = \nabla^d Y_t$, $x_t = \nabla^d X_t$, and $n_t = \nabla^d N_t$ are stationary processes with
zero means. Then, on multiplying throughout in (11.2.3) by x_{t-k} for $k \geqslant 0$,
we obtain

$$x_{t-k} y_t = v_0 x_{t-k} x_t + v_1 x_{t-k} x_{t-1} + \cdots + x_{t-k} n_t \qquad (11.2.4)$$

If we make the further assumption that x_{t-k} is uncorrelated with n_t for all k,
taking expectations in (11.2.4) yields the set of equations

$$\gamma_{xy}(k) = v_0 \gamma_{xx}(k) + v_1 \gamma_{xx}(k-1) + \cdots \qquad k = 0, 1, 2 \ldots \quad (11.2.5)$$

Suppose that the weights v_j are effectively zero beyond $k = K$. Then the
first $K + 1$ of the equations (11.2.5) can be written

$$\gamma_{xy} = \Gamma_{xx} v \qquad (11.2.6)$$

where

$$\gamma_{xy} = \begin{bmatrix} \gamma_{xy}(0) \\ \gamma_{xy}(1) \\ \vdots \\ \gamma_{xy}(K) \end{bmatrix} \quad \Gamma_{xx} = \begin{bmatrix} \gamma_{xx}(0) & \gamma_{xx}(1) & \cdots & \gamma_{xx}(K) \\ \gamma_{xx}(1) & \gamma_{xx}(0) & \cdots & \gamma_{xx}(K-1) \\ \vdots & \vdots & \cdots & \vdots \\ \gamma_{xx}(K) & \gamma_{xx}(K-1) & \cdots & \gamma_{xx}(0) \end{bmatrix} \quad v = \begin{bmatrix} v_0 \\ v_1 \\ \vdots \\ v_K \end{bmatrix}$$

Substituting estimates $r_{xx}(k)$ of the autocorrelation function of the input
and estimates $r_{xy}(k)$ of the cross correlation function between the input and
output, (11.2.6) provides $K + 1$ linear equations for the first $K + 1$ weights.
However, these equations which do not in general provide efficient estimates
are cumbersome to solve and in any case require knowledge of the point K
beyond which v_j is effectively zero.

11.2.1 Identification of transfer function models by prewhitening the input

Considerable simplification in the identification process would occur if
the input to the system were white noise. Indeed, as is discussed in more
detail in Section 11.6, when the choice of the input is at our disposal, there
is much to recommend such an input. When the original input follows some
other stochastic process, simplification is possible by "prewhitening."

Suppose the suitably differenced input process x_t is stationary and is
capable of representation by some member of the general linear class of
autoregressive-moving average models. Then, given a set of data, we can

carry out our usual identification and estimation methods to obtain a model for the x_t process

$$\phi_x(B)\theta_x^{-1}(B)x_t = \alpha_t \tag{11.2.7}$$

which, to a close approximation, transforms the correlated input series x_t to the uncorrelated white noise series α_t. At the same time, we can obtain an estimate s_α^2 of σ_α^2 from the sum of squares of the $\hat{\alpha}$'s. If we now apply this same transformation to y_t to obtain

$$\beta_t = \phi_x(B)\theta_x^{-1}(B)y_t$$

then the model (11.2.3) may be written

$$\beta_t = v(B)\alpha_t + \varepsilon_t \tag{11.2.8}$$

where ε_t is the transformed noise series defined by

$$\varepsilon_t = \phi_x(B)\theta_x^{-1}(B)n_t \tag{11.2.9}$$

On multiplying (11.2.8) on both sides by α_{t-k} and taking expectations, we obtain

$$\gamma_{\alpha\beta}(k) = v_k\sigma_\alpha^2 \tag{11.2.10}$$

where $\gamma_{\alpha\beta}(k) = E[\alpha_{t-k}\beta_t]$ is the cross covariance at lag $+k$ between α and β. Thus

$$v_k = \frac{\gamma_{\alpha\beta}(k)}{\sigma_\alpha^2}$$

or in terms of the cross correlations,

$$v_k = \frac{\rho_{\alpha\beta}(k)\sigma_\beta}{\sigma_\alpha} \qquad k = 0, 1, 2, \ldots \tag{11.2.11}$$

Hence, after "prewhitening" the input, the cross correlation function between the prewhitened input and correspondingly transformed output is directly proportional to the impulse response function. We note that the effect of prewhitening is to convert the nonorthogonal set of equations (11.2.6) into the orthogonal set (11.2.10).

In practice, we do not know the theoretical cross correlation function $\rho_{\alpha\beta}(k)$, so we must substitute estimates in (11.2.11) to give

$$\hat{v}_k = \frac{r_{\alpha\beta}(k)s_\beta}{s_\alpha} \qquad k = 0, 1, 2, \ldots \tag{11.2.12}$$

The preliminary estimates $\hat{v}_k$ so obtained are again, in general, statistically inefficient but can provide a rough basis for selecting suitable operators $\delta(B)$ and $\omega(B)$ in the transfer function model. We now illustrate this identification and preliminary estimation procedure with an actual example.

11.2.2 An example of the identification of a transfer function model

In an investigation on adaptive optimization [83], a gas furnace was employed in which air and methane combined to form a mixture of gases containing CO_2 (carbon dioxide). The air feed was kept constant, but the methane feed rate could be varied in any desired manner and the resulting CO_2 concentration in the off gases measured. The continuous data of Figure 11.1 were collected to provide information about the dynamics of the system over a region of interest where it was known that an approximately linear steady state relationship applied. The continuous stochastic input series $X(t)$ shown in the top half of Figure 11.1 was generated by passing white noise through a linear filter. The process had mean zero and, during the realization that was used for this experiment, varied from -2.5 to $+2.5$. It was desired that the actual methane gas feedrate should cover a range from 0.5 to 0.7 cubic feet per minute. To ensure this, the input gas feedrate was caused to follow the process

$$\text{Methane Gas Input Feed} = 0.60 - 0.04X(t)$$

For simplicity we shall work throughout with the "coded" input $X(t)$. The final transfer function expressed in terms of the actual feedrate is readily obtained by substitution. Series J in the collection of time series at the end of this volume shows 296 successive pairs of observations (X_t, Y_t) read off from the continuous records at 9 second intervals. In this particular experiment the nature of the input disturbance was known because it was deliberately induced. However, we proceed as if it were not. The estimated auto- and cross-correlation functions of X_t and Y_t damped out fairly quickly, confirming that no differencing was necessary. The usual identification and fitting procedure applied to the input X_t indicated that it is well described by a third-order autoregressive process

$$(1 - \phi_1 B - \phi_2 B^2 - \phi_3 B^3)X_t = \alpha_t$$

with $\hat{\phi}_1 = 1.97$, $\hat{\phi}_2 = -1.37$, $\hat{\phi}_3 = 0.34$, and $s_\alpha^2 = 0.0353$. Hence the transformations

$$\alpha_t = (1 - 1.97B + 1.37B^2 - 0.34B^3)X_t$$
$$\beta_t = (1 - 1.97B + 1.37B^2 - 0.34B^3)Y_t$$

were applied to the input and output series to yield the series α_t and β_t with $s_\alpha = 0.188$, $s_\beta = 0.358$. The estimated cross correlation function between α_t and β_t is shown in Table 11.1, together with the estimate (11.2.12) of the impulse response function

$$\hat{v}_k = \frac{0.358}{0.188} r_{\alpha\beta}(k)$$

TABLE 11.1   Estimated cross correlation function after prewhitening and approximate impulse
response function for gas furnace data

k	0	1	2	3	4	5	6	7	8	9	10
$r_{\alpha\beta}(k)$	−0.01	0.05	−0.03	−0.28	−0.33	−0.46	−0.27	−0.17	−0.03	0.03	−0.05
$\hat{\sigma}(r)$	0.06	0.06	0.06	0.05	0.06	0.05	0.06	0.06	0.06	0.06	0.06
$\hat{v}_k$	−0.02	0.10	−0.06	−0.53	−0.63	−0.88	−0.52	−0.32	−0.06	0.06	−0.10

The approximate standard errors for the cross correlations $r_{\alpha\beta}(k)$ shown in
Table 11.1 are the square roots of the variances obtained from (11.1.7):—
(a) with cross correlations up to lag + 2 and from lag + 8 onwards assumed
    equal to zero
(b) with autocorrelations $r_{\alpha\alpha}(k)$ assumed zero for $k > 0$
(c) with autocorrelations $r_{\beta\beta}(k)$ assumed zero for $k > 4$
(d) with estimated correlations from Table 11.1 replacing theoretical
    values.
The estimated cross correlations together with one and two standard error
limits centered on zero are plotted in Figure 11.5. For this example the
standard errors differ very little from the approximate values $n^{-\frac{1}{2}} = 0.06$
appropriate to the hypothesis that the series are uncorrelated.

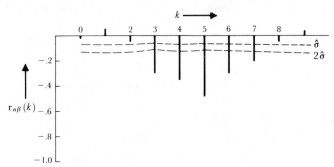

FIG. 11.5   Estimated cross correlation function for coded gas furnace data after
prewhitening

The values $\hat{v}_0$, $\hat{v}_1$, and $\hat{v}_2$ are small compared with their standard errors
suggesting that $b = 3$ (that there are two whole periods of delay). Using the
results of Section 11.2.1 the subsequent pattern of the $\hat{v}$'s might be accounted
for by a model with (r, s, b) either equal to $(1, 2, 3)$ or to $(2, 2, 3)$. The first
model would imply that v_3 and v_4 were preliminary values following no
fixed pattern and that v_5 provided the starting value for an exponential decay
determined by the difference equation $v_j - \delta v_{j-1} = 0$, $j > 5$. The second
model would imply that v_3 was a single preliminary value and that v_4 and v_5
provided the starting values for a pattern of double exponential decay
determined by the difference equation $v_j - \delta_1 v_{j-1} - \delta_2 v_{j-2} = 0$, $j > 5$.

Thus the preliminary identification suggests a transfer function model

$$(1 - \delta_1 B - \delta_2 B^2)Y_t = (\omega_0 - \omega_1 B - \omega_2 B^2)X_{t-b} \qquad (11.2.13)$$

or some simplification of it, probably with $b = 3$.

Preliminary estimates. Assuming the model (11.2.13) with $b = 3$, the equations (10.2.8) for the impulse response function are

$$v_j = 0 \qquad j < 3$$

$$v_3 = \omega_0$$

$$v_4 = \delta_1 v_3 - \omega_1$$

$$v_5 = \delta_1 v_4 + \delta_2 v_3 - \omega_2 \qquad\qquad (11.2.14)$$

$$v_6 = \delta_1 v_5 + \delta_2 v_4$$

$$v_7 = \delta_1 v_6 + \delta_2 v_5$$

Substituting the estimates $\hat{v}_k$ from Table 11.1 in the last two of these equations, we obtain

$$-0.88\hat{\delta}_1 - 0.63\hat{\delta}_2 = -0.52$$

$$-0.52\hat{\delta}_1 - 0.88\hat{\delta}_2 = -0.32$$

which give preliminary estimates $\hat{\delta}_1 = 0.57$, $\hat{\delta}_2 = 0.02$. If these values are now substituted in the second, third, and fourth of the equations (11.2.14), we obtain

$$\hat{\omega}_0 = \hat{v}_3 = -0.53$$

$$\hat{\omega}_1 = \hat{\delta}_1 \hat{v}_3 - \hat{v}_4 = (0.57)(-0.53) + 0.63 = 0.33$$

$$\hat{\omega}_2 = \hat{\delta}_1 \hat{v}_4 + \hat{\delta}_2 \hat{v}_3 - \hat{v}_5 = (0.57)(-0.63) + (0.02)(-0.53) + 0.88 = 0.51$$

Thus, the preliminary identification suggests a tentative transfer function model

$$(1 - 0.57B - 0.02B^2)Y_t = -(0.53 + 0.33B + 0.51B^2)X_{t-3}$$

The estimates so obtained can be used as starting values for the more efficient iterative estimation methods which will be described in Section 11.3. Note that the estimate $\hat{\delta}_2$ is very small and suggests that this parameter may be omitted, but we shall retain it for the time being.

11.2.3 *Identification of the noise model*

Reverting to the general case, suppose that (where necessary, after suitable differencing) the model could be written

$$y_t = v(B)x_t + n_t$$

where

$$n_t = \nabla^d N_t$$

Given that a preliminary estimate $\hat{v}(B)$ of the transfer function has been obtained in the manner discussed in the last section, then an estimate of the noise series is provided by

$$\hat{n}_t = y_t - \hat{v}(B)x_t$$

that is

$$\hat{n}_t = y_t - \hat{v}_0 x_t - \hat{v}_1 x_{t-1} - \hat{v}_2 x_{t-2} - \cdots$$

Alternatively, $\hat{v}(B)$ may be replaced by the tentative transfer function model $\hat{\delta}^{-1}(B)\hat{\omega}(B)B^b$ determined by preliminary identification. Thus

$$\hat{n}_t = y_t - \hat{\delta}^{-1}(B)\hat{\omega}(B)x_{t-b}$$

and $\hat{n}_t$ may be computed from

$$\hat{n}_t = y_t + \hat{\delta}_1(\hat{n}_{t-1} - y_{t-1}) + \cdots + \hat{\delta}_r(\hat{n}_{t-r} - y_{t-r}) - \hat{\omega}_0 x_{t-b}$$
$$+ \hat{\omega}_1 x_{t-b-1} + \cdots + \hat{\omega}_s x_{t-b-s}$$

In either case, study of the estimated autocorrelation function of $\hat{n}_t$ can lead to identification of the noise model.

It is also possible to identify the noise using the correlation functions for the input and output, after prewhitening, as follows:

Suppose that the input could be exactly prewhitened to give

$$\beta_t = v(B)\alpha_t + \varepsilon_t \tag{11.2.15}$$

where the known relationship

$$\varepsilon_t = \phi_x(B)\theta_x^{-1}(B)n_t \tag{11.2.16}$$

would link ε_t and n_t. If a stochastic model could be found for ε_t, then, using (11.2.16), a model could be deduced for n_t and hence for N_t. If we now write $v(B)\alpha_t = u_t$, so that $\beta_t = u_t + \varepsilon_t$, and provided our independence assumption concerning x_t and n_t, and hence concerning u_t and ε_t, is justified, then we can write

$$\gamma_{\beta\beta}(k) = \gamma_{uu}(k) + \gamma_{\varepsilon\varepsilon}(k) \tag{11.2.17}$$

Since α_t is white noise, $\gamma_{uu}(k)$ may be obtained using the result (3.1.8) which gives the autocorrelation function of a linear process.

Thus

$$\gamma_{uu}(k) = \sigma_\alpha^2 \sum_{j=0}^{\infty} v_j v_{j+k}$$

$$= \frac{1}{\sigma_\alpha^2} \sum_{j=0}^{\infty} \gamma_{\alpha\beta}(j)\gamma_{\alpha\beta}(j+k)$$

using (11.2.10). Hence, using (11.2.17),

$$\gamma_{\varepsilon\varepsilon}(k) = \gamma_{\beta\beta}(k) - \frac{1}{\sigma_\alpha^2} \sum_{j=0}^{\infty} \gamma_{\alpha\beta}(j)\gamma_{\alpha\beta}(j + k)$$

$$\gamma_{\varepsilon\varepsilon}(0) = \gamma_{\beta\beta}(0) - \frac{1}{\sigma_\alpha^2} \sum_{j=0}^{\infty} \gamma_{\alpha\beta}^2(j)$$

and

$$\rho_{\varepsilon\varepsilon}(k) = \frac{\rho_{\beta\beta}(k) - \sum_{j=0}^{\infty} \rho_{\alpha\beta}(j)\rho_{\alpha\beta}(j + k)}{1 - \sum_{j=0}^{\infty} \rho_{\alpha\beta}^2(j)} \qquad (11.2.18)$$

Now, in practice, it is necessary to *estimate* the prewhitening transformation. Having made the approximate prewhitening transformation, rough values for $\rho_{\varepsilon\varepsilon}(k)$ may be obtained from (11.2.18) by substituting the estimates $r_{\alpha\beta}(j)$ of the cross correlation function between transformed input and output and $r_{\beta\beta}(j)$ of the autocorrelation function of the transformed output.

Application to the gas furnace example. Table 11.2 shows the first few values of $r_{\beta\beta}(k)$ and $r_{\alpha\beta}(k)$ and the corresponding values of $r_{\varepsilon\varepsilon}(k)$ computed from (11.2.18). In this table we have replaced small correlations by zeroes, supposing in particular that $r_{\beta\beta}(k) = 0$, $k > 4$, and $r_{\alpha\beta}(k) = 0$, $k < 3$.

TABLE 11.2   Calculation of autocorrelation function of noise in gas furnace data

k	0	1	2	3	4	5	6	7
$r_{\beta\beta}(k)$	1.000	0.223	0.359	0.126	0.081	0.000	0.000	0.000
$r_{\alpha\beta}(k)$	0.000	0.000	0.000	−0.283	−0.331	−0.456	−0.268	−0.168
$r_{\varepsilon\varepsilon}(k)$	1.000	−0.382	0.126	−0.011	0.064	0.000	0.000	0.000

The autocorrelation function of the noise might be represented by a first-order autoregressive process or by a first-order moving average process. Since the estimate $r_{\varepsilon\varepsilon}(1)$ is not very large, it will probably make little difference which is chosen. However, we consider the two possibilities separately.

First, if the process were AR (1), then setting $\hat{\phi}_1 = r_{\varepsilon\varepsilon}(1) = -0.38$, the model for ε_t would be

$$(1 + 0.38B)\varepsilon_t = a_t$$

Now, since we know that

$$(1 - 1.97B + 1.37B^2 - 0.34B^3)N_t = \varepsilon_t$$

the identification implies that the noise model is approximately

$$(1 + 0.38B)(1 - 1.97B + 1.37B^2 - 0.34B^3)N_t = a_t$$

or

$$(1 - 1.59B + 0.62B^2 + 0.18B^3 - 0.13B^4)N_t = a_t \qquad (11.2.19)$$

Since the last two coefficients are small, it seems sensible to suppose tentatively that N_t might be represented by a second-order autoregressive process which is approximately $(1 - 1.6B + 0.6B^2)N_t = a_t$.

Alternatively, assuming that ε_t follows the MA (1) process $\varepsilon_t = (1 - \theta B)a_t$, then entering Table A in the collection of tables and charts at the end of the book with a value of $r_1 = -0.38$, we obtain a rough estimate 0.46 for $\hat{\theta}$. Hence the model for N_t would be

$$(1 - 1.97B + 1.37B^2 - 0.34B^3)N_t = (1 - 0.46B)a_t$$

On dividing by $(1 - 0.46B)$, we obtain very nearly

$$(1 - 1.51B + 0.68B^2)N_t = a_t$$

which is very similar to the model previously suggested.

Thus, the analysis of this Section and Section 11.2.2 suggests the identification

$$Y_t = \frac{(\omega_0 - \omega_1 B - \omega_2 B^2)}{1 - \delta_1 B - \delta_2 B^2} X_{t-3} + \frac{1}{1 - \phi_1 B - \phi_2 B^2} a_t \qquad (11.2.20)$$

for the gas furnace model. Furthermore, the initial estimates $\hat{\omega}_0 = -0.53$, $\hat{\omega}_1 = 0.33$, $\hat{\omega}_2 = 0.51$, $\hat{\delta}_1 = 0.57$, $\hat{\delta}_2 = 0.02$, $\hat{\phi}_1 = 1.51$, $\hat{\phi}_2 = -0.68$ can be used as rough starting values for the nonlinear estimation procedures to be described in Section 11.3.

11.2.4 Some general considerations in identifying transfer function models

Some general remarks can now be made concerning the procedures for identifying transfer function and noise models which we have just described.

(1) For many practical situations, when the effect of noise is appreciable, a delayed first- or second-order system such as that given by (11.2.13), or some simplification of it, would often provide as elaborate a model as could be justified for the data.

(2) Efficient estimation is only possible assuming the model *form* to be known. The estimates $\hat{v}_k$ given by (11.2.12) are in general *necessarily* inefficient therefore. They are employed at the identification stage because they are easily computed and can indicate a form of model worthy to be fitted by more elaborate means.

(3) Even if these were efficient estimates, the number of $\hat{v}$'s required to fully trace out the impulse response function would usually be considerably larger than the number of parameters in a transfer function model. In cases where the δ's and ω's in an adequate transfer function model could

be estimated accurately, nevertheless, the estimates of the corresponding v's could have large variances and be highly correlated.

(4) The variance of

$$r_{\alpha\beta}(k) = \hat{v}_k \frac{s_\alpha}{s_\beta}$$

is of order $1/n$. Thus, we can expect that the estimates $r_{\alpha\beta}(k)$, and hence the $\hat{v}_k$, will be buried in noise unless σ_α is reasonably large compared with the residual noise, or unless n is large. Thus the identification procedure requires that the variation in the input X shall be reasonably large compared with the variation due to the noise and/or that a large volume of data is available. These requirements are satisfied by the gas furnace data for which, as we show in Section 11.3, the initial identification is remarkably good. When these requirements are not satisfied, the identification procedure may fail. Usually, this will mean that only very rough estimates are possible with the available data. However, some kind of rudimentary modelling may be possible by postulating a plausible but simple transfer function/noise model, fitting directly by the least squares procedures of the next section, and applying diagnostic checks leading to elaboration of the model when this proves necessary.

Lack of uniqueness of the model.   Suppose a particular dynamic system is represented by the model

$$Y_t = \delta^{-1}(B)\omega(B)X_{t-b} + \varphi^{-1}(B)\theta(B)a_t$$

Then it could equally well be represented by

$$L(B)Y_t = L(B)\delta^{-1}(B)\omega(B)X_{t-b} + L(B)\varphi^{-1}(B)\theta(B)a_t$$

Again, if the arbitrary operator $L(B)$ were set equal to $\varphi(B)$, we would have

$$L(B)Y_t = L(B)\delta^{-1}(B)\omega(B)X_{t-b} + \theta(B)a_t$$

The chance that we may iterate towards a model of unnecessarily complicated form is reduced if we base our strategy on the following considerations:

(1) Since rather simple transfer function models of first or second order, with or without delay, are often adequate, iterative model building should begin with a fairly simple model, looking for further simplification if this is possible, and reverting to more complicated models only as the need is demonstrated.

(2) One should be always on the look out for the possibility of removing a factor common to two or more of the operators on Y_t, X_t, and a_t. We have illustrated such a factorization on the noise model identification in Section 11.2.3. In practice, we shall be dealing with estimated coefficients, which may be subject to rather large errors, so that only approximate factorization can be expected, and considerable imagination may

need to be exerted to spot a possible factorization. The factored model may be refitted and checked to show whether the simplification can be justified.

(3) When simplification by factorization is possible, but is overlooked, the least squares estimation procedure may become extremely unstable since the minimum will tend to lie on a line or surface in the parameter space rather than at a point. Conversely, instability in the solution can point to the possibility of simplification of the model. As previously emphasized, one reason for carrying out the identification procedure before fitting the model is to avoid redundancy or, conversely, to achieve *parsimony* in parameterization.

An alternative method of identifying transfer function models, which is capable of ready generalization to deal with multiple inputs, is given in Appendix A11.1.

11.3 FITTING AND CHECKING TRANSFER FUNCTION MODELS

11.3.1 The conditional sum of squares function

We now consider the problem of efficiently and simultaneously estimating the parameters b, δ, ω, ϕ, and θ in the tentatively identified model

$$y_t = \delta^{-1}(B)\omega(B)x_{t-b} + n_t \tag{11.3.1}$$

where $y_t = \nabla^d Y_t$, $x_t = \nabla^d X_t$, $n_t = \nabla^d N_t$ are all stationary processes and

$$n_t = \phi^{-1}(B)\theta(B)a_t \tag{11.3.2}$$

It is assumed that $n = N - d$ pairs of values are available for the analysis and that Y_t and X_t, (y_t and x_t if $d > 0$), denote deviations from expected values. These expected values may be estimated along with the other parameters, but for the lengths of time series normally worth analyzing it will usually be sufficient to use the sample means as estimates. When $d > 0$ it will frequently be true that expected values for y_t and x_t are zero.

If starting values $\mathbf{x}_0$, $\mathbf{y}_0$, and $\mathbf{a}_0$ prior to the commencement of the series were available, then given the data, for any choice of the parameters $(b, \delta, \omega, \phi, \theta)$ and of the starting values $(\mathbf{x}_0, \mathbf{y}_0, \mathbf{a}_0)$, we could calculate, successively, values of $a_t = a_t(b, \delta, \omega, \phi, \theta | \mathbf{x}_0, \mathbf{y}_0, \mathbf{a}_0)$ for $t = 1, 2, \ldots, n$. Under the Normal assumption for the a's, a close approximation to the maximum likelihood estimates of the parameters can be obtained by minimizing the *conditional sum of squares function*

$$S_0(b, \delta, \omega, \phi, \theta) = \sum_{t=1}^{n} a_t^2(b, \delta, \omega, \phi, \theta | \mathbf{x}_0, \mathbf{y}_0, \mathbf{a}_0) \tag{11.3.3}$$

Three stage procedure for calculating the a's. Given appropriate starting values, the generation of the *a*'s *for any particular choice of the parameter values* may be accomplished using the following three stage procedure.

First, the output y_t from the transfer function model may be computed from

$$y_t = \delta^{-1}(B)\omega(B)x_{t-},$$

that is from

$$\delta(B)y_t = \omega(B)x_{t-b}$$

or from

$$y_t - \delta_1 y_{t-1} - \cdots - \delta_r y_{t-r} = \omega_0 x_{t-b} - \omega_1 x_{t-b-1} - \cdots - \omega_s x_{t-b-s}$$

$$(11.3.4)$$

Having calculated the y_t series, then using (11.3.1), the noise series n_t can be obtained from

$$n_t = y_t - y_t \qquad (11.3.5)$$

Finally, the *a*'s can be obtained from (11.3.2) written in the form

$$a_t = \theta^{-1}(B)\phi(B)n_t$$

that is

$$a_t = \theta_1 a_{t-1} + \cdots + \theta_q a_{t-q} + n_t - \phi_1 n_{t-1} - \cdots - \phi_p n_{t-p} \qquad (11.3.6)$$

Starting values. As discussed in Section 7.1.3 for stochastic model estimation, the effect of transients can be minimized if the difference equations are started off from a value of t for which all previous x's and y's are known. Thus y_t in (11.3.4) is calculated from $t = u + 1$ onwards, where u is the larger of r and $s + b$. This means that n_t will be available from n_{u+1} onwards; hence, if unknown *a*'s are set equal to their unconditional expected values of zero, the *a*'s may be calculated from a_{u+p+1} onwards. Thus, the conditional sum of squares function is

$$S_0(b, \delta, \omega, \phi, \theta) = \sum_{t=u+p+1}^{n} a_t^2(b, \delta, \omega, \phi, \theta | x_0, y_0, a_0) \qquad (11.3.7)$$

Example using the gas furnace data. For this data the model (11.2.20), namely

$$Y_t = \frac{\omega_0 - \omega_1 B - \omega_2 B^2}{1 - \delta_1 B - \delta_2 B^2} X_{t-3} + \frac{1}{1 - \phi_1 B - \phi_2 B^2} a_t$$

has been identified. Equations (11.3.4), (11.3.5), and (11.3.6) then become

$$\mathcal{Y}_t = \delta_1 \mathcal{Y}_{t-1} + \delta_2 \mathcal{Y}_{t-2} + \omega_0 X_{t-3} - \omega_1 X_{t-4} - \omega_2 X_{t-5} \quad (11.3.8)$$

$$N_t = Y_t - \mathcal{Y}_t \quad (11.3.9)$$

$$a_t = N_t - \phi_1 N_{t-1} - \phi_2 N_{t-2} \quad (11.3.10)$$

Thus, (11.3.8) can be used to generate $\mathcal{Y}_t$ from $t = 6$ onwards and (11.3.10) to generate a_t from $t = 8$ onwards. The slight loss of information which results will not be important for a sufficiently long length of series. For example, since $N = 296$ for the gas furnace data, the loss of seven values at the beginning of the series is of little practical consequence. For illustration, Table 11.3 shows the calculation of the first few values of a_t for the coded gas furnace data with

$$b = 3 \quad \delta_1 = 0.1 \quad \delta_2 = 0.1 \quad \omega_0 = 0.1 \quad \omega_1 = -0.1 \quad \omega_2 = -0.1$$

$$\phi_1 = 0.1 \quad \phi_2 = 0.1$$

The X_t and Y_t values in columns 2 and 4 were obtained by subtracting the means $\overline{X} = -0.057$ and $\overline{Y} = 53.51$ from the values of the series given in the Collection of Time Series at the end of this volume.

In the above we have assumed $b = 3$. To estimate b, the values of δ, ω, ϕ, and θ, which minimize the conditional sum of squares, can be calculated for each value of b in the likely range and the overall minimum with respect to b, δ, ω, ϕ, and θ obtained.

TABLE 11.3   Calculation of first few values a_t for gas furnace data when $b = 3$, $\delta_1 = 0.1$, $\delta_2 = 0.1$, $\omega_0 = 0.1$, $\omega_1 = -0.1$, $\omega_2 = -0.1$, $\phi_1 = 0.1$, $\phi_2 = 0.1$

t	X_t	$\mathcal{Y}_t$	Y_t	N_t	a_t
1	−0.052	—	0.29	—	—
2	0.057	—	0.09	—	—
3	0.235	—	−0.01	—	—
4	0.396	—	−0.01	—	—
5	0.430	—	−0.11	—	—
6	0.498	0.024	−0.41	−0.434	—
7	0.518	0.071	−0.81	−0.881	—
8	0.405	0.116	−1.11	−1.226	−1.094
9	0.184	0.151	−1.31	−1.461	−1.250
10	−0.123	0.171	−1.51	−1.681	−1.412

11.3.2 Nonlinear estimation

A nonlinear least squares algorithm, analogous to that given for fitting the stochastic model in Section 7.2.4, can be used to obtain the least squares estimates and their approximate standard errors. The algorithm will behave well when the sum of squares function is very roughly quadratic. However, the procedure can sometimes run into trouble, in particular if the parameters are very highly correlated (if, for example, the model approaches singularity due to near-factorization), or in some cases, if estimates are near a boundary of the permissible parameter space. In difficult cases the estimation situation may be clarified by plotting sums of squares contours for selected two-dimensional sections of the parameter space.

The algorithm may be derived as follows: At any stage of the iteration, and for some fixed value of the delay parameter b, let the best guesses available for the remaining parameters be denoted by

$$\boldsymbol{\beta}_0' = (\delta_{1,0}, \ldots, \delta_{r,0}; \omega_{0,0}, \ldots, \omega_{s,0}; \phi_{1,0}, \ldots, \phi_{p,0}; \theta_{1,0}, \ldots, \theta_{q,0})$$

Now let $a_{t,0}$ be that value computed from the model, as in Section 11.3.1, for the guessed parameter values $\boldsymbol{\beta}_0$ and denote the negative of the derivatives of a_t with respect to the parameters as follows:

$$d_{i,t}^{(\delta)} = -\frac{\partial a_t}{\partial \delta_i}\bigg|_{\boldsymbol{\beta}_0} \qquad d_{j,t}^{(\omega)} = -\frac{\partial a_t}{\partial \omega_j}\bigg|_{\boldsymbol{\beta}_0} \qquad d_{g,t}^{(\phi)} = -\frac{\partial a_t}{\partial \phi_g}\bigg|_{\boldsymbol{\beta}_0} \qquad d_{h,t}^{(\theta)} = -\frac{\partial a_t}{\partial \theta_h}\bigg|_{\boldsymbol{\beta}_0}$$

$$(11.3.11)$$

Then a Taylor series expansion of $a_t = a_t(\boldsymbol{\beta})$ about parameter values $\boldsymbol{\beta} = \boldsymbol{\beta}_0$ can be rearranged in the form

$$a_{t,0} \simeq \sum_{i=1}^{r} (\delta_i - \delta_{i,0})d_{i,t}^{(\delta)} + \sum_{j=0}^{s} (\omega_j - \omega_{j,0})d_{j,t}^{(\omega)}$$

$$+ \sum_{g=1}^{p} (\phi_g - \phi_{g,0})d_{g,t}^{(\phi)} + \sum_{h=1}^{q} (\theta_h - \theta_{h,0})d_{h,t}^{(\theta)} + a_t \qquad (11.3.12)$$

We proceed as in Section 7.2 to obtain adjustments $\delta_i - \delta_{i,0}$, $\omega_j - \omega_{j,0}$, etc. by fitting this linearized equation by standard linear least squares. By adding the adjustments to the first guesses $\boldsymbol{\beta}_0$, a set of second guesses can be formed and the process repeated until convergence is reached.

As with stochastic models (see Chapter 7 and especially Section 7.2.3) the derivatives may be computed recursively. However it seems simplest to work with a standard nonlinear least squares computer program in which derivatives are determined numerically and an option is available of "constrained iteration" to prevent instability (see Chapter 7). It is then only necessary to program the computation of a_t itself.

The covariance matrix of the estimates may be obtained from the converged value of the matrix $(X_{\hat{\boldsymbol{\beta}}}' X_{\hat{\boldsymbol{\beta}}})^{-1}\hat{\sigma}_a^2$ as described in Section 7.2.2. If b, which is an integer, needs to be estimated, then the iteration may be run to convergence for a series of values of b and that value of b giving the minimum sum of squares, selected.

11.3.3 Use of residuals for diagnostic checking

Serious model inadequacy can usually be detected by examining

(a) the autocorrelation function $r_{\hat{a}\hat{a}}(k)$ of the residuals $\hat{a}_t = a_t(\hat{b}, \hat{\delta}, \hat{\omega}, \hat{\phi}, \hat{\theta})$ from the fitted model, and

(b) certain cross correlation functions involving input and residuals: in particular the cross correlation function $r_{\alpha\hat{a}}(k)$ between prewhitened input α_t and the residuals $\hat{a}_t$.

Suppose, if necessary after suitable differencing, that the model can be written

$$y_t = \delta^{-1}(B)\omega(B)x_{t-b} + \phi^{-1}(B)\theta(B)a_t$$

$$= v(B)x_t + \psi(B)a_t \tag{11.3.13}$$

Now, suppose that we select an incorrect model leading to residuals a_{0t}, where

$$y_t = v_0(B)x_t + \psi_0(B)a_{0t}$$

Then

$$a_{0t} = \psi_0^{-1}(B)\{v(B) - v_0(B)\}x_t + \psi_0^{-1}(B)\psi(B)a_t \tag{11.3.14}$$

whence it is apparent in general that, if a wrong model is selected, the a_{0t}'s will be autocorrelated and the a_{0t}'s will be cross correlated with the x_t's and hence with the α_t's which generate the x_t's.

Now consider what happens in the two special cases

(a) when the transfer function model is correct, but the noise model is incorrect;

(b) when the transfer function model is incorrect.

Transfer function model correct—noise model incorrect. If $v_0(B) = v(B)$ but $\psi_0(B) \neq \psi(B)$, then (11.3.14) becomes

$$a_{0t} = \psi_0^{-1}(B)\psi(B)a_t \tag{11.3.15}$$

Therefore, the a_{0t}'s would *not* be cross correlated with x_t's or with α_t's. However, the a_{0t} process would be autocorrelated, and the form of the autocorrelation function could indicate appropriate modification of the noise structure.

Transfer function model incorrect. From (11.3.14) it is apparent that, if the transfer function model were incorrect, not only would the a_{0t}'s be cross correlated with the x_t's (and α_t's), but *also the a_{0t}'s would be autocorrelated.* This would be true even if the noise model were correct, for then (11.3.14) would become

$$a_{0t} = \psi^{-1}(B)\{v(B) - v_0(B)\}x_t + a_t \tag{11.3.16}$$

Whether or not the noise model was correct, a cross correlation analysis could indicate the modifications needed in the transfer function model. This aspect is clarified by considering the model after prewhitening. If the output and the input are assumed to be transformed so that the input is white noise, then, as in (11.2.8), we may write the model as

$$\beta_t = v(B)\alpha_t + \varepsilon_t$$

where $\beta_t = \phi_x(B)\theta_x^{-1}(B)y_t$, $\varepsilon_t = \phi_x(B)\theta_x^{-1}(B)n_t$. Now, consider the quantities

$$\varepsilon_{0t} = \beta_t - v_0(B)\alpha_t$$

Since $\varepsilon_{0t} = \{v(B) - v_0(B)\}\alpha_t + \varepsilon_t$, arguing as in Section 11.2.1, the cross correlations between the ε_{0t}'s and the α_t's measure the discrepancy between the correct and incorrect impulse functions. Specifically, as in (11.2.11),

$$v_k - v_{0k} = \frac{\rho_{\alpha\varepsilon_0}(k)\sigma_{\varepsilon_0}}{\sigma_\alpha} \qquad k = 0, 1, 2, \dots \qquad (11.3.17)$$

11.3.4 Specific checks applied to the residuals

In practice, we do not know the process parameters exactly but must apply our checks to residuals $\hat{a}_t$ computed after least squares fitting. Even if the functional form of the fitted model were adequate, the parameter estimates would differ somewhat from the true values and the distribution of the correlations of the residual $\hat{a}_t$'s would also differ to some extent from that of the autocorrelations of the a_t's. Therefore, some caution is necessary in using the results of the previous sections to suggest the behavior of residual correlations. The brief discussion which follows is based on a more detailed study given in [84].

Autocorrelation checks. Suppose that, a transfer function-noise model having been fitted by least squares and the residual $\hat{a}_t$'s calculated by substituting least squares estimates for the parameters, the estimated auto-correlation function $r_{\hat{a}\hat{a}}(k)$ of these residuals is computed. Then, as we have seen

(a) if the autocorrelation function $r_{\hat{a}\hat{a}}(k)$ shows marked correlation patterns, this suggests model inadequacy;

(b) if the cross correlation checks do not indicate inadequacy of the transfer function model, the inadequacy is probably in the fitted noise model $n_t = \psi_0(B)a_t$.

In the latter case, identification of a subsidiary model

$$\hat{a}_{0t} = T(B)a_t$$

to represent the correlation of the residuals from the primary model can, in accordance with (11.3.15), indicate roughly the form

$$n_t = \psi_0(B)T(B)a_t$$

to take for the modified noise model. However, in making assessments of whether an apparent discrepancy of estimated autocorrelations from zero is, or is not, likely to point to a nonzero theoretical value, certain facts must be borne in mind analogous to those discussed in Section 8.2.1.

Suppose that, after allowing for starting values, $m = n - u - p$ values of the $\hat{a}_t$'s are actually available for this computation. Then if the model was correct in functional form and the *true parameter values were substituted*, the residuals would be white noise and the estimated autocorrelations would be distributed mutually independently about zero with variance $1/m$. When estimates are substituted for the parameter values, the distributional properties of the correlations at low lag are affected. In particular, the variance of these estimated low lag correlations can be considerably less than $1/m$, and the values can be highly correlated. Thus, with k small, comparison of an estimated autocorrelation $r_{\hat{a}\hat{a}}(k)$ with a "standard error" $1/\sqrt{m}$ could greatly underestimate its significance. Also, ripples in the estimated autocorrelation function at low lags can arise simply because of the high induced correlation between these estimates. If the amplitude of such low lag ripples is small compared with $1/\sqrt{m}$, they could have arisen by chance alone and need not be indicative of some real pattern in the theoretical autocorrelations.

A helpful overall check, which takes account of these distributional effects produced by fitting, is as follows. Consider the first K estimated autocorrelations $r_{\hat{a}\hat{a}}(1), \ldots, r_{\hat{a}\hat{a}}(K)$ and let K be taken sufficiently large so that, if the model is written as $y_t = v(B)x_t + \psi(B)a_t$, the weights ψ_j can be expected to be negligible for $j > K$. Then, if the functional form of the model is adequate, the quantity

$$Q = m \sum_{k=1}^{K} r_{\hat{a}\hat{a}}^2(k) \qquad (11.3.18)$$

is approximately distributed as χ^2 with $K - p - q$ degrees of freedom. Note that the degrees of freedom in χ^2 depends on the number of parameters in the noise model but not on the number of parameters in the transfer function model. By referring Q to a table of percentage points of χ^2, we can obtain an approximate test of the hypothesis of model adequacy.

Cross correlation check. As we have seen in the last section:

(1) A pattern of markedly nonzero cross correlations $r_{x\hat{a}}(k)$ suggests inadequacy of the transfer function model.

(2) A somewhat different cross correlation analysis can suggest the *type* of modification needed. Specifically, if the fitted transfer function is $\hat{v}_0(B)$ and we consider the cross correlations between quantities $\hat{\varepsilon}_{0t} = \beta_t - \hat{v}_0(B)\alpha_t$ and α_t, then rough estimates of the discrepancies $v_k - v_{0k}$ are given by

$$\frac{r_{\alpha\hat{\varepsilon}_0}(k)s_{\hat{\varepsilon}_0}}{s_\alpha}$$

Suppose the model were of the correct functional form and *true* parameter values had been substituted. The residuals would be white noise uncorrelated with the n's and, using (11.1.11), the variance of the $r_{x\alpha}(k)$ for an effective length of series m would be approximately $1/m$. However, unlike the auto-correlations $r_{aa}(k)$, these cross correlations will not be approximately uncorrelated. In general, if the x's are autocorrelated, then so are the cross correlations $r_{xa}(k)$. In fact as has been seen in (11.1.12), on the assumption that the x's and the a's have no cross correlation, the correlation coefficient between $r_{xa}(k)$ and $r_{xa}(k + l)$ is

$$\rho[r_{xa}(k), r_{xa}(k + l)] \simeq \rho_{xx}(l) \qquad (11.3.19)$$

That is, approximately, the cross correlations have *the same* autocorrelation function as does the original input series x_t. Thus, when the x_t's are auto-correlated, a perfectly adequate transfer function model will give rise to cross correlations $r_{x\hat{a}}(k)$ which although small in magnitude may show *pronounced patterns*. This effect is eliminated if the check is made by comput-ing cross correlations $r_{\alpha\hat{a}}(k)$ with the *prewhitened* input α_t.

As with the autocorrelations, when estimates are substituted for parameter values, the distributional properties of the autocorrelations are affected. However, a rough overall test of the hypothesis of model adequacy, similar to the autocorrelation test, can be obtained based on the sizes of the cross correlations. To employ the check, the cross correlations $r_{\alpha\hat{a}}(k)$ for $k = 0, 1, 2, \ldots, K$ between the input α_t in *prewhitened* form and the residuals $\hat{a}_t$ are estimated, and K is chosen sufficiently large so that the weights v_j and ψ_j in (11.3.13) can be expected to be negligible for $j > K$. The effects resulting from the use of estimated parameters in calculating residuals are, as before, principally confined to correlations of low order whose variances are con-siderably less than m^{-1} and which may be highly correlated even when the input is white noise.

However, it is true [84] that

$$S = m \sum_{k=0}^{K} r_{\alpha\hat{a}}^2(k) \qquad (11.3.20)$$

is approximately distributed as χ^2 with $K + 1 - (r + s + 1)$ degrees of freedom, where $(r + s + 1)$ is the number of parameters fitted in the transfer function model. Note that the number of degrees of freedom is independent of the number of parameters fitted in the noise model.

11.4 SOME EXAMPLES OF FITTING AND CHECKING TRANSFER FUNCTION MODELS

11.4.1 Fitting and checking of the gas furnace model

We now illustrate the approach described in Section 11.3 to the fitting of the model

$$Y_t = \frac{\omega_0 - \omega_1 B - \omega_2 B^2}{1 - \delta_1 B - \delta_2 B^2} X_{t-3} + \frac{1}{1 - \phi_1 B - \phi_2 B^2} a_t$$

which was identified for the gas furnace data in Sections 11.2.2 and 11.2.3.

Nonlinear estimation.   Using the initial estimates $\hat{\omega}_0 = -0.53, \hat{\omega}_1 = 0.33,$ $\hat{\omega}_2 = 0.51, \hat{\delta}_1 = 0.57, \hat{\delta}_2 = 0.02, \hat{\phi}_1 = 1.51, \hat{\phi}_2 = -0.68$ derived in Sections 11.2.2 and 11.2.3 with the conditional least squares algorithm described in Section 11.3.2, least squares values, to two decimals, were achieved in 4 iterations. To test whether the iteration would converge in unfavorable circumstances, Table 11.4 shows how the iteration proceeded with all starting values taken to be either $+0.1$ or -0.1. The fact that, even then, convergence was achieved in 10 iterations with as many as 7 parameters in the model is encouraging.

TABLE 11.4   Convergence of nonlinear least squares fit of gas furnace data

Iteration	ω_0	ω_1	ω_2	δ_1	δ_2	ϕ_1	ϕ_2	Sum of Squares
0	0.10	−0.10	−0.10	0.10	0.10	0.10	0.10	13,601
1	−0.46	0.63	0.60	0.14	0.27	1.33	−0.27	273.1
2	−0.52	0.45	0.31	0.40	0.52	1.37	−0.43	92.5
3	−0.63	0.60	0.01	0.12	0.73	1.70	−0.76	31.8
4	−0.54	0.50	0.29	0.24	0.42	1.70	−0.81	19.7
5	−0.50	0.31	0.51	0.63	0.09	1.56	−0.68	16.84
6	−0.53	0.38	0.53	0.54	0.01	1.54	−0.64	16.60
7	−0.53	0.37	0.51	0.56	0.01	1.53	−0.63	16.60
8	−0.53	0.37	0.51	0.56	0.01	1.53	−0.63	16.60
9	−0.53	0.37	0.51	0.57	0.01	1.53	−0.63	16.60
Preliminary estimates	−0.53	0.33	0.51	0.57	0.02	1.51	−0.68	

The last line in Table 11.4 shows the rough preliminary estimates obtained at the identification stage in Sections 11.2.2 and 11.2.3. It is seen that, for this example, they are in close agreement with the least squares estimates given on the previous line.

Thus the final fitted transfer function model is

$$(1 - 0.57B - 0.01B^2)Y_t = -(0.53 + 0.37B + 0.51B^2)X_{t-3} \quad (11.4.1)$$
$$(\pm 0.21) \ (\pm 0.14) \qquad (\pm 0.08)(\pm 0.15)\ (\pm 0.16)$$

and the fitted noise model is

$$(1 - 1.53B + 0.63B^2)N_t = a_t \qquad (11.4.2)$$
$$(\pm 0.05)\ (\pm 0.05)$$

with $\hat{\sigma}_a^2 = 0.0561$, where the limits in brackets are the $\pm$ one standard error limits obtained from the nonlinear estimation procedure.

Diagnostic checking. Before accepting the above model as an adequate representation of the system, autocorrelation and cross correlation checks should be applied, as described in Section 11.3.4. The first 36 lags of the residual autocorrelations are tabulated in Table 11.5, together with the upper bound $1/\sqrt{m}$ for their standard errors ($m = 289$) assuming that the model is adequate. There seems to be no evidence of model inadequacy from the behavior of individual autocorrelations. This is confirmed by calculating the Q criterion (11.3.18) which is

$$Q = 289 \sum_{k=1}^{36} r_{\hat{a}\hat{a}}^2(k) = 41.7$$

Comparison of Q with the χ^2 table for $K - p - q = 36 - 2 - 0 = 34$ degrees of freedom provides no grounds for questioning model adequacy.

The first 36 lags of the cross correlation function $r_{x\hat{a}}(k)$ between the input and estimated residuals are given in Table 11.6(a), together with the upper bound $1/\sqrt{m}$ for their standard errors. It will be seen that, although the cross correlations $r_{x\hat{a}}(k)$ are not especially large compared with the upper bound of their standard errors, they are themselves highly autocorrelated. This is to be expected because, as indicated by (11.3.19), the cross correlations follow the same stochastic process as does the input x_t and, as we have already seen, for this example the input was highly autocorrelated.

The corresponding cross correlations between $\hat{a}_t$ and the prewhitened input α_t are given in Table 11.6(b).

The criterion (11.3.20) yields

$$S = 289 \sum_{k=0}^{35} r_{\alpha\hat{a}}^2(k) = 29.4$$

Comparison of S with the χ^2 table for $K + 1 - (r + s + 1) = 36 - 5 = 31$ degrees of freedom again provides no evidence that the model is inadequate.

Step and impulse responses. The estimate $\hat{\delta}_2 = 0.01$ in (11.4.1) is very small when compared with its standard error ± 0.14, and the parameter δ_2 can in fact be omitted from the model without affecting the estimates of the remaining parameters to the accuracy considered. The final form of the combined transfer function-noise model for the gas furnace data is

$$Y_t = \frac{-(0.53 + 0.37B + 0.51B^2)}{1 - 0.57B} X_{t-3} + \frac{1}{1 - 1.53B + 0.63B^2} a_t$$

The step and impulse response functions corresponding to the transfer function model

$$(1 - 0.57B)Y_t = -(0.53 + 0.37B + 0.51B^2)X_{t-3}$$

TABLE 11.5 Estimated autocorrelation function $r_{\hat{a}\hat{a}}(k)$ of residuals from fitted gas furnace model

Lag k	$r_{\hat{a}\hat{a}}(k)$												Upper bound to Standard Error
1–12	0.02	0.06	−0.07	−0.05	−0.05	0.12	0.03	0.03	−0.08	0.05	0.02	0.10	±0.06
13–24	−0.04	0.05	−0.09	−0.01	−0.08	0.00	−0.12	0.00	−0.01	0.08	0.02	−0.01	±0.06
25–36	0.04	−0.02	0.02	0.09	−0.12	0.06	−0.03	−0.06	−0.11	0.02	0.03	0.06	±0.06

TABLE 11.6(a) Estimated cross correlation function $r_{x\hat{a}}(k)$ between the input and output residuals for gas furnace data

Lag k	$r_{x\hat{a}}(k)$												Upper bound to Standard Error
0–11	0.00	0.00	0.00	0.00	0.00	0.00	−0.01	−0.02	−0.03	−0.05	−0.06	−0.05	±0.06
12–23	−0.03	−0.03	−0.03	−0.07	−0.10	−0.12	−0.12	−0.10	−0.04	−0.01	−0.01	−0.02	±0.06
24–35	−0.04	−0.04	−0.04	−0.02	−0.01	0.02	0.04	0.05	0.06	0.07	0.07	0.06	±0.06

TABLE 11.6(b) Estimated cross correlation function $r_{\alpha\hat{a}}(k)$ between the prewhitened input and output residuals for gas furnace data

Lag k	$r_{\alpha\hat{a}}(k)$												Upper bound to Standard Error
0–11			−0.06	0.03	−0.01	0.01	0.01	−0.04	0.02	0.07	−0.03	−0.02	±0.06
12–23			−0.03	−0.11	0.02	0.04	0.01	−0.15	−0.03	−0.07	−0.08	0.02	±0.06
24–35			−0.01	0.02	0.05	0.00	−0.15	0.04	0.03	−0.02	0.00	0.03	±0.06

are given in Figure 11.6. Using (10.2.5), the steady state gain of the coded data is

$$g = \frac{-(0.53 + 0.37 + 0.51)}{1 - 0.57} = -3.3$$

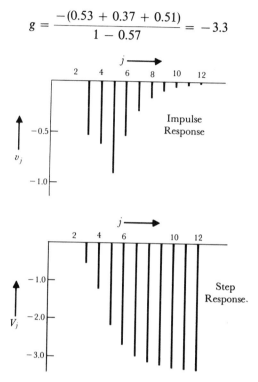

FIG. 11.6   Impulse and step responses for transfer function model $(1 - 0.57B)Y_t = -(0.53 + 0.37B + 0.51B^2)X_{t-3}$ fitted to coded gas furnace data

The results agree very closely with those obtained by cross spectral analysis [27].

Choice of sampling interval. When a choice is available, the sampling interval should be taken as fairly short compared with the time constants expected for the system. When in doubt the analysis can be repeated with several trial sampling intervals. In the choice of sampling interval it is the noise at the output that is important, and its variance should approach a minimum value as the interval is shortened. Thus, in the gas furnace example which we have used for illustration, a pen recorder was used to provide a continuous record of input and output. The discrete data which we have actually analyzed were obtained by reading off values from this continuous record at points separated by 9 second intervals. This interval was chosen because inspection of the traces shown in Figure 11.1 suggested that it ought to be adequate to allow all the variation (apart from slight pen-chatter) which occurred in input and output to be taken account of. The use of this kind of commonsense is usually a reliable guide in choosing the interval. The estimated mean square error for the gas furnace data (obtained by dividing $\sum(Y - \hat{Y})^2$ by the appropriate number of degrees of freedom, with $\hat{Y}$ the fitted value) is shown for various time intervals in Table 11.7. These values are also plotted in Figure 11.7. Little change in mean square error occurs until the interval is almost 40 seconds, when a very rapid rise occurs. There is little difference in the mean square error, or indeed the plotted step response, for the 9, 18, and 27 second intervals, but a considerable change occurs when the 36 second interval is used. It will be seen that the 9 second interval we have used in this example is, in fact, conservative.

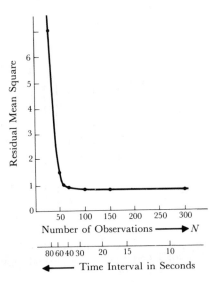

FIG. 11.7   Mean square error at the output for various choices of sampling interval

TABLE 11.7   Mean square error at the output for various choices of the sampling interval

Interval (seconds)	9	18	27	36	45	54	72
Number of data points N	296	148	98	74	59	49	37
M.S. Error	0.71	0.78	0.74	0.95	0.97	1.56	7.11

11.4.2 A simulated example with two inputs

The fitting of models involving more than one input series involves no difficulty in principle, except for the increase in the number of parameters which has to be handled.

For example, for two inputs we can write the model as

$$y_t = \delta_1^{-1}(B)\omega_1(B)x_{1,t-b_1} + \delta_2^{-1}(B)\omega_2(B)x_{2,t-b_2} + n_t$$

with

$$n_t = \phi^{-1}(B)\theta(B)a_t$$

where $y_t = \nabla^d Y_t$, $x_{1,t} = \nabla^d X_{1,t}$, $x_{2,t} = \nabla^d X_{2,t}$ and $n_t = \nabla^d N_t$ are stationary processes. To compute the a_t's, we first calculate for specified values of the parameters b_1, δ_1, ω_1,

$$\mathscr{Y}_{1,t} = \delta_1^{-1}(B)\omega_1(B)x_{1,t-b_1} \tag{11.4.3}$$

and for specified values of b_2, δ_2, ω_2,

$$\mathscr{Y}_{2,t} = \delta_2^{-1}(B)\omega_2(B)x_{2,t-b_2} \tag{11.4.4}$$

Then the noise n_t can be calculated from

$$n_t = y_t - \mathscr{Y}_{1,t} - \mathscr{Y}_{2,t} \tag{11.4.5}$$

and finally, a_t from

$$a_t = \theta^{-1}(B)\phi(B)n_t \tag{11.4.6}$$

A simulated example.   It is clear that even simple situations can lead to the estimation of a large number of parameters. The example below, with two imput variables and delayed first-order models, has eight unknown parameters. To determine whether the iterative nonlinear least squares procedure described in Section 11.3.2 could be used to obtain estimates of the parameters in such models, an experiment was performed using manufactured data, details of which are given in [85]. The data were generated from the model written in ∇ form as

$$Y_t = \beta + g_1 \frac{(1 + \eta_1\nabla)}{(1 + \xi_1\nabla)} X_{1,t-1} + g_2 \frac{(1 + \eta_2\nabla)}{(1 + \xi_2\nabla)} X_{2,t-1} + \frac{1}{(1 - \phi_1 B)} a_t \tag{11.4.7}$$

with $\beta = 60$, $g_1 = 13.0$, $\eta_1 = -0.6$, $\xi_1 = 4.0$, $g_2 = -5.5$, $\eta_2 = -0.6$, $\xi_2 = 4.0$, $\phi_1 = 0.5$, and $\sigma_a^2 = 9.0$. The input variables X_1 and X_2 were changed according to a randomized 2^2 factorial design replicated three times. Each input condition was supposed held fixed for 5 minutes and output observations taken every minute. The data are plotted in Figure 11.8 and appear as Series K in the Collection of Time Series at the end of the volume.

The constrained iterative nonlinear least squares program, described in Chapter 7, was used to obtain the least squares estimates, so that it was only necessary to calculate the a's. Thus, for specified values of the parameters $g_1, g_2, \xi_1, \xi_2, \eta_1, \eta_2$, the values $\mathscr{Y}_{1,t}$, and $\mathscr{Y}_{2,t}$ can be obtained from

$$(1 + \xi_1\nabla)\mathscr{Y}_{1,t} = g_1(1 + \eta_1\nabla)X_{1,t-1}$$

$$(1 + \xi_2\nabla)\mathscr{Y}_{2,t} = g_2(1 + \eta_2\nabla)X_{2,t-1}$$

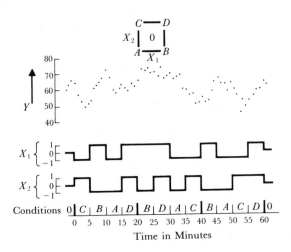

FIG. 11.8   Data for simulated two-input example (Series K)

and can be used to calculate

$$N_t = Y_t - \mathscr{Y}_{1,t} - \mathscr{Y}_{2,t}$$

Finally, for specified values of ϕ_1, a_t can be calculated from

$$a_t = N_t - \phi_1 N_{t-1}$$

It was assumed that the process inputs had been maintained at their center conditions for some time before the start of the experiment, so that $\mathscr{Y}_{1,t}$, $\mathscr{Y}_{2,t}$, and hence N_t, may be computed from $t = 0$ onwards and a_t from $t = 1$.

Two runs were made of the nonlinear least squares procedure using two different sets of initial values. In the first the parameters were chosen as representing what a person reasonably familiar with the process might guess for initial values. In the second, the starting value for β was chosen to be the mean $\bar{Y}$ of all observations and all other starting values were set equal to 0.1. Thus, the second run represents a much more extreme situation than would normally arise in practice. Table 11.8 shows that, with the first set of initial values, convergence occurs after 5 iterations, and Table 11.9 shows that convergence with the second set occurs after 9 iterations. These results suggest that, in realistic circumstances, multiple inputs can be handled without serious estimation difficulties.

11.5 FORECASTING USING LEADING INDICATORS

Frequently, forecasts of a time series $Y_t, Y_{t-1}, \ldots$ may be considerably improved by using information coming from some associated series

TABLE 11.8 Convergence of nonlinear least squares procedure for two inputs using guessed initial estimates

Iteration	β	g_1	g_2	η_1	η_2	ξ_1	ξ_2	ϕ_1	Sum of squares
0	59.19	10.00	−7.00	−0.50	−0.50	1.00	1.00	0.10	2,046.8
1	59.20	9.07	−6.37	−0.58	−0.56	1.33	1.31	0.24	1,085.4
2	59.24	8.38	−5.35	−0.70	−0.59	2.03	1.75	0.39	621.5
3	59.35	9.24	−3.98	−0.75	−0.55	3.45	1.95	0.36	503.5
4	59.41	11.90	−3.40	−0.75	−0.56	5.21	1.66	0.22	463.7
5	59.39	12.03	−3.52	−0.80	−0.57	4.99	1.76	0.21	461.8
6	59.39	12.08	−3.53	−0.79	−0.56	5.03	1.77	0.21	461.8
7	59.39	12.07	−3.53	−0.79	−0.56	5.03	1.77	0.21	461.8
8	59.39	12.07	−3.53	−0.79	−0.56	5.03	1.77	0.21	461.8

TABLE 11.9 Convergence of nonlinear least squares procedure for two inputs using extreme initial values

Iteration	β	g_1	g_2	η_1	η_2	ξ_1	ξ_2	ϕ_1	Sum of squares
0	59.19	0.10	0.10	0.10	0.10	0.10	0.10	0.10	2,496.4
1	59.19	0.24	−0.07	−1.57	0.48	1.77	−0.28	0.15	2,190.5
2	59.22	1.62	−0.29	−2.09	−2.24	−0.07	0.26	0.29	1,473.6
3	59.21	1.80	−0.77	−1.75	0.58	0.20	−0.10	0.56	1,016.8
4	59.21	3.01	−1.31	−1.15	−0.83	0.91	0.22	0.72	743.1
5	59.31	6.17	−2.82	−0.93	−0.65	3.03	1.20	0.67	611.4
6	59.61	15.83	−3.25	−0.70	−0.66	8.88	1.64	0.26	534.2
7	59.47	10.31	−3.48	−0.74	−0.56	3.52	1.63	0.23	501.9
8	59.41	11.89	−3.41	−0.74	−0.58	5.01	1.65	0.20	462.8
9	59.39	12.07	−3.52	−0.79	−0.57	5.04	1.76	0.21	461.8
10	59.39	12.07	−3.53	−0.79	−0.56	5.03	1.77	0.21	461.8
11	59.39	12.07	−3.53	−0.79	−0.56	5.03	1.77	0.21	461.8

$X_t, X_{t-1}, \ldots$. This is particularly true if changes in Y tend to be *anticipated* by changes in X, in which case economists call X a "leading indicator" for Y.

In order to obtain an optimal forecast using information from both Y and X, we first build a transfer function-noise model connecting Y and X in the manner already outlined.

Suppose, using previous notations, an adequate model is

$$Y_t = \delta^{-1}(B)\omega(B)X_{t-b} + \varphi^{-1}(B)\theta(B)a_t \qquad b \geqslant 0 \qquad (11.5.1)$$

In general, the noise component of this model, which is assumed statistically independent of the input X_t, is nonstationary with

$$\varphi(B) = \phi(B)\nabla^d$$

so that if

$$\nabla^d Y_t = y_t \quad \text{and} \quad \nabla^d X_t = x_t$$

$$y_t = \delta^{-1}(B)\omega(B)x_{t-b} + \phi^{-1}(B)\theta(B)a_t$$

Also, we shall assume that an adequate stochastic model for the leading series is

$$X_t = \varphi_x^{-1}(B)\theta_x(B)\alpha_t \tag{11.5.2}$$

so that with

$$\varphi_x(B) = \phi_x(B)\nabla^d$$

$$x_t = \phi_x^{-1}(B)\theta_x(B)\alpha_t$$

11.5.1 The minimum mean square error forecast

Now (11.5.1) may be written

$$Y_t = v(B)\alpha_t + \psi(B)a_t \tag{11.5.3}$$

with the a's and the α's statistically independent. Arguing as in Section 5.1.1, suppose the forecast $\hat{Y}_t(l)$ of Y_{t+l} made at origin t is of the form

$$\hat{Y}_t(l) = \sum_{j=0}^{\infty} v_{l+j}^0 \alpha_{t-j} + \sum_{j=0}^{\infty} \psi_{l+j}^0 a_{t-j}$$

Then

$$Y_{t+l} - \hat{Y}_t(l) = \sum_{i=0}^{l-1} (v_i \alpha_{t+l-i} + \psi_i a_{t+l-i})$$

$$+ \sum_{j=0}^{\infty} \{(v_{l+j} - v_{l+j}^0)\alpha_{t-j} + (\psi_{l+j} - \psi_{l+j}^0)a_{t-j}\}$$

and

$$E\{Y_{t+l} - \hat{Y}_t(l)\}^2 = (v_0^2 + v_1^2 + \cdots + v_{l-1}^2)\sigma_\alpha^2 + (1 + \psi_1^2 + \cdots + \psi_{l-1}^2)\sigma_a^2$$

$$+ \sum_{j=0}^{\infty} \{(v_{l+j} - v_{l+j}^0)^2\sigma_\alpha^2 + (\psi_{l+j} - \psi_{l+j}^0)^2\sigma_a^2\}$$

which is minimized only if $v_{l+j}^0 = v_{l+j}$ and $\psi_{l+j}^0 = \psi_{l+j}$. Thus the minimum mean square error forecast $\hat{Y}_t(l)$ of Y_{t+l} at origin t is given by the conditional expectation of Y_{t+l} at time t. Theoretically, this expectation is conditional on knowledge of the series from the infinite past up to the present origin t. As in Chapter 5 such results are of practical use because, usually, the forecasts depend appreciably only on *recent* past values of X and Y.

Computation of the forecast. Now (11.5.1) may be written

$$\varphi(B)\delta(B)Y_t = \varphi(B)\omega(B)X_{t-b} + \delta(B)\theta(B)a_t,$$

which we shall write as

$$\delta^*(B)Y_t = \omega^*(B)X_{t-b} + \theta^*(B)a_t$$

Then, using square brackets to denote conditional expectations at time t, and writing $p^* = p + d$, we have for the lead-l forecast

$$\hat{Y}_t(l) = [Y_{t+l}] = \delta_1^*[Y_{t+l-1}] + \cdots + \delta_{p^*+r}^*[Y_{t+l-p^*-r}] + \omega_0^*[X_{t+l-b}]$$
$$- \cdots - \omega_{p^*+s}^*[X_{t+l-b-p^*-s}] + [a_{t+1}] - \theta_1^*[a_{t+1-1}]$$
$$- \cdots - \theta_{q+r}^*[a_{t+l-q-r}] \tag{11.5.4}$$

where

$$[Y_{t+j}] = \begin{cases} Y_{t+j} & j \leqslant 0 \\ \hat{Y}_t(j) & j > 0 \end{cases}$$

$$[X_{t+j}] = \begin{cases} X_{t+j} & j \leqslant 0 \\ \hat{X}_t(j) & j > 0 \end{cases} \tag{11.5.5}$$

$$[a_{t+j}] = \begin{cases} a_{t+j} & j \leqslant 0 \\ 0 & j > 0 \end{cases}$$

and a_t is calculated from (11.5.1), or if $b \geqslant 1$, from

$$a_t = Y_t - \hat{Y}_{t-1}(1)$$

Thus, by appropriate substitutions, the minimum square error forecast is readily computed directly using (11.5.4) and (11.5.5). The forecasts $\hat{X}_t(j)$ are obtained in the usual way (see Section 5.2) utilizing the model (11.5.2).

The variance of the forecast. The v weights and the ψ weights of (11.5.3) may be obtained explicitly by equating coefficients in

$$\delta(B)\varphi_x(B)v(B) = \omega(B)\theta_x(B)B^b$$

and in

$$\varphi(B)\psi(B) = \theta(B)$$

The variance of the lead-l forecast error is then given by

$$V\{l\} = E\{Y_{t+l} - \hat{Y}_t(l)\}^2 = \sigma_\alpha^2 \sum_{j=b}^{l-1} v_j^2 + \sigma_a^2 \sum_{j=0}^{l-1} \psi_j^2 \tag{11.5.6}$$

Forecasts as a weighted aggregate of previous observations. It is instructive to consider, for any given example, precisely how the forecasts of future values of Y utilize the previous values of the X and Y series.

We have seen in Section 5.3.3 how the forecasts may be written as linear aggregates of previous values of the series. Thus, for forecasts of the leading indicator, we could write

$$\hat{X}_t(l) = \sum_{j=1}^{\infty} \pi_j^{(l)} X_{t-j+1} \tag{11.5.7}$$

The weights $\pi_j^{(1)} = \pi_j$ arise when the model (11.5.2) is written in the form

$$\alpha_t = X_t - \pi_1 X_{t-1} - \pi_2 X_{t-2} - \cdots$$

and may thus be obtained by explicitly equating coefficients in

$$\varphi_x(B) = (1 - \pi_1 B - \pi_2 B^2 - \cdots)\theta_x(B)$$

Also, using (5.3.9),

$$\pi_j^{(l)} = \pi_{j+l-1} + \sum_{h=1}^{l-1} \pi_h \pi_j^{(l-h)} \tag{11.5.8}$$

In a similar way, we can write the transfer function model (11.5.1) in the form

$$a_t = Y_t - \sum_{j=1}^{\infty} P_j Y_{t-j} - \sum_{j=1}^{\infty} Q_j X_{t-j} \quad b > 0 \tag{11.5.9}$$

(It should be noted that if the transfer function between the leading indicator X and the output Y is such that $v_j = 0$ for $j < b$, then $Q_1, Q_2, \ldots, Q_{b-1}$ in (11.5.9) will be zero.)

Now (11.5.9) may be written

$$a_t = \left(1 - \sum_{j=1}^{\infty} P_j B^j\right) Y_t - \sum_{j=1}^{\infty} Q_j B^j X_t$$

Comparison with (11.5.1) shows that the P and Q weights may be obtained by equating coefficients in the expressions

$$\theta(B)\left(1 - \sum_{j=1}^{\infty} P_j B^j\right) = \varphi(B)$$

$$\theta(B)\delta(B) \sum_{j=1}^{\infty} Q_j B^j = \varphi(B)\omega(B)B^b$$

On substituting $t + l$ for t in (11.5.9), and taking conditional expectations at origin t, we have the lead l forecast in the form

$$\hat{Y}_t(l) = \sum_{j=1}^{\infty} P_j[Y_{t+l-j}] + \sum_{j=1}^{\infty} Q_j[X_{t+l-j}] \tag{11.5.10}$$

Now the lead-1 forecast is

$$\hat{Y}_t(1) = \sum_{j=1}^{\infty} P_j Y_{t+1-j} + \sum_{j=1}^{\infty} Q_j X_{t+1-j}$$

Also, the quantities in squared brackets in (11.5.10) are either known values of X and Y series or forecasts which are linear functions of these known values.

Thus, we can write the forecast in terms of the values of the series which have already occurred at time t in the form

$$\hat{Y}_t(l) = \sum_{j=1}^{\infty} P_j^{(l)} Y_{t+1-j} + \sum_{j=1}^{\infty} Q_j^{(l)} X_{t+1-j} \qquad (11.5.11)$$

where the coefficients $P_j^{(l)}$, $Q_j^{(l)}$ may be computed recursively as follows:

$$\left.\begin{array}{l} P_j^{(1)} = P_j \quad Q_j^{(1)} = Q_j \\[2mm] P_j^{(l)} = P_{j+l-1} + \sum_{h=1}^{l-1} P_h P_j^{(l-h)} \\[2mm] Q_j^{(l)} = Q_{j+l-1} + \sum_{h=1}^{l-1} \{P_h Q_j^{(l-h)} + Q_h \pi_j^{(l-h)}\} \end{array}\right\} \qquad (11.5.12)$$

11.5.2 Forecast of CO_2 output from gas furnace

For illustration, consider the gas furnace data shown in Figure 11.1. For this example, the fitted model (see Section 11.4.1) was

$$Y_t = \frac{-(0.53 + 0.37B + 0.51B^2)}{1 - 0.57B} X_{t-3} + \frac{a_t}{1 - 1.53B + 0.63B^2}$$

and $(1 - 1.97B + 1.37B^2 - 0.34B^3)X_t = \alpha_t$.

The forecast function, written in the form (11.5.4), is thus

$$\begin{aligned} \hat{Y}_t(l) = [Y_{t+l}] = {}& 2.1[Y_{t+l-1}] - 1.5021[Y_{t+l-2}] + 0.3591[Y_{t+l-3}] \\ & - 0.53[X_{t+l-3}] + 0.4409[X_{t+l-4}] - 0.2778[X_{t+l-5}] \\ & + 0.5472[X_{t+l-6}] - 0.3213[X_{t+l-7}] \\ & + [a_{t+l}] - 0.57[a_{t+l-1}] \end{aligned}$$

Figure 11.9 shows the forecasts for lead times $l = 1, 2, \ldots, 12$ made at origin $t = 206$.

The π, P and Q weights for the model are given in Table 11.10.

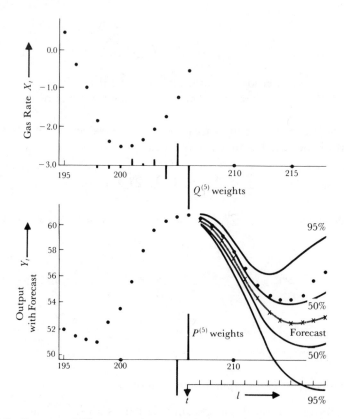

FIG. 11.9   Forecast of CO_2 output from a gas furnace using input and output series

TABLE 11.10 π, P and Q weights for gas furnace model

j	π_j	P_j	Q_j	j	π_j	P_j	Q_j
1	1.97	1.53	0	7	0	0	-0.07
2	-1.37	-0.63	0	8	0	0	-0.04
3	0.34	0	-0.53	9	0	0	-0.02
4	0	0	0.14	10	0	0	-0.01
5	0	0	-0.20	11	0	0	-0.01
6	0	0	0.43				

Figure 11.9 shows the weights $P_j^{(5)}$ and $Q_j^{(5)}$ appropriate to the lead-5 forecast.

The weights v_i and ψ_i of (11.5.3) are listed in Table 11.11.

TABLE 11.11 v and ψ weights for gas furnace model

i	v_i	ψ_i	i	v_i	ψ_i
0	0	1	6	−5.33	0.89
1	0	1.53	7	−6.51	0.62
2	0	1.71	8	−6.89	0.39
3	−0.53	1.65	9	−6.57	0.20
4	−1.72	1.45	10	−5.77	0.06
5	−3.55	1.18	11	−4.73	−0.03

Using estimates $\hat{\sigma}_\alpha^2 = 0.0353$ and $\hat{\sigma}_a^2 = 0.0561$, obtained in Sections 11.2.2 and 11.4.1, respectively, (11.5.6) may be employed to obtain variances of the forecast errors and the 50% and 95% probability limits shown in Figure 11.9.

To illustrate the advantages of using a leading indicator in forecasting, assume that only the Y series is available. The usual identification and fitting procedure applied to this series indicated that it is well described by an ARMA (4, 2) process

$$(1 - 2.42B + 2.38B^2 - 1.16B^3 + 0.23B^4)Y_t = (1 - 0.31B + 0.47B^2)\varepsilon_t$$

with $\hat{\sigma}_\varepsilon^2 = 0.1081$.

Table 11.12 shows estimated standard deviations of forecast errors made with and without the leading indicator.

TABLE 11.12 Estimated standard deviations of forecast errors
made with and without the leading indicator

l	with leading indicator	without leading indicator	l	with leading indicator	without leading indicator
1	0.23	0.33	7	1.52	2.74
2	0.43	0.77	8	1.96	2.86
3	0.59	1.30	9	2.35	2.95
4	0.72	1.82	10	2.65	3.01
5	0.86	2.24	11	2.87	3.05
6	1.12	2.54	12	3.00	3.08

As might be expected, for short lead times use of the leading indicator can produce forecasts of considerably greater accuracy.

11.5.3 Forecast of nonstationary sales data using a leading indicator

As a second illustration, consider the data on sales Y_t in relation to a leading indicator X_t, plotted in Figure 11.10 and listed as Series M in the collection

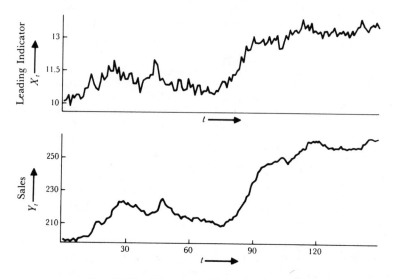

FIG. 11.10   Sales data with leading indicator

of time series at the end of this volume. The data is typical of that arising in business forecasting and is well fitted by the nonstationary model*

$$y_t = 0.035 + \frac{4.82x_{t-3}}{1 - 0.72B} + (1 - 0.54B)a_t$$

$$x_t = (1 - 0.32B)\alpha_t$$

with y_t and x_t first differences of the series. The forecast function, in the form (11.5.4), is then

$$\hat{Y}_t(l) = [Y_{t+l}] = 1.72[Y_{t+l-1}] - 0.72[Y_{t+l-2}] + 0.0098 + 4.82[X_{t+l-3}]$$

$$- 4.82[X_{t+l-4}] + [a_{t+l}] - 1.26[a_{t+l-1}]$$

$$+ 0.3888[a_{t+l-2}]$$

Figure 11.11 shows the forecasts for lead times $l = 1, 2, \ldots, 12$ made at origin $t = 89$. The weights v_j and ψ_j are given in Table 11.13.

TABLE 11.13 v and ψ weights for nonstationary model

j	v_j	ψ_j	j	v_j	ψ_j
0	0	1	6	9.14	0.46
1	0	0.46	7	9.86	0.46
2	0	0.46	8	10.37	0.46
3	4.82	0.46	9	10.75	0.46
4	6.75	0.46	10	11.02	0.46
5	8.14	0.46	11	11.21	0.46

*Using data, the latter part of which, is listed as Series M.

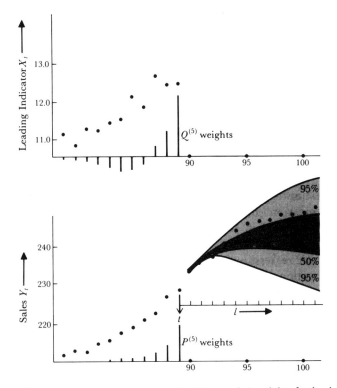

FIG. 11.11   Forecast of sales at origin $t = 89$ with P and Q weights for lead-5 forecast

Using the estimates $\hat{\sigma}_\alpha^2 = 0.0676$ and $\hat{\sigma}_a^2 = 0.0484$, obtained in fitting the above model, the variance of the forecast error may be found from (11.5.6). The 50% and 95% probability limits are shown in Figure 11.11. It will be seen that, in this particular example, the use of the leading indicator allows very accurate forecasts to be obtained for lead times $l = 1, 2$ and 3.

The π, P, and Q weights for this model are given in Table 11.14.

TABLE 11.14 π, P, and Q weights for nonstationary model

j	π_j	P_j	Q_j	j	π_j	P_j	Q_j
1	0.68	0.46	0	9	0.00	0.00	-0.74
2	0.22	0.25	0	10	0.00	0.00	-0.59
3	0.07	0.13	4.82	11	0.00	0.00	-0.29
4	0.02	0.07	1.25	12	0.00	0.00	-0.13
5	0.01	0.04	-0.29	13	0.00	0.00	-0.06
6	0.00	0.02	-0.86	14	0.00	0.00	-0.02
7	0.00	0.01	-0.97	15	0.00	0.00	0.00
8	0.00	0.01	-0.89				

The weights $P_j^{(5)}$ and $Q_j^{(5)}$ appropriate to the lead-5 forecasts are shown in Figure 11.11.

11.6 SOME ASPECTS OF THE DESIGN OF EXPERIMENTS TO ESTIMATE TRANSFER FUNCTIONS

In some engineering applications the form of the input X_t can be deliberately chosen so as to obtain good estimates of the parameters in the transfer function-noise model

$$Y_t = \delta^{-1}(B)\omega(B)X_{t-b} + N_t$$

The estimation of the transfer function is equivalent to estimation of a dynamic "regression" model and the methods which can be used are very similar to those used in ordinary non-dynamic regression. As might be expected, the same problems [86] face us.

As with static regression, it is very important to be clear on the objective of the investigation. In some situations we want to answer the question

"If the input X is merely observed (but not interfered with) what can this tell us of the present and future behavior of the output Y under *normal* conditions of process operation?"

In other situations the appropriate question is

"If the input X is *changed* in some specific way, what *change* will be induced in the present and future behavior of the output Y?"

The types of data we need to answer these two questions are different.

To answer the first question unambiguously, we must use data, obtained by observing, *but not interfering with*, the normal operation of the system.

By contrast, the second question can only be answered unambiguously from data in which *deliberate* changes have been induced into the input of the system, that is the data must be specially generated by a *designed experiment*.

Clearly, if X is to be used as a control variable, that is a variable which may be used to manipulate the output, then we need to answer the second question. To understand how we can design experiments to obtain valid estimates of the parameters of a cause and effect relationship, it is necessary to examine the assumptions of the analysis.

A critical assumption is that the X_t's are distributed independently of the N_t's. When this assumption is violated:

(1) The estimates we obtain are, in general, not even consistent. Specifically, as the sample size is made large, the estimates converge not on the true values but on other values differing from the true values by an unknown amount.

(2) The violation of this assumption is not detectable by examining the data. Therefore, the possibility that in any particular situation the independence assumption may not be true is a particularly disturbing one. The only way it is possible to guarantee its truth is by deliberately *designing* the experiment rather than using data which have simply "happened." Specifically, we must deliberately generate and feed into the process a disturbance X_t, which we know to be uncorrelated with N_t because we have generated it by some external random process.

The disturbance X_t can, of course, be autocorrelated; it is necessary only that it should not be cross-correlated with N_t. To satisfy this requirement, we could, for example, draw a set of random deviates α_t and use them to generate any desired input process $X_t = \psi_x(B)\alpha_t$.

Alternatively, we can choose a fixed "design," for example the factorial design used in Section 11.4.2, and randomize the order in which the runs are made. Appendix A11.2 contains a preliminary discussion of some elementary design problems, and it is sufficient to expose some of the difficulties in the practical selection of an "optimal" stochastic input. In particular, as is true in a wider context:

(1) It is difficult to decide what is a sensible criterion for optimality.

(2) The choice of "optimal" input depends on the values of the unknown parameters which are to be optimally estimated.

In general, a white noise input has distinct advantages in simplifying identification, and if nothing very definite were known about the system under study, would provide a sensible initial choice of input.

APPENDIX A11.1 USE OF CROSS SPECTRAL ANALYSIS FOR TRANSFER FUNCTION MODEL IDENTIFICATION

In this Appendix we show that an alternative method for identifying transfer function models, which does not require prewhitening of the input, can be based on spectral analysis. Furthermore, it is easily generalized to multiple inputs.

A11.1.1 Identification of single input transfer function models

Suppose the transfer function $v(B)$ is *defined* so as to allow the possibility of non-zero impulse response weights v_j for j a negative integer, so that

$$v(B) = \sum_{k=-\infty}^{\infty} v_k B^k$$

Then if, corresponding to (11.2.3), the transfer function-noise model is

$$y_t = v(B)x_t + n_t$$

the equations (11.2.5) become

$$\gamma_{xy}(k) = \sum_{j=-\infty}^{\infty} v_j \gamma_{xx}(k-j) \qquad k = 0, \pm 1, \pm 2, \ldots \qquad \text{(A11.1.1)}$$

We now define a *cross covariance generating function*

$$\gamma^{xy}(B) = \sum_{k=-\infty}^{\infty} \gamma_{xy}(k) B^k \qquad \text{(A11.1.2)}$$

which is analogous to the autocovariance generating function (3.1.10). On multiplying throughout in (A11.1.1) by B^k and summing, we obtain

$$\gamma^{xy}(B) = v(B)\gamma^{xx}(B) \qquad \text{(A11.1.3)}$$

If we now substitute $B = e^{-i2\pi f}$ in (A11.1.2), we obtain the cross spectrum $p_{xy}(f)$ between input and output. Making the same substitution in (A11.1.3) yields

$$v(e^{-i2\pi f}) = \frac{p_{xy}(f)}{p_{xx}(f)} \qquad -\tfrac{1}{2} \leqslant f < \tfrac{1}{2} \qquad \text{(A11.1.4)}$$

where

$$v(e^{-i2\pi f}) = G(f)\, e^{i2\pi\phi(f)} = \sum_{k=-\infty}^{\infty} v_k\, e^{-i2\pi fk} \qquad \text{(A11.1.5)}$$

is called the *frequency response function* of the system and is the Fourier transform of the impulse response function. Since $v(e^{-i2\pi f})$ is complex, we write it as a product involving a *gain function* $G(f)$ and a *phase function* $\phi(f)$. (A11.1.4) shows that the frequency response function is the ratio of the cross spectrum to the input spectrum. Methods for estimating the frequency response function $v(e^{-i2\pi f})$ are described in [27]. Knowing $v(e^{-i2\pi f})$, the impulse response function v_k can then be obtained from

$$v_k = \int_{-1/2}^{1/2} v(e^{-i2\pi f})\, e^{i2\pi fk}\, df \qquad \text{(A11.1.6)}$$

Using a similar approach, the autocovariance generating function of the noise n_t is

$$\gamma^{nn}(B) = \gamma^{yy}(B) - \frac{\gamma^{xy}(B)\gamma^{xy}(F)}{\gamma^{xx}(B)} \qquad \text{(A11.1.7)}$$

On substituting $B = e^{-i2\pi f}$ in (A11.1.7), we obtain the expression

$$p_{nn}(f) = p_{yy}(f)[1 - \kappa_{xy}^2(f)] \qquad \text{(A11.1.8)}$$

for the spectrum of the noise, where

$$\kappa_{xy}^2(f) = \frac{|p_{xy}(f)|^2}{p_{xx}(f)p_{yy}(f)}$$

and $\kappa_{xy}(f)$, the *coherency spectrum*, behaves like a correlation coefficient at each frequency f. Knowing the noise spectrum, the noise autocovariance function $\gamma_{nn}(k)$ may then be obtained from

$$\gamma_{nn}(k) = 2 \int_0^{1/2} p_{nn}(f) \cos 2\pi f \, k \, df$$

By substituting estimates of the spectra such as are described in [27], estimates of the impulse response weights v_k and noise autocorrelation function are obtained. These can be used to identify the transfer function model and noise model as described in Sections 11.2.1 and 6.2.1.

A11.1.2 Identification of multiple input transfer function models

We now generalize the model

$$Y_t = v(B)X_{t-b} + N_t$$

$$= \delta^{-1}(B)\omega(B)X_{t-b} + N_t$$

to allow for several inputs $X_{1,t}, X_{2,t}, \ldots, X_{m,t}$. Thus

$$Y_t = v_1(B)X_{1,t} + \cdots + v_m(B)X_{m,t} + N_t \tag{A11.1.9}$$

$$= \delta_1^{-1}(B)\omega_1(B)X_{1,t-b_1} + \cdots + \delta_m^{-1}(B)\omega_m(B)X_{m,t-b_m} + N_t$$

$$\tag{A11.1.10}$$

where $v_j(B)$ is the generating function of the impulse response weights relating $X_{j,t}$ to the output. We assume, as before, that after differencing, (A11.1.9) may be written

$$y_t = v_1(B)x_{1,t} + \cdots + v_m(B)x_{m,t} + n_t$$

Multiplying throughout by $x_{1,t-k}, x_{2,t-k}, \ldots, x_{m,t-k}$ in turn, taking expectations and forming the generating functions, we obtain

$$\gamma^{x_1 y}(B) = v_1(B)\gamma^{x_1 x_1}(B) + v_2(B)\gamma^{x_1 x_2}(B) + \cdots + v_m(B)\gamma^{x_1 x_m}(B)$$

$$\gamma^{x_2 y}(B) = v_1(B)\gamma^{x_2 x_1}(B) + v_2(B)\gamma^{x_2 x_2}(B) + \cdots + v_m(B)\gamma^{x_2 x_m}(B) \quad \text{(A11.1.11)}$$

$$\vdots \qquad\qquad\qquad\qquad \vdots$$

$$\gamma^{x_m y}(B) = v_1(B)\gamma^{x_m x_1}(B) + v_2(B)\gamma^{x_m x_2}(B) + \cdots + v_m(B)\gamma^{x_m x_m}(B)$$

On substituting $B = e^{-i2\pi f}$, the spectral equations are obtained. For example, with $m = 2$

$$p_{x_1 y}(f) = H_1(f)p_{x_1 x_1}(f) + H_2(f)p_{x_1 x_2}(f)$$

$$p_{x_2 y}(f) = H_1(f)p_{x_2 x_1}(f) + H_2(f)p_{x_2 x_2}(f)$$

and the frequency response functions $H_1(f) = v_1(e^{-i2\pi f})$, $H_2(f) = v_2(e^{-i2\pi f})$ can be calculated as described in [27].

The impulse response weights can then be obtained using the inverse transformation (A11.1.6).

APPENDIX A11.2 CHOICE OF INPUT TO PROVIDE OPTIMAL PARAMETER ESTIMATES

Suppose the input to a dynamic system can be made to follow an imposed stochastic process which is at choice. For example, it might be an autoregressive process, a moving average process, or white noise. To illustrate the problems involved in the optimal selection of this stochastic process, it is sufficient to consider an elementary example.

A11.2.1 Design of optimal inputs for a simple system

Suppose a system is under study for which the transfer function-noise model is assumed to be

$$Y_t = \beta_1 Y_{t-1} + \beta_2 X_{t-1} + a_t \qquad |\beta_1| < 1 \qquad \text{(A11.2.1)}$$

where a_t is white noise. It is supposed also that the input and output processes are stationary and that X_t, Y_t denote deviations of these processes from their respective means. For large samples, and associated with any fixed probability, the approximate area of the Bayesian HPD region for β_1 and β_2, and also of the corresponding confidence region, is proportional to $\Delta^{-1/2}$, where

$$\Delta = \begin{vmatrix} E[Y_t^2] & E[Y_t X_t] \\ E[Y_t X_t] & E[X_t^2] \end{vmatrix}$$

We shall proceed by attempting to find the design minimizing the area of the region and thus maximizing Δ. Now

$$E[Y_t^2] = \sigma_Y^2 = \sigma_X^2 \beta_2^2 \frac{(1 + 2q)}{1 - \beta_1^2} + \frac{\sigma_a^2}{1 - \beta_1^2} \qquad \text{(A11.2.2)}$$

$$E[Y_t X_t] = \sigma_X^2 \frac{\beta_2}{\beta_1} q$$

$$E[X_t^2] = \sigma_X^2$$

where

$$q = \sum_{i=1}^{\infty} \beta_1^i \rho_i \qquad \sigma_X^2 \rho_i = E[X_t X_{t-i}]$$

The value of the determinant may be written in terms of σ_X^2 as

$$\Delta = \frac{\sigma_X^2 \sigma_a^2}{1 - \beta_1^2} + \frac{\beta_2^2 \sigma_X^4}{(1 - \beta_1^2)^2} - \frac{\sigma_X^4 \beta_2^2}{\beta_1^2} \left\{ q - \frac{\beta_1^2}{1 - \beta_1^2} \right\}^2 \qquad \text{(A11.2.3)}$$

Thus, as might be expected, the area of the region can be made small by making σ_X^2 large (that is, by varying the input variable over a wide range). In practice, there may be limits to the amount of variation that can be allowed in X. Let us proceed by first supposing that σ_X^2 is to be held fixed at some specified value.

The solution with σ_X^2 fixed. With $(1 - \beta_1^2) > 0$, and for any fixed σ_X^2, we see from (A11.2.3) that Δ is maximized by setting

$$q = \frac{\beta_1^2}{1 - \beta_1^2}$$

that is

$$\beta_1 \rho_1 + \beta_1^2 \rho_2 + \beta_1^3 \rho_3 + \cdots = \beta_1^2 + \beta_1^4 + \beta_1^6 + \cdots$$

There are an infinity of ways in which, for given β_1, this equality could be achieved. One solution is

$$\rho_i = \beta_1^i$$

Thus one way to maximize Δ, for fixed σ_X^2, would be to force the input to follow the autoregressive process

$$(1 - \beta_1 B)X_t = \varepsilon_t$$

where ε_t is a white noise process with variance $\sigma_\varepsilon^2 = \sigma_X^2(1 - \beta_1^2)$.

The solution with σ_Y^2 fixed. So far we have supposed that σ_Y^2 is unrestricted. In some cases we might wish to avoid too great a variation in the output rather than in the input. Suppose that σ_Y^2 is held equal to some fixed acceptable value but that σ_X^2 is unrestricted. Then the value of the determinant Δ can be written in terms of σ_Y^2 as

$$\Delta = \frac{\sigma_Y^4}{\beta_2^2} \left\{ \frac{(\sigma_Y^2 - \sigma_a^2)}{\sigma_Y^2} - \frac{\beta_1^2}{s^2} \left(\frac{q + s}{1 + 2q} \right)^2 \right\} \qquad \text{(A11.2.4)}$$

where

$$s = \frac{\beta_1^2 t}{(1 + \beta_1^2 t)} \qquad \text{(A11.2.5)}$$

and

$$t = \frac{\sigma_Y^2}{(\sigma_Y^2 - \sigma_a^2)} \qquad \text{(A11.2.6)}$$

The maximum is achieved by setting

$$q = -s = -\beta_1^2 t/(1 + \beta_1^2 t) \qquad \text{(A11.2.7)}$$

that is

$$\beta_1 \rho_1 + \beta_1^2 \rho_2 + \beta_1^3 \rho_3 + \cdots = -\beta_1^2 t + \beta_1^4 t^2 - \beta_1^6 t^3 + \cdots$$

There are again an infinity of ways of satisfying this equality. In particular we can make

$$\rho_i = (-\beta_1 t)^i \qquad \text{(A11.2.8)}$$

by forcing the input to follow the autoregressive process

$$(1 + \beta_1 tB)X_t = \varepsilon_t \qquad \text{(A11.2.9)}$$

where ε_t is a white noise process with variance $\sigma_\varepsilon^2 = \sigma_X^2(1 - \beta_1^2 t^2)$. Since t is essentially positive, the sign of the parameter $\beta_1 t$ of this autoregressive process is opposite to that obtained for the optimal input with σ_X^2 fixed.

The solution with $\sigma_Y^2 \times \sigma_X^2$ fixed. In practice it might happen that excessive variations in input and output were both to be avoided. If it were true that a given *percentage* decrease in the variance of X was equally as desirable as the same *percentage* decrease in the variance of Y, then it would be sensible to maximize Δ subject to a fixed value of the product $\sigma_X^2 \times \sigma_Y^2$. The determinant is

$$\Delta = \sigma_X^2\sigma_Y^2 - (\sigma_X^4\beta_2^2 q^2/\beta_1^2) \qquad \text{(A11.2.10)}$$

which is maximized for fixed $\sigma_X^2\sigma_Y^2$ only if $q = 0$. Once again there is an infinity of solutions. However, by using a white noise input, Δ is maximized *whatever the value of β_1.* For such an input, using (A11.2.2), σ_X^2 is the positive root of

$$\sigma_X^4\beta_2^2 + \sigma_X^2\sigma_a^2 - k(1 - \beta_1^2) = 0 \qquad \text{(A11.2.11)}$$

where $k = \sigma_X^2\sigma_Y^2$.

A11.2.2 A numerical example

Suppose we were studying the first-order dynamic system (A11.2.1) with $\beta_1 = 0.50$, $\beta_2 = 1.00$, so that

$$Y_t = 0.50Y_{t-1} + 1.00X_{t-1} + a_t$$

where $\sigma_a^2 = 0.2$.

σ_X^2 *fixed, σ_Y^2 unrestricted.* Suppose at first that the design is chosen to maximize Δ with $\sigma_X^2 = 1.0$. Then one optimal choice for the input X_t will be the autoregressive process

$$(1 - 0.5B)X_t = \varepsilon_t$$

where the white noise process ε_t would have variance $\sigma_\varepsilon^2 = \sigma_X^2(1 - \beta_1^2) = 0.75$. Using (A11.2.2), the variance σ_Y^2 of the output would be 2.49, and the scheme will achieve a Bayesian region for β_1 and β_2 whose area is proportional to $\Delta^{-1/2} = 0.70$.

σ_Y^2 *fixed*, σ_X^2 *unrestricted.*   The above scheme is optimal under the assumption that the input variance is $\sigma_X^2 = 1$ and the output variance is unrestricted. This output variance then turns out to be $\sigma_Y^2 = 2.49$. If, instead, the input variance were unrestricted, then with a *fixed* output variance of 2.49, we could, of course, do considerably better. In fact, using (A11.2.6), $t = 1.087$, so that from (A11.2.9), one optimal choice for the unrestricted input would be the autoregressive process

$$(1 + 0.54B)X_t = \varepsilon_t$$

where in this case ε_t is a white noise process with $\sigma_\varepsilon^2 = 2.05$. The variance σ_X^2 of the input would now be increased to 2.91, and $\Delta^{-1/2}$, which measures the area of the Bayesian region, would be reduced to $\Delta^{-1/2} = 0.42$.

Product $\sigma_Y^2 \times \sigma_X^2$ *fixed.*   Finally, we consider a scheme which attempts to control both σ_Y^2 and σ_X^2 by minimizing Δ with $\sigma_Y^2 \times \sigma_X^2$ fixed. In the previous example in which σ_Y^2 was fixed, we found that $\Delta^{-1/2} = 0.42$ with $\sigma_X^2 = 2.91$ and $\sigma_Y^2 = 2.49$, so that the product $2.91 \times 2.49 = 7.25$. If our objective had been to minimize $\Delta^{-1/2}$ while keeping this product equal to 7.25 we could have made an optimal choice *without knowledge of* β_1 by choosing a white noise input $X_t = \varepsilon_t$. Using (A11.2.11), $\sigma_X^2 = \sigma_\varepsilon^2 = 2.29$, $\sigma_Y^2 = 3.16$ and in this case, as expected, $\Delta^{-1/2} = 0.37$ is slightly smaller than in the previous example.

It is worth considering this example in terms of spectral ideas. To optimize with σ_X^2 fixed we have used an autoregressive input with ϕ positive which has high power at low frequencies. Since the gain of the system is high at low frequencies, this achieves maximum transfer from X to Y and so induces large variations in Y. When σ_Y^2 is fixed, we have introduced an input which is an autoregressive process with ϕ negative. This has high power at high frequencies. Since there is minimum transfer from X to Y at high frequencies, the disturbance in X must now be made large at these frequencies. When the product $\sigma_X^2 \times \sigma_Y^2$ is fixed, the "compromise" input white noise is indicated and does not require knowledge of β_1. This final maximization is equivalent to minimizing the correlation between the estimates $\hat{\beta}_1$ and $\hat{\beta}_2$, and in fact the correlation between these estimates is zero when a white noise input is used.

Conclusions.   This investigation shows

(1) The optimal choice of design rests heavily on how we define "optimal."
(2) Both in the case where σ_X^2 is held fixed and in that where σ_Y^2 is held fixed, the optimal choices require specific stochastic processes whose parameters are functions of the *unknown* dynamic parameters. Thus, we are in

the familiar paradoxical situation where we can do a better job of data gathering only to the extent that we already know something about the answer we seek. A sequential approach, where we improve the design as we find out more about the parameters, is a possibility worth further investigation. In particular, a pilot investigation using a possibly non-optimal input disturbance, say white noise, could be used to generate data from which preliminary estimates of the dynamic parameters could be obtained. These estimates could then be used to specify a further input disturbance using one of our previous criteria.

(3) The use of white noise is shown, *for the simple case investigated*, to be optimal for a sensible criterion of optimality, and its use as an input requires no prior knowledge of the parameters.

Part IV

Design of Discrete Control Schemes

In earlier chapters we studied the modelling of discrete time series and dynamic systems. We saw how, once adequate models have been obtained, they may be put to use to yield forecasts of time series and to characterize the transfer function of a dynamic system. However, the models and the methods for their manipulation are of much wider importance than even these applications indicated. The ideas we have outlined are of importance in the analysis of a wide class of stochastic-dynamic systems occurring for example, in economics, engineering, commerce and in organizational studies.

It is obviously impossible to illustrate every application. Rather, it is hoped that the theory and examples of this book may help the reader to adapt the general methods to his own particular problems. In doing this, the dynamic and stochastic models we have discussed will often act as *building blocks* which can be linked together to represent the particular system under study. Also, techniques of identification, estimation and diagnostic checking, similar to those we have illustrated, will be needed to establish the model. Finally, recursive calculations and the ideas considered under the general heading of forecasting will have wider application in working out the consequences of a model once it has been fitted.

We shall conclude this book by illustrating these possibilities in one further application—the design of optimal feedforward and feedback control schemes. In working through the following chapters, it is the exercise of bringing together the previously discussed ideas in a fresh application, quite as much as the detailed results, which we hope will be of value.

12

Design of Feedforward and Feedback Control Schemes

A common control problem is how to maintain some output variable as close as possible to a target value in a system subject to disturbances. We now consider this problem using the previously discussed stochastic and transfer function models to describe disturbances and system dynamics.

We shall continue to assume that data is available at discrete equispaced time intervals, when opportunity can also be taken to make adjustments. It is assumed also that no appreciable extra cost is associated with corrective action. This is the case for many industrial processes subject to manual or automatic control. It is sensible then to seek control schemes which minimize some overall measure of error at the output. The overall error measure we use is the *mean square error*.

In some instances, one or more *sources* of disturbance may be measured and these measurements used to compensate potential deviations in the output. Such action is called *feedforward control*. In other situations, the only evidence we have of the existence of the disturbance is the deviation from target it produces in the output. When this deviation itself is used as a basis for adjustment, this action is *feedback control*. In some instances, a combination of the two modes of control is desirable, and this is referred to as *feedforward-feedback control*.

In this chapter we first show how one can design control schemes of the various types described above to yield minimum square error at the output. Finally we show how process data and, in particular, data collected during the operation of a pilot control scheme may be used to obtain better estimates of the model and its parameters. This allows us to employ an *iterative approach* in arriving at an optimal control scheme.

12.1 FEEDFORWARD CONTROL

We now consider the design of discrete feedforward control schemes which give minimum mean square error at the output. A situation arising in the manufacture of a polymer is illustrated in Figure 12.1. The viscosity Y_t of the product is known to vary in part due to fluctuations in the feed concentration z_t, which can be observed but not changed. The steam pressure X_t is a control variable which is measured, can be manipulated, and is potentially available to alter the viscosity by any desired amount and hence compensate potential deviations from target. The total effect in the output viscosity of all *other* sources of disturbance at time t is denoted by N_t.

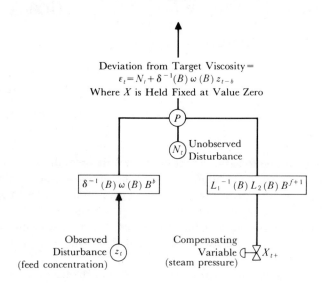

FIG. 12.1   A system at time t subject to an observed disturbance z_t and unobserved disturbance N_t, with potential compensating variable X_t held fixed at $X_t = 0$

12.1.1 Feedforward control to minimize mean square error at the output

We can suppose that z_t, X_t, N_t are deviations from reference values, which are such that if the conditions $z = 0$, $X = 0$, $N = 0$ were continuously maintained, then the process would remain in an equilibrium state such that the output was exactly on the target value $Y = 0$.

The transfer function model which connects the observed disturbance z_t (feed concentration) and the output Y_t (viscosity) is assumed to be

$$Y_t = \delta^{-1}(B)\omega(B)B^b z_t$$

Now, changes will be made in X at times $t, t - 1, t - 2, \ldots$ immediately after the observations $z_t, z_{t-1}, z_{t-2}, \ldots$ are taken. Hence we obtain a "pulsed" input, and we denote the level of X in the interval t to $t + 1$ by X_{t+}. For this pulsed input, it is assumed that the transfer function model which connects the compensating variable X_t (steam pressure) and the output Y_t (viscosity) is

$$Y_t = L_1^{-1}(B)L_2(B)B^{f+1}X_{t+}$$

where $L_1(B)$ and $L_2(B)$ are polynomials in B. Then if no control is exerted (the potential compensating variable X_t is held fixed at $X_t = 0$), the total error in the output viscosity will be

$$\varepsilon_t = N_t + \delta^{-1}(B)\omega(B)z_{t-b}$$

Clearly, it ought to be possible to compensate the effect of the measured parts of the overall disturbance by manipulating X_t. Now at time t, and at the point P in Figure 12.1:

(1) The total effect of the disturbance (z) is

$$\delta^{-1}(B)\omega(B)z_{t-b}$$

(2) The total effect of the compensation (X) is

$$L_1^{-1}(B)L_2(B)X_{t-f-1+}$$

Then the effect of the observed disturbance z will be cancelled if we set

$$L_1^{-1}(B)L_2(B)X_{t-f-1+} = -\delta^{-1}(B)\omega(B)z_{t-b}$$

Thus, the control action at time t should be such that

$$L_1^{-1}(B)L_2(B)X_{t+} = -\delta^{-1}(B)\omega(B)z_{t-(b-f-1)} \qquad (12.1.1)$$

Case 1: $b \geqslant f + 1$. Now at time t, the values $z_{t+1}, z_{t+2} \ldots$ are unknown. The control action (12.1.1) is directly realizable, therefore, only if $(b - f - 1) \geqslant 0$, in which case the desired control action at time t is to set the manipulated variable X to the level

$$X_{t+} = -\frac{L_1(B)\omega(B)}{L_2(B)\delta(B)}z_{t-(b-f-1)}$$

Alternatively, it is often more convenient to define the control action in terms of the *change* $x_t = X_{t+} - X_{t-1+}$ which is to be made in the level of X immediately after the observation z_t has come to hand. This is

$$x_t = -\frac{L_1(B)\omega(B)}{L_2(B)\delta(B)}\{z_{t-(b-f-1)} - z_{t-1-(b-f-1)}\} \qquad (12.1.2)$$

The situation is illustrated in Figure 12.2. The effect at P from the control action is $-\delta^{-1}(B)\omega(B)z_{t-b}$, and this exactly cancels the effect at P of the disturbance. The component of the deviation from target due to z_t is (theoretically at least) exactly eliminated at the observation times, and only the component N_t due to the unobserved disturbance remains.

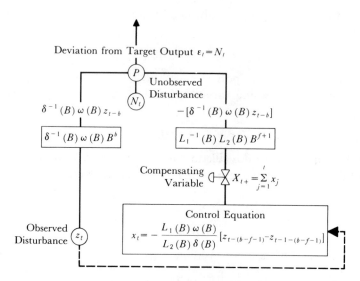

Fig. 12.2   Feedforward control scheme at time t when $b \geqslant f + 1$

Case 2: $(b - f - 1)$ *negative.*   It can happen that $f + 1 > b$. This means that an observed disturbance reaches the output before it is possible for compensating action to become effective. In this case the action

$$X_{t+} = -L_1(B)L_2^{-1}(B)\omega(B)\delta^{-1}(B)z_{t+f+1-b} \qquad (12.1.3)$$

is not realizable because at time t, when the action is to be taken, the relevant value $z_{t+(f+1-b)}$ of the disturbance is not yet available. One would usually avoid this situation if one could (if some quicker acting compensating variable could be used instead of X), but sometimes such an alternative is not available.

Now with $z_t' = \omega(B)\delta^{-1}(B)z_t$ represented by the linear model [117]

$$z_t' = \left\{ 1 + \sum_{i=1}^{\infty} \psi_i B^i \right\} a_t$$

where, as before, a_t is a white noise process with mean zero and variance σ_a^2, then

$$z_{t+f+1-b}' = \hat{z}_t'(f + 1 - b) + e_t'(f + 1 - b)$$

In this expression

$$e_t'(f + 1 - b) = a_{t+f+1-b} + \psi_1 a_{t+f-b} + \cdots + \psi_{f-b} a_{t+1}$$

is the forecast error. Then we can write the right-hand side of (12.1.3) in the form

$$-L_1(B)L_2^{-1}(B)\hat{z}_t'(f+1-b) - L_1(B)L_2^{-1}(B)e_t'(f+1-b)$$

Now, $e_t'(f+1-b)$ is a function of the uncorrelated random deviates a_{t+h} ($h \geq 1$) which have not yet occurred at time t and which are uncorrelated with any variable known at time t (and are therefore unforecastable). It follows that the optimal action is achieved by setting

$$X_{t+} = -\frac{L_1(B)}{L_2(B)}\hat{z}_t'(f+1-b) \tag{12.1.4}$$

that is by making the *change* in the compensating variable at time t equal to

$$x_t = -\frac{L_1(B)}{L_2(B)}\{\hat{z}_t'(f+1-b) - \hat{z}_{t-1}'(f+1-b)\} \tag{12.1.5}$$

This results in an additional component in the deviation ε_t from the target, which now becomes

$$\varepsilon_t = N_t + e_{t-f-1}'(f+1-b)$$

If the disturbance model is $\varphi(B)z_t = \theta(B)a_t$, then that for z_t' can be written $\varphi'(B)z_t' = \theta'(B)a_t$, with $\varphi'(B) = \varphi(B)\delta(B)$ and $\theta'(B) = \theta(B)\omega(B)$. The needed forecast $\hat{z}_t'(f+1-b)$, obtained as in Chapter 5, can then be written conveniently in terms of previous z's and a's obtainable from the z series itself.

12.1.2 An Example—Control of the specific gravity of an intermediate product

In the manufacture of an intermediate product, used for the production of a synthetic resin, the specific gravity Y_t of the product had to be maintained as close as possible to the value 1.260. This was actually achieved by a mixed scheme of feedforward and feedback control. We consider the complete scheme later and discuss here only the feedforward part. The process has rather slow dynamics, and also the disturbance is known to change slowly, so that observations and adjustments are made at two-hourly intervals. The uncontrolled disturbance which is fed forward is the feed concentration z_t, which is measured from an origin of 30 grams per liter. The relation between specific gravity and feed concentration over the range of normal operation is

$$Y_t = 0.0016\, z_t$$

where Y_t is measured from the target value 1.260.

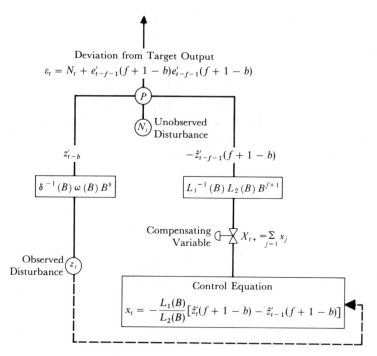

Deviation from Target Output
$$\varepsilon_t = N_t + e'_{t-f-1}(f + 1 - b)e'_{t-f-1}(f + 1 - b)$$

FIG. 12.3   Feedforward control scheme at time t when $f + 1 > b$

This relation contains "no dynamics" because the feed concentration can only be measured at the inlet to the reactor, so that in our general notation

$$\delta(B) = 1 \qquad \omega(B) = 0.0016 \qquad b = 0$$

Control is achieved by varying pressure which is referred to a convenient origin of 25 psi. The transfer function model relating specific gravity and pressure was estimated as

$$(1 - 0.7B)Y_t = 0.0024 \, X_{t-1+}$$

so that

$$L_1(B) = (1 - 0.7B) \qquad L_2(B) = 0.0024 \qquad f = 0$$

So far as could be ascertained, the effects of pressure and feed concentration were approximately additive in the region of normal operation. Therefore, the control equation (12.1.4) is used, since $b - f - 1$ is negative, and yields

$$X_{t+} = -\frac{(1 - 0.7B)0.0016}{0.0024} \hat{z}_t(1) \qquad (12.1.6)$$

for, in this particular example, $z'_t = 0.0016z_t$ and hence $\hat{z}'_t(1) = 0.0016\hat{z}_t(1)$.

Study of the feed concentration showed that it could be represented by the linear stochastic model of order $(0, 1, 1)$

$$\nabla z_t = (1 - \theta B)a_t$$

with $\theta = 0.5$. For such a process,

$$\hat{z}_t(1) = (1 - \theta)z_t + \theta\hat{z}_{t-1}(1)$$

that is

$$(1 - \theta B)\hat{z}_t(1) = (1 - \theta)z_t$$

or

$$\hat{z}_t(1) = \frac{(1 - \theta)}{(1 - \theta B)}z_t$$

Thus, the control equation (12.1.6) can be written finally as

$$X_{t+} = -\frac{(1 - 0.7B)0.0016(0.5)}{0.0024(1 - 0.5B)}z_t$$

or

$$X_{t+} = 0.5X_{t-1+} - 0.33\{z_t - 0.7z_{t-1}\}. \qquad (12.1.7)$$

Table 12.1 shows the calculation of the first few of a series of settings of the pressure required to compensate the variations in feed concentration, given the starting conditions for time $t = 0$ of $z_0 = 1.6$, $X_{0+} = -0.63$.

TABLE 12.1   Calculation of adjustments for feedforward control scheme (12.1.7)

| | Concentration | | | Pressure | |
t	$z_t + 30$	z_t	X_{t+}	$X_{t+} + 25$	x_t
0	31.6	1.6	−0.63	24.4	
1	31.1	1.1	−0.31	24.7	0.3
2	34.4	4.4	−1.36	23.6	−1.1
3	32.0	2.0	−0.32	24.7	1.1
4	28.2	−1.8	0.90	25.9	1.2

Once the calculation has been started off, it is sometimes more convenient to work directly with the change x_t to be made at time t using

$$x_t = 0.5x_{t-1} - 0.33\{\nabla z_t - 0.7\nabla z_{t-1}\} \qquad (12.1.8)$$

Figure 12.4 shows a section of the input disturbance and the corresponding output after applying feedforward control. The lower graph shows the calculated output (specific gravity) which would have resulted if no control

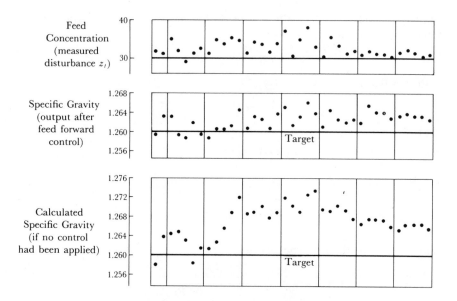

FIG. 12.4   Measured disturbance and output from feedforward control scheme

had been applied. These values Y_t are, of course, not directly available but
may be obtained in general from the values Y'_t which actually occurred using

$$Y_t = Y'_t + \hat{z}'_{t-f-1}(f + 1 - b).$$

For this example then

$$Y_t = Y'_t + \frac{0.0008z_{t-1}}{1 - 0.5B}$$

that is

$$Y_t = 0.5Y_{t-1} + Y'_t - 0.5Y'_{t-1} + 0.0008z_{t-1}$$

As a result of feedforward control, the root mean square error deviation of
the output from the target value over the sample record shown is 0.003.
Over the same period the root mean square error of the uncorrected series
would have been 0.008. The improvement is marked and extremely worthwhile. However, it appears that other unidentified sources of disturbance
exist in the process, as evidenced by the drift away from target. This kind of
tendency is frequently met in pure feedforward schemes but may be compensated by the addition of feedback control, as discussed in Section 12.2.

12.1.3 A nomogram for feedforward control

Control action is effected in whatever manner is most suited to the situation.
If changes are made infrequently, and if the control equation is fairly simple,

the theory we have outlined may be used to obtain optimal *manual* control. It is then convenient to use some form of control chart or nomogram which can be easily understood by the process operator.

For illustration, we design a nomogram to indicate the appropriate feedforward control action for the previous example. The control equation (12.1.7) is

$$(1 - 0.5B)X_{t+} = -0.33(1 - 0.7B)z_t$$

and since

$$(1 - \delta B) = (1 - \delta)\left(1 + \frac{\delta}{1 - \delta}\nabla\right)$$

this may be written in difference notation as

$$(1 + \nabla)X_{t+} = -0.20(1 + 2.33\nabla)z_t \qquad (12.1.9)$$

To design a nomogram which allows us to compute the value of

$$r_t = (1 + \xi\nabla)X_{t+} = X_{t+} + \xi(X_{t+} - X_{t-1+})$$

we construct three vertical scales to accommodate X_{t-1+}, X_{t+} and r_t, like the scales A, B, and C in Figure 12.5, and mark them off in units of X_t and space them so that $BC/AB = \xi$. Then, by simple geometry, it is evident that the value of r_t may be obtained by projecting a line through the points corresponding to X_{t-1+} and X_{t+} on the A and B scales onto the C scale—the scale of r_t.

To achieve the control action of equation (12.1.9), we must equate two expressions of this type, and hence we need five scales as shown in Figure 12.5. Four of these, namely A, B, D, and E, are to accommodate X_{t-1+}, X_{t+}, z_t, and z_{t-1}, respectively, and a further scale C allows the right side of the equation to be equated to the left. The scales are arranged so that

(1) The origins, pressure $= 25$ psi: feed concentration $= 30$ grams per liter, are in the same horizontal line, and so that 1 unit of z_t equals -0.2 units of X_t

(2) $\dfrac{BC}{AB} = 1$, $\quad \dfrac{CD}{DE} = 2.33$.

The data of Table 12.1 may be used to show the calculation of the action appropriate at time $t = 2$. We project onto the C scale at P the line joining the previous feed concentration $(z_1 + 30) = 31.1$ on the E scale and the present feed concentration $(z_2 + 30) = 34.4$ on the D scale. We then join this projected point P to the points marking the previous pressure $X_{1+} + 25 = 24.7$ on the A scale and read off on scale B the value $X_{2+} + 25 = 23.6$ to which the pressure must now be adjusted and held for the next two hours.

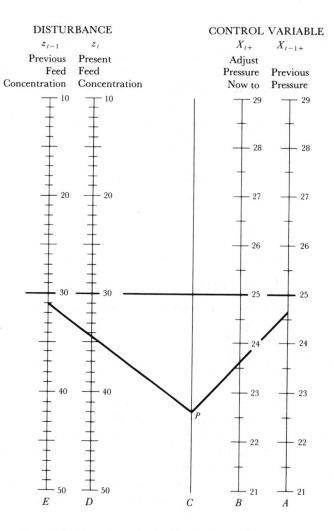

FIG. 12.5 Nomogram for feedforward control scheme

12.1.4 *Feedforward control with multiple inputs*

No difficulty arises in principle when the effect of several additive input disturbances $z_1, z_2, \ldots, z_m$ are to be compensated by changes in X using feedforward control. Suppose the combined effect at the output of all the disturbances is given by

$$Y_t = \sum_{j=1}^{m} \delta_j^{-1}(B)\omega_j(B)B^{b_j}z_{j,t} = \sum_{j=1}^{m} B^{b_j}z'_{j,t}$$

and, as before, the transfer function model for the compensating variable is given by

$$Y_t = L_1^{-1}(B)L_2(B)B^{f+1}X_{t+}$$

Then, proceeding precisely as before, the required control action is to change X at time t by an amount

$$x_t = -L_1(B)L_2^{-1}(B) \sum_{j=1}^{m} \left[z'_{j,t+f+1-b_j} - z'_{j,t+f-b_j} \right] \qquad (12.1.10)$$

where

$$\left[z'_{j,t+f+1-b_j} - z'_{j,t+f-b_j} \right]$$

$$= \begin{cases} z'_{j,t+f+1-b_j} - z'_{j,t+f-b_j} & f+1-b_j \leqslant 0 \\ \hat{z}'_{j,t}(f+1-b_j) - \hat{z}'_{j,t-1}(f+1-b_j) & f+1-b_j > 0 \end{cases} \qquad (12.1.11)$$

If, as before, N_t is an unmeasurable disturbance, then the error at the output will be

$$\varepsilon_t = N_t + \sum_{j=1}^{m} e'_{j,t-f-1}(f+1-b_j) \qquad (12.1.12)$$

where $e'_{j,t-f-1}(f+1-b_j) = 0$ if $f+1-b_j \leqslant 0$.

On the one hand, feedforward control allows us to take prompt action to cancel the effect of disturbance variables, and if $f+1-b_j \leqslant 0$, to anticipate completely such disturbances, at least in theory. On the other hand, to use this type of control we must be able to measure the disturbing variables and possess complete knowledge—or at least a good estimate—of the relationship between each disturbance variable and the output. In practice, we could never measure *all* of the disturbances that affected the system. The remaining disturbances, which we have denoted by N_t, and which are not affected by feedforward control, could of course increase the variance at the output or cause the process to wander off target, as in fact occurred in the example discussed in Section 12.1.2.

Clearly, we should be able to prevent this from happening by using the error ε_t itself to indicate an appropriate adjustment, that is, by using feedback control.

12.2 FEEDBACK CONTROL

Consider the feedback scheme shown in Figure 12.6. Here N_t measures the joint effect at the output of unobserved disturbances and is defined as the deviation from target that would occur in the output at time t if no control action were taken. It is assumed to follow some linear stochastic process defined by

$$N_t = \varphi^{-1}(B)\theta(B)a_t \qquad (12.2.1)$$

or by

$$N_t = \left\{1 + \sum_{i=1}^{\infty} \psi_i B^i\right\}a_t \qquad (12.2.2)$$

where a_t is a white noise process. As in Section 12.1, the transfer function model linking the controllable variable and the output is

$$Y_t = L_1^{-1}(B)L_2(B)B^{f+1}X_{t+} \qquad (12.2.3)$$

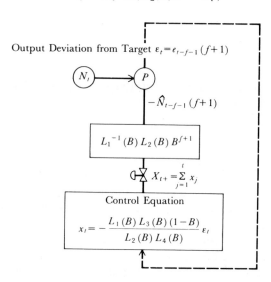

FIG. 12.6   Feedback control scheme at time t

Arguing precisely as in Section 12.1.1, we have at the point P for time t:

Total effect of disturbance $= N_t$

Total effect of compensation $= L_1^{-1}(B)L_2(B)X_{t-f-1+}$

12.2.1 Feedback control to minimize output mean square error

The effect of the disturbance would be cancelled if it were possible to set

$$X_{t+} = -L_1(B)L_2^{-1}(B)N_{t+f+1}$$

Since $f + 1$ is positive, this is not possible, but we can obtain minimum mean square control error by replacing N_{t+f+1} by the forecast $\hat{N}_t(f + 1)$, that is, by taking the control action

$$X_{t+} = -L_1(B)L_2^{-1}(B)\hat{N}_t(f+1) \tag{12.2.4}$$

Thus, the *change* or adjustment to be made in the manipulated variable is

$$x_t = -L_1(B)L_2^{-1}(B)\{\hat{N}_t(f+1) - \hat{N}_{t-1}(f+1)\} \tag{12.2.5}$$

in which case the error at the output at time t will be the forecast error at lead time $f+1$ for the N_t process, that is

$$\varepsilon_t = N_t - \hat{N}_{t-f-1}(f+1) = e_{t-f-1}(f+1)$$

Now $\hat{N}_t(f+1) - \hat{N}_{t-1}(f+1)$ is not known directly but it can nevertheless be deduced from the error sequence $\varepsilon_t, \varepsilon_{t-1}, \varepsilon_{t-2}, \ldots$ which is observed. This follows from the fact that

$$N_{t+f+1} = \left\{1 + \sum_{i=1}^{\infty} \psi_i B^i\right\} a_{t+f+1}$$

$$= \{a_{t+f+1} + \psi_1 a_{t+f} + \cdots + \psi_f a_{t+1}\} + \{\psi_{f+1} a_t + \psi_{f+2} a_{t-1} + \cdots\}$$

$$= e_t(f+1) + \hat{N}_t(f+1)$$

Now both $e_t(f+1)$ and $\hat{N}_t(f+1)$ are linear functions of the a's, so that we can write the preceding equation as

$$N_{t+f+1} = L_4(B)a_{t+f+1} + L_3(B)a_t$$

Knowing the model $N_t = \varphi^{-1}(B)\theta(B)a_t = \psi(B)a_t$ for the stochastic process, we can deduce the operators $L_3(B)$ and $L_4(B)$ in the relations

$$e_{t-f-1}(f+1) = L_4(B)a_t \qquad \hat{N}_t(f+1) = L_3(B)a_t \tag{12.2.6}$$

and hence the relations

$$\hat{N}_t(f+1) = \frac{L_3(B)}{L_4(B)} e_{t-f-1}(f+1) = \frac{L_3(B)}{L_4(B)} \varepsilon_t$$

Finally then, the feedback control equation (12.2.4) resulting in smallest mean square error at the output may be written

$$X_{t+} = -\frac{L_1(B)L_3(B)}{L_2(B)L_4(B)} \varepsilon_t \tag{12.2.7}$$

Alternatively, if as is frequently convenient, we define the control action in terms of the *adjustment* $x_t = X_{t+} - X_{t-1+}$ to be made at time t, then

$$x_t = -\frac{L_1(B)L_3(B)(1-B)}{L_2(B)L_4(B)} \varepsilon_t \tag{12.2.8}$$

In practice, the nature of $L_3(B)(1 - B)$ is often best deduced by noting that this is the operator which occurs in the updating formula,

$$\hat{N}_t(f + 1) - \hat{N}_{t-1}(f + 1) = L_3(B)(1 - B)a_t \qquad (12.2.9)$$

12.2.2 Application of the control equation: relation with three-term controller

In this book we are concerned principally with the derivation of the optimal control *equation* which indicates how the manipulated variable should be changed to maintain the controlled variable close to some target value. In practice, the actual measuring, and the computing and carrying out of the required action, can be done in a number of ways. At one end of the scale of sophistication, one may have electrical measuring instruments, the results from which are fed to a computer which calculates the required control action and directly activates transducers which carry it into effect. At the other end of this scale, one may have a plant operator who periodically takes a measurement, reads off the required action from a simple chart or nomogram and carries it out by hand. The theory which we have described has been used successfully in both kinds of situations. We go to some pains to describe in detail some of the manual applications because we feel that the use of elementary control ideas to assist the plant operator to do his job well has been somewhat neglected in the past. Although, undisputably, more and more schemes of automatic control are coming into use, there is still a great deal of manual operation, and this is likely to continue for a long period of time.

Three-term controller. A type of automatic control device which has been in use for many years is the "three-term controller." Controllers of this kind may operate through either mechanical, pneumatic, hydraulic, or electrical means, and their operation is based on continuous rather than discrete measurement and adjustment. If ε_t is the error at the output at time t, control action may be made proportional to ε itself, its integral with respect to time or its derivative with respect to time. A three-term controller uses a linear combination of all of these, so that if X_t indicates the level of the manipulated variable at time t, the control equation is of the form

$$X_t = k_D \frac{d\varepsilon_t}{dt} + k_P \varepsilon_t + k_I \int \varepsilon_t \, dt$$

where k_D, k_P, and k_I are constants.

In some situations, only one or two of these three modes of action are used. Thus, we find instances of simple proportional control ($k_D = 0, k_I = 0$), of simple integral control ($k_D = 0, k_P = 0$), of proportional-integral control ($k_D = 0$), and of proportional-derivative control ($k_I = 0$).

The discrete analogue of this continuous control equation is

$$X_{t+} = k_D \nabla \varepsilon_t + k_P \varepsilon_t + k_I S \varepsilon_t$$

or in terms* of the adjustments to be made,

$$x_t = k_D \nabla^2 \varepsilon_t + k_P \nabla \varepsilon_t + k_I \varepsilon_t$$

We shall find that many of the simple situations which we meet do lead to control equations containing terms of these types. For example, if the noise can be represented by a $(0, 1, 1)$ process $\nabla N_t = (1 - \theta B)a_t$, while the dynamics can be represented by the first-order system $(1 + \xi \nabla)Y_t = gX_{t-1+}$, (12.2.7) reduces to

$$X_{t+} = -\frac{(1 - \theta)\xi}{g}\varepsilon_t - \frac{(1 - \theta)}{g}S\varepsilon_t$$

Thus the action called for is the discrete analogue of proportional-integral control.

However, it is clear that not all control actions that might be called for by (12.2.7) could be produced by a three-term controller, and rather simple examples occur where other modes of control are required. We now consider some specific examples.

12.2.3 Examples of discrete feedback control

Example 1.   In a scheme to control the viscosity Y of a polymer employed in the manufacture of a synthetic fiber, the controlled variable, viscosity, was checked every hour and adjusted by manipulating the catalyst formulation X. The desired target value for viscosity was 47 units. The transfer function model between X and Y was adequately described by the simple first-order system

$$(1 - \delta B)Y_t = (1 - \delta)gX_{t-1+}$$

Furthermore, the time constant of the system was short compared with the sampling interval. Specifically, $\delta \simeq 0.04$, so that an estimated 96% of the eventual change occurred in the sampling interval of one hour. Therefore, to a sufficient approximation, we can set $\delta = 0$. Furthermore, catalyst formulation changes were, by custom, scaled in terms of the effect they were expected to produce. Thus, one unit of formulation increase was such as would *decrease* viscosity by one unit. Hence $g = -1$, and the transfer function model was

---

* In previous publications [14], [15], we have used a different nomenclature. For example an adjustment $x_t = k_I \varepsilon_t$ was there referred to as proportional control. This control action is equivalent to $X_{t+} = k_I S \varepsilon_t$, that is to *integral* action in the *level set-point* X of the manipulated variable. It is this latter nomenclature which has been used traditionally by control engineers and which we adopt here.

taken to be

$$Y_t = gX_{t-1+}$$

with $g = -1$, or in terms of the general model (12.2.3),

$$L_1(B) = 1 \qquad L_2(B) = g = -1 \qquad f = 0$$

The disturbance N_t at the output, which it will be recalled is defined as the variation in viscosity if no control were exerted, was adequately described by the stochastic process of order $(0, 1, 1)$

$$\nabla N_t = (1 - \theta B)a_t$$

with $\theta = 0.53$, $\lambda = (1 - \theta) = 0.47$, so that

$$\hat{N}_t(1) - \hat{N}_{t-1}(1) = \lambda a_t$$

whence using (12.2.9), $L_3(B)(1 - B) = \lambda = 0.47$. Furthermore, since $\varepsilon_t = e_{t-f-1}(f + 1) = e_{t-1}(1) = a_t$, then, using (12.2.6),

$$L_4(B) = 1$$

Finally then, the adjustment called for at time t is

$$x_t = -\frac{L_1(B)L_3(B)(1 - B)}{L_2(B)L_4(B)}\varepsilon_t = -\frac{\lambda}{g}\varepsilon_t$$

that is,

$$x_t = 0.47\varepsilon_t \quad \text{or} \quad X_{t+} = 0.47S\varepsilon_t$$

In this situation then, where the time constant of the system is small compared with the sampling interval, optimal control requires the discrete analogue of simple integral control action. We can derive the required control action for this type of example somewhat more directly as follows:

$$\text{Predicted change at the output} = \hat{N}_t(1) - \hat{N}_{t-1}(1) = \lambda a_t$$

$$\text{Effect of adjustment} = gx_t$$

Therefore, the adjustment required to compensate is such that $gx_t = -\lambda a_t$. But with this adjustment, the error at the output is $\varepsilon_t = a_t$. Thus the optimal feedback control equation is $x_t = -(\lambda/g)\varepsilon_t$.

The efficiency of control action of this kind is insensitive to moderate changes in parameter values and, to a sufficient approximation, we can take

$$x_t = 0.5\varepsilon_t$$

A convenient chart for use when, as in this example, manual control action is employed, is shown in Figure 12.7(a). On this chart the output (viscosity) scale and the action scale are arranged so that the output target is aligned with zero action, and so that one unit of output is matched by $-\lambda/g$ units of action. To employ the chart, the plant operator simply plots the latest output (viscosity) value and reads off the appropriate adjustment on the action scale.

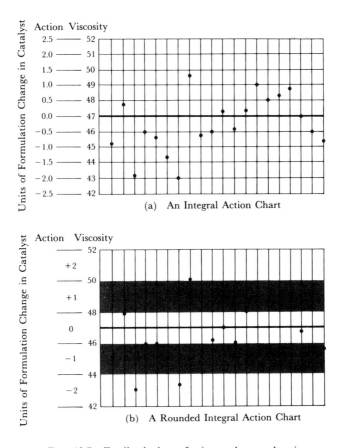

(a)   An Integral Action Chart

(b)   A Rounded Integral Action Chart

FIG. 12.7   Feedback charts for integral control action

On this particular process, control had previously been carried out using a chart based somewhat arbitrarily on a sequential significance testing scheme. It had turned out in this connection that it was convenient to add or subtract from the catalyst formulation in standard steps. Possible actions were: no action, ± one step, or ± two steps of catalyst formulation.

Significance testing procedures have little relevance in the present context. However, the previous scheme did have the advantages (i) that it had not been necessary to make changes every time and (ii) when changes were called for they were of one of five definite types, making the procedure easy to apply and supervise. However, these features can easily be included in the present control scheme, with very little increase in the error, by using a "rounded" action chart.

Rounded charts. A rounded chart is easily constructed from the original chart by dividing the action scale into bands. The adjustment made when an observation falls within the band is that appropriate to the middle point of the band on an ordinary chart. Figure 12.7(b) shows a rounded chart in which possible action is limited to -2, -1, 0, 1, or 2 catalyst formulation changes. Figures 12.7(a) and (b) have been constructed by back calculating the values of a_t from a set of operating data and reconstructing the charts that would have resulted from using an unrounded and a rounded scheme. The increase in mean square error (less than 5% for this example), which results from using the rounded scheme, is often outweighed by the convenience of working with a small number of standard adjustments. The effect of rounding is discussed in more detail in Section 13.1.1.

Example 2. At a later stage of manufacture of the polymer, the objective was to maintain the output viscosity Y as close as possible to a target value of 92 by adjusting the gas rate. Hourly determinations of viscosity were used for making regular adjustments to the gas rate every hour. We shall discuss here the planning of the pilot control scheme for this process. It will be described later how the data collected during the running of this preliminary pilot study was used to re-estimate parameters and to arrive thereby at an improved control scheme.

At this stage of the investigation some data were available which showed the variation which occurred in viscosity when no control was applied (that is when the gas rate was held fixed). These data came from a previous period of operation, during which compensations for variations in viscosity were made at a later stage. These data are in fact the 310 observations of Series D, listed in the Collection of Time Series at the end of the volume. We have shown in Chapter 6 and 7 that this series could be described fairly well by the IMA of order (0, 1, 1), $\nabla N_t = (1 - \theta B)a_t$, with θ close to zero (that is $\lambda = 1 - \theta$ close to unity). There was good evidence that, over the range of operation, the steady state relation between gas rate and viscosity was linear and that a unit change in gas rate produced 0.20 units of change in viscosity, so that the steady state gain was taken to be $g = 0.20$. Experimental evidence of questioned reliability indicated simple exponential dynamics with no dead time such that about half of the eventual change occurred in one hour.

Thus we have, tentatively, for the transfer function model connecting viscosity Y and gas rate X,

$$(1 - 0.5B)Y_t = 0.10X_{t-1+}$$

so that $L_1(B) = 1 - 0.5B$, $L_2(B) = 0.10$, $f = 0$. Also, using the disturbance model

$$\nabla N_t = a_t$$

we have $L_4(B) = 1$, $L_3(B)(1 - B) = 1$ and the appropriate feedback control equation (12.2.8) is

$$x_t = -\frac{L_1(B)L_3(B)(1 - B)}{L_2(B)L_4(B)}\varepsilon_t = -\frac{(1 - 0.5B)}{0.10}\varepsilon_t$$

or

$$x_t = -10\varepsilon_t + 5\varepsilon_{t-1}$$

where ε_t is the output deviation from target at time t.

If the action is expressed in terms of the backward difference operator ∇, we have

$$x_t = -5(1 + \nabla)\varepsilon_t$$

or

$$X_{t+} = -5\varepsilon_t - 5S\varepsilon_t$$

so that what we have is a combination of "integral" and proportional control.

A projection chart. The situation in which the disturbance N_t can be represented by a linear model of order $(0, 1, 1)$

$$\nabla N_t = (1 - \theta B)a_t$$

and the transfer function model is of the simple first order form

$$Y_t = g(1 + \xi\nabla)^{-1}X_{t-1+}$$

is of sufficiently common occurrence to warrant special mention. In general, the control adjustment (12.2.8) will be

$$x_t = -\frac{(1 - \theta)}{g}(1 + \xi\nabla)\varepsilon_t \qquad\qquad (12.2.10)$$

and the set point of the control variable is

$$X_{t+} = -\left\{\frac{(1 - \theta)\xi}{g}\varepsilon_t + \frac{(1 - \theta)}{g}S\varepsilon_t\right\}$$

With manual control, this proportional-integral action is conveniently indicated by a suitable "projection" chart. That shown in Figure 12.8(a), which was, in fact, used to implement the control action in the example described above, will illustrate the general mode of construction. The deviation from the central target line, when read on the viscosity scale, corresponds to the deviation ε_t from target. A second scale is also shown indicating the control action x_t to be taken, with zero action ($x_t = 0$) aligned with the target value. The scales are arranged so that one unit in the output viscosity (ε_t) scale corresponds to $-[(1-\theta)/g]$ units on the control action scale.

The appropriate action at time t can be read off by projecting ξ time units ahead a line through ε_t and ε_{t-1} (or equivalently through the last two viscosity measurements). For the present pilot scheme, $\xi = 1$, so we must project one time unit ahead. The control action at time $t = 2$, for example, is found by joining the viscosity values at time $t = 1$ and $t = 2$ by a line projecting one step ahead and reading off the value -30 on the action scale.

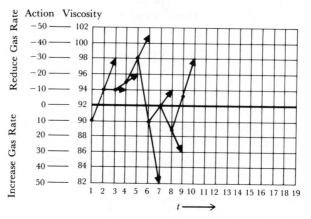

(a) A Proportional-Integral Action Chart

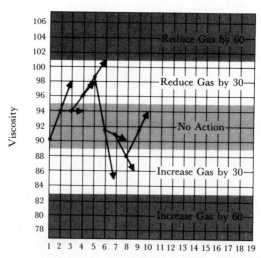

(b) A Rounded Proportional-Integral Action Chart

FIG. 12.8   Feedback charts for proportional-integral action

This indicates that the gas rate should be decreased by 30 units and held at the new value until further information becomes available at time $t = 3$.

A rounded chart. As we have mentioned earlier in this Section, exception is sometimes taken to control schemes based on charts like the one above because they require that action be taken after *each* observation. It may be felt that action ought to be taken "only when it is necessary." Two different kinds of reasoning may underlie this feeling, one having a more valid basis than the other.

(1) The practitioner who is familiar with statistical significance tests and standard quality control charts may be persuaded that he ought to have real evidence that "the process has deviated from target" before any action is taken. When, as in the mass production metal working industries (where standard quality control procedures have traditionally been used), *an additional cost* is incurred every time a change is made, it is possible to agree with the consequences of this thinking if not its customary justification [15]. However, within the process industries, the process operator (or the controlling computer) is usually going to be on duty anyway to check the process periodically, so that there is no additional cost in making a change. In this latter case, it is appropriate simply to minimize some measure of deviation from target such as the mean square error, and this is what we do here.

(2) A second and more sensible argument might be that, in any industrial operation, it is always advantageous to simplify as much as possible the actions that the plant operator is expected to take. If a chart could be devised which, without much loss, required him to take one of a small number of distinct actions, this would be an advantage.

As we have seen before, this objective is easily gained by the use of a "rounded" chart. A suitable "rounded" chart for the present example is shown in Figure 12.8(b). In this chart the action scale has been divided into 5 bands, each 30 gas rate units in width. The bands correspond to the 5 actions: reduce gas rate by 60, reduce gas rate by 30, no action, increase gas rate by 30, increase gas rate by 60. The viscosity is plotted and the points projected exactly as before, but the action is "rounded" and corresponds to the central value of the band in which the projected point falls. The chance of a projected point falling outside an outer band is small, and such points are treated as having fallen within the appropriate outer band. In other words, the outer bands are extended to stretch to plus and minus infinity.

Of course, the result of using a rounded chart is to increase somewhat the variance of the output viscosity about target. However, even with such severe rounding, as is illustrated here, the increase is usually not very great. In Section 13.1.1 we discuss the general question of the effect of added noise in the input of the process. Using the derivation given there, it turns out that the increase in the standard deviation of viscosity about the target, produced by the rounding illustrated in Figure 12.8(b), is about 7%. The points which have been placed on the rounded chart in Figure 12.8(b) for illustration were, in fact, back calculated assuming that the same disturbance is present as for the unrounded chart in Figure 12.8(a). It is shown in Chapter 13 that, provided δ is not too close to 1, (that is, provided the time constant of the system is not too long compared with the sampling interval) a rounding interval as wide as one standard deviation of x may be used without causing a large increase in the variance of the output. The approximate effect of such rounding, applied to the control equation (12.2.10), is now considered.

Let the rounding interval be denoted by $R\sigma_x$. Then for the particular choice $R = 1$, and assuming a Normal distribution, we would have the following distribution of actions for a chart such as 12.8(b)

Zone $\left\{ \begin{array}{c} \\ \\ \end{array} \right.$	$\begin{array}{c} -\infty \\ -1.5\sigma \end{array}$	$\begin{array}{c} -1.5\sigma_x \\ -0.5\sigma_x \end{array}$	$\begin{array}{c} -0.5\sigma_x \\ 0.5\sigma_x \end{array}$	$\begin{array}{c} 0.5\sigma_x \\ 1.5\sigma_x \end{array}$	$\begin{array}{c} 1.5\sigma_x \\ \infty \end{array}$
"Central" value of zone	$-2\sigma_x$	$-\sigma_x$	0	σ_x	$2\sigma_x$
Probability (%) of falling in zone	6.7	24.2	38.3	24.2	6.7

Strictly speaking, the theoretical results of Section 13.1.1 concerning the increase in output variance, due to rounding, assume that there are also zones centered on $3\sigma_x$, $-3\sigma_x$, $4\sigma_x$, $-4\sigma_x$, and so on. However, the total probability of a point falling into these outer zones would be only 1.24%, and the effect of combining them all into the $\pm 2\sigma_x$ zones is assumed small.

Specifically, it can be shown that with these assumptions the standard deviation of the output is increased by the factor F, where

$$F^2 \simeq 1 + \frac{R^2}{12} \frac{(1 + \theta\delta)(1 - \theta)(1 + \delta^2)}{(1 - \theta\delta)(1 + \theta)(1 - \delta^2)} \qquad (12.2.11)$$

For the chart in Figure 12.8(b), $\theta = 0$, $\delta = 0.5$, $R \simeq 1$, so that $F \simeq 1.07$.

Example 3. For further illustration, we consider the slightly more complicated situation which occurs when the transfer function model may be represented by a first-order system with dead time (delay). Thus, with

$$\nabla Y_t = g(1 + \xi\nabla)^{-1} \{(1 - v)x_{t-f-1} + vx_{t-f-2}\}$$

then in terms of the general model (12.2.3),

$$L_1(B)/L_2(B) = (1 + \xi\nabla) \{g(1 - v\nabla)\}^{-1}$$

If the disturbance N_t is represented, as before, by a process of order $(0, 1, 1)$

$$\nabla N_t = (1 - \theta B)a_t$$

$$\hat{N}_t(f + 1) - \hat{N}_{t-1}(f + 1) = (1 - \theta)a_t$$

$$e_t(f + 1) = [1 + (1 - \theta)\{B + B^2 + \cdots + B^f\}]a_{t+f+1}$$

so that

$$L_3(B)(1 - B) = (1 - \theta), \quad L_4(B) = 1 + (1 - \theta)\{B + B^2 + \cdots + B^f\}$$

Hence, using (12.2.8), the optimal action is given by making an adjustment x_t given by

$$\{1 - v\nabla\}\{1 + (1 - \theta)(B + B^2 + \cdots + B^f)\}x_t = -\frac{(1 - \theta)}{g}(1 + \xi\nabla)\varepsilon_t$$

$$\text{(12.2.12)}$$

that is

$$x_t = -(1 - \theta)(X_{t-1+} - X_{t-f-1+}) - \frac{(1 - \theta)(1 + \xi\nabla)}{g(1 - v\nabla)}\varepsilon_t \quad \text{(12.2.13)}$$

We notice that the introduction of delay into the transfer function model results in a mode of control in which the present adjustment depends on past *action* over the period of the delay as well as on present and past errors ε_t. In particular, in the common situation where $f = 0$, we obtain

$$x_t = v\nabla x_t - \frac{(1 - \theta)}{g}(1 + \xi\nabla)\varepsilon_t$$

A delay nomogram. Using the same argument as in Section 12.1.3, it is easy to design a nomogram to compute the required action

$$(1 - v\nabla)x_t = -\frac{(1 - \theta)}{g}(1 + \xi\nabla)\varepsilon_t \quad \text{(12.2.14)}$$

appropriate in the special case where $f = 0$. Suppose we had a situation with the same background as in Example 2 where it was desired to maintain viscosity at the value 92 as nearly as possible. However, suppose now that

$$\theta = 0.5 \qquad \xi = 0.7 \qquad v = 0.25 \qquad g = 0.20$$

Then, substituting in (12.2.14), the required adjustment is

$$x_t = 0.25\nabla x_t - 2.50\varepsilon_t - 1.75\nabla\varepsilon_t$$

that is

$$x_t = -0.33x_{t-1} - 5.67\varepsilon_t + 2.33\varepsilon_{t-1}$$

This action is computed using the nomogram of Figure 12.9 with scales A, B, E, D, indicating, respectively, $\varepsilon_t, \varepsilon_{t-1}, x_t, x_{t-1}$ and a scale C used to

equate the two sides of the control equation. The scales are arranged so that :
(1) Zero action and target value are aligned.
(2) One unit on the viscosity scale is equal to $-(1 - \theta)/g = -2.5$ units in the gas rate scale.
(3) The distances between the scales are such that $AC/AB = \xi = 0.7$, $CE/DE = v = 0.25$.

On the nomogram shown in Figure 12.9, a value of 92 for the viscosity has just come to hand. A straight line joining this to the previous viscosity reading of 96 is projected to cut the C scale at a point marked P. A line drawn through P and the value -32, corresponding to the previous adjustment, cuts the action scale at 20. This tells us that the present optimal adjustment is to increase the gas rate by 20 units.

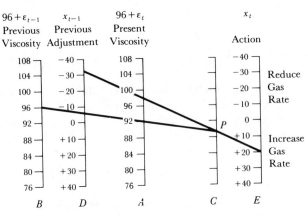

FIG. 12.9   Control nomogram for a simple delayed system

Note that, in this particular example, the current value of viscosity is on target. Nevertheless, taking into account the previous behavior of the process and its dynamic-stochastic characteristics, corrective action is still necessary. The plant operator must increase the gas rate by 20 units if he is to follow a policy which will minimize the mean square deviation from target viscosity.

As before, if it were desired to simplify the control action, a "rounded" nomogram with the action scale divided up into a suitable number of zones could be used.

12.3 FEEDFORWARD-FEEDBACK CONTROL

A combined feedforward-feedback scheme provides for the elimination of identifiable disturbances by feedforward control and for the reduction of the remaining disturbance by feedback control. Figure 12.10 shows part of a combined feedforward-feedback scheme in which m identifiable disturbances $z_1, z_2, \ldots, z_m$ are fed forward. It is supposed that N'_t is a further unidentified disturbance and that the *augmented noise* N_t is made up of N'_t plus that part of the feedforward disturbance that cannot be predicted at time t. Thus, using (12.1.12),

$$N_t = N'_t + \sum_{j=1}^{m} e'_{j,t-f-1}(f + 1 - b_j)$$

with $e'_{j,t-f-1}(f + 1 - b_j) = 0$ if $f + 1 - b_j \leqslant 0$, and includes any further contributions from errors in forecasting the identifiable inputs. It is assumed that N_t can be represented by a linear stochastic process so that, as in (12.2.9), the relationship between the forecasts of this noise process and the forecast errors may be written

$$\frac{L_3(B)(1 - B)}{L_4(B)}\varepsilon_t = \hat{N}_t(f + 1) - \hat{N}_{t-1}(f + 1)$$

where $\varepsilon_t = e_{t-f-1}(f + 1)$.

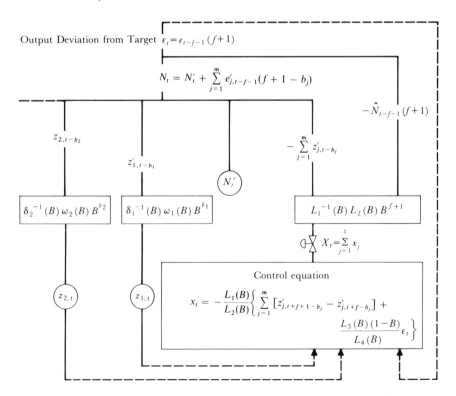

FIG. 12.10 Mixed feedforward-feedback control scheme at time t

12.3.1 Feedforward-feedback control to minimize output mean square error

Arguing as in (12.1.10) and (12.2.8), the optimal control action is

$$x_t = -\frac{L_1(B)}{L_2(B)}\left\{\sum_{j=1}^{m}\left[z'_{j,t+f+1-b_j} - z'_{j,t+f-b_j}\right] + \frac{L_3(B)(1-B)}{L_4(B)}\varepsilon_t\right\}$$

$$(12.3.1)$$

where

$$z'_{j,t+f+1-b_j} - z'_{j,t+f-b_j}$$

$$= \begin{cases} z'_{j,t+f+1-b_j} - z'_{j,t+f-b_j} & f+1-b_j \leqslant 0 \\ \hat{z}'_{j,t}(f+1-b_j) - \hat{z}'_{j,t-1}(f+1-b_j) & f+1-b_j > 0 \end{cases}$$

$$(12.3.2)$$

The first term in the control equation (12.3.1) is the same as (12.1.10) and compensates for changes in the feedforward variables. The second term in (12.3.1) is the same as (12.2.8) and compensates for that part N'_t of the augmented noise which can be predicted at time t. In Figure 12.10, the output from the right-hand box is split into two parts only for diagrammatic convenience.

12.3.2 An example of feedforward-feedback control

We illustrate by discussing further the example used in Section 12.1.2, where it was desired to control specific gravity as close as possible to a target value 1.260. Study of the deviations from target occurring *after feedforward control* showed that they could be represented by the IMA (0, 1, 1) process

$$\nabla N_t = (1 - 0.5B)a_t$$

where a_t is a white noise process. Thus

$$\frac{L_3(B)(1-B)}{L_4(B)}a_t = \hat{N}_t(1) - \hat{N}_{t-1}(1) = 0.5a_t$$

and

$$\varepsilon_t = e_{t-1}(1) = a_t$$

As in Section 12.1.2, the remaining parameters are

$$\delta^{-1}(B)\omega(B) = 0.0016 \qquad\qquad b = 0$$

$$L_2^{-1}(B)L_1(B) = \frac{(1-0.7B)}{0.0024} \qquad f = 0$$

and

$$\hat{z}_t(1) - \hat{z}_{t-1}(1) = \frac{0.5}{1-0.5B}(z_t - z_{t-1})$$

Using (12.3.1), the optimal adjustment incorporating feedforward and feedback control is

$$x_t = -\frac{(1 - 0.7B)}{0.0024}\left[\frac{(0.0016)(0.5)}{1 - 0.5B}(z_t - z_{t-1}) + 0.5\varepsilon_t\right] \tag{12.3.3}$$

that is

$$x_t = 0.5x_{t-1} - 0.33(1 - 0.7B)(z_t - z_{t-1}) - 208(1 - 0.7B)(1 - 0.5B)\varepsilon_t$$

or

$$x_t = 0.5x_{t-1} - 0.33z_t + 0.56z_{t-1} - 0.23z_{t-2} - 208\varepsilon_t + 250\varepsilon_{t-1} - 73\varepsilon_{t-2}$$

$$\tag{12.3.4}$$

Figure 12.11 shows the section of record previously given in Figure 12.4, when only feedforward control was employed, and the corresponding calculated variation that would have occurred if no control had been applied. This is now compared with a record from a scheme using both feedforward and feedback control. The introduction of feedback control resulted in a further substantial reduction in mean square error and corrected the tendency to drift from target which was experienced with the feedforward scheme.

Note that with a feedback scheme, the correction employs a forecast having lead time $f + 1$, whereas with a feedforward scheme the forecast has lead time $f + 1 - b$ and no forecasting is involved if $f + 1 - b$ is zero or negative. Thus, feedforward control gains in the immediacy of possible adjustment whenever b is greater than zero.

The example we have quoted is exceptional in that $b = 0$, and consequently no advantage of immediacy is, in this case, gained by feedforward control. It might be true in this case that equally good control could have been obtained by feedback alone. In practice, possibly because of error transmission problems, the mixed scheme did rather better than the pure feedback system.

12.3.3 Advantages and disadvantages of feedforward and feedback control

With feedback control it is the total disturbance, as evidenced by the error at the output, that actuates compensation. Therefore, it is not necessary to be able to identify and measure the sources of disturbance. All that is needed is that we *characterize* the disturbance N_t at the output by an appropriate stochastic process. Because we are not relying on "dead reckoning," unexpected disturbances and moderate errors in estimating the system's characteristics will normally result only in greater variation about the target value and not (as may occur with feedforward control) in a consistent drift away from the target value. On the other hand, especially if the delay $f + 1$ is large, the

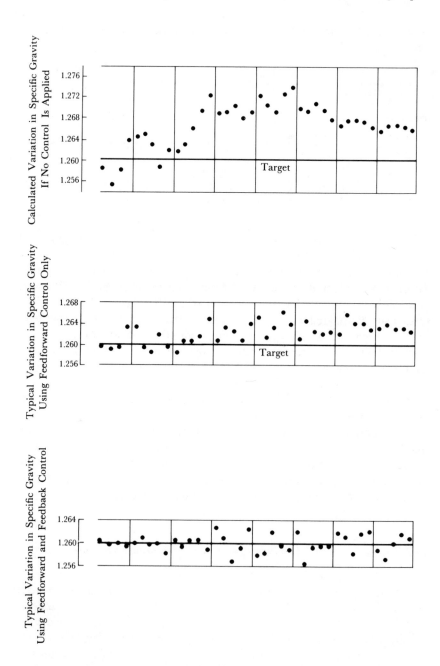

FIG. 12.11   Typical variation in specific gravity with: no control, feedforward control
        only, and feedforward-feedback control

errors about the target (since they are then the errors of a remote forecast) may be large although they have zero mean. Clearly, if identifiable sources of disturbance can be partially or wholly eliminated by feedforward control, then this should be done. Then, only the unidentifiable error has to be dealt with by feedback control.

In summary, although we can design a feedback scheme which is optimal, in the sense that it is the best possible feedback scheme, it will not usually be as good as a combined feedforward-feedback scheme in which sources of error which can be, are, eliminated before the feedback loop.

12.4 FITTING TRANSFER FUNCTION-NOISE MODELS USING OPERATING DATA

12.4.1 Iterative model building

It is desirable that the parameters of a control system be estimated from data collected under as nearly as possible the conditions which will apply when the control scheme is in actual operation. The calculated control action, using estimates so obtained, properly takes account of noise in the system, which will be characterized as if it entered at the point provided for in the model (see Section 13.1.3). This being so, it is desirable to proceed iteratively in the development of a control scheme. Using technical knowledge of the process, together with whatever can be gleaned from past operating data, preliminary transfer function and noise models are postulated and used to design a pilot control scheme. The operation of this pilot scheme can then be used to supply further data, which may be analyzed to give improved estimates of the transfer function and noise models, and then used to plan an improved scheme.

12.4.2 Estimation from operating data

It will be sufficient to consider a feedforward-feedback scheme with a single feedforward input.

$b - f - 1$ *nonnegative.*   If we suppose that $b - f - 1$ is nonnegative, then, whatever the inputs z_t and X_{t+}, it will be true that the error is given by

$$\varepsilon_t = \delta^{-1}(B)\omega(B)z_{t-b} + L_1^{-1}(B)L_2(B)X_{t-f-1+} + N_t \qquad (12.4.1)$$

We see that (12.4.1) is of similar form to the open loop models considered in Section 11.4.2 and contains two inputs z_t and X_{t+}. The model (12.4.1) can thus be fitted exactly as described in Chapter 11.

If it is assumed, as in Chapter 11, that the noise may be described by an ARIMA (p, d, q) model

$$N_t = \phi^{-1}(B)\nabla^{-d}\theta(B)a_t$$

the model (12.4.1) may be rewritten

$$\nabla^d \varepsilon_t = \mathscr{y}_{1,t} + \mathscr{y}_{2,t} + n_t \qquad (12.4.2)$$

where

$$\mathscr{y}_{1,t} = \delta^{-1}(B)\omega(B)\nabla^d z_{t-b} \qquad (12.4.3)$$

$$\mathscr{y}_{2,t} = L_1^{-1}(B)L_2(B)\nabla^d X_{t-f-1+} \qquad (12.4.4)$$

$$n_t = \nabla^d N_t = \phi^{-1}(B)\theta(B)a_t \qquad (12.4.5)$$

It is supposed that data is available, in the form of simultaneous series for $\varepsilon_t, z_t,$ and X_{t+}, during a fairly long period of actual plant operation. Usually, although not necessarily, this would be a period during which some pre-liminary pilot control scheme was being operated. Then, proceeding as in Chapter 11, for specified values of the parameters, $\mathscr{y}_{1,t}$ can be generated from z_t and $\mathscr{y}_{2,t}$ from X_{t+}. Then the differenced noise n_t can be calculated from

$$n_t = \nabla^d \varepsilon_t - \mathscr{y}_{1,t} - \mathscr{y}_{2,t} \qquad (12.4.6)$$

and finally a_t from

$$a_t = \theta^{-1}(B)\phi(B)n_t \qquad (12.4.7)$$

Equation (12.4.7) allows the a_t's to be calculated for any chosen values of the parameters. To estimate these parameters, we need only program the recursive calculation of the a_t's and insert this subroutine into the general nonlinear estimation program which computes the derivatives numerically and which automatically proceeds with the iteration, as already described in Chapters 7 and 11.

Feedback Control. Consider now a pure feedback system which, with $e_t = \nabla^d \varepsilon_t,$ $x_t = \nabla^d X_t,$ and $n_t = \nabla^d N_t = \psi(B)a_t,$ may be represented in the form

$$e_t = v(B)x_t + \psi(B)a_t \qquad (12.4.8)$$

$$x_t = c(B)e_t\{+d_t\} \qquad (12.4.9)$$

where $c(B)$ is the transfer function of the controller, not necessarily optimal, and d_t is either a mistake representing the difference between $c(B)e_t$, the calculated adjustment, and x_t the adjustment known to have actually been applied, or an added "dither" signal (see Section 12.4.4) which has been deliberately introduced. The curly brackets in (12.4.9) emphasize that the added term may or may not be present. In either case, in the fitting process, equation (12.4.8) may be used to compute the a_t's for any given values of the parameters of $v(B) = L_1^{-1}(B)L_2(B)B^{f+1}$ and $\psi(B) = \theta(B)/\phi(B)$ as is done in the example which follows.

12.4.3 An example

In the second feedback control example in Section 12.2.3, the objective was to maintain the viscosity of a polymer as close as possible to the target value of 92 by hourly readings of viscosity and adjustment of the gas rate. The previous discussion was concerned with the design of a pilot control scheme based on information of questionable accuracy. Essentially the pilot scheme assumed that the noise and transfer-function models were

$$\nabla N_t = (1 - \theta B)a_t$$

$$(1 - \delta B)Y_t = g(1 - \delta)X_{t-1+}$$

with $\theta = 0$, $\delta = 0.5$, $g = 0.20$.

These models led to the control equation $x_t = -10\varepsilon_t + 5\varepsilon_{t-1}$ as defining the optimal adjustment at time t. Part of the actual operating record using this pilot scheme is shown in Figure 12.12. The changes in gas rate x_t and the corresponding deviations from target ε_t now supply the data from which new estimates may be obtained. We proceed on the assumption that the form of model is adequate but that the estimates of the parameters θ, δ, and g may be in error. In this case, equations (12.4.4), (12.4.6) and (12.4.7) become, respectively,

$$y_t = \delta y_{t-1} + g(1 - \delta)x_{t-1} \tag{12.4.10}$$

where $x_t = \nabla X_{t+}$,

$$n_t = \nabla \varepsilon_t - y_t$$

$$a_t = \theta a_{t-1} + n_t$$

$$= \theta a_{t-1} + \nabla \varepsilon_t - y_t \tag{12.4.11}$$

For illustration, a set of eight pairs of values of x_t and ε_t are given in Table 12.2. These are the initial values of a series of 312 observations made during 13 days of running of the pilot scheme and fully listed as Series L, "Pilot Scheme Data" in the list of series and data at the end of this volume.

TABLE 12.2   Eight pairs of values of (x_t, ε_t) series from pilot scheme

t	1	2	3	4	5	6	7	8
x_t	30	0	-10	0	-40	0	-10	10
ε_t	-4	-2	0	0	4	2	2	0

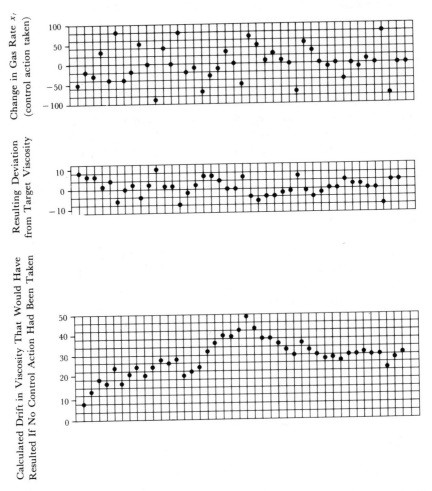

Fig. 12.12   Pilot scheme for control of viscosity: part of the operating record with reconstructed disturbance

Table 12.3 shows the beginning of the recursive calculation of a_t^0 for the parameter values $\theta = 0.2$, $\delta = 0.6$, $g = 0.25$. For these values, equations (12.4.11) and (12.4.10) become

$$a_t^0 = 0.2a_{t-1}^0 + \nabla\varepsilon_t - y_t^0 \tag{12.4.12}$$

$$y_t^0 = 0.6y_{t-1}^0 + 0.1x_{t-1} \tag{12.4.13}$$

The data are given in columns (1), (2), and (3) of Table 12.3. The entries in column (4) are obtained using (12.4.13) and represent the changes at the

TABLE 12.3 Recursive calculation of a_t^0 for data from pilot scheme for parameter values $\theta = 0.2$, $\delta = 0.6$, $g = 0.25$

t	x_t	ε_t	$y_t^0 = 0.6y_{t-1}^0 + 0.1x_{t-1}$	$\nabla \varepsilon_t$	$\nabla \varepsilon_t - y_t^0$	$a_t^0 = 0.2a_{t-1}^0 + (\nabla \varepsilon_t - y_t^0)$
			y_1^0			a_1^0
1	30	-4				
2	0	-2	$0.60y_1^0 + 3.00$	2	$-1.00 - 0.60y_1^0$	$-1.00 + 0.20a_1^0 - 0.60y_1^0$
3	-10	0	$0.36y_1^0 + 1.80$	2	$0.20 - 0.36y_1^0$	$0.00 + 0.04a_1^0 - 0.48y_1^0$
4	0	0	$0.22y_1^0 + 0.08$	0	$-0.08 - 0.22y_1^0$	$-0.08 + 0.01a_1^0 - 0.31y_1^0$
5	-40	4	$0.13y_1^0 + 0.05$	4	$3.95 - 0.13y_1^0$	$3.93 - 0.19y_1^0$
6	0	2	$0.08y_1^0 - 3.97$	-2	$1.97 - 0.08y_1^0$	$2.76 - 0.12y_1^0$
7	-10	2	$0.05y_1^0 - 2.38$	0	$2.38 - 0.05y_1^0$	$2.93 - 0.07y_1^0$
8	10	0	$0.03y_1^0 - 2.43$	-2	$0.43 - 0.03y_1^0$	$1.02 - 0.04y_1^0$

output which are produced by the changes x_t. Columns (5) and (6) are obtained by simple arithmetic, and column (7) from (12.4.12). In this table, y_1^0 and a_1^0 have been inserted for the unknown starting values. The entries in the table show the influence which the choice of these values has on subsequent calculations.

A number of points are clarified by the table.

(1) We notice that the choices of a_1^0 and y_1^0 influence only the first few values of a_t^0. This will be true more generally, except for parameter values in ranges for which the weight functions for the noise model or for the transfer function model are very slow to die out. With the approach we adopt here, the true values of the parameters are unlikely to be within these critical ranges.

(2) We can substitute guesses for a_1 and y_1 and when, as in this example, data is cheap, throw away the first few values of a_t^0 to allow transients arising from non-optimal choice of a_1 and y_1 to die out.

(3) The values of the a_t's with those starting values a_1^0 and y_1^0 which give a minimum sum of squares *conditional* on the choice of the "main" parameters may be computed and employed in subsequent least squares calculations. Some further refinements along the lines of Section 7.1 are possible but will not be further discussed here.

We illustrate this final point with the data of Table 12.3 where the calculation is particularly simple. The values a_1^0 and y_1^0 which minimize $\Sigma(a_j^0)^2$ for the particular choice of parameters $\theta = 0.2$, $\delta = 0.6$, $g = 0.25$ are found by "regressing" column (a) on columns (b) and (c) in Table 12.4.

TABLE 12.4   Calculation of maximum likelihood
estimates of starting values

(a)	(b)	(c)
0.00	−1.00	0.00
−1.00	−0.20	0.60
0.00	−0.04	0.48
−0.08	−0.01	0.31
3.93	0.00	0.19
2.76	0.00	0.12
2.93	0.00	0.07
1.02	0.00	0.04

The elements in the table are all taken from the extreme right-hand column of Table 12.3. The elements in column (a) are the terms independent of a_1 and y_1, and the elements of columns (b) and (c) are the coefficients of $-a_1^0$ and $-y_1^0$, respectively. Because the coefficients in columns (b) and (c) rapidly die out, for the purpose of computing $\hat{a}_1^0$ and $\hat{y}_1^0$ we need be concerned only with the first values of the series. In fact, for the particular case considered above, we need only take account of the first eight entries. The normal equations are then

$$0.2008 = 1.0417a_1^0 - 0.1423\hat{y}_1^0$$

$$0.6990 = -0.1423a_1^0 + 0.7435\hat{y}_1^0$$

yielding solutions $\hat{a}_1^0 = 0.33$, $\hat{y}_1^0 = 1.00$, for the starting values.

The nature of the sum of squares surface for this example can be seen from Figure 12.13. The contours were obtained by interpolating in a grid of computed values. In each case, starting values were obtained in the manner described above. The approximate three dimensional 95% confidence region is indicated by the shaded areas in the figure.

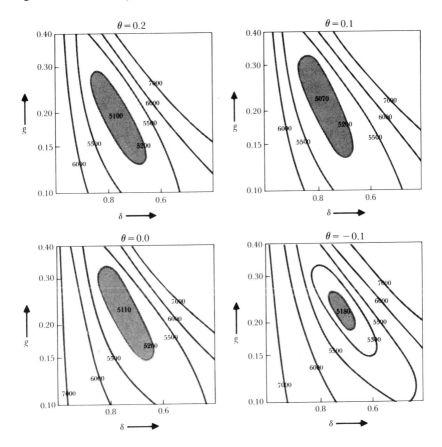

FIG. 12.13   Sums of squares contours and approximate 95% confidence region for (θ, g, δ) using data from pilot control scheme (g is plotted on a log scale)

As an additional check, the nonlinear least squares routine was run using as initial parameter values the rough estimates employed in the pilot control scheme. The iteration proceeded as shown in Table 12.5.

TABLE 12.5   Convergence of parameters in simultaneous fitting of transfer function and noise models

Iteration	θ	$(1 - \delta)g$	δ	Sum of squares
0	0.01	0.10	0.50	6,247.6
1	−0.06	0.09	0.53	5,661.3
2	−0.11	0.08	0.61	5,275.9
3	−0.02	0.06	0.71	5,115.9
4	0.08	0.05	0.77	5,067.6
5	0.10	0.05	0.77	5,065.2
6	0.11	0.05	0.77	5,065.1
7	0.11	0.05	0.77	5,065.1

In this example it is clear that the estimates $\hat{\theta} = 0$, $\hat{g} = 0.20$ used in the pilot scheme were about right, but the value $\hat{\delta} = 0.5$ was too low for the estimate of the dynamic parameter, a value of $\hat{\delta} = 0.77$ now being indicated. For the re-estimated values of the parameters, the optimal control scheme is

$$x_t = -17.8\varepsilon_t + 13.7\varepsilon_{t-1}$$

which may be compared with the pilot scheme

$$x_t = -10\varepsilon_t + 5\varepsilon_{t-1}$$

12.4.4   Model fitting under feedback conditions

Following [109] and [110], to better understand the nature of the fitting procedure used above, we may substitute (12.4.9) in (12.4.8) to obtain

$$\{1 - v(B)c(B)\}e_t = \psi(B)a_t\{+v(B)d_t\} \qquad (12.4.14)$$

First, consider the case where d_t is *zero*. Because, from (12.4.9), x_t is then a deterministic function of the e_t's, the model, (which appears in (12.4.8) to be of the transfer function form), is seen in (12.4.14) to be equivalent to an ARMA model whose coefficients are functions of the known parameters of $c(B)$ and of the unknown dynamic and stochastic parameters of the model, which are to be estimated by minimizing Σa_t^2. It is then apparent that, with d_t absent, all dynamic and stochastic model forms $v_0(B)$ and $\psi_0(B)$ which are such that

$$\psi_0^{-1}(B)\{1 - v_0(B)c(B)\} = \psi^{-1}(B)\{1 - v(B)c(B)\} \qquad (12.4.15)$$

will fit equally well. This does not lead to estimation difficulties in the example above where the model is of a simple form in which the degrees of each of the polynomials in the rational functions $v(B)$ and $\psi(B)$ are known and the pilot controller is *not* very close to optimality in the sense of minimizing mean square errors. More generally however difficulties can arise. In particular, consider the case where the dynamic and stochastic models are of the form $v(B) = \omega(B)B/\delta(B)$, $\psi(B) = \theta(B)/\phi(B)$. Then it may be shown [111], that as the pilot controller used during the generation of the data approaches near-optimality, then near-singularities occur in the sum of squares surface in the $r + s + 1 + p + q$ dimensional space of $\delta, \omega, \phi, \theta$. The individual parameters may then be estimated only very imprecisely or in the limit will be non-estimable. However, in these circumstances, accurate estimates of those functions of the parameters which are the constants of the control equation may be obtainable. Thus while data collected under feedback conditions may be inadequate for estimating the *individual* dynamic and stochastic parameters of the system, it may nevertheless be used for updating the estimates of the constants of a control equation whose mathematical form is assumed known. The situation can be much improved by the deliberate introduction during data generation of a random signal d_t as in (12.4.9). To achieve this the action $c(B)e_t$ is first computed according to the control equation and then d_t is added on. The added signal can, for example, be a random Normal deviate or a random binary digit and should have mean zero and variance sufficiently small so as not to unduly upset the process. The computations of the a_t's needed in the fitting may be carried through using equation (12.4.8) exactly as before. We see, however, from (12.4.14) that with d_t present this procedure now employs a genuine transfer model form in which e_t depends on the random input d_t as well as on the shocks a_t. Thus with d_t present the fitting procedure tacitly employs, not only information arising from the autocorrelation of the e's, but also additional information associated with the cross correlations of the e's and the d's.

Identification under feedback conditions. In the example above, data from a pilot scheme was used to reestimate parameters with the model form *already identified* from open loop data and from previous knowledge of the system. Considerable care is needed in using closed loop data in the identification process itself. In the first place, if d_t is absent, it is apparent from (12.4.9) that cross correlation of the 'output' e_t and the 'input' x_t with or without prewhitening will tell us (what we already know) about $c(B)$ and not, as might appear if (12.4.8) were treated as defining an open loop system, about $v(B)$. Furthermore, since the autocorrelations of the e_t's will be the same for all model forms satisfying (12.4.15) unique identification is not possible if nothing is known about the form of either $\psi(B)$ or $v(B)$. On the other hand if either of $\psi(B)$ or $v(B)$ is known the autocorrelation function can be used for the identification of the other. With d_t present, the form of (12.4.14) is that of the transfer function model considered in Chapter 11 and corresponding methods may be used for identification.

13

Some Further Problems in Control

In this chapter we consider three further problems which arise in the design of discrete control schemes. First, we consider the effect of additional noise on the design of a control scheme (a) by calculating the extent to which the performance of the control scheme is lowered if the added noise is ignored and (b) by computing the optimal action when the effect of the noise is allowed for. Second, we show how to construct optimal schemes when the adjustment variance of the control variable is restricted by practical considerations. Finally, we show how the choice of sampling interval affects the degree of control that is possible.

13.1 EFFECT OF ADDED NOISE IN FEEDBACK SCHEMES

In Chapter 12 we emphasized the importance of estimating the parameters of the system under, as nearly as possible, the actual control conditions which will obtain in the final scheme. The main reason for this is to ensure that all sources of noise are taken account of. If we estimate the system parameters under working control conditions, then automatically we shall estimate the noise *as if* it all originated at the source provided for it in the model. The effect of this will be that parameter estimates will be obtained which will give near optimal control action under actual working conditions.

By contrast, suppose the stochastic and transfer function models were estimated "piecemeal." For example, we might use records which indicated the noise *actually originating* at P in Figure 13.1 to estimate the noise model for N_t. Provided the amount of additional noise was not excessive, the control scheme obtained using this estimate might still be reasonably good. However, the ignoring of large additional noise sources could lead to inefficient control action.

In the sections that follow, we investigate, for a feedback scheme, the following problems:
(1) The effect of ignoring added noise.
(2) "Rounding" the control action as a source of added noise.
(3) Differences in optimal action produced by added noise.
(4) Effective transference of the noise origin which occurs when data are

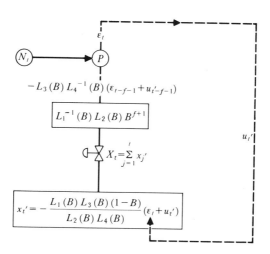

FIG. 13.1   Feedback control with error in the loop

collected under operating conditions similar to those obtained in the final
control scheme.

13.1.1 Effect of ignoring added noise—rounded schemes

Consider the feedback control loop of Figure 13.1 in which the noise *actually*
originating at P is N_t and $\nabla^d N_t = \phi^{-1}(B)\theta(B)a_t$. As shown in Section 12.2.1,
on the assumption that N_t is the only noise component, optimal action results
from the control equation

$$X_{t+} = -\frac{L_1(B)L_3(B)}{L_2(B)L_4(B)}\varepsilon_t \tag{13.1.1}$$

with

$$\hat{N}_t(f+1) = L_3(B)a_t \qquad e_{t-f-1}(f+1) = L_4(B)a_t \qquad \varepsilon_t = e_{t-f-1}(f+1)$$

Suppose now that an additional source of noise u'_t exists, whose effect is to
change the error signal from ε_t to $\varepsilon_t + u'_t$. Then the action actually taken is

$$X'_{t+} = -\frac{L_1(B)L_3(B)}{L_2(B)L_4(B)}(\varepsilon_t + u'_t) \tag{13.1.2}$$

so that the adjustment is

$$x'_t = -\frac{L_1(B)L_3(B)(1-B)}{L_2(B)L_4(B)}(\varepsilon_t + u'_t)$$

Then at P in Figure 13.1,

$$\varepsilon_t = N_t - L_3(B)L_4^{-1}(B)B^{f+1}(\varepsilon_t + u_t')$$

or

$$L_3(B)L_4^{-1}(B)u_{t-f-1}' - N_t = -\{L_3(B)L_4^{-1}(B)B^{f+1} + 1\}\varepsilon_t \quad (13.1.3)$$

However, using the results of Section 12.2.1,

$$N_t = \hat{N}_{t-f-1}(f + 1) + e_{t-f-1}(f + 1)$$

and

$$\hat{N}_{t-f-1}(f + 1) = L_3(B)L_4^{-1}(B)B^{f+1}e_{t-f-1}(f + 1)$$

Hence

$$N_t = \{L_3(B)L_4^{-1}(B)B^{f+1} + 1\}e_{t-f-1}(f + 1) \quad (13.1.4)$$

Adding (13.1.3) and (13.1.4), we obtain

$$L_3(B)L_4^{-1}(B)u_{t-f-1}' = \{L_3(B)L_4^{-1}(B)B^{f+1} + 1\}(e_{t-f-1}(f + 1) - \varepsilon_t) \quad (13.1.5)$$

Now, since

$$\nabla^d N_t = \phi^{-1}(B)\theta(B)a_t$$

$$= \phi^{-1}(B)\theta(B)L_4^{-1}(B)e_{t-f-1}(f + 1) \quad (13.1.6)$$

substituting (13.1.4) in (13.1.6) gives

$$\nabla^d(L_3(B)L_4^{-1}(B)B^{f+1} + 1) = \phi^{-1}(B)\theta(B)L_4^{-1}(B)$$

It follows that (13.1.5) may be rewritten

$$L_3(B)L_4^{-1}(B)\nabla^d u_{t-f-1}' = \phi^{-1}(B)\theta(B)L_4^{-1}(B)\{e_{t-f-1}(f + 1) - \varepsilon_t\}$$

so that

$$\varepsilon_t = e_{t-f-1}(f + 1) - L_3(B)\phi(B)\theta^{-1}(B)\nabla^d u_{t-f-1}' \quad (13.1.7)$$

Now

$$e_{t-f-1}(f + 1) = L_4(B)a_t = a_t + \psi_1 a_{-1} + \cdots + \psi_f a_{t-f}$$

is statistically uncorrelated with u_{t-f-1}' provided only that the cross covariances $\gamma_{u'a}(k)$ are zero for $k \geqslant 1$. In what follows, we assume this condition to be satisfied.

    If the additional noise u_t' is represented by the stochastic process

$$\phi_1(B)\nabla^{d_1}u_t' = \theta_1(B)b_t'$$

with b_t' a white noise process, then (13.1.7) becomes

$$\varepsilon_t = L_4(B)a_t - L_3(B)\phi(B)\theta^{-1}(B)\phi_1^{-1}(B)\theta_1(B)\nabla^{d-d_1}b_{t-f-1}' \quad (13.1.8)$$

and provided $d \geqslant d_1$, ε_t will be a stationary process. For any choice of the parametric models for the noise at P, the additional noise in the system, and the transfer function, the variance of ε_t at the output can now be calculated.

Errors in x_t. If we wish to think of the ignored error as occurring in the adjustments x_t, we can write the control equation as

$$x_t = -\frac{L_1(B)}{L_2(B)} \frac{L_3(B)(1-B)}{L_4(B)} \varepsilon_t + u_t$$

where

$$u_t = -\frac{L_1(B)L_3(B)(1-B)}{L_2(B)L_4(B)} u_t'$$

Equation (13.1.7) then becomes

$$\varepsilon_t = e_{t-f-1}(f+1) + L_1^{-1}(B)L_2(B)L_4(B)\phi(B)\theta^{-1}(B)\nabla^{d-1}u_{t-f-1} \quad (13.1.9)$$

and if the errors in x_t follow a stochastic process

$$\phi_2(B)\nabla^{d_2}u_t = \theta_2(B)b_t \quad (13.1.10)$$

then, on substituting (13.1.10) in (13.1.9) and writing $e_{t-f-1}(f+1) = L_4(B)a_t$, we obtain

$$\varepsilon_t = L_4(B)a_t + L_1^{-1}(B)L_2(B)L_4(B)\phi(B)\theta^{-1}(B)\phi_2^{-1}(B)\theta_2(B)\nabla^{d-d_2-1}b_{t-f-1}$$

$$(13.1.11)$$

Provided then that $d > d_2$, ε_t will follow a stationary process, and its variance may be calculated for any given choice of the parameters.

Ignoring observational errors in x_t for a simple control scheme. For illustration, we now study the effect of ignoring observational errors in x_t for an important but simple control scheme of the type considered before in Section 12.2. The noise and the transfer function are defined, respectively, by

$$\nabla N_t = (1 - \theta B)a_t$$

and

$$\nabla Y_t = g\frac{(1-\delta)}{(1-\delta B)}x_{t-1}$$

and the optimal control adjustment (12.2.8), assuming no errors in the loop, is

$$x_t = -\frac{(1-\theta)}{(1-\delta)g}(1-\delta B)\varepsilon_t$$

with $\varepsilon_t = a_t$. We suppose that the adjustment actually made is

$$x'_t = x_t + u_t$$

with the adjustment errors $u_t, u_{t-1}, u_{t-2} \ldots$ uncorrelated and having variance σ_u^2. Then $L_1(B)L_2^{-1}(B) = (1 - \delta B)/(1 - \delta)g$, $f = 0$, $L_3(B)(1 - B) = (1 - \theta)$, $L_4(B) = 1$, $\phi^{-1}(B)\theta(B) = 1 - \theta B$, $\phi_2^{-1}(B)\theta_2(B) = 1$, $d = 1$, $d_2 = 0$. Substituting these values in (13.1.11), we obtain

$$\varepsilon_t = a_t + \frac{g(1 - \delta)}{(1 - \delta B)(1 - \theta B)} u_{t-1}$$

$$\sigma_\varepsilon^2 = \sigma_a^2 + \frac{g^2(1 - \delta)^2(1 + \theta\delta)}{(1 - \theta\delta)(1 - \theta^2)(1 - \delta^2)} \sigma_u^2$$

To make comparison simpler, it is convenient to express σ_u as a multiple $k\sigma_x$ of the standard deviation σ_x of x when no additional noise is present. Then

$$\sigma_u^2 = k^2 \sigma_x^2 = k^2 \frac{(1 - \theta)^2(1 + \delta^2)}{g^2(1 - \delta)^2} \sigma_a^2 \qquad (13.1.12)$$

Finally, if the additional noise in x raises the variance to $(1 + k^2)\sigma_x^2$, then the variance of the deviation from target output is increased according to

$$\sigma_\varepsilon^2 = \sigma_a^2 \left\{ 1 + k^2 \frac{(1 + \theta\delta)(1 - \theta)(1 + \delta^2)}{(1 - \theta\delta)(1 + \theta)(1 - \delta^2)} \right\} \qquad (13.1.13)$$

Rounding error in the adjustment. In particular, (13.1.13) allows us to obtain, approximately, the effect of "rounding" the adjustments x_t as is done, for example, in the chart of Figure 12.8(b). Suppose that the rounding interval is $R\sigma_x$. Very approximately we can represent the effect of rounding by adding an error u_t to x_t which is uniformly distributed over the interval $R\sigma_x$. Moreover, although there will be some autocorrelation among the u_t's, for most practically occurring cases this will be slight and so we assume them to be uncorrelated. With these approximations,

$$\sigma_\varepsilon^2 = \sigma_a^2 \left\{ 1 + \frac{R^2}{12} \frac{(1 + \theta\delta)(1 - \theta)(1 + \delta^2)}{(1 - \theta\delta)(1 + \theta)(1 - \delta^2)} \right\}$$

which gives the formula (12.2.12) previously quoted. Thus, for the chart of 12.8(b), $\theta = 0$, $\delta = 0.5$, $R \simeq 1$, so that

$$\sigma_\varepsilon^2 \simeq \sigma_a^2 \left\{ 1 + \tfrac{5}{36} \right\}$$

$$\sigma_\varepsilon \simeq 1.067 \sigma_a$$

13.1.2 Optimal action when there are observational errors in the adjustments x_t

Equation (13.1.11) makes it possible to calculate the *effect* of added noise in x_t when the optimal scheme which assumes no added noise is used. It is of interest also to derive the *optimal* scheme for specified added noise and to see how it differs from the scheme which assumes no added noise. We use for illustration the example considered in Section 13.1.1.

Suppose the control action actually taken is

$$x_t = -\frac{(1 - \theta)(1 - \delta B)}{g(1 - \delta)}L(B)\varepsilon_t + u_t$$

where again $u_t, u_{t-1}, \ldots$ are uncorrelated with variance σ_u^2 and that the noise N_t can be represented by an IMA process of order $(0, 1, 1)$. We wish to choose $L(B)$ so as to minimize σ_ε^2.

Considering, as before, the situation at the point P in the feedback loop, we obtain

$$\nabla\varepsilon_t = g(1 - \delta)(1 - \delta B)^{-1}x_{t-1} + \nabla N_t$$

whence

$$(1 - B)\varepsilon_t = -(1 - \theta)L(B)\varepsilon_{t-1} + (1 - \theta B)a_t + g(1 - \delta)(1 - \delta B)^{-1}u_{t-1}$$

that is

$$(1 - \delta B)\{1 - B + (1 - \theta)BL(B)\}\varepsilon_t = (1 - \delta B)(1 - \theta B)a_t + g(1 - \delta)u_{t-1}$$

$$(13.1.14)$$

Now the right-hand side of (13.1.14) is a representation of a second-order moving average process with added white noise and can therefore (see Section A4.4.1) be represented by another second-order moving average process

$$(1 - \pi_1 B - \pi_2 B^2)b_t$$

where b_t is a white noise process. Therefore, the problem is reduced to that of choosing $L(B)$ so that var $[\varepsilon_t]$ is minimized, where

$$(1 - \delta B)\{1 - B + (1 - \theta)BL(B)\}\varepsilon_t = (1 - \pi_1 B - \pi_2 B^2)b_t$$

Alternatively, we can write this equality in the form

$$\varepsilon_t = (1 + \psi_1 B + \psi_2 B^2 + \cdots)b_t$$

so that

$$\sigma_\varepsilon^2 = (1 + \psi_1^2 + \psi_2^2 + \cdots)\sigma_b^2$$

and σ_ε^2 is minimized only if $0 = \psi_1 = \psi_2 = \psi_3 = \ldots$. We require then that

$$(1 - \delta B)\{1 - B + (1 - \theta)BL(B)\} = 1 - \pi_1 B - \pi_2 B^2 \quad (13.1.15)$$

that is

$$L(B) = \frac{(1 + \delta - \pi_1) - (\delta + \pi_2)B}{(1 - \theta)(1 - \delta B)}$$

Therefore, the optimal adjustment is

$$x_{0t} = -\left\{ \frac{(1 + \delta - \pi_1) - (\delta + \pi_2)B}{g(1 - \delta)} \right\} \varepsilon_t \qquad (13.1.16)$$

Now, substituting (13.1.15) in (13.1.14), we obtain

$$(1 - \pi_1 B - \pi_2 B^2)\varepsilon_t = (1 - \delta B)(1 - \theta B)a_t + g(1 - \delta)u_{t-1}$$

whence π_1 and π_2 may be found by equating covariances of lags 0, 1, 2. Writing $r = \sigma_\varepsilon^2/\sigma_a^2$, we obtain

$$\left. \begin{array}{c} (1 + \pi_1^2 + \pi_2^2)r = 1 + (\delta + \theta)^2 + (\delta\theta)^2 + g^2(1 - \delta)^2\dfrac{\sigma_u^2}{\sigma_a^2} \\[2mm] \pi_1(1 - \pi_2)r = (\delta + \theta)(1 + \delta\theta) \\[2mm] -\pi_2 r = \delta\theta \end{array} \right\} \qquad (13.1.17)$$

Optimal rounded control scheme. For illustration, consider again the rounded chart of Figure 12.8(b). Making the same approximations as before, we consider what would have been the optimal control scheme given that the additional rounding error is to be taken account of.

Suppose, as in the earlier discussion at the end of Section 13.1.1, that $g = 0.2$, $\theta = 0$, $\delta = 0.5$, $R = 1$, and hence that

$$\frac{\sigma_u^2}{\sigma_a^2} = \frac{1}{12}\frac{(1 - \theta)^2(1 + \delta^2)}{g^2(1 - \delta)^2} = \frac{125}{12}$$

Then using (13.1.17)

$$\pi_2 = 0 \qquad \pi_1 = \frac{0.5}{r}$$

$$r + \frac{0.25}{r} = 1 + 0.5^2 + \frac{1.25}{12} = 1.3542$$

Hence

$$r = 1.134 \qquad \pi_1 = 0.43 \qquad \pi_2 = 0$$

Substituting these values in (13.1.16), we now find that the optimal control adjustment is

$$x_{0t} = -10.68\varepsilon_t + 5.00\varepsilon_{t-1}$$

with $\sigma_\varepsilon = 1.065\sigma_a$. This may be compared with the scheme

$$x_t = -10.00\varepsilon_t + 5.00\varepsilon_{t-1}$$

with $\sigma_\varepsilon = 1.067\sigma_a$, which was actually used and which is optimal on the assumption that there is no added error. Clearly, in this case, the choice of optimal control equation is not much effected by the added noise.

Changes in the optimal adjustment induced by noise in the input. If, as before, we write

$$\sigma_u^2 = k^2\sigma_x^2 = k^2 \frac{(1-\theta)^2(1+\delta^2)}{g^2(1-\delta)^2}\sigma_o^2$$

then, from equations (13.1.17), we obtain

$$\pi_2 = -\frac{\delta\theta}{r} \qquad\qquad (13.1.18a)$$

$$\pi_1 = \frac{(\delta+\theta)(1+\delta\theta)}{r+\delta\theta} \qquad\qquad (13.1.18b)$$

$$r\left\{1 + \frac{(\delta+\theta)^2(1+\delta\theta)^2}{(r+\delta\theta)^2} + \left(\frac{\delta\theta}{r}\right)^2\right\} = 1 + (\delta+\theta)^2 + (\delta\theta)^2$$

$$+ k^2(1-\theta)^2(1+\delta^2)$$

$$(13.1.18c)$$

where, as before, $r = \sigma_\varepsilon^2/\sigma_a^2$. In practice, when relating r to k^2, it is easiest to solve (13.1.18c) in terms of k^2 for a series of suitably chosen values of r and then obtain the corresponding values of π_1 and π_2 by substituting in (13.1.18a) and (13.1.18b).

In this example then, a moderate amount of additional noise (due to severe rounding) did not greatly increase σ_ε^2, nor was the optimal scheme which took account of the added noise much better than the scheme which ignored it. This kind of conclusion applies for moderate added noise levels over wide ranges of the parameters. However, it does not apply when δ approaches unity (the system has a time constant which is large compared with the sampling interval) and for very large components of added noise in the loop. To shed some further light on these questions, we consider some examples. In each case we take $k^2 = R^2/12$, with $R = 1$, so that $\sigma_u/\sigma_x = 0.29$. Then this corresponds to adding noise u with standard deviation σ_u the same as that of the rounding error, with the rounding interval equal to σ_x (where σ_x is the standard deviation of x for the no noise case).

We now consider two cases.

Case 1: $g = 1$ $\theta = 0.5$ $\delta = 0.5$
Case 2: $g = 1$ $\theta = 0.5$ $\delta = 0.9$

The optimal control schemes corresponding to these parameters are summarized in Table 13.1. To obtain a fuller understanding of the results of Table 13.1, we notice that if instead of writing the control equation in terms

TABLE 13.1 Behavior of particular control schemes with added noise at the input

Case 1: $g = 1, \theta = 0.5, \delta = 0.5$

	Control equation for adjustment x_t	Variance at output σ_ε^2
Optimal scheme. No added noise	$-x_t = 1.00\varepsilon_t - 0.50\varepsilon_{t-1}$ $= 0.50(1 + 1.00\nabla)\varepsilon_t$	$1.000\sigma_a^2$
Effect on "no added noise" scheme of noise at input with $\sigma_u/\sigma_x = 0.29$	As above	$1.077\sigma_a^2$
Optimal scheme with added noise	$-x_t = 1.11\varepsilon_t - 0.53\varepsilon_{t-1}$ $= 0.58(1 + 0.93\nabla)\varepsilon_t$	$1.072\sigma_a^2$

Case 2: $g = 1, \theta = 0.5, \delta = 0.9$

	Control equation for adjustment x_t	Variance at output σ_ε^2
Optimal scheme. No added noise	$-x_t = 5.00\varepsilon_t - 4.50\varepsilon_{t-1}$ $= 0.50(1 + 9.00\nabla)\varepsilon_t$	$1.000\sigma_a^2$
Effect on "no added noise" scheme of noise at input with $\sigma_u/\sigma_x = 0.29$	As above	$1.697\sigma_a^2$
Optimal scheme with added noise	$-x_t = 7.25\varepsilon_t - 5.50\varepsilon_{t-1}$ $= 1.77(1 + 3.1\nabla)\varepsilon_t$	$1.278\sigma_a^2$

of the adjustment $x_t = X_{t+} - X_{t-1+}$, we write it in terms of the level X_{t+} at which the manipulated variable is maintained from time t to $t + 1$, then all of the schemes in Table 13.1 would be of the form

$$-X_{t+} = k_P\varepsilon_t + k_I S\varepsilon_t$$

calling for proportional-integral control action.
The adjustment equation is then

$$-x_t = k_I\left\{1 + \frac{k_P}{k_I}\nabla\right\}\varepsilon_t \qquad (13.1.19)$$

We see from the table that with $\delta = 0.5$ (the time constant of the system of moderate size compared with the sampling interval), the ratio of proportional to integral control $k_P/k_I = 1.0$ and that the introduction of the noise does not change the nature of the optimal control very much. However, when $\delta = 0.9$ (so that the time constant of the system is very large compared with the sampling interval), the ratio of proportional to integral control is large $(k_P/k_I = 9.0)$. The optimal scheme accommodates to the added noise by increasing the amount k_I and drastically cutting back on the ratio k_P/k_I of proportional to integral control. We can use the ratio

$$E = \frac{\text{Variance of optimal ``added noise'' scheme}}{\text{Variance of optimal ``no added noise'' scheme}} \times 100$$

to measure the efficiency of the optimal "no added noise" scheme in the noisy situation. Thus, for the schemes considered above,

$$E = 99.54\% \quad \text{for } \delta = 0.5$$

and

$$E = 75.31\% \quad \text{for } \delta = 0.9$$

For further illustration, Figures 13.2(a) and (b) show the changes in the efficiency factor E and the values of k_I and k_P/k_I as more and more noise is introduced into the loop for the two cases $(\theta = 0.5, \delta = 0.5)$ and $(\theta = 0.5, \delta = 0.9)$ previously considered. In inspecting these figures it should be borne in mind that:
(1) In industrial control applications, even a 10% error in the input might be rather unusual and certainly in the range $0 < 100\sigma_u/\sigma_x < 10$, even with $\delta = 0.9$, the efficiency of the scheme which assumes no added noise is quite good.
(2) If the parameters are estimated from operating data, the added noise will have already been taken account of in the basic scheme.
  Nevertheless, if the parameters had not been estimated in this way and if there was a great deal of added noise in the input which had been ignored in designing the scheme, then control could be very inefficient. For these examples, the optimal schemes for added noise involve a greater use of integral action and a smaller ratio of proportional to integral action.

13.1.3 Transference of the noise origin

It is instructive to consider the derivation of (13.1.15) in the previous section from a different point of view. We supposed there that, although the intended action was

$$x_{0t} = -\frac{(1 - \theta)(1 - \delta B)}{g(1 - \delta)} L(B)\varepsilon_t$$

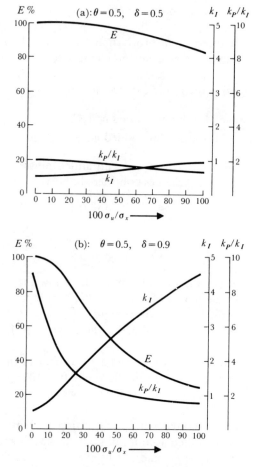

Fig. 13.2 Characteristics of control schemes for various proportions of added noise at the input

because of the error u_t, the action actually taken was

$$x_t = -\frac{(1 - \theta)(1 - \delta B)}{g(1 - \delta)}L(B)\varepsilon_t + u_t$$

The appropriate operator $L(B)$ to give optimal control was derived in these circumstances.

Now the effect of the additional noise u_t is that after being acted upon by the dynamics of the process, an additional component is produced at P in Figure 12.6. We could equally well regard this component *as part of* the noise source at P. In fact, the situation is *as if* the noise entering at P was such that

$$\nabla N'_t = \nabla N_t + \frac{g(1 - \delta)}{1 - \delta B} u_{t-1}$$

In that case

$$\nabla N'_t = (1 - \theta B)a_t + \frac{g(1 - \delta)}{1 - \delta B} u_{t-1}$$

$$(1 - \delta B)\nabla N'_t = (1 - \delta B)(1 - \theta B)a_t + g(1 - \delta)u_{t-1}$$

$$= (1 - \pi_1 B - \pi_2 B^2)b_t$$

where π_1, π_2, and b_t are defined precisely as before. We can now apply the general equation (12.2.8) for optimal adjustment

$$x_t = -\frac{L_1(B)L_3(B)}{L_2(B)L_4(B)}(1 - B)\varepsilon_t$$

with $\varepsilon_t = e_{t-f-1}(f + 1)$. The total noise at P is now represented by the process of order $(1, 1, 2)$

$$\nabla N'_{t+1} = \frac{(1 - \pi_1 B - \pi_2 B^2)}{1 - \delta B} b_{t+1}$$

so that

$$\hat{N}'_t(1) - \hat{N}'_{t-1}(1) = \left\{ \frac{(1 - \pi_1 B - \pi_2 B^2)}{1 - \delta B} - (1 - B) \right\} b_{t+1}$$

$$= \left\{ \frac{(1 + \delta - \pi_1) - (\delta + \pi_2)B}{1 - \delta B} \right\} b_t = \frac{L_3(B)(1 - B)}{L_4(B)} b_t$$

and $f = 0$, so that $\varepsilon_t = b_t$. Also $L_1(B)/L_2(B) = (1 - \delta B)/g(1 - \delta)$.

Thus, optimal adjustment is obtained, as before, by setting

$$x_{0t} = -\left\{ \frac{(1 + \delta - \pi_1) - (\delta + \pi_2)B}{g(1 - \delta)} \right\} \varepsilon_t$$

This device of transference of the noise origin can be applied more generally to obtain optimal control action with additional noise entering the system at any point.

Implications for estimation of transfer function-noise model. The fact that the noise origin can be transferred in the manner described above has a very important practical implication which has already been referred to. Provided the model parameters are estimated from actual operating records when closed loop control is being applied, the estimates will automatically take account of added noise, and a control scheme based on these parameters will be optimal for the actual situation in which added noise occurs. On the other hand, a scheme based on estimating the *actual* noise N_t which really originates at the point P in Figure 12.6 could fail to give optimal control. For example, consider again the simple scheme with added noise in the input x discussed in Section 13.1.3. In practice, to use such a scheme, we would need to know the form of the appropriate noise and transfer function models and to have estimates of the parameters. Specifically, if we were successful in characterizing the actual noise at P by, for example, performing an experiment in which the process was run with the manipulated variable X held fixed, we would be led to the noise model $\nabla N_t = (1 - \theta B)a_t$. If, under normal operating conditions, there was really a great deal of additional noise entering the system from observational errors in x which were not present under the conditions of the experiment, then the scheme ignoring this additional noise could be rather inefficient. On the other hand, if data collected during the actual running of a closed loop control scheme, necessarily not optimal, were used to estimate parameters, added white noise u_t in the adjustments x_t would lead to the noise at P being represented by

$$\nabla N_t' = (1 - \delta B)^{-1}(1 - \pi_1 B - \pi_2 B^2)b_t$$

and could lead to the design of an optimal scheme.

13.2 FEEDBACK CONTROL SCHEMES WHERE THE ADJUSTMENT VARIANCE IS RESTRICTED

The discrete feedback control schemes previously discussed were designed to produce minimum mean square error at the output. It was tacitly supposed that there was no restriction in the amount of adjustment x_t that could be tolerated to achieve this. It sometimes happens that we are not able to employ these schemes because the amount of variation which can be allowed in x_t is restricted by practical limitations. Therefore, we consider how a particular class of feedback control schemes would need to be modified if a constraint was placed on var $[x_t]$, supposing x_t to be a stationary process.

We consider again the important case in which the disturbance N_t at the output can be represented by a model

$$\nabla N_t = (1 - \theta B)a_t \qquad -1 < \theta < 1 \qquad (13.2.1)$$

of order $(0, 1, 1)$, while the output and input are related by a first-order transfer function model, such that

$$\frac{(1 - \delta B)}{1 - \delta} y_t = g x_{t-1} \qquad (13.2.2)$$

where $\nabla Y_t = y_t$. It will be recalled that $1 - \delta$ may be interpreted as the proportion of the total response to a step input that occurs in the first time interval. As we have seen in Section 12.2.3, the control equation yielding minimum output variance is

$$x_t = -\frac{\lambda}{g}\frac{(1 - \delta B)}{1 - \delta}\varepsilon_t \qquad (13.2.3)$$

where $\lambda = 1 - \theta$ and $\varepsilon_t = a_t$.

If δ is negligibly small, optimal control is obtained from $x_t = -(\lambda/g)\varepsilon_t$ and in that case, let us write var $[x_t] = (\lambda^2/g^2)\sigma_a^2 = k$. It follows that when δ is *not* negligible, var $[x_t] = k[(1 + \delta^2)/(1 - \delta)^2]$. If δ is near its upper limit of unity, var $[x_t]$ can become very large. For example, if $\delta = 0.9$ (so that only one tenth of the eventual change produced by a step input is experienced in the first interval), then var $[x_t] = 181\, k$. In fact, as δ approaches unity, the control action

$$x_t = -\frac{\lambda}{g(1 - \delta)}(\varepsilon_t - \delta\varepsilon_{t-1})$$

takes on more and more of an "alternating" character, the adjustment made at time t reversing a substantial portion of the adjustment made at time $t - 1$. Now a value of $\delta = 0.9$ corresponds to a time constant for the system of over nine sampling intervals (see for example Table 10.4). The occurrence of such a value would immediately raise the question as to whether the sampling interval was being taken too short; whether in fact, the inertia of the process was so large that little would be lost by less frequent surveillance.

Now, (see Section 13.3) the question of the choice of sampling interval must depend on the nature of the noise which infects the system. Because the properties of the noise usually reflect system inertia as well, in many cases it would be concluded that the sampling interval should be increased. Nevertheless, cases have occurred in practice [105] where a sensible sampling interval has been used and yet the excessive size of var $[x_t]$ has rendered a scheme which minimizes output variance impossible to operate.

Consider now the situation where the models for the noise and system dynamics are again given by (13.2.1) and (13.2.2), but some restriction of the input variance var $[x_t]$ is necessary. The unrestricted optimal scheme has the property that the errors in the output $\varepsilon_t, \varepsilon_{t-1}, \varepsilon_{t-2}, \ldots$ are the uncorrelated random variables $a_t, a_{t-1}, a_{t-2}, \ldots$ and the variance of the output σ_ε^2 has the minimum possible value σ_a^2. With the restricted schemes, the variance σ_ε^2 will necessarily be greater than σ_a^2, and the errors $\varepsilon_t, \varepsilon_{t-1}, \varepsilon_{t-2}, \ldots$ at the output will be correlated.

We shall pose our problem in the following form: Given that σ_ε^2 be allowed to increase to some value $\sigma_\varepsilon^2 = (1 + c)\sigma_a^2$, where c is a positive constant, to find that control scheme which produces the minimum value for var $[x_t]$.

13.2.1 Derivation of optimal adjustment

Let the optimal adjustment, *expressed in terms of the a_t's, be*

$$x_t = -\frac{1}{g} L(B) a_t \tag{13.2.4}$$

where

$$L(B) = l_0 + l_1 B + l_2 B^2 + \cdots$$

Then, referring to Figure 13.3, we see that the error ε_t at the output is given by

$$\varepsilon_t = a_t + \left\{ \lambda - \frac{L(B)(1 - \delta))}{1 - \delta B} \right\} S a_{t-1} \tag{13.2.5}$$

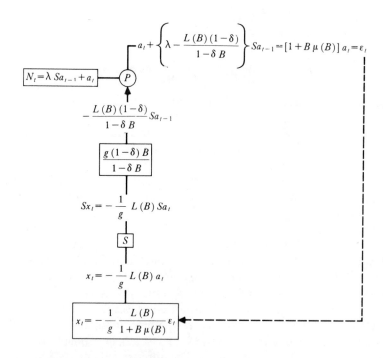

FIG. 13.3 A feedback control scheme for first-order transfer function model and (0, 1, 1) noise model when the input variance is constrained

The coefficient of a_t in this expression is unity, so that we can write

$$\varepsilon_t = \{1 + B\mu(B)\} a_t \tag{13.2.6}$$

where

$$\mu(B) = \mu_1 + \mu_2 B + \mu_3 B^2 + \cdots$$

Furthermore, in practice, control would need to be exerted in terms of the observed output errors ε_t rather than in terms of the a_t's, so that the control equation actually used would be of the form

$$x_t = -\frac{1}{g}\frac{L(B)}{1 + B\mu(B)}\varepsilon_t \tag{13.2.7}$$

Equating (13.2.5) and (13.2.6), we obtain

$$(1 - \delta)L(B) = \{\lambda - (1 - B)\mu(B)\}(1 - \delta B) \tag{13.2.8}$$

Since δ, g, and σ_a^2 are constants, we can proceed conveniently by finding an unrestricted minimum of

$$\frac{(1 - \delta)^2 g^2 V[x_t]}{\sigma_a^2} + v\left\{\frac{V[\varepsilon_t]}{\sigma_a^2} - (1 + c)\right\} \tag{13.2.9}$$

where, for example,

$$V[x_t] = \mathrm{var}\,[x_t]$$

Equivalently, using covariance generating functions, we require an unrestricted minimum of the coefficient of $B^0 = 1$ in the expression

$$G(B) = (1 - \delta)^2 L(B)L(F) + v\{1 + B\mu(B)\}\{1 + F\mu(F)\}$$

that is, in

$$G(B) = (1 - \delta B)(1 - \delta F)\{\lambda - (1 - B)\mu(B)\}\{\lambda - (1 - F)\mu(F)\}$$
$$+ v\{1 + B\mu(B)\}\{1 + F\mu(F)\} \tag{13.2.10}$$

where $F = B^{-1}$. This we can obtain by differentiating $G(B)$ with respect to each $\mu_i\ (i = 1, 2, \ldots)$, selecting the coefficients of $B^0 = 1$ in the resulting expression, equating them to zero and solving the resulting equations. Thus

$$\frac{\partial}{\partial \mu_i}G(B) = (1 - \delta B)(1 - \delta F)[-\lambda\{(1 - B)B^{i-1} + (1 - F)F^{i-1}\}$$
$$+ (1 - B)(1 - F)\{\mu(B)F^{i-1} + \mu(F)B^{i-1}\}] \tag{13.2.11}$$
$$+ v[B^i + F^i + B^{i-1}\mu(F) + F^{i-1}\mu(B)]$$

After selecting the coefficients of $B^0 = 1$ for $i = 1, 2, 3, \ldots$ and setting each of these equal to zero, we obtain the following equations

$$(i = 1): -\lambda(1 + \delta + \delta^2) + 2(1 + \delta + \delta^2)\mu_1 - (1 + \delta)^2\mu_2$$
$$+ \delta\mu_3 + v\mu_1 = 0 \tag{13.2.12}$$
$$(i = 2): \lambda\delta - (1 + \delta)^2\mu_1 + 2(1 + \delta + \delta^2)\mu_2 - (1 + \delta)^2\mu_3$$
$$+ \delta\mu_4 + v\mu_2 = 0 \tag{13.2.13}$$

$$(i > 2): \{\delta B^2 - (1 + \delta)^2 B + 2(1 + \delta + \delta^2) - (1 + \delta)^2 F$$
$$+ \delta F^2 + v\}\mu_i = 0 \qquad (13.2.14)$$

The case where δ is negligible. Consider first the simpler case where δ is negligibly small and can be set equal to zero. Then the above equations can be written

$$(i = 1): \qquad -(\lambda - \mu_1) + (\mu_1 - \mu_2) + v\mu_1 = 0 \qquad (13.2.15)$$

$$(i > 1): \qquad \{B - (2 + v) + F\}\mu_j = 0 \qquad (13.2.16)$$

These difference equations have a solution of the form

$$\mu_i = A_1 \kappa_1^i + A_2 \kappa_2^i$$

where κ_1 and κ_2 are the roots of the characteristic equation

$$B^2 - (2 + v)B + 1 = 0 \qquad (13.2.17)$$

that is, of

$$B + B^{-1} = 2 + v$$

Evidently, if κ is a root, then so is κ^{-1}. Thus the solution is of the form $\mu_i = A_1 \kappa^i + A_2 \kappa^{-i}$. Now if κ has modulus less than or equal to 1, then κ^{-1} has modulus greater than or equal to 1, and since $\varepsilon_t = \{1 + B\mu(B)\}a_t$ must have finite variance, A_2 must be zero with $|\kappa| < 1$. By substituting the solution $\mu_i = A_1 \kappa^i$ in (13.2.15), we find that $A_1 = \lambda$.

Finally, then, $\mu_i = \lambda \kappa^i$ and since μ_i and λ must be real, then so must the root κ. Hence

$$\mu(B) = \frac{\lambda \kappa}{1 - \kappa B} \qquad\qquad 0 < \kappa < 1 \quad (13.2.18)$$

$$1 + B\mu(B) = 1 + \frac{\lambda \kappa B}{1 - \kappa B} = \frac{1 - \theta \kappa B}{1 - \kappa B} \qquad (13.2.19)$$

where $\theta = 1 - \lambda$. Thus

$$\varepsilon_t = \frac{1 - \theta \kappa B}{1 - \kappa B} a_t$$

so that

$$\frac{V[\varepsilon_t]}{\sigma_a^2} = 1 + \frac{\lambda^2 \kappa^2}{1 - \kappa^2} \qquad (13.2.20)$$

Also, using (13.2.8) with $\delta = 0$,

$$L(B) = \lambda - \frac{(1 - B)\lambda\kappa}{1 - \kappa B} = \frac{\lambda(1 - \kappa)}{1 - \kappa B} \tag{13.2.21}$$

Thus

$$x_t = -\frac{\lambda}{g} \frac{(1 - \kappa)}{1 - \kappa B} a_t$$

and

$$\frac{V[x_t]}{\sigma_a^2} = \frac{\lambda^2}{g^2} \frac{(1 - \kappa)^2}{1 - \kappa^2} = \frac{\lambda^2}{g^2} \frac{(1 - \kappa)}{(1 + \kappa)} \tag{13.2.22}$$

Using (13.2.7) with (13.2.19) and (13.2.21), we now find that the optimal control action, in terms of the observed output error ε_t, is

$$x_t = -\frac{1}{g} \frac{\lambda(1 - \kappa)}{1 - \theta\kappa B} \varepsilon_t$$

that is

$$x_t = (1 - \lambda)\kappa x_{t-1} - \frac{1}{g}\lambda(1 - \kappa)\varepsilon_t \tag{13.2.23}$$

Note that the constrained control equation differs from the unconstrained one in two respects:
(1) A new factor $(1 - \lambda)\kappa x_{t-1}$ is introduced, thus making present action depend partly on previous action.
(2) The constant determining the amount of integral control is reduced by a factor $1 - \kappa$.
We have supposed that the output variance is allowed to increase to some value $\sigma_a^2(1 + c)$. It follows from (13.2.20) that

$$c = \frac{\lambda^2\kappa^2}{1 - \kappa^2}$$

that is

$$\kappa = \sqrt{\frac{c}{\lambda^2 + c}}$$

where the positive square root is to be taken. It is convenient to write $Q = c/\lambda^2$. Then $Q = \kappa^2/(1 - \kappa^2)$ and $\kappa^2 = Q/(1 + Q)$ and the output variance becomes $\sigma_a^2(1 + \lambda^2 Q)$.

In summary, supposing we are prepared to tolerate an increase in variance in the output to some value $\sigma_a^2(1 + \lambda^2 Q)$, then

(1) We compute $\kappa = \sqrt{\dfrac{Q}{1 + Q}}$

(2) Optimal control will be achieved by taking action

$$x_t = (1 - \lambda)\kappa x_{t-1} - \frac{1}{g}\lambda(1 - \kappa)\varepsilon_t$$

(3) The variance of the input will be reduced to

$$V[x_t] = \frac{\lambda^2}{g^2}\frac{1 - \kappa}{1 + \kappa}\sigma_a^2$$

that is, it will reduce to a value that is $W\%$ of that for the unconstrained scheme, where

$$W = 100\left(\frac{1 - \kappa}{1 + \kappa}\right).$$

Table 13.2 shows κ and W for values of Q between 0.1 and 1.0.

TABLE 13.2   Values of parameters for a simple constrained control scheme

$\dfrac{c}{\lambda^2} = Q$	0.10	0.20	0.30	0.40	0.50	0.60	0.70	0.80	0.90	1.00
κ	0.302	0.408	0.480	0.535	0.577	0.612	0.641	0.667	0.688	0.707
W	53.7	42.0	35.1	30.3	26.8	24.0	21.9	20.0	18.5	17.2

For illustration, suppose $\lambda = 0.4$. Then the optimal unconstrained scheme will employ the control action

$$x_t = -\frac{0.4}{g}\varepsilon_t$$

with $\varepsilon_t = a_t$. The variance of x_t would be $V[x_t] = (\sigma_a^2/g^2)0.16$. Suppose it was desired to reduce this by a factor of four to the value $(\sigma_a^2/g^2)0.04$. Thus, we require W to be 25%. Table 13.2 shows that a reduction of the input variance to 24% of its unconstrained value is possible with $Q = 0.60$ and $\kappa = 0.612$. If we use this scheme, the output variance will be

$$\sigma_\varepsilon^2 = \sigma_a^2\{1 + 0.16 \times 0.60\} = 1.10\sigma_a^2$$

Thus, by the use of the control action

$$x_t = 0.37x_{t-1} - \frac{1}{g}0.16\varepsilon_t$$

instead of

$$x_t = -\frac{0.4}{g}\varepsilon_t$$

the variance of the input is reduced to about a quarter of its previous value, while the variance of the output is increased by only 10%.

Case where δ is not negligible. Consider now the more general situation where δ is not negligible and the system dynamics must be taken account of. The difference equation (13.2.14) is of the form

$$(\alpha B^{-2} + \beta B^{-1} + \gamma + \beta B + \alpha B^2)\mu_i = 0$$

and if κ is a root of the characteristic equation, then so is κ^{-1}. Suppose the roots are $\kappa_1, \kappa_2, \kappa_1^{-1}, \kappa_2^{-1}$ and suppose that κ_1 and κ_2 are a pair of roots with modulus < 1. Then in the solution

$$\mu_i = A_1\kappa_1^i + A_2\kappa_2^i + A_3\kappa_1^{-i} + A_4\kappa_2^{-i}$$

A_3 and A_4 must be zero, because ε_t is required to have a finite variance. Hence the solution is of the form

$$\mu_i = A_1\kappa_1^i + A_2\kappa_2^i \qquad |\kappa_1| < 1 \qquad |\kappa_2| < 1$$

The A's satisfying the initial conditions, defined by (13.2.12) and (13.2.13), are obtained by substitution to give

$$A_1 = \frac{\lambda\kappa_1(1 - \kappa_2)}{\kappa_1 - \kappa_2} \qquad A_2 = -\frac{\lambda\kappa_2(1 - \kappa_1)}{\kappa_1 - \kappa_2}$$

If we write $k_0 = \kappa_1 + \kappa_2 - \kappa_1\kappa_2$, $k_1 = \kappa_1\kappa_2$, then

$$\mu(B) = \lambda\left\{\frac{k_0 - k_1 B}{1 - (k_0 + k_1)B + k_1 B^2}\right\} \qquad (13.2.24)$$

and

$$1 + B\mu(B) = \frac{1 - k_1 B - (1 - \lambda)(k_0 B - k_1 B^2)}{1 - (k_0 + k_1)B + k_1 B^2} \qquad (13.2.25)$$

Now substituting (13.2.24) in (13.2.8),

$$L(B) = \frac{\lambda(1 - \delta B)(1 - k_0)}{(1 - \delta)[1 - (k_0 + k_1)B + k_1 B^2]} \qquad (13.2.26)$$

and

$$\frac{L(B)}{1 + B\mu(B)} = \frac{\lambda(1 - \delta B)(1 - k_0)}{(1 - \delta)\{1 - k_1 B - (1 - \lambda)(k_0 B - k_1 B^2)\}}$$

Therefore, using (13.2.7) we find that the optimal control action in terms of the error ε_t is

$$x_t = -\frac{\lambda}{g}\frac{(1 - \delta B)(1 - k_0)}{(1 - \delta)\{1 - k_1 B - (1 - \lambda)(k_0 B - k_1 B^2)\}}\varepsilon_t \quad (13.2.27)$$

or

$$x_t = (k_1 + (1 - \lambda)k_0)x_{t-1} - (1 - \lambda)k_1 x_{t-2} - \frac{\lambda(1 - k_0)(1 - \delta B)}{g(1 - \delta)}\varepsilon_t$$

$$(13.2.28)$$

Thus the modified control scheme makes x_t depend on both x_{t-1} and x_{t-2} (only on x_{t-1} if $\lambda = 1$) and reduces the standard integral and proportional action by a factor $1 - k_0$.

The variances of output and input. The actual variances for the output and input are readily found since

$$\varepsilon_t = a_t + \lambda\frac{(k_0 - k_1 B)}{1 - (k_0 + k_1)B + k_1 B^2}a_{t-1}$$

The second term on the right defines a mixed autoregressive-moving average process of order $(2, 0, 1)$, the variance for which is readily obtained to give

$$\frac{V[\varepsilon_t]}{\sigma_a^2} = 1 + \lambda^2\left\{\frac{(k_0 + k_1)^2(1 - k_1) - 2k_1(k_0 - k_1^2)}{(1 - k_1)\{(1 + k_1)^2 - (k_0 + k_1)^2\}}\right\} = 1 + \lambda^2 Q$$

$$(13.2.29)$$

Also,

$$\frac{V[x_t]}{\sigma_a^2} = \frac{\lambda^2}{g^2(1 - \delta)^2}\frac{(1 - k_0)\{(1 + \delta^2)(1 + k_1) - 2\delta(k_0 + k_1)\}}{(1 + k_0 + 2k_1)(1 - k_1)} \quad (13.2.30)$$

Computation of k_0 and k_1. Returning to the difference equations (13.2.14), the characteristic equation may be written

$$B^4 - MB^3 + NB^2 - MB + 1 = 0$$

where $M = (1 + \delta)^2/\delta$ and $N = \{(1 + \delta)^2 + (1 + \delta^2) + v_1^2\}/\delta$. It may also be written in the form

$$(B^2 - TB + P)(B^2 - P^{-1}TB + P^{-1}) = 0$$

where

$$T = \kappa_1 + \kappa_2 \text{ and } P = \kappa_1\kappa_2$$

Equating coefficients of B,

$$T + P^{-1}T = M \quad \text{that is} \quad T = \frac{PM}{1 + P}$$

$$P + P^{-1} + P^{-1}T^2 = N$$

Thus $P + P^{-1} + PM^2/(1 + P)^2 = N$, that is

$$(P + 2 + P^{-1})(P + P^{-1}) + M^2 = N(P + 2 + P^{-1})$$

$$(P + P^{-1})^2 + (2 - N)(P + P^{-1}) + M^2 - 2N = 0$$

For suitable values of v, this quadratic equation will have two real roots

$$u_1 = \kappa_1\kappa_2 + \kappa_1^{-1}\kappa_2^{-1} \qquad u_2 = \kappa_1\kappa_2^{-1} + \kappa_1^{-1}\kappa_2$$

the root u_1 being the larger. The required quantity P is now the smaller root of the quadratic equation

$$P^2 - u_1P + 1 = 0$$

and T is given by

$$T = \{P(u_2 + 2)\}^{1/2}$$

Table of optimal values for constrained schemes—Construction of the table. Table 13.3 is provided to facilitate the selection of an optimal control scheme. The tabled values were obtained as follows for each chosen value of the parameter δ in the transfer function model:

(1) Compute $M = \dfrac{(1 + \delta)^2}{\delta}$ and $N = \dfrac{(1 + \delta)^2 + (1 + \delta^2) + v}{\delta}$ for a series of

values of v chosen to provide a suitable range for Q.

(2) Compute $\quad u_1 = \dfrac{1}{2}(N - 2) + \left\{ \left(\dfrac{N - 2}{2}\right)^2 + 2N - M^2 \right\}^{1/2}$

and $\qquad u_2 = \dfrac{1}{2}(N - 2) - \left\{ \left(\dfrac{N - 2}{2}\right)^2 + 2N - M^2 \right\}^{1/2}$

(3) Compute $k_1 = P = \dfrac{1}{2}u_1 - \left\{ \left(\dfrac{1}{2}u_1\right)^2 - 1 \right\}^{1/2}$

and $\qquad k_0 = T - P = \{k_1(u_2 + 2)\}^{1/2} - k_1$

(4) Compute $\quad Q = \dfrac{(k_0 + k_1)^2(1 - k_1) - 2k_1(k_0 - k_1^2)}{(1 - k_1)\{(1 + k_1)^2 - (k_0 + k_1)^2\}}$

(5) Compute $\quad W = \dfrac{(1 - k_0)\{(1 + \delta^2)(1 + k_1) - 2\delta(k_0 + k_1)\}}{(1 + k_0 + 2k_1)(1 - k_1)(1 + \delta^2)}$

TABLE 13.3   Table to facilitate the calculation of optimal constrained control schemes

δ		20	40	60	80	100
				$100Q$		
0.9	$100W$	21.7	11.3	6.7	4.5	3.1
	k_0	0.44	0.585	0.68	0.74	0.78
	k_1	0.18	0.27	0.34	0.39	0.44
0.8	$100W$	22.0	11.7	7.2	4.8	3.4
	k_0	0.44	0.585	0.68	0.74	0.78
	k_1	0.18	0.27	0.33	0.38	0.43
0.7	$100W$	22.7	12.4	8.0	5.6	4.1
	k_0	0.44	0.585	0.68	0.74	0.78
	k_1	0.17	0.25	0.32	0.36	0.40
0.6	$100W$	24.1	13.6	9.0	6.6	5.0
	k_0	0.44	0.58	0.67	0.73	0.78
	k_1	0.16	0.24	0.29	0.33	0.365
0.5	$100W$	26.5	15.5	10.5	7.9	6.2
	k_0	0.43	0.58	0.67	0.72	0.77
	k_1	0.15	0.21	0.26	0.29	0.32
0.4	$100W$	28.5	17.7	12.7	9.8	7.9
	k_0	0.43	0.57	0.66	0.72	0.76
	k_1	0.13	0.18	0.22	0.245	0.265
0.3	$100W$	31.5	20.5	15.2	12.0	9.9
	k_0	0.43	0.57	0.65	0.71	0.75
	k_1	0.105	0.145	0.17	0.19	0.20
0.2	$100W$	34.8	23.6	18.0	14.5	12.2
	k_0	0.42	0.56	0.64	0.69	0.73
	k_1	0.07	0.10	0.12	0.13	0.14
0.1	$100W$	38.2	26.7	21.0	17.3	14.6
	k_0	0.42	0.55	0.63	0.68	0.72
	k_1	0.04	0.05	0.06	0.065	0.07

(6) Interpolate among the W, k_0, k_1 values at convenient values of Q.

Use of the table.   Table 13.3 may be used as follows. The value of δ is entered in the vertical margin. Using the fact that $V[\varepsilon_t] = (1 + \lambda^2 Q)\sigma_a^2$, the percentage increase in output variance is $100Q\lambda^2$. A suitable value of Q is entered in the horizontal margin. The entries in the table are then

(a) $100W$, the percentage reduction in the variance of x_t;
(b) k_0;
(c) k_1.

For illustration, suppose $\lambda = 0.6$, $\delta = 0.5$, $g = 1$. The optimal uncon-strained control equation is then

$$x_t = -1.2(1 - 0.5B)\varepsilon_t = -1.2(1 - 0.5B)a_t$$

and var $[x_t] = 1.80\sigma_a^2$. Suppose that this amount of variation in the input variable produces difficulties in process operation and it is desired to cut var $[x_t]$ to about $0.50\sigma_a^2$, that is, to about 28% of the value for the uncon-strained scheme. Inspection of Table 13.3 in the column labelled $\delta = 0.5$ shows that a reduction to 26.5% can be achieved by using a control scheme with constants $k_0 = 0.43$, $k_1 = 0.15$, that is, by employing the control equation (13.2.28) to give

$$x_t = 0.32x_{t-1} - 0.06x_{t-2} - (0.57 \times 1.2)(1 - 0.5B)\varepsilon_t$$

This solution corresponds to a value $Q = 0.20$. Therefore, the variance at the output will be increased by a factor of $1 + \lambda^2 Q = 1 + 0.6^2(0.2) = 1.072$, that is by about 7%.

13.2.2 A constrained scheme for the viscosity/gas rate example

In the second example in Section 12.2.3, we considered a chemical process in which viscosity was controlled to a target value of 92 by varying the gas rate. For the pilot control scheme, $\lambda = 1.0$, $(\theta = 0)$, $\delta = 0.5$, so that the optimal control action was

$$x_t = -\frac{1}{g}(2\varepsilon_t - \varepsilon_{t-1})$$

with $\varepsilon_t = a_t$. We later showed (Section 12.4.3) that this model was somewhat in error. However, for the purposes of the present illustration we suppose it correct. The variance of x_t would then be

$$\sigma_x^2 = \frac{1}{g^2} 5\sigma_a^2$$

that is

$$g\frac{\sigma_x}{\sigma_a} = \sqrt{5} = 2.24$$

Figure 13.4 shows the reduction in $g(\sigma_x/\sigma_a)$ possible for various values of $\sigma_\varepsilon/\sigma_a$, together with the accompanying optimal control parameters. We see, in particular, that for a 10% increase in the standard deviation of the output, the standard deviation of the input can be halved.

Figure 13.5 illustrates this point further. A set of twenty-four successive observations showing the values of inputs (gas rate) and outputs (viscosity) are reproduced in the left-hand diagrams as they were actually recorded using the optimal unrestricted scheme $x_t = -(1/g)(2\varepsilon_t - \varepsilon_{t-1})$ with $g = 0.2$. Also shown is the reconstructed noise. Supposing the scheme to be initially on target, this reconstructed noise is the (computed) drift away from target that would have occurred if no control action had been taken. The right-hand diagrams show the calculated behavior that would have occurred with the same noise if the control equation

$$x_t = 0.15x_{t-1} - \frac{0.55}{g}(2\varepsilon_t - \varepsilon_{t-1})$$

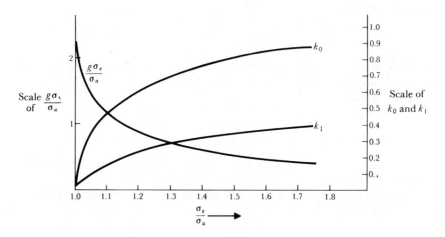

FIG. 13.4   Control of viscosity by varying gas rate. Values of W, $g\sigma_x/\sigma_a$, k_0, k_1 for a range of values of $\sigma_\varepsilon/\sigma_a$

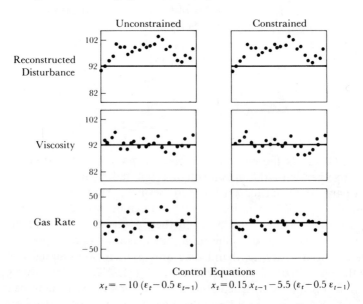

FIG. 13.5   Behavior of unconstrained and constrained control schemes for viscosity/gas rate example

had been used. This is the scheme which provides smallest input standard deviation when the output standard deviation is allowed to increase by 10%. As expected, very little increase is observed in the output standard deviation, but σ_x is roughly halved.

Constrained schemes with delayed dynamics. A general analysis for constrained schemes of the sort mentioned above has recently been presented by G. T. Wilson [107] using the Wiener–Hopf argument [106] as described in Whittle [49]. Using Wilson's results, J. F. MacGregor has made a preliminary investigation, partially reported below, of the interesting situation where there is delay (or dead time) in the feedback loop. For illustration, we again consider the important case for which the unconstrained solution is given in Example 3 in Section 12.2.3.

The model assumes delayed first order dynamics between the output Y_t and the input X_t, with a disturbance N_t which can be represented by a process of order $(0, 1, 1)$. Thus,

$$(1 - \delta B)Y_t = g(1 - \delta)\{(1 - v) + vB\}X_{t-f-1+}$$

$$\nabla N_t = (1 - \theta B)a_t$$

The optimal constrained controller is of the form, when $f \geqslant 1$,

$$-x_t = c_0 x_{t-1} + c_1 x_{t-2} + \cdots + c_f x_{t-f-1} + c(1 - \delta B)\varepsilon_t$$

of which the optimal unconstrained controller (12.2.13) is a special case.

In particular, suppose that $f = 1$. The dynamic model will then parallel a continuous system with dead time equal to one whole period plus some fraction of a period. Then the optimal constrained controller is

$$-x_t = c_0 x_{t-1} + c_1 x_{t-2} + c(1 - \delta B)\varepsilon_t$$

where

$$c_0 = \gamma_0^{-1}\{\gamma_1 + (1 - \theta)\gamma_0\}$$

$$c_1 = \gamma_0^{-1}\{(1 - \theta)(1 - \delta)v + \theta\gamma_2\}$$

$$c = \gamma_0^{-1}g^{-1}(1 - \theta)$$

and

$$\gamma_0 + \gamma_1 + \gamma_2 = (1 - \delta)$$

$$\gamma_0\gamma_1 + \gamma_1\gamma_2 = v(1 - v)(1 - \delta)^2 - (1 + \delta)^2 G$$

$$\gamma_0\gamma_2 = \delta G$$

where G is an undetermined multiplier.

Also, the variance of x_t and ε_t are given by

$$\frac{V(x_t)}{\sigma_a^2} = \frac{(1 - \theta)^2\{(\gamma_0 + \gamma_2)(1 + \delta^2) + 2\gamma_1\delta\}}{(\gamma_0 - \gamma_2)\{(\gamma_0 + \gamma_2)^2 - \gamma_1^2\}}$$

$$\frac{V(\varepsilon_t)}{\sigma_a^2} = 1 + \frac{(1 - \theta)^2\{(\gamma_0 + \gamma_2)[\gamma_0^2 + (v - v\delta - \gamma_2)^2] - 2\gamma_0\gamma_1(v - v\delta - \gamma_2)\}}{(\gamma_0 - \gamma_2)\{(\gamma_0 + \gamma_2)^2 - \gamma_1^2\}}$$

The case in which the system is delayed is of special interest, because unconstrained minimum mean square error schemes often call for impractically large alternating adjustments. To illustrate the dramatic reductions possible in the adjustment variance $V(x_t)$ when schemes of this kind are used, calculations were made for the case $g = 1, f = 1, v = 0.4, \theta = 0.6$ (a) with $\delta = 0.5$ and (b) with $\delta = 0.9$.

The characteristics of the unconstrained schemes are as follows. For $\delta = 0.5$,

$$-x_t = 1.07x_{t-1} + 0.27x_{t-2} + 1.33(\varepsilon_t - 0.5\varepsilon_{t-1})$$

$$\frac{V(\varepsilon_t)}{\sigma_a^2} = 1.16 \qquad \frac{V(x_t)}{\sigma_a^2} = 6.13$$

and for $\delta = 0.9$,

$$-x_t = 1.07x_{t-1} + 0.27x_{t-2} + 1.33(\varepsilon_t - 0.9\varepsilon_{t-1})$$

$$\frac{V(\varepsilon_t)}{\sigma_a^2} = 1.16 \qquad \frac{V(x_t)}{\sigma_a^2} = 9.63$$

Various optimal constrained schemes are shown in Table 13.4.

13.3 CHOICE OF THE SAMPLING INTERVAL

By comparison with continuous systems, discrete systems of control, such as are discussed here, can be very efficient provided that the sampling interval is suitably chosen. Roughly speaking, we want the interval to be such that not too much change can occur during the sampling interval. Usually, the behavior of the disturbance which has to pass through all or part of the system reflects the inertia or dynamic properties of the system, so that the sampling interval will often be chosen tacitly or explicitly to be proportional to the time constant or constants of the system. In chemical processes involving reaction and mixing of liquids, where time constants of 2 or 3 hours are common, rather infrequent sampling, say at hourly intervals and possibly with operator surveillance and manual adjustment, will be sufficient. By contrast, where reactions between gases are involved, a suitable sampling interval may be measured in seconds and automatic monitoring and adjustment may be essential.

TABLE 13.4 Comparison of constrained and unconstrained feedback schemes

δ	% increase in $V(\varepsilon_t)/\sigma_a^2$	% decrease in $V(x_t)/\sigma_a^2$	Controller Parameters		
			c_0	c_1	c
			(unconstrained scheme)		
	0	0	1.07	0.27	1.33
0.5	0.2	49.1	0.88	0.24	1.19
	0.7	69.8	0.69	0.23	1.06
	2.8	90.0	0.24	0.20	0.74
			(unconstrained scheme)		
	0	0	1.07	0.27	1.33
0.9	0.1	40.5	0.93	0.25	1.23
	0.7	71.7	0.68	0.23	1.05
	2.4	89.4	0.29	0.21	0.78

It will be noticed that, for example, with $\delta = 0.9$, an 89% reduction in var(x_t) to a value $1.02\sigma_a^2$ is obtained with an accompanying increase of only 2.4% in var(ε_t).

In some cases, experimentation may be needed to arrive at a satisfactory sampling interval, and in others rather simple calculations will show how the choice of sampling interval will affect the degree of control that is possible.

13.3.1 An illustration of the effect of reducing sampling frequency

To illustrate the kind of calculation that is helpful, suppose again that we have a simple system in which, using a particular sampling interval, the noise is represented by a $(0, 1, 1)$ process $\nabla N_t = (1 - \theta B)a_t$ and the transfer function model by the first-order system $(1 - \delta B)y_t = g(1 - \delta)x_{t-1}$.

In this case, if we employ the optimal adjustment

$$x_t = -\frac{(1 - \theta)}{g(1 - \delta)}(1 - \delta B)\varepsilon_t \qquad (13.3.1)$$

then the deviation from target is $\varepsilon_t = a_t$ and has variance $\sigma_a^2 = \sigma_1^2$ say.

In practice, the question has often arisen: How much worse off would we be if we took samples less frequently? To answer this question, we must consider the effect of sampling the stochastic process involved.

13.3.2 Sampling an IMA (0, 1, 1) process

Suppose that, with observations being made at some "unit" interval, we have a noise model

$$\nabla N_t = (1 - \theta_1 B) a_t$$

with var $[a_t] = \sigma_a^2 = \sigma_1^2$, where the subscript 1 is used in this context to denote the choice of sampling interval. Then, for the differences ∇N_t, the autocovariances γ_k are given by

$$\gamma_0 = (1 + \theta_1^2) \sigma_1^2$$
$$\gamma_1 = -\theta_1 \sigma_1^2 \qquad\qquad (13.3.2)$$
$$\gamma_j = 0 \qquad j \geqslant 2$$

Writing $\zeta = (\gamma_0 + 2\gamma_1)/\gamma_1$, we obtain

$$\zeta = -(1 - \theta_1)^2/\theta_1$$

so that, given γ_0 and γ_1, the parameter λ of the IMA process may be obtained by solving the quadratic equation

$$(1 - \theta_1)^2 - \zeta(1 - \theta_1) + \zeta = 0$$

selecting that root for which $-1 < \theta_1 < 1$. Also

$$\sigma_1^2 = -\gamma_1/\theta_1 \qquad\qquad (13.3.3)$$

Suppose now that the process N_t is observed at intervals of h units (where h is a positive integer) and the resulting process is denoted by M_t. Then

$$\nabla M_t = N_t - N_{t-h} = (a_t + a_{t-1} + \cdots + a_{t-h+1})$$
$$- \theta_1(a_{t-1} + a_{t-2} + \cdots + a_{t-h})$$
$$\nabla M_{t-h} = N_{t-h} - N_{t-2h} = (a_{t-h} + a_{t-h-1} + \cdots + a_{t-2h+1})$$
$$- \theta_1(a_{t-h-1} + \cdots + a_{t-2h})$$

and so on. Then, for the differences ∇M_t, the autocovariances $\gamma_k(h)$ are

$$\gamma_0(h) = \{(1 + \theta_1^2) + (h - 1)(1 - \theta_1)^2\}\sigma_1^2$$
$$\gamma_1(h) = -\theta_1 \sigma_1^2 \qquad\qquad (13.3.4)$$
$$\gamma_j(h) = 0 \qquad j \geqslant 2$$

It follows that the process M_t is also an IMA process of order (0, 1, 1),

$$\nabla M_t = (1 - \theta_h B) e_t$$

where e_t is a white noise process with variance σ_h^2. Now

$$\frac{\gamma_0(h) + 2\gamma_1(h)}{\gamma_1(h)} = -\frac{h(1 - \theta_1)^2}{\theta_1}$$

so that

$$\frac{h(1 - \theta_1)^2}{\theta_1} = \frac{(1 - \theta_h)^2}{\theta_h} \tag{13.3.5}$$

Also, since $\gamma_1(h) = -\theta_h\sigma_h^2 = -\theta_1\sigma_1^2$, it follows that

$$\frac{\sigma_h^2}{\sigma_1^2} = \frac{\theta_1}{\theta_h} \tag{13.3.6}$$

Therefore, we have shown that the sampling of an IMA process of order $(0, 1, 1)$ at interval h produces another IMA process of order $(0, 1, 1)$. From (13.3.5), we can obtain the value of the parameter θ_h for the sampled process and from (13.3.6) we can obtain the variance σ_h^2 of that process in terms of the parameters θ_1 and σ_1^2 of the original process.

In Figure 13.6, θ_h is plotted against log h, a scale of h being appended.

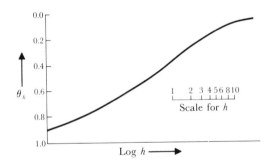

FIG. 13.6   Sampling of IMA $(0, 1, 1)$ process. Parameter θ_h plotted against log h

The graph enables one to find the effect of increasing the sampling interval of a $(0, 1, 1)$ process by any given multiple. For illustration, suppose we have a process for which $\theta_1 = 0.5$ and $\sigma_1^2 = 1$. Let us use the graph to find the values of the corresponding parameters θ_2, θ_4, σ_2^2, σ_4^2 when the sampling interval is (a) doubled, (b) quadrupled. Marking on the edge of a piece of paper the points $h = 1$, $h = 2$, $h = 4$ from the scale on the graph, we set the paper horizontally so that $h = 1$ corresponds to the point on the curve for which $\theta_1 = 0.5$. We then read off the ordinates for θ_2 and θ_4 corresponding to $h = 2$ and $h = 4$. We find

$$\theta_1 = 0.5 \qquad \theta_2 = 0.38 \qquad \theta_4 = 0.27$$

Using (13.3.6), the variances are in inverse proportion to the values of θ, so that

$$\sigma_1^2 = 1.00 \qquad \sigma_2^2 = 1.32 \qquad \sigma_4^2 = 2.17$$

Suppose now that, for the original scheme with unit interval, the dynamic constant was δ_1 (again we shall use the subscript to denote the sampling interval). Then, since in real time the same fixed time constant $T = -h/\ln \delta$ applies to all the schemes, we have

$$\delta_2 = \delta_1^2 \qquad \delta_4 = \delta_1^4$$

The scheme giving minimum mean square error for a *particular* sampling interval h would be

$$x_t(h) = -\frac{(1 - \theta_h)}{g(1 - \delta_1^h)}(1 - \delta_1^h B)\varepsilon_t(h)$$

or

$$x_t(h) = -\frac{(1 - \theta_h)}{g}\left\{1 + \frac{\delta_1^h}{1 - \delta_1^h}\nabla\right\}\varepsilon_t(h) \qquad (13.3.7)$$

Suppose, for example, with $\theta_1 = 0.5$ as above, $\delta_1 = 0.8$, so that $\delta_2 = 0.64$, $\delta_4 = 0.41$. Then the optimal schemes would be

$$h = 1 : x_t(1) = -\frac{0.5}{g}(1 + 4\nabla)\varepsilon_t(1) \qquad \sigma_\varepsilon^2 = 1.00 \qquad g^2\sigma_x^2 = 10.25$$

$$h = 2 : x_t(2) = -\frac{0.62}{g}(1 + 1.78\nabla)\varepsilon_t(2) \qquad \sigma_\varepsilon^2 = 1.32 \qquad g^2\sigma_x^2 = 5.50$$

$$h = 4 : x_t(4) = -\frac{0.73}{g}(1 + 0.69\nabla)\varepsilon_t(4) \qquad \sigma_\varepsilon^2 = 2.17 \qquad g^2\sigma_x^2 = 3.84$$

In accordance with expectation, as the sampling interval is increased and the dynamics of the system have relatively less importance, the amount of "integral" control is increased and the ratio of proportional to integral control is markedly reduced. We have noted earlier that in some cases an excessively large adjustment variance σ_x^2 would be a disadvantage. The values of $g^2\sigma_x^2$ are indicated to show how the schemes differ in this respect. The smaller value for σ_x^2 would not of itself, of course, justify the choice $h = 4$. Using an optimal constrained scheme, such as is described in Section 13.2 with $h = 1$ a very large reduction in σ_x would be produced with only a small increase in the output variance. For example, entering Table 13.3 with $\delta = 0.8$, $100Q = 20$, we find that, for a 5% increase of output variance to the value $(1 + \lambda^2 Q)\sigma_1^2 = 1.05\sigma_1^2$, the input variance for the scheme with $h = 1$ could be reduced to 22% of its unconstrained value, so that

$g^2\sigma_x^2 = 10.25 \times 0.22 = 2.26$. Using (13.2.28), we obtain for the constrained scheme with $h = 1$,

$$x_t = 0.40x_{t-1} - 0.09x_{t-2} - 0.56\left\{\frac{0.5}{g}(1 + 4\nabla)\right\}\varepsilon_t(1)$$

$$\sigma_\varepsilon^2 = 1.05 \qquad g^2\sigma_x^2 = 2.26$$

In practice, various alternative schemes could be set out with their accompanying characteristics and an economic choice made to suit the particular problem. In general, the increase in output variance which comes with the larger interval would have to be balanced off against the economic advantage, if any, of less frequent surveillance.

Part V

This part of the book is a collection of auxiliary material useful in the analysis of time series. This includes a description of computer programs needed for time series analysis, tables and charts for obtaining preliminary estimates of autoregressive-moving average models, together with the usual tail-area tables of the Normal, χ^2 and t distributions. This is followed by a complete listing of all the Time Series analyzed in the book.

Description of Computer Programs

Program 1 Univariate Stochastic Model Identification (USID)
Program 2 Univariate Stochastic Model Preliminary Estimation (USPE)
Program 3 Univariate Stochastic Model Estimation (USES)
Program 4 Univariate Stochastic Model Forecasting (USFO)
Program 5 Univariate Transfer Function Model Identification (UTID)
Program 6 Univariate Transfer Function Model Preliminary Estimation (UTPE)
Program 7 Univariate Transfer Function Model Estimation (UTES)

USE OF THE COMPUTER IN MODEL BUILDING

The computer has great speed and precision, compared with which the human mind is slow and inaccurate. However, the mind has powers of critical and inductive reasoning, relative to which, those of the present-day computer are minuscule. This makes possible a happy division of labor in which mind and machine each contributes what it does better. Models built entirely automatically by the computer without the intervention of intellectual imagination and restraint are unlikely to be very good, and could be very bad. Programs such as those outlined below are not intended to provide automatic model-building. Rather, when properly used, they allow the investigator great freedom to experiment with his data, and can spark the ideas necessary to creative iteration. Once the *form* of the tentative model has been decided, it is often possible to obtain remarkably rapid convergence using the iterative nonlinear estimation routine, *even when rather remote initial parameter values are employed*. It may therefore be possible to dispense with the preliminary estimation phase embodied in Programs 2 or 6. In suitable cases, after tentative identification of the model *form* with Programs 1 or 5, the investigator may proceed directly with the estimation Programs 3 or 7.

ACKNOWLEDGMENT

Programs described below were written in ALGOL at the University of Lancaster by John Hampton, Granville Wilson, Elaine Hodgkinson and

Patricia Blant. They are presented here in the form of computational algorithms. Although necessary input and desirable output is fully specified, no discussion of program organization is attempted since this will generally depend upon local conditions and the requirements of individual users.

PROGRAM 1

UNIVARIATE STOCHASTIC MODEL IDENTIFICATION (USID)

1.1 General description

The program inputs a time series, appropriately transforms each observation, applies a differencing operation and then calculates the following quantities:
(1) Mean $\bar{w}$ and variance s_w^2.
(2) Autocovariance function (*acvf*) c_k.
(3) Autocorrelation function (*acf*) r_k.
(4) Partial autocorrelation function (*pacf*) $\hat{\phi}_{ll}$.

1.2 Input

The minimum information required for computation includes:

Time series values:	$\{z_t\}, t = 1, 2, \ldots, N$
Number of observations:	N
Degree of nonseasonal differencing:	$d \geqslant 0$
Degree of seasonal differencing:	$D \geqslant 0$
Period of seasonality:	$s \geqslant 1 \quad N - d - sD > 1$
Maximum lag of *acvf, acf*:	K
Maximum lag of *pacf*:	$L \leqslant K$
Transformation parameters:	λ, m

1.3 Computation

Transforming and differencing. For each set of values (d, D, λ, m), the time series is transformed as follows:

$$z_t' = \begin{cases} (z_t + m)^\lambda & \lambda \neq 0 \\ \log_e (z_t + m) & \lambda = 0 \end{cases}$$

where, if $\lambda \neq 0$, the parameter m is chosen so that $z_t + m$ is positive for all t and, if $\lambda = 1$, m is set equal to zero so that the z's remain unchanged.

The series is then differenced so that

$$w_t = \nabla^d \nabla_s^D z_t' \qquad t = 1, 2, \ldots, n$$

where

$$\nabla w_t = w_t - w_{t-1}$$

$$\nabla_s w_t = w_t - w_{t-s}$$

$$n = N - d - sD$$

Then the following quantities are calculated:
Mean and variance.

$$\bar{w} = \frac{1}{n} \sum_{t=1}^{n} w_t$$

$$s_w^2 = c_0$$

where c_0 is defined below.
Autocovariance function.

$$c_k = \frac{1}{n} \sum_{t=1}^{n-k} (w_t - \bar{w})(w_{t+k} - \bar{w})$$

where $k = 0, 1, \ldots, K$.
Autocorrelation function.

$$r_k = \frac{c_k}{c_0}$$

where $k = 0, 1, \ldots, K$.
Partial autocorrelation function.

$$\hat{\phi}_{ll} = \begin{cases} r_1 & l = 1 \\[2ex] \dfrac{r_l - \sum_{j=1}^{l-1} \hat{\phi}_{l-1,j} r_{l-j}}{1 - \sum_{j=1}^{l-1} \hat{\phi}_{l-1,j} r_j} & l = 2, 3, \ldots, L \end{cases}$$

where

$$\hat{\phi}_{lj} = \hat{\phi}_{l-1,j} - \hat{\phi}_{ll}\hat{\phi}_{l-1,l-j} \qquad j = 1, 2, \ldots, l-1$$

Alternatively, the partial autocorrelation function may be evaluated more precisely, using Program 3 (USES), by calculating the least squares estimate of the last autoregressive parameter $\hat{\phi}_{ll}$ in successive $AR(l)$ process, $l = 1, 2, \ldots$ fitted to the data.

1.4 Output

This should include all input information and
Transformation parameters: λ, m

Values of differenced and transformed series: $\quad \{w_t\}, t = 1, 2, \ldots, n$
Number of w_t values: $\quad n$
Mean of w series: $\quad \bar{w}$
Variance of w series: $\quad s_w^2$
acvf of w series: $\quad c_k, k = 0, 1, \ldots, K$
acf of w series: $\quad r_k, k = 0, 1, \ldots, K$
pacf of w series: $\quad \hat{\phi}_{ll}, l = 1, 2, \ldots, L$

1.5 Implementation

To give widest generality, control parameters should be employed to steer program activity.

PROGRAM 2

UNIVARIATE STOCHASTIC MODEL PRELIMINARY ESTIMATION (USPE)

2.1 General description

Using the autocovariances, calculated using Program 1, this program computes for the nonseasonal model

$$w_t = \phi_1 w_{t-1} + \cdots + \phi_p w_{t-p} + \theta_0 + a_t - \theta_1 a_{t-1} - \cdots - \theta_q a_{t-q}$$

identified for the suitably differenced and transformed series, the following quantities:
(1) Initial estimates $\hat{\boldsymbol{\phi}}_0 = (\hat{\phi}_{10}, \hat{\phi}_{20}, \ldots, \hat{\phi}_{p0})$ of the nonseasonal autoregressive parameters.
(2) Initial estimates $\hat{\boldsymbol{\theta}}_0 = (\hat{\theta}_{10}, \hat{\theta}_{20}, \ldots, \hat{\theta}_{q0})$ of the nonseasonal moving average parameters.
(3) Initial estimate $\hat{\theta}_{00}$ of the overall constant term.
(4) Initial estimate $\hat{\sigma}_a^2$ of the white noise variance.

2.2 Input

The minimum information for input includes:
Number of autoregressive parameters: $\quad p \geqslant 0$
Number of moving average parameters: $\quad q \geqslant 0, p + q > 0$
Mean of w_t series: $\quad \bar{w}$
Autocovariances of w_t series: $\quad c_k, k = 0, 1, \ldots, K$
$\quad K \geqslant p + q$
where $\bar{w}$ and c_k are computed as in Program 1.

2.3 Computation

Estimates $\hat{\boldsymbol{\phi}}$ *of autoregressive parameters.* If $p > 0$, solve the set of p linear equations

$$\mathbf{A}\boldsymbol{\phi}_0 = \mathbf{x}$$

where

$$A_{ij} = c_{|q+i-j|}$$

$$x_i = c_{q+i}$$

$$i, j = 1, 2, \ldots, p$$

Estimates $\hat{\boldsymbol{\theta}}$ *of moving average parameters.* (1) Using the autocovariances c_k of the w_t series, the following modified covariance sequence c'_j is calculated:

$$c'_j = \begin{cases} \sum_{i=0}^{p} \sum_{k=0}^{p} \hat{\phi}_{i0}\hat{\phi}_{k0}c_{|j+i-k|} & p > 0 \quad (\hat{\phi}_{00} = -1) \\ c_j & p = 0 \end{cases}$$

where $j = 0, 1, \ldots, q$.
(2) Then the Newton–Raphson algorithm

$$\boldsymbol{\tau}^{i+1} = \boldsymbol{\tau}^i - \mathbf{h}$$

where

$$\mathbf{T}^i\mathbf{h} = \mathbf{f}^i$$

is used to calculate the vector $\boldsymbol{\tau}^{i+1}$ at the $(i + 1)$st iteration from its value $\boldsymbol{\tau}^i$ at the ith iteration, where

$$\boldsymbol{\tau} = (\tau_0, \tau_1, \ldots, \tau_q)$$

$$f_j = \sum_{i=0}^{q-j} \tau_i\tau_{i+j} - c'_j$$

$$\mathbf{f} = (f_0, f_1, \ldots, f_q)$$

$$\mathbf{T} = \begin{bmatrix} \tau_0 & \tau_1 & \cdots & & \tau_q \\ \tau_1 & \tau_2 & \cdots & \tau_q \\ \vdots & & & \\ & & 0 & \\ \tau_q & & & \end{bmatrix} + \begin{bmatrix} \tau_0 & \tau_1 & \cdots & & \tau_q \\ & \tau_0 & \tau_1 & \cdots & \tau_{q-1} \\ & & & \vdots \\ & & 0 & \\ & & & \tau_0 \end{bmatrix}$$

with starting values $\tau_0 = \sqrt{c'_0}, \tau_1 = \tau_2 = \cdots = \tau_q = 0$.
(3) When $|f_j| < \varepsilon, j = 0, 1, \ldots, q$, for some prescribed value ε, the process is considered to have converged and the parameter estimates are obtained

from the final τ values according to

$$\hat{\theta}_{j0} = -\tau_j/\tau_0 \qquad j = 1, 2, \ldots, q$$

Estimate $\hat{\theta}_{00}$ of overall constant.

$$\hat{\theta}_{00} = \begin{cases} \bar{w}\left(1 - \sum_{i=1}^{p} \hat{\phi}_{i0}\right) & p > 0 \\ \bar{w} & p = 0 \end{cases}$$

Estimate $\hat{\sigma}_a^2$ of white noise variance.

$$\hat{\sigma}_a^2 = \begin{cases} \tau_0^2 & q > 0 \\ c_0 - \sum_{i=1}^{p} \hat{\phi}_i c_i & q = 0 \end{cases}$$

where τ_0 is calculated as in (2) above.

2.4 Output

This should include all input information and
Initial estimates of nonseasonal
autoregressive parameters: $\hat{\mathbf{\phi}}_0 = (\hat{\phi}_{10}, \hat{\phi}_{20}, \ldots, \hat{\phi}_{p0})$
Initial estimates of nonseasonal
moving average parameters: $\hat{\mathbf{\theta}}_0 = (\hat{\theta}_{10}, \hat{\theta}_{20}, \ldots, \hat{\theta}_{q0})$
Initial estimate of overall constant term: $\hat{\theta}_{00}$
Initial estimate of white noise variance: $\hat{\sigma}_a^2$
The computation and output may be organized for several degrees of differencing and for all models (p, q) from $(1, 0)$ and $(0, 1)$ up to (p, q) inclusive.

2.5 Implementation

To give the widest generality, control parameters should be employed to steer program activity. In practice, the computational algorithms for this program and for Program 1 may be combined in a single systems program to allow both stochastic model identification and preliminary estimation as desired.

PROGRAM 3

UNIVARIATE STOCHASTIC MODEL ESTIMATION (USES)

3.1 General description

The program inputs initial estimates of the parameters. These could be those from Program 2 if the model is nonseasonal and as described in Chapter 9 if the model is seasonal. It then calculates

(1) The least squares estimates of the parameters μ, ϕ, Φ, θ, Θ and σ_a^2 in the seasonal model

$$\phi(B)\Phi(B^s)(w_t - \mu) = \theta(B)\Theta(B^s)a_t$$

where $w_t = \nabla^d\nabla_s^D z_t'$,

z_t' is the transformed time series as defined in Program 1,

μ is the mean value of the w series

(2) The standard errors of the estimates and an estimate of their correlation matrix.

(3) The autocorrelation function of the residuals corresponding to the least squares estimates and the associated chi-square statistic.

3.2 Input parameters

The minimum information required for computation includes

$\{z_t\}, N, d, D, s, K, \lambda, m$:	as defined in Program 1
p, q:	as defined in Program 2
Number of seasonal autoregressive parameters:	$P \geqslant 0$
Number of seasonal moving average parameters:	$Q \geqslant 0, P + Q > 0$
Control parameter for mean μ:	$M = 1$ (μ included)
	$M = 0$ (μ omitted)
Initial estimate of μ:	$\bar{w}$ if $M = 1$; zero if $M = 0$
Initial estimates of nonseasonal autoregressive parameters:	$\hat{\phi} = (\hat{\phi}_{10}, \hat{\phi}_{20}, \ldots, \hat{\phi}_{p0})$
Initial estimates of seasonal autoregressive parameters:	$\hat{\Phi} = (\hat{\Phi}_{10}, \hat{\Phi}_{20}, \ldots, \hat{\Phi}_{P0})$
Initial estimates of nonseasonal moving average parameters:	$\hat{\theta} = (\hat{\theta}_{10}, \hat{\theta}_{20}, \ldots, \hat{\theta}_{q0})$
Initial estimates of seasonal moving average parameters:	$\hat{\Theta} = (\hat{\Theta}_{10}, \hat{\Theta}_{20}, \ldots, \hat{\Theta}_{Q0})$
Maximum number of iterations to be attempted:	it

3.3 Computation

Calculation of residual sum of squares. (1) Having backforecast initial w's, the residuals for a specified set of values of the parameters are calculated by the two stage process:

$$\alpha_t = (w_t - \mu) - \sum_{i=1}^{p} \phi_i (w_{t-i} - \mu) + \sum_{j=1}^{q} \theta_j \alpha_{t-j}$$

$$a_t = \alpha_t - \sum_{i=1}^{P} \Phi_i \alpha_{t-is} + \sum_{j=1}^{Q} \Theta_j a_{t-js}$$

where we here employ the shortened notation a_t for $[a_t|\mu, \phi, \Phi, \theta, \Theta, w]$,

    μ only appears if $M = 1$,

    $t = Q', Q' + 1, \ldots, n$ and

    Q' is a *negative* origin for t beyond which the back forecasts are negligible.

(2) For starting the forward recursion in (1), a procedure from Program 4 is used to back-forecast the values $w_0, w_{-1}, \ldots$, the process being terminated when $w_t - \hat{\mu}$ becomes negligibly small.

(3) For given values of the parameters $(\mu, \phi, \Phi, \theta, \Theta)$, the residual sum of squares is calculated from

$$S(\mu, \phi, \Phi, \theta, \Theta) = \sum_{t=Q'}^{n} a_t^2$$

Calculation of least squares estimates. The values of the parameters which minimize the residual sum of squares are obtained by a constrained optimization method, proposed by Marquardt [63], which is described at the end of this program. The input parameter "it" defines the maximum number of such iterations to be attempted.

Standard errors and correlation matrix. The estimate of the residual variance is obtained from the value of the sum of squares function at convergence using

$$\hat{\sigma}_a^2 = \frac{1}{n - p - q - P - Q - M} S(\hat{\mu}, \hat{\phi}, \hat{\Phi}, \hat{\theta}, \hat{\Theta})$$

and the covariance matrix $\mathbf{V}$ of the estimates from

$$\mathbf{V} = \{V_{ij}\} = (\mathbf{X'X})^{-1} \hat{\sigma}_a^2$$

where $\mathbf{X}$ is the regression matrix in the linearised model, calculated at the last iteration of the Marquardt procedure.

The standard errors are

$$s_i = \sqrt{V_{ii}} \qquad i = 1, 2, \ldots, p + q + P + Q + M$$

and the elements R_{ij} of the correlation matrix are obtained from

$$R_{ij} = V_{ij}/\sqrt{V_{ii}V_{jj}}.$$

Finally an estimate $\hat{\theta}_0$ of the overall constant term is

$$\hat{\theta}_0 = \hat{\mu}G$$

where

$$G = \left(1 - \sum_{i=1}^{p} \hat{\phi}_i\right)\left(1 - \sum_{j=1}^{P} \hat{\Phi}_j\right)$$

Diagnostic checks. Using the residuals $\hat{a}_t$ corresponding to the least square estimates, the residual autocorrelations are obtained from

$$r_{\hat{a}\hat{a}}(k) = c_{\hat{a}\hat{a}}(k)/c_{\hat{a}\hat{a}}(0)$$

where

$$c_{\hat{a}\hat{a}}(k) = \frac{1}{n} \sum_{t=1}^{n-k} (\hat{a}_t - \bar{a})(\hat{a}_{t+k} - \bar{a})$$

$$\bar{a} = \frac{1}{n} \sum_{t=1}^{n} \hat{a}_t$$

and

$$k = 0, 1, \ldots, K$$

Finally, the chi-square statistic is calculated from

$$\chi^2 = n \sum_{k=1}^{K} r_{\hat{a}\hat{a}}^2(k)$$

and is compared with a chi-square distribution with

$$v = K - M - p - q - P - Q$$

degrees of freedom.

3.4 Output

This should include all input information and:

$$\hat{\mu}, \hat{\boldsymbol{\phi}}, \hat{\boldsymbol{\Phi}}, \hat{\boldsymbol{\theta}}, \hat{\boldsymbol{\Theta}}, S(\hat{\mu}, \hat{\boldsymbol{\phi}}, \hat{\boldsymbol{\Phi}}, \hat{\boldsymbol{\theta}}, \hat{\boldsymbol{\Theta}})$$

at each iteration and the following information at the final iteration (that is, following convergence, failure of the search routine, or when convergence is not obtained after the maximum number, "it", of iterations)

Residuals corresponding to least squares estimates:	$\{\hat{a}_t\}$	$t = 1, 2, \ldots, n$
Residual (white noise) variance estimate:	$\hat{\sigma}_a^2$	
Covariance matrix of estimates:	$\mathbf{V}$	
Standard errors of estimates:	$\mathbf{s} = (s_1, s_2, \ldots, s_{M+p+q+P+Q})$	
Correlation matrix of estimates:	$\mathbf{R}$	
Overall constant estimate:	$\hat{\theta}_0$	
Residual autocorrelations:	$r_{\hat{a}\hat{a}}(k)$	$k = 1, 2, \ldots, K$
Chi-square statistic:	χ^2	

3.5 Implementation

To give widest generality, control parameters should be employed to steer program activity. Tabulations of $S(\hat{\mu}, \hat{\phi}, \hat{\Phi}, \hat{\theta}, \hat{\Theta})$ around the minimum may be included by computing this function on a grid centred on the least squares values.

MARQUARDT ALGORITHM FOR NONLINEAR LEAST SQUARES*

1. Supplied Quantities

Denoting by $\beta = (\beta_1, \beta_2, \ldots, \beta_k)$ all the parameters in the model, that is $\beta = (\mu, \phi, \Phi, \theta, \Theta)$, starting values β_0 are specified together with parameters π and F_2, which constrain the search, and a convergence parameter ε. During the search, the values $a_t = [a_t | \beta, w]$ and the derivatives

$$x_{i,t} = -\frac{\partial a_t}{\partial \beta_i}$$

need to be evaluated at each stage of the iterative process.

2. Calculation of derivatives

Using the residuals, calculated as described in Section 3.3 of Program 3, the derivatives are obtained from

$$x_{i,t} = \{a_t(\beta_{1,0}, \ldots, \beta_{i,0}, \ldots, \beta_{k,0}) - a_t(\beta_{1,0}, \ldots, \beta_{i,0} + \delta_i, \ldots, \beta_{k,0})\}/\delta_i$$

3. The Iteration

STAGE (1).   With a_t, $x_{i,t}$ supplied from the current parameter values, the following quantities are formed:
(1) The $k \times k$ matrix

$$\mathbf{A} = \{A_{ij}\}$$

where

$$A_{ij} = \sum_{t=Q'}^{n} x_{i,t} x_{j,t}$$

(2) The vector $\mathbf{g}$ with elements $g_1, g_2, \ldots, g_k$, where

$$g_i = \sum_{x=Q'}^{n} x_{i,t} a_t$$

(3) The scaling quantities $D_i = \sqrt{A_{ii}}$

*Slightly modified by G. T. Wilson.

STAGE (2). The modified (scaled and constrained) linearized equations

$$\mathbf{A}^*\mathbf{h}^* = \mathbf{g}^*$$

are constructed according to

$$A_{ij}^* = A_{ij}/D_i D_j \qquad i \neq j$$
$$A_{ii}^* = 1 + \pi$$
$$g_i^* = g_i/D_i$$

The equations are solved for $\mathbf{h}^*$, which is scaled back to give the parameter corrections h_j, where

$$h_j = h_j^*/D_j$$

Then the new parameter values are constructed from

$$\boldsymbol{\beta} = \boldsymbol{\beta}_0 + \mathbf{h}$$

and the sum of squares of residuals $S(\boldsymbol{\beta})$ evaluated.

STAGE (3). (1) If $S(\boldsymbol{\beta}) < S(\boldsymbol{\beta}_0)$, the parameter corrections $\mathbf{h}$ are tested. If all are smaller than ε, convergence is assumed and the $k \times k$ matrix $\mathbf{A}^{-1}$ is used to calculate the covariance matrix of the estimates as described in Program 3; otherwise, $\boldsymbol{\beta}_0$ is reset to the value $\boldsymbol{\beta}$, π is reduced by a factor F_2 and computation returns to Stage (1).
(2) If $S(\boldsymbol{\beta}) > S(\boldsymbol{\beta}_0)$, the constraint parameter π is increased by a factor F_2 and computation resumed at Stage (2). In all but exceptional cases, a reduced sum of squares will eventually be found. However, an upper bound is placed on π, and if this bound is exceeded, the search is terminated.
When convergence has occurred, either according to the criterion in (1) of Stage (3), or it is assumed to have taken place after a specified number of iterations, the residual variance and the covariance matrix of the estimates are calculated as described in Program 3.

PROGRAM 4

UNIVARIATE STOCHASTIC MODEL FORECASTING (USFO)

4.1 General description

The program inputs the least squares estimates $(\hat{\theta}_0, \hat{\boldsymbol{\phi}}, \hat{\boldsymbol{\Phi}}, \hat{\boldsymbol{\theta}}, \hat{\boldsymbol{\Theta}})$ of the parameters in the general model, fitted using Program 2, together with the values of the time series and computes
(1) The forecast function $\hat{z}_{N-b}(l)$, $l = 1, 2, \ldots, L$ for each origin $b = 0, 1, \ldots, B$ time units before the end of the series. The forecasts are first formed for

the transformed series z'_t and then converted into forecasts for the original time series z_t.

(2) Upper and lower probability limits $z_t(-)$, $z_t(+)$ such that probability $P\{z_t(-) < z_t < z_t(+)\}$ of some future value z_t of the time series lying between these limits is equal to a specified value α.

4.2 Input parameters

The minimum information for computation includes:

$\{z_t\}$, N, d, D, s, K:	as defined in Program 1
p, P, q, Q:	as defined in Program 3
Control parameter for constant term:	$M = 1$ (constant term included)
	$M = 0$ (constant term omitted)
Maximum lead time of forecast:	L
Maximum value of backward origin:	B
Least squares estimate of constant term:	$\hat{\theta}_0$
Least squares estimates of non-seasonal autoregressive parameters:	$\hat{\phi}_1, \hat{\phi}_2, \ldots, \hat{\phi}_p$
Least squares estimates of seasonal autoregressive parameters:	$\hat{\Phi}_1, \hat{\Phi}_2, \ldots, \hat{\Phi}_P$
Least squares estimates of non-seasonal moving average parameters:	$\hat{\theta}_1, \hat{\theta}_2, \ldots, \hat{\theta}_q$
Least squares estimates of seasonal moving average parameters:	$\hat{\Theta}_1, \hat{\Theta}_2, \ldots, \hat{\Theta}_Q$
Least squares estimate of residual variance:	$\hat{\sigma}_a^2$

4.3 Computation

Unscrambling the operators. For forecasting purposes, the seasonal model is written in the form

$$z_t = \Phi_1^* z_{t-1} + \cdots + \Phi_{p*}^* z_{t-p*} + \theta_0 + a_t - \Theta_1^* a_{t-1} - \cdots - \Theta_{Q*}^* a_{t-Q*}$$

where in practice estimates are substituted for unknown parameters. To generate the Φ_i^*, first the parameters $\Phi'_1, \ldots, \Phi'_{p'}$ in the operator

$$\Phi'(B) = \Phi(B^s)\phi(B) = 1 - \Phi'_1 B - \cdots - \Phi'_{p'} B^{p'}$$

are derived, where $p' = p + sP$, using

$$\Phi'_k = -\sum_i \sum_{\substack{j \\ j+si=k}} \Phi_i \phi_j$$

the double sum being evaluated by ranging over all $j = 0, \ldots, p;\ i = 0, \ldots, P$ and accumulating the products $\Phi_i \phi_j$ appropriately. The constants Φ_0 and ϕ_0 are both taken as -1. Second, the parameters $\Phi_1^*, \ldots, \Phi_{p*}^*$ in the operator

$$\Phi^*(B) = \nabla^d \nabla_s^D \Phi'(B) = 1 - \Phi_1^* B - \cdots - \Phi_{p*}^* B^{p*}$$

with $p^* = p' + d + sD$, are obtained by the following three stage procedure:

(1)
$$\Phi_j'' = \begin{cases} \Phi_j' & 0 \leqslant j < s \\ \Phi_j' - \Phi_{j-s}' & s \leqslant j \leqslant p' \\ -\Phi_{j-s}' & p' < j \leqslant p' + s \end{cases}$$

(2) Apply (1) repeatedly D times, replacing Φ_j' with Φ_j'', p' with $p'' = p' + s$ at each iteration to obtain the coefficients in $\nabla_s^D \Phi'(B)$.
(3) Similarly, continue by applying (1), with $s = 1$, repeatedly d times to the values obtained from (2) to obtain eventually the coefficients $\Phi_1^*, \ldots, \Phi_{p*}^*$.

The moving average parameters Θ_k^* are obtained in the same way as the Φ' parameters using

$$\Theta_k^* = - \sum_i \sum_{\substack{j \\ j+si=k}} \Theta_i \theta_j$$

where $i = 0, 1, \ldots, Q, j = 0, 1, \ldots, q$, with $\Theta_0 = -1$, $\theta_0 = -1$.

Generation of forecasts. (1) For the transformed series z_t', the forecasts $\hat{z}_{N-b}'(l)$ are obtained from

$$\hat{z}_{N-b}'(l) = \theta_0 + \sum_{i=1}^{p+sP+d+sD} \Phi_i^* [z_{N-b-i+l}'] - \sum_{j=1}^{q+SQ} \Theta_j^* [a_{N-b-j+l}]$$

where

$$[z_{N-b-i+l}'] = \begin{cases} \hat{z}_{N-b}'(l - i) & l > i \\ z_{N-b-i+l}' & l \leqslant i \end{cases}$$

$$[a_{N-b-j+l}] = \begin{cases} 0 & l > j \\ z_{N-b-j+l}' - \hat{z}_{N-b-j+l-1}'(1) & l \leqslant j \end{cases}$$

with $l = 1, 2, \ldots, L$ and z_t' as defined in Program 1. The forecasts are obtained for each backward origin $b = 0, 1, \ldots, B$.
(2) Then the forecasts $\hat{z}_{N-b}(l)$ for the original series are obtained from

$$\hat{z}_{N-b}(l) = \begin{cases} \hat{z}_{N-b}'(l) & \lambda = 1 \\ \{\hat{z}_{N-b}'(l)\}^{1/\lambda} - m & \lambda \neq 0 \\ \exp\{\hat{z}_{N-b}'(l)\} - m & \lambda = 0 \end{cases}$$

Accuracy of forecasts. (1) For the transformed forecasts, the upper and lower probability limits are:

$$z'_{N-b+l}(\pm) = \hat{z}'_{N-b}(l) \pm u\sqrt{V(l)}$$

where $u = 0.68, 1.65, 1.96$ or 2.58 depending on whether the probability that a future value lies in the interval is $0.50, 0.90, 0.95$ or 0.99, respectively. The variance function is

$$V(l) = \hat{\sigma}_a^2 \sum_{j=0}^{l-1} \psi_j^2$$

where

$$\psi_j = \begin{cases} 1 & j = 0 \\ \sum_{i=1}^{j} \Phi_i^* \psi_{j-i} - \Theta_j^* & j \geq 1 \end{cases}$$

for $j = 1, 2, \ldots, L$ and

$$\Phi_i^* = 0 \qquad i > p + sP + d + sD$$

$$\Theta_j^* = 0 \qquad j > q + sQ$$

(2) For the transformed forecasts, the upper and lower limits are

$$z_{N-b+l}(\pm) = \begin{cases} z'_{N-b+l}(\pm) & \lambda = 1 \\ \{z'_{N-b+l}(\pm)\}^{1/\lambda} - m & \lambda \neq 0 \\ \exp\{z'_{N-b+l}(\pm)\} - m & \lambda = 0 \end{cases}$$

4.4 Output

This should include all input information and:

Parameters in generalized autoregressive operator: $\mathbf{\Phi}^*$

Parameters in generalized moving average operator: $\mathbf{\Theta}^*$

ψ weights: $\psi_j, j = 1, 2, \ldots, L$

For a given probability level, forecasts and their upper and lower probability limits:
   $z_{N-b+l}(+)$
   $\hat{z}_{N-b}(l)$
   $z_{N-b+l}(-)$

for $l = 1, 2, \ldots, L$, and for selected origins decided by choice of $b = 0, 1, \ldots, B$.

4.5 Implementation

To give widest generality, control parameters should be employed to steer program activity. In practice, the forecasts should appear in the output as a suitably displayed matrix so that every element in each column corresponds to the z_t it forecasts.

PROGRAM 5

UNIVARIATE TRANSFER FUNCTION MODEL IDENTIFICATION (UTID)

5.1 General description

The program inputs a pair of time series, differences each series, applies a prewhitening transformation to each series to produce another pair of time series (α_t, β_t) and then calculates the following:

(1) The autocorrelation function $r_{\alpha\alpha}(k)$ and the cross correlation function $(ccf)\ r_{\alpha\beta}(k)$ of the prewhitened series.

(2) Estimates $\hat{v}_k$ of the impulse response weights in the model

$$y_t = v_0 x_t + v_1 x_{t-1} + \cdots + v_h x_{t-h} + n_t$$

where n_t is the noise component.

(3) The noise variance s_n^2, autocorrelation function $r_{nn}(k)$ and partial auto-correlation function $\hat{\phi}_{ll}$.

5.2 Input

The minimum information required for computation includes:

"Input" time series values:	$\{X_t\}, t = 1, 2, \ldots, N$
"Output" time series values:	$\{Y_t\}, t = 1, 2, \ldots, N$
Number of observations in each series:	N
Degree of differencing:*	d
Number of parameters in autoregressive prewhitening operator:	p
Number of parameters in moving average prewhitening operator:	q
Parameter values for autoregressive operator:	$\phi = (\phi_1, \phi_2, \ldots, \phi_p)$
Parameter values for moving average operator:	$\theta = (\theta_1, \theta_2, \ldots, \theta_q)$
Number of v_k weights to be estimated:	h
Number of v_k weights for generating noise series n_t:	g
Transformation parameters:	$\lambda_x, \lambda_y, m_x, m_y$

*More recent versions of programs 5, 6, and 7 allow for different degrees of differencing in the X and Y series.

5.3 Computation

Differencing and prewhitening. The input series X_t and output series Y_t may first be transformed to X_t', Y_t' as in Program 1 and then differenced to give $n = N - d$ values of

$$x_t = \begin{cases} \nabla^d X_t' & d > 0 \\ X_t' - \bar{X}' & d = 0 \end{cases}$$

$$y_t = \begin{cases} \nabla^d Y_t' & d > 0 \\ Y_t' - \bar{Y}' & d = 0 \end{cases}$$

where $\bar{X}'$, $\bar{Y}'$ are the arithmetic means of the X_t' and Y_t' series. The differenced series are then prewhitened to give $n' = n - p$ values of the α_t, β_t series according to

$$\alpha_t = x_t - \sum_{i=1}^{p} \phi_i x_{t-i} + \sum_{j=1}^{q} \theta_j \alpha_{t-j}$$

$$\beta_t = y_t - \sum_{i=1}^{p} \phi_i y_{t-i} + \sum_{j=1}^{q} \theta_j \beta_{t-j}$$

Prewhitened output autocorrelation function.

$$r_{\beta\beta}(k) = \frac{\sum_{j=1}^{n'-k} (\beta_j - \bar{\beta})(\beta_{j+k} - \bar{\beta})}{\sum_{j=1}^{n'} (\beta_j - \bar{\beta})^2} \qquad k = 0, 1, \ldots, h$$

Prewhitened input-output cross-correlation function.

$$r_{\alpha\beta}(k) = \frac{c_{\alpha\beta}(k)}{s_\alpha s_\beta}$$

where

$$c_{\alpha\beta}(k) = \frac{1}{n} \sum_{j=1}^{n'-k} (\alpha_j - \bar{\alpha})(\beta_{j+k} - \bar{\beta}) \qquad k = 0, 1, \ldots, h$$

$$c_{\alpha\beta}(-k) = c_{\beta\alpha}(k) \qquad\qquad\qquad k = 1, 2, \ldots, h$$

$$s_\alpha = \sqrt{c_{\alpha\alpha}(0)}$$

$$s_\beta = \sqrt{c_{\beta\beta}(0)}$$

Impulse response function estimate.

$$\hat{v}_k = \frac{s_\beta}{s_\alpha} r_{\alpha\beta}(k) \qquad k = 0, 1, \ldots, h$$

Noise variance and autocorrelation function. Using the estimates $\hat{v}_k$ of the impulse response weights, the noise series n_t is regenerated from

$$n_t = y_t - \hat{v}_0 x_t - \hat{v}_1 x_{t-1} - \cdots - \hat{v}_g x_{t-g}$$

and then the variance, autocorrelation and partial autocorrelation functions calculated as in Program 1 from the values $n_t, t = 1, 2, \ldots, (n - g)$.

5.4 Output

This should include all input information and the following:

Prewhitened output autocorrelation
function: $\quad\quad r_{\beta\beta}(k), k = 0, 1, \ldots, h$

Prewhitened input–output cross
correlation function: $\quad\quad r_{\alpha\beta}(k), k = 0, \pm 1, \ldots, \pm h$

Prewhitened input and output standard
deviations: $\quad\quad s_\alpha, s_\beta$

Estimated impulse response weights: $\quad\quad \hat{v}_k, k = 0, 1, \ldots, h$

Noise variance: $\quad\quad s_n^2$

Noise autocorrelation function: $\quad\quad r_{nn}(k), k = 0, 1, \ldots, h$

Noise partial autocorrelation function: $\quad\quad \hat{\phi}_{ll}, l = 1, 2, \ldots, h$

Values of noise series: $\quad\quad \{n_t\}, t = 1, 2, \ldots, (n - g)$

5.5 Implementation

To give widest generality, control parameters should be employed to steer program activity.

PROGRAM 6

UNIVARIATE TRANSFER FUNCTION MODEL PRELIMINARY ESTIMATION (UTPE)

6.1 General description

With y_t and x_t defined as in Program 5 and using estimates of the impulse response weights and noise autocorrelation function obtained from that program, Program 6 computes for the transfer function-noise model

$$y_t = \delta^{-1}(B)\omega(B)x_{t-b} + \phi^{-1}(B)\theta(B)a_t$$

the following quantities:

(1) initial estimates $\boldsymbol{\delta}_0 = (\delta_{10}, \delta_{20}, \ldots, \delta_{r0})$ of the "left-hand side" parameters in the transfer function model,

(2) initial estimates $\boldsymbol{\omega}_0 = (\omega_{00}, \omega_{10}, \ldots, \omega_{s0})$ of the "right-hand side" parameters in the transfer function model,

(3) initial estimates $\boldsymbol{\phi}_0 = (\phi_{10}, \phi_{20}, \ldots, \phi_{p0})$ of the autoregressive parameters in the noise model,

(4) initial estimates $\boldsymbol{\theta}_0 = (\theta_{10}, \theta_{20}, \ldots, \theta_{q0})$ of the moving average parameters in the noise model,

(5) initial estimate $\hat{\sigma}_a^2$ of the white noise variance.

6.2 Input

The minimum information required for computation includes:

Number of "left-hand side" transfer function parameters:	r
Number of "right-hand side" transfer function parameters:	$s+1$
Number of noise autoregressive parameters:	p
Number of noise moving average parameters:	q
Delay parameter:	b
Number of impulse response weights:	$f \geqslant b + s + r$
Impulse response weights:	$\hat{v}_k, k = 1, \ldots, f$
Noise autocorrelations:	$r_{nn}(k), k = 1, \cdots, (p + q)$
Maximum number of iterations attempted:	Iter

6.3 Computation

Estimates $\hat{\boldsymbol{\delta}}_0$ of "left-hand side" transfer function parameters. If $r > 0$, solve the set of equations

$$\mathbf{A}\hat{\boldsymbol{\delta}}_0 = \mathbf{h}$$

where

$$A_{ij} = \begin{cases} \hat{v}_{b+s+i-j} & s + i \geqslant j \\ 0 & s + i < j \end{cases}$$

$$h_i = \hat{v}_{b+s+i}$$

$$i, j = 1, 2, \ldots, r$$

Estimates $\boldsymbol{\omega}_0$ of "right-hand side" transfer function parameters.

$$\hat{\omega}_{00} = \hat{v}_b$$

If $r \geqslant s$,

$$\hat{\omega}_{j0} = \sum_{i=1}^{j} \hat{\delta}_i \hat{v}_{b+j-i} - \hat{v}_{b+j}$$

If $r < s$,

$$\hat{\omega}_{j0} = \sum_{i=1}^{j} \hat{\delta}_i \hat{v}_{b+j-i} - \hat{v}_{b+j} \qquad j = 1, 2, \ldots, r$$

$$\hat{\omega}_{j0} = \sum_{i=1}^{r} \hat{\delta}_i \hat{v}_{b+j-i} - \hat{v}_{b+j} \qquad j = r+1, \ldots, s$$

The noise model parameters (3), (4) and (5) are calculated from the noise autocorrelations $r_{nn}(k)$ in the same way as in Program 2.

6.4 Output

This should include all input information and the following:

Initial estimates of "left-hand side" transfer
function parameters: $\hat{\delta}_{10}, \hat{\delta}_{20}, \ldots, \hat{\delta}_{r0}$
Initial estimates of "right-hand side" transfer
function parameters: $\hat{\omega}_{00}, \hat{\omega}_{10}, \ldots, \hat{\omega}_{s0}$
Initial estimates of noise autoregressive
parameters: $\hat{\phi}_{10}, \hat{\phi}_{20}, \ldots, \hat{\phi}_{p0}$
Initial estimates of noise moving average
parameters: $\hat{\theta}_{10}, \hat{\theta}_{20}, \ldots, \hat{\theta}_{q0}$

Output may be arranged at each iteration of the moving average parameter estimation to examine the rate of convergence. It may also be organized for several degrees of differencing and for all models up to (r, s, b, p, q), inclusively.

6.5 Implementation

To give the widest generality, control parameters should be employed to steer program activity. In practice, the computational algorithms for this program and for Program 5 may be combined in a single systems program to allow both transfer function-noise model identification and preliminary estimation as required.

PROGRAM 7

UNIVARIATE TRANSFER FUNCTION MODEL ESTIMATION (UTES)

7.1 General description

The program inputs initial estimates of the parameters. These could be those obtained from Program 6. It then calculates

(1) The least squares estimates of the parameters $\delta, \omega, \phi, \theta$ and σ_a^2 in the transfer function-noise model defined in Program 6.

(2) The standard errors of the estimates and an estimate of their correlation matrix.

(3) The autocorrelation of the residuals $\hat{a}_t$ corresponding to the least squares estimates and the associated chi-square statistic.

(4) The cross correlation function between the residuals $\hat{a}_t$ and the pre-whitened input and the associated chi-square statistic.

7.2 Input parameters

The minimum information required for input includes:

$\{X_t\}, \{Y_t\}, N, d, \lambda_x, \lambda_y, m_x, m_y$:	as defined in Program 5
r, s, b, p, q:	as defined in Program 6
Maximum lag of *acf* and *ccf*:	K
Initial estimates of "left-hand side" transfer function parameters:	$\hat{\delta}_{10}, \hat{\delta}_{20}, \ldots, \hat{\delta}_{r0}$
Initial estimates of "right-hand side" transfer function parameters:	$\hat{\omega}_{00}, \hat{\omega}_{10}, \ldots, \hat{\omega}_{s0}$
Initial estimates of noise autoregressive parameters:	$\hat{\phi}_{10}, \hat{\phi}_{20}, \ldots, \hat{\phi}_{p0}$
Initial estimates of noise moving average parameters:	$\hat{\theta}_{10}, \hat{\theta}_{20}, \ldots, \theta_{q0}$

In addition, the following parameters

$$(p', d', q')$$

$$\phi'_1, \ldots, \phi'_{p'}$$

$$\theta'_1, \ldots, \theta'_{q'}$$

specify an estimated noise model for the input series $\{x_t\}$.

7.3 Computation

Calculation of residual sum of squares. Given the x_t and y_t series, for given values of the parameters $(\delta, \omega, \phi, \theta)$, the residuals a_t are calculated by the three stage process

(a) $\quad \mathscr{y}_t = \delta_1 \mathscr{y}_{t-1} + \cdots + \delta_r \mathscr{y}_{t-r} + \omega_0 x_{t-b} - \omega_1 x_{t-b-1} - \cdots - \omega_s x_{t-b-s}$

for $t \geqslant b + s + 1$, and values of $\mathscr{y}_t$ previous to this point set equal to zero.

(b) $\qquad\qquad\qquad\qquad n_t = y_t - \mathscr{y}_t$

(c) $\quad a_t = n_t - \phi_1 n_{t-1} - \cdots - \phi_p n_{t-p} + \theta_1 a_{t-1} + \cdots + \theta_q a_{t-q}$

for $t \geqslant s + b + p + 1$, and values of a_t previous to this point, set equal to zero.

Calculation of least squares estimates. As described in Program 3.

Standard errors and correlation matrix. The estimate of the residual variance is obtained from the value of the sum of squares function at convergence using

$$\hat{\sigma}_a^2 = \frac{1}{n - r - 2s - b - 2p - q - 1} S(\hat{\delta}, \hat{\omega}, \hat{\phi}, \hat{\theta})$$

Then the covariance and correlation matrix of the estimates are obtained as described in the "Standard Errors and Correlation Matrix" in Program 3.

Autocorrelation diagnostic check. Using the residuals corresponding to the least squares estimates, the residual autocorrelations are obtained from

$$r_{\hat{a}\hat{a}}(k) = c_{\hat{a}\hat{a}}(k)/c_{\hat{a}\hat{a}}(0) \qquad k = 0, 1, \ldots, K$$

where

$$(n - s - b - p - 1)\, c_{\hat{a}\hat{a}}(k) = \sum_{t = s + b + p + 1}^{n - k} (\hat{a}_t - \bar{\hat{a}})(\hat{a}_{t+k} - \bar{\hat{a}})$$

The chi-square statistic is calculated from

$$P = (n - s - b - p) \sum_{k=1}^{K} r_{\hat{a}\hat{a}}^2(k)$$

and is compared with a chi-square distribution with $K - p - q$ degrees of freedom.

Cross correlation diagnostic check. The cross correlations between the prewhitened input series α_t obtained from

$$\alpha_t = x_t - \phi_1' x_{t-1} - \cdots - \phi_{p'}' x_{t-p'} + \theta_1' \alpha_{t-1} + \theta_2' \alpha_{t-2} + \cdots + \theta_{q'}' a_{t-q'}$$

for $t \geqslant p' + 1$, with values of α_t previous to this point set to zero, and the residuals $\hat{a}_t$ are calculated from

$$r_{\alpha\hat{a}}(k) = c_{\alpha\hat{a}}(k)/\sqrt{c_{\hat{a}\hat{a}}(0)c_{\alpha\alpha}(0)} \qquad k = 0, 1, \ldots, K$$

where

$$c_{\alpha\hat{a}} = \frac{1}{n - v} \sum_{t = v + 1}^{n - k} (\alpha_t - \bar{\alpha})(\hat{a}_{t+k} - \bar{\hat{a}})$$

$$c_{\alpha\alpha}(0) = \frac{1}{n - p'} \sum_{t = p' + 1}^{n} (\alpha_t - \bar{\alpha})^2$$

$$v = \max{(s + b + p, p')}$$

The chi-square statistic is calculated from

$$Q = (n - v) \sum_{k=0}^{K} r_{\alpha\hat{a}}^2(k)$$

and is compared with a chi-square distribution with $K - r - s$ degrees of freedom.

7.4 Output

This should include all input information and:

$$\hat{\delta}, \hat{\omega}, \hat{\phi}, \hat{\theta} \qquad S(\hat{\delta}, \hat{\omega}, \hat{\phi}, \theta)$$

and the following information at convergence (see Program 3):

Residual variance estimate:	$\hat{\sigma}_a^2$
Covariance matrix of estimates:	$\mathbf{V}$
Standard errors of estimates:	$\mathbf{s} = (s_1, s_2, \ldots, s_{r+s+p+q+1})$.
Correlation matrix of estimates:	$\mathbf{R}$
Residuals corresponding to least squares estimates:	$\{\hat{a}_t\}, t = s + b + p + 1, \ldots, n$
Residual autocorrelations:	$r_{\hat{a}\hat{a}}(k), k = 1, 2, \ldots, K$
Chi-square statistic:	P
Degrees of freedom:	$K - p - q$
Prewhitened input-residual cross correlations:	$r_{\alpha\hat{a}}(k), k = 0, 1, \ldots, K$
Chi-square statistic:	Q
Degrees of freedom:	$K - r - s$

7.5 Implementation

To give widest generality, control parameters should be employed to steer program activity. Tabulations of $S(\delta, \omega, \phi, \theta)$ around the minimum may be included by computing this function on a grid centred on the least squares values. The program is readily extended to make provision for the simultaneous estimation of a constant term θ_0.

Collection of Tables and Charts

Charts B, C and D are adapted and reproduced by permission of the author from reference [33]. Tables E, F and G are condensed and adapted from *Biometrika Tables for Statisticians Volume I*, with permission from the trustees of Biometrika.

TABLE A   TABLE RELATING ρ_1 TO θ FOR A FIRST-ORDER MOVING AVERAGE PROCESS

θ	ρ_1	θ	ρ_1
0.00	0.000	0.00	0.000
0.05	−0.050	−0.05	0.050
0.10	−0.099	−0.10	0.099
0.15	−0.147	−0.15	0.147
0.20	−0.192	−0.20	0.192
0.25	−0.235	−0.25	0.235
0.30	−0.275	−0.30	0.275
0.35	−0.315	−0.35	0.315
0.40	−0.349	−0.40	0.349
0.45	−0.374	−0.45	0.374
0.50	−0.400	−0.50	0.400
0.55	−0.422	−0.55	0.422
0.60	−0.441	−0.60	0.441

TABLE A   *Continued*

θ	ρ_1	θ	ρ_1
0.65	−0.457	−0.65	0.457
0.70	−0.468	−0.70	0.468
0.75	−0.480	−0.75	0.480
0.80	−0.488	−0.80	0.488
0.85	−0.493	−0.85	0.493
0.90	−0.497	−0.90	0.497
0.95	−0.499	−0.95	0.499
1.00	−0.500	−1.00	0.500

The table may be used to obtain first estimates of the parameters in the $(0, d, 1)$ process $w_t = (1 - \theta B)a_t$, where $w_t = \nabla^d z_t$, by substituting $r_1(w)$ for ρ_1.

CHART B   CHART RELATING ρ_1 AND ρ_2 TO ϕ_1 AND ϕ_2 FOR A SECOND-ORDER AUTOREGRESSIVE PROCESS

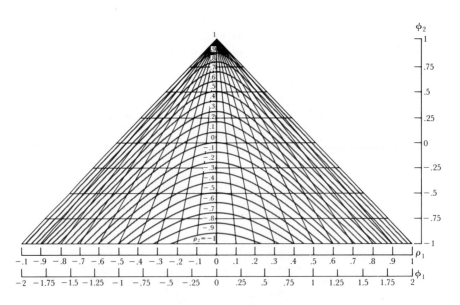

The chart may be used to obtain estimates of the parameters in the $(2, d, 0)$ process: $(1 - \phi_1 B - \phi_2 B^2)w_t = a_t$, where $w_t = \nabla^d z_t$, by substituting $r_1(w)$ and $r_2(w)$ for ρ_1 and ρ_2.

Chart C CHART RELATING ρ_1 AND ρ_2 TO θ_1 AND θ_2 FOR A SECOND-ORDER MOVING AVERAGE PROCESS

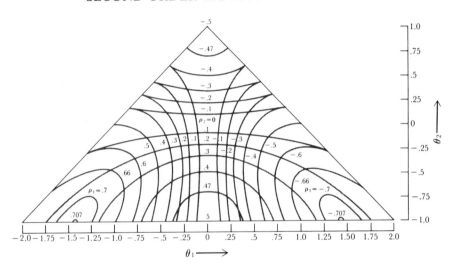

The chart may be used to obtain first estimates of the parameters in the $(0, d, 2)$ process $w_t = (1 - \theta_1 B - \theta_2 B^2)a_t$, where $w_t = \nabla^d z_t$, by substituting $r_1(w)$ and $r_2(w)$ for ρ_1 and ρ_2.

CHART D   CHART RELATING ρ_1 AND ρ_2 TO ϕ AND θ FOR A MIXED FIRST-ORDER AUTOREGRESSIVE-MOVING AVERAGE PROCESS

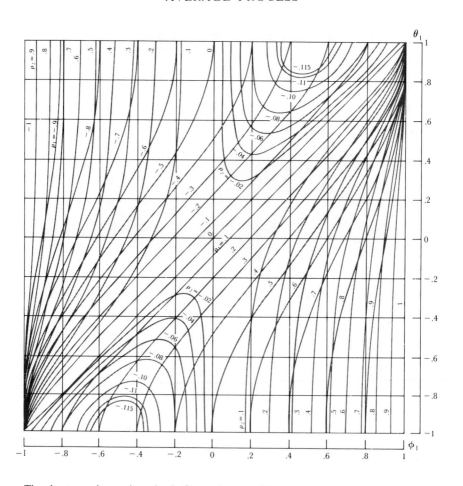

The chart may be used to obtain first estimates of the parameters in the $(1, d, 1)$ process $(1 - \phi B)w_t = (1 - \theta B)a_t$, where $w_t = \nabla^d z_t$, by substituting $r_1(w)$ and $r_2(w)$ for ρ_1 and ρ_2.

TABLE E TAIL AREAS AND ORDINATES OF UNIT NORMAL DISTRIBUTION

u_ε	ε	$p(u_\varepsilon)$	u_ε	ε	$p(u_\varepsilon)$
0.0	0.500	0.3989	1.6	0.055	0.1109
0.1	0.460	0.3969	1.7	0.045	0.0940
0.2	0.421	0.3910	1.8	0.036	0.0790
0.3	0.382	0.3814	1.9	0.029	0.0656
0.4	0.345	0.3683	2.0	0.023	0.0540
0.5	0.309	0.3521	2.1	0.018	0.0440
0.6	0.274	0.3322	2.2	0.014	0.0355
0.7	0.242	0.3123	2.3	0.011	0.0283
0.8	0.212	0.2897	2.4	0.008	0.0224
0.9	0.184	0.2661	2.5	0.006	0.0175
1.0	0.159	0.2420	2.6	0.005	0.0136
1.1	0.136	0.2179	2.7	0.003	0.0104
1.2	0.115	0.1942	2.8	0.003	0.0079
1.3	0.097	0.1714	2.9	0.002	0.0059
1.4	0.081	0.1497	3.0	0.001	0.0044
1.5	0.067	0.1295			

Table showing the values of the unit Normal deviate u_ε such that $\Pr\{u > u_\varepsilon\} = \varepsilon$. Also shown are the ordinates $p(u = u_\varepsilon)$.

TABLE F   TAIL AREAS OF THE CHI-SQUARE DISTRIBUTION

p	0.995	0.99	0.975	0.95	0.9	0.75	0.5	0.25	0.1	0.05	0.025	0.01	0.005	0.001	p
1	—	—	—	—	0.016	0.102	0.455	1.32	2.71	3.84	5.02	6.63	7.88	10.8	1
2	0.010	0.020	0.051	0.103	0.211	0.575	1.39	2.77	4.61	5.99	7.38	9.21	10.6	13.8	2
3	0.072	0.115	0.216	0.352	0.584	1.21	2.37	4.11	6.25	7.81	9.35	11.3	12.8	16.3	3
4	0.207	0.297	0.484	0.711	1.06	1.92	3.36	5.39	7.78	9.49	11.1	13.3	14.9	18.5	4
5	0.412	0.554	0.831	1.15	1.61	2.67	4.35	6.63	9.24	11.1	12.8	15.1	16.7	20.5	5
6	0.676	0.872	1.24	1.64	2.20	3.45	5.35	7.84	10.6	12.6	14.4	16.8	18.5	22.5	6
7	0.989	1.24	1.69	2.17	2.83	4.25	6.35	9.04	12.0	14.1	16.0	18.5	20.3	24.3	7
8	1.34	1.65	2.18	2.73	3.49	5.07	7.34	10.2	13.4	15.5	17.5	20.1	22.0	26.1	8
9	1.73	2.09	2.70	3.33	4.17	5.90	8.34	11.4	14.7	16.9	19.0	21.7	23.6	27.9	9
10	2.16	2.56	3.25	3.94	4.87	6.74	9.34	12.5	16.0	18.3	20.5	23.2	25.2	29.6	10
11	2.60	3.05	3.82	4.57	5.58	7.58	10.3	13.7	17.3	19.7	21.9	24.7	26.8	31.3	11
12	3.07	3.57	4.40	5.23	6.30	8.44	11.3	14.8	18.5	21.0	23.3	26.2	28.3	32.9	12
13	3.57	4.11	5.01	5.89	7.04	9.30	12.3	16.0	19.8	22.4	24.7	27.7	29.8	34.5	13
14	4.07	4.66	5.63	6.57	7.79	10.2	13.3	17.1	21.1	23.7	26.1	29.1	31.3	36.1	14
15	4.60	5.23	6.26	7.26	8.55	11.0	14.3	18.2	22.3	25.0	27.5	30.6	32.8	37.7	15
16	5.14	5.81	6.91	7.96	9.31	11.9	15.3	19.4	23.5	26.3	28.8	32.0	34.3	39.3	16
17	5.70	6.41	7.56	8.67	10.1	12.8	16.3	20.5	24.8	27.6	30.2	33.4	35.7	40.8	17
18	6.26	7.01	8.23	9.39	10.9	13.7	17.3	21.6	26.0	28.9	31.5	34.8	37.2	42.3	18
19	6.84	7.63	8.91	10.1	11.7	14.6	18.3	22.7	27.2	30.1	32.9	36.2	38.6	43.8	19
20	7.43	8.26	9.59	10.9	12.4	15.5	19.3	23.8	28.4	31.4	34.2	37.6	40.0	45.3	20
21	8.03	8.90	10.3	11.6	13.2	16.3	20.3	24.9	29.6	32.7	35.5	38.9	41.4	46.8	21
22	8.64	9.54	11.0	12.3	14.0	17.2	21.3	26.0	30.8	33.9	36.8	40.3	42.8	48.3	22
23	9.26	10.2	11.7	13.1	14.8	18.1	22.3	27.1	32.0	35.2	38.1	41.6	44.2	49.7	23
24	9.89	10.9	12.4	13.8	15.7	19.0	23.3	28.2	33.2	36.4	39.4	43.0	45.6	51.2	24
25	10.5	11.5	13.1	14.6	16.5	19.9	24.3	29.3	34.4	37.7	40.6	44.3	46.9	52.6	25
26	11.2	12.2	13.8	15.4	17.3	20.8	25.3	30.4	35.6	38.9	41.9	45.6	48.3	54.1	26
27	11.8	12.9	14.6	16.2	18.1	21.7	26.3	31.5	36.7	40.1	43.2	47.0	49.6	55.5	27
28	12.5	13.6	15.3	16.9	18.9	22.7	27.3	32.6	37.9	41.3	44.5	48.3	51.0	56.9	28
29	13.1	14.3	16.0	17.7	19.8	23.6	28.3	33.7	39.1	42.6	45.7	49.6	52.3	58.3	29
30	13.8	15.0	16.8	18.5	20.6	24.5	29.3	34.8	40.3	43.8	47.0	50.9	53.7	59.7	30

Table of $\chi^2_\varepsilon(p)$ such that $\Pr\{\chi^2(p) > \chi^2_\varepsilon(p)\} = \varepsilon$, where p is the number of degrees of freedom

TABLE G TAIL AREAS OF THE t-DISTRIBUTION

v	ε					
	0.25	0.10	0.05	0.025	0.01	0.005
1	1.00	3.08	6.31	12.71	31.82	63.66
2	0.82	1.89	2.92	4.30	6.96	9.92
3	0.76	1.64	2.35	3.18	4.54	5.84
4	0.74	1.53	2.13	2.78	3.75	4.60
5	0.73	1.48	2.02	2.57	3.36	4.03
6	0.72	1.44	1.94	2.45	3.14	3.71
7	0.71	1.42	1.90	2.36	3.00	3.50
8	0.71	1.40	1.86	2.31	2.90	3.36
9	0.70	1.38	1.83	2.26	2.82	3.25
10	0.70	1.37	1.81	2.23	2.76	3.17
11	0.70	1.36	1.80	2.20	2.72	3.11
12	0.70	1.36	1.78	2.18	2.68	3.06
13	0.69	1.35	1.77	2.16	2.65	3.01
14	0.69	1.34	1.76	2.14	2.62	3.00
15	0.69	1.34	1.75	2.13	2.60	2.95
16	0.69	1.34	1.75	2.12	2.58	2.92
17	0.69	1.33	1.74	2.11	2.57	2.90
18	0.69	1.33	1.73	2.10	2.55	2.88
19	0.69	1.33	1.73	2.09	2.54	2.86
20	0.69	1.33	1.72	2.09	2.53	2.84
30	0.68	1.31	1.70	2.04	2.46	2.75
40	0.68	1.30	1.68	2.02	2.42	2.70
60	0.68	1.30	1.67	2.00	2.39	2.66
120	0.68	1.29	1.66	1.98	2.36	2.62
∞	0.67	1.28	1.64	1.96	2.33	2.58

Table of $t_\varepsilon(v)$ such that $\Pr\{t(v) > t_\varepsilon(v)\} = \varepsilon$, where v is the number of degrees of freedom

Collection of Time Series Used for Examples in the Text

Series A   Chemical process concentration readings: every two hours.

Series B   IBM common stock closing prices: Daily, 17th May 1961–2nd November 1962.

Series B′   IBM common stock closing prices: Daily, 29th June 1959–30th June 1960.

Series C   Chemical process temperature readings; every minute.

Series D   Chemical process viscosity readings: every hour.

Series E   Wölfer sunspot numbers: yearly.

Series F   Series of 70 consecutive yields from a batch chemical process (tabulated in Table 2.1 in the text).

Series G   International airline passengers: Monthly totals (thousands of passengers) January 1949–December 1960.

Series J   Gas furnace data.

Series K   Simulated dynamic data with two inputs.

Series L   Pilot scheme data.

Series M   Sales data with leading indicator.

SERIES A CHEMICAL PROCESS CONCENTRATION READINGS: EVERY TWO HOURS*

1	17.0	41	17.6	81	16.8	121	16.9	161	17.1
2	16.6	42	17.5	82	16.7	122	17.1	162	17.1
3	16.3	43	16.5	83	16.4	123	16.8	163	17.1
4	16.1	44	17.8	84	16.5	124	17.0	164	17.4
5	17.1	45	17.3	85	16.4	125	17.2	165	17.2
6	16.9	46	17.3	86	16.6	126	17.3	166	16.9
7	16.8	47	17.1	87	16.5	127	17.2	167	16.9
8	17.4	48	17.4	88	16.7	128	17.3	168	17.0
9	17.1	49	16.9	89	16.4	129	17.2	169	16.7
10	17.0	50	17.3	90	16.4	130	17.2	170	16.9
11	16.7	51	17.6	91	16.2	131	17.5	171	17.3
12	17.4	52	16.9	92	16.4	132	16.9	172	17.8
13	17.2	53	16.7	93	16.3	133	16.9	173	17.8
14	17.4	54	16.8	94	16.4	134	16.9	174	17.6
15	17.4	55	16.8	95	17.0	135	17.0	175	17.5
16	17.0	56	17.2	96	16.9	136	16.5	176	17.0
17	17.3	57	16.8	97	17.1	137	16.7	177	16.9
18	17.2	58	17.6	98	17.1	138	16.8	178	17.1
19	17.4	59	17.2	99	16.7	139	16.7	179	17.2
20	16.8	60	16.6	100	16.9	140	16.7	180	17.4
21	17.1	61	17.1	101	16.5	141	16.6	181	17.5
22	17.4	62	16.9	102	17.2	142	16.5	182	17.9
23	17.4	63	16.6	103	16.4	143	17.0	183	17.0
24	17.5	64	18.0	104	17.0	144	16.7	184	17.0
25	17.4	65	17.2	105	17.0	145	16.7	185	17.0
26	17.6	66	17.3	106	16.7	146	16.9	186	17.2
27	17.4	67	17.0	107	16.2	147	17.4	187	17.3
28	17.3	68	16.9	108	16.6	148	17.1	188	17.4
29	17.0	69	17.3	109	16.9	149	17.0	189	17.4
30	17.8	70	16.8	110	16.5	150	16.8	190	17.0
31	17.5	71	17.3	111	16.6	151	17.2	191	18.0
32	18.1	72	17.4	112	16.6	152	17.2	192	18.2
33	17.5	73	17.7	113	17.0	153	17.4	193	17.6
34	17.4	74	16.8	114	17.1	154	17.2	194	17.8
35	17.4	75	16.9	115	17.1	155	16.9	195	17.7
36	17.1	76	17.0	116	16.7	156	16.8	196	17.2
37	17.6	77	16.9	117	16.8	157	17.0	197	17.4
38	17.7	78	17.0	118	16.3	158	17.4		
39	17.4	79	16.6	119	16.6	159	17.2		
40	17.8	80	16.7	120	16.8	160	17.2		

* 197 Observations

Series B IBM COMMON STOCK CLOSING PRICES: DAILY, 17TH MAY 1961–2ND NOVEMBER 1962*

460	471	527	580	551	523	333	394	330
457	467	540	579	551	516	330	393	340
452	473	542	584	552	511	336	409	339
459	481	538	581	553	518	328	411	331
462	488	541	581	557	517	316	409	345
459	490	541	577	557	520	320	408	352
463	489	547	577	548	519	332	393	346
479	489	553	578	547	519	320	391	352
493	485	559	580	545	519	333	388	357
490	491	557	586	545	518	344	396	
492	492	557	583	539	513	339	387	
498	494	560	581	539	499	350	383	
499	499	571	576	535	485	351	388	
497	498	571	571	537	454	350	382	
496	500	569	575	535	462	345	384	
490	497	575	575	536	473	350	382	
489	494	580	573	537	482	359	383	
478	495	584	577	543	486	375	383	
487	500	585	582	548	475	379	388	
491	504	590	584	546	459	376	395	
487	513	599	579	547	451	382	392	
482	511	603	572	548	453	370	386	
479	514	599	577	549	446	365	383	
478	510	596	571	553	455	367	377	
479	509	585	560	553	452	372	364	
477	515	587	549	552	457	373	369	
479	519	585	556	551	449	363	355	
475	523	581	557	550	450	371	350	
479	519	583	563	553	435	369	353	
476	523	592	564	554	415	376	340	
476	531	592	567	551	398	387	350	
478	547	596	561	551	399	387	349	
479	551	596	559	545	361	376	358	
477	547	595	553	547	383	385	360	
476	541	598	553	547	393	385	360	
475	545	598	553	537	385	380	366	
475	549	595	547	539	360	373	359	
473	545	595	550	538	364	382	356	
474	549	592	544	533	365	377	355	
474	547	588	541	525	370	376	367	
474	543	582	532	513	374	379	357	
465	540	576	525	510	359	386	361	
466	539	578	542	521	335	387	355	
467	532	589	555	521	323	386	348	
471	517	585	558	521	306	389	343	

* 369 Observations (Read downwards).

SERIES B′   IBM COMMON STOCK CLOSING PRICES: DAILY, 29TH JUNE 1959–30TH JUNE 1960*

445	425	406	441	415	461
448	421	407	437	420	463
450	414	410	427	420	463
447	410	408	423	424	461
451	411	408	424	426	465
453	406	409	428	423	473
454	406	410	428	423	473
454	413	409	431	425	475
459	411	405	425	431	499
440	410	406	423	436	485
446	405	405	420	436	491
443	409	407	426	440	496
443	410	409	418	436	504
440	405	407	416	443	504
439	401	409	419	445	509
435	401	425	418	439	511
435	401	425	416	443	524
436	414	428	419	445	525
435	419	436	425	450	541
435	425	442	421	461	531
435	423	442	422	471	529
433	411	433	422	467	530
429	414	435	417	462	531
428	420	433	420	456	527
425	412	435	417	464	525
427	415	429	418	463	519
425	412	439	419	465	514
422	412	437	419	464	509
409	411	439	417	456	505
407	412	438	419	460	513
423	409	435	422	458	525
422	407	433	423	453	519
417	408	437	422	453	519
421	415	437	421	449	522
424	413	444	421	447	522
414	413	441	419	453	
419	410	440	418	450	
429	405	441	421	459	
426	410	439	420	457	
425	412	439	413	453	
424	413	438	413	455	
425	411	437	408	453	
425	411	441	409	450	
424	409	442	415	456	

* 255 Observations (Read downwards).

Series C CHEMICAL   PROCESS   TEMPERATURE   READINGS:
EVERY MINUTE*

26.6	19.6	24.4	21.1	24.4
27.0	19.6	24.4	20.9	24.2
27.1	19.6	24.4	20.8	24.2
27.1	19.6	24.4	20.8	24.1
27.1	19.6	24.5	20.8	24.1
27.1	19.7	24.5	20.8	24.0
26.9	19.9	24.4	20.9	24.0
26.8	20.0	24.3	20.8	24.0
26.7	20.1	24.2	20.8	23.9
26.4	20.2	24.2	20.7	23.8
26.0	20.3	24.0	20.7	23.8
25.8	20.6	23.9	20.8	23.7
25.6	21.6	23.7	20.9	23.7
25.2	21.9	23.6	21.2	23.6
25.0	21.7	23.5	21.4	23.7
24.6	21.3	23.5	21.7	23.6
24.2	21.2	23.5	21.8	23.6
24.0	21.4	23.5	21.9	23.6
23.7	21.7	23.5	22.2	23.5
23.4	22.2	23.7	22.5	23.5
23.1	23.0	23.8	22.8	23.4
22.9	23.8	23.8	23.1	23.3
22.8	24.6	23.9	23.4	23.3
22.7	25.1	23.9	23.8	23.3
22.6	25.6	23.8	24.1	23.4
22.4	25.8	23.7	24.6	23.4
22.2	26.1	23.6	24.9	23.3
22.0	26.3	23.4	24.9	23.2
21.8	26.3	23.2	25.1	23.3
21.4	26.2	23.0	25.0	23.3
20.9	26.0	22.8	25.0	23.2
20.3	25.8	22.6	25.0	23.1
19.7	25.6	22.4	25.0	22.9
19.4	25.4	22.0	24.9	22.8
19.3	25.2	21.6	24.8	22.6
19.2	24.9	21.3	24.7	22.4
19.1	24.7	21.2	24.6	22.2
19.0	24.5	21.2	24.5	21.8
18.9	24.4	21.1	24.5	21.3
18.9	24.4	21.0	24.5	20.8
19.2	24.4	20.9	24.5	20.2
19.3	24.4	21.0	24.5	19.7
19.3	24.4	21.0	24.5	19.3
19.4	24.3	21.1	24.5	19.1
19.5	24.4	21.2	24.4	19.0
				18.8

* 226 Observations (Read downwards).

SERIES D   CHEMICAL PROCESS VISCOSITY READINGS: EVERY HOUR*

8.0	8.6	9.3	9.8	9.4	9.6	9.4
8.0	8.4	9.5	9.6	9.6	9.6	10.0
7.4	8.3	9.4	9.6	9.6	9.6	10.0
8.0	8.4	9.0	9.4	9.6	9.6	10.0
8.0	8.3	9.0	9.4	10.0	9.6	10.2
8.0	8.3	8.8	9.4	10.0	9.0	10.0
8.0	8.1	9.0	9.4	9.6	9.4	10.0
8.8	8.2	8.8	9.6	9.2	9.4	9.6
8.4	8.3	8.6	9.6	9.2	9.4	9.0
8.4	8.5	8.6	9.4	9.2	9.6	9.0
8.0	8.1	8.0	9.4	9.0	9.4	8.6
8.2	8.1	8.0	9.0	9.0	9.6	9.0
8.2	7.9	8.0	9.4	9.6	9.6	9.6
8.2	8.3	8.0	9.4	9.8	9.8	9.6
8.4	8.1	8.6	9.6	10.2	9.8	9.0
8.4	8.1	8.0	9.4	10.0	9.8	9.0
8.4	8.1	8.0	9.2	10.0	9.8	8.9
8.6	8.4	8.0	8.8	10.0	9.6	8.8
8.8	8.7	7.6	8.8	9.4	9.2	8.7
8.6	9.0	8.6	9.2	9.2	9.6	8.6
8.6	9.3	9.6	9.2	9.6	9.2	8.3
8.6	9.3	9.6	9.6	9.7	9.2	7.9
8.6	9.5	10.0	9.6	9.7	9.6	8.5
8.6	9.3	9.4	9.8	9.8	9.6	8.7
8.8	9.5	9.3	9.8	9.8	9.6	8.9
8.9	9.5	9.2	10.0	9.8	9.6	9.1
9.1	9.5	9.5	10.0	10.0	9.6	9.1
9.5	9.5	9.5	9.4	10.0	10.0	9.1
8.5	9.5	9.5	9.8	8.6	10.0	
8.4	9.5	9.9	8.8	9.0	10.4	
8.3	9.9	9.9	8.8	9.4	10.4	
8.2	9.5	9.5	8.8	9.4	9.8	
8.1	9.7	9.3	8.8	9.4	9.0	
8.3	9.1	9.5	9.6	9.4	9.6	
8.4	9.1	9.5	9.6	9.4	9.8	
8.7	8.9	9.1	9.6	9.6	9.6	
8.8	9.3	9.3	9.2	10.0	8.6	
8.8	9.1	9.5	9.2	10.0	8.0	
9.2	9.1	9.3	9.0	9.8	8.0	
9.6	9.3	9.1	9.0	9.8	8.0	
9.0	9.5	9.3	9.0	9.7	8.0	
8.8	9.3	9.1	9.4	9.6	8.4	
8.6	9.3	9.5	9.0	9.4	8.8	
8.6	9.3	9.4	9.0	9.2	8.4	
8.8	9.9	9.5	9.4	9.0	8.4	
8.8	9.7	9.6	9.4	9.4	9.0	
8.6	9.1	10.2	9.6	9.6	9.0	

* 310 Observations (Read downwards).

SERIES E   WÖLFER SUNSPOT NUMBERS: YEARLY*

1770	101	1795	21	1820	16	1845	40
1771	82	1796	16	1821	7	1846	62
1772	66	1797	6	1822	4	1847	98
1773	35	1798	4	1823	2	1848	124
1774	31	1799	7	1824	8	1849	96
1775	7	1800	14	1825	17	1850	66
1776	20	1801	34	1826	36	1851	64
1777	92	1802	45	1827	50	1852	54
1778	154	1803	43	1828	62	1853	39
1779	125	1804	48	1829	67	1854	21
1780	85	1805	42	1830	71	1855	7
1781	68	1806	28	1831	48	1856	4
1782	38	1807	10	1832	28	1857	23
1783	23	1808	8	1833	8	1858	55
1784	10	1809	2	1834	13	1859	94
1785	24	1810	0	1835	57	1860	96
1786	83	1811	1	1836	122	1861	77
1787	132	1812	5	1837	138	1862	59
1788	131	1813	12	1838	103	1863	44
1789	118	1814	14	1839	86	1864	47
1790	90	1815	35	1840	63	1865	30
1791	67	1816	46	1841	37	1866	16
1792	60	1817	41	1842	24	1867	7
1793	47	1818	30	1843	11	1868	37
1794	41	1819	24	1844	15	1869	74

* 100 Observations

SERIES F   YIELDS FROM BATCH CHEMICAL PROCESS*

47	44	50	62	68
64	80	71	44	38
23	55	56	64	50
71	37	74	43	60
38	74	50	52	39
64	51	58	38	59
55	57	45	59	40
41	50	54	55	57
59	60	36	41	54
48	45	54	53	23
71	57	48	49	
35	50	55	34	
57	45	45	35	
40	25	57	54	
58	59	50	45	

* 70 Observations (Read downwards). This Series also appears in Table 2.1

Series G INTERNATIONAL AIRLINE PASSENGERS: MONTHLY TOTALS (THOUSANDS OF PASSENGERS) JANUARY 1949–DECEMBER 1960*

	Jan.	Feb.	Mar.	Apr.	May	June	July	Aug.	Sept.	Oct.	Nov.	Dec.
1949	112	118	132	129	121	135	148	148	136	119	104	118
1950	115	126	141	135	125	149	170	170	158	133	114	140
1951	145	150	178	163	172	178	199	199	184	162	146	166
1952	171	180	193	181	183	218	230	242	209	191	172	194
1953	196	196	236	235	229	243	264	272	237	211	180	201
1954	204	188	235	227	234	264	302	293	259	229	203	229
1955	242	233	267	269	270	315	364	347	312	274	237	278
1956	284	277	317	313	318	374	413	405	355	306	271	306
1957	315	301	356	348	355	422	465	467	404	347	305	336
1958	340	318	362	348	363	435	491	505	404	359	310	337
1959	360	342	406	396	420	472	548	559	463	407	362	405
1960	417	391	419	461	472	535	622	606	508	461	390	432

* 144 Observations

Series J   Gas Furnace Data*

t	X_t	Y_t	t	X_t	Y_t	t	X_t	Y_t
1	−0.109	53.8	51	1.608	46.9	101	−0.288	51.0
2	0.000	53.6	52	1.905	47.8	102	−0.153	51.8
3	0.178	53.5	53	2.023	48.2	103	−0.109	52.4
4	0.339	53.5	54	1.815	48.3	104	−0.187	53.0
5	0.373	53.4	55	0.535	47.9	105	−0.255	53.4
6	0.441	53.1	56	0.122	47.2	106	−0.229	53.6
7	0.461	52.7	57	0.009	47.2	107	−0.007	53.7
8	0.348	52.4	58	0.164	48.1	108	0.254	53.8
9	0.127	52.2	59	0.671	49.4	109	0.330	53.8
10	−0.180	52.0	60	1.019	50.6	110	0.102	53.8
11	−0.588	52.0	61	1.146	51.5	111	−0.423	53.3
12	−1.055	52.4	62	1.155	51.6	112	−1.139	53.0
13	−1.421	53.0	63	1.112	51.2	113	−2.275	52.9
14	−1.520	54.0	64	1.121	50.5	114	−2.594	53.4
15	−1.302	54.9	65	1.223	50.1	115	−2.716	54.6
16	−0.814	56.0	66	1.257	49.8	116	−2.510	56.4
17	−0.475	56.8	67	1.157	49.6	117	−1.790	58.0
18	−0.193	56.8	68	0.913	49.4	118	−1.346	59.4
19	0.088	56.4	69	0.620	49.3	119	−1.081	60.2
20	0.435	55.7	70	0.255	49.2	120	−0.910	60.0
21	0.771	55.0	71	−0.280	49.3	121	−0.876	59.4
22	0.866	54.3	72	−1.080	49.7	122	−0.885	58.4
23	0.875	53.2	73	−1.551	50.3	123	−0.800	57.6
24	0.891	52.3	74	−1.799	51.3	124	−0.544	56.9
25	0.987	51.6	75	−1.825	52.8	125	−0.416	56.4
26	1.263	51.2	76	−1.456	54.4	126	−0.271	56.0
27	1.775	50.8	77	−0.944	56.0	127	0.000	55.7
28	1.976	50.5	78	−0.570	56.9	128	0.403	55.3
29	1.934	50.0	79	−0.431	57.5	129	0.841	55.0
30	1.866	49.2	80	−0.577	57.3	130	1.285	54.4
31	1.832	48.4	81	−0.960	56.6	131	1.607	53.7
32	1.767	47.9	82	−1.616	56.0	132	1.746	52.8
33	1.608	47.6	83	−1.875	55.4	133	1.683	51.6
34	1.265	47.5	84	−1.891	55.4	134	1.485	50.6
35	0.790	47.5	85	−1.746	56.4	135	0.993	49.4
36	0.360	47.6	86	−1.474	57.2	136	0.648	48.8
37	0.115	48.1	87	−1.201	58.0	137	0.577	48.5
38	0.088	49.0	88	−0.927	58.4	138	0.577	48.7
39	0.331	50.0	89	−0.524	58.4	139	0.632	49.2
40	0.645	51.1	90	0.040	58.1	140	0.747	49.8
41	0.960	51.8	91	0.788	57.7	141	0.900	50.4
42	1.409	51.9	92	0.943	57.0	142	0.993	50.7
43	2.670	51.7	93	0.930	56.0	143	0.968	50.9
44	2.834	51.2	94	1.006	54·7	144	0.790	50.7
45	2.812	50.0	95	1.137	53.2	145	0.399	50.5
46	2.483	48.3	96	1.198	52.1	146	−0.161	50.4
47	1.929	47.0	97	1.054	51.6	147	−0.553	50.2
48	1.485	45.8	98	0.595	51.0	148	−0.603	50.4
49	1.214	45.6	99	−0.080	50.5	149	−0.424	51.2
50	1.239	46.0	100	−0.314	50.4	150	−0.194	52.3

* X : 0.60–0.04 (input gas rate in cu. ft/min) Y: % CO_2 in outlet gas
  Sampling interval 9 seconds $N = 296$ pairs of data points

t	X_t	Y_t	t	X_t	Y_t	t	X_t	Y_t
151	−0.049	53.2	201	−2.473	55.6	251	0.185	56.3
152	0.060	53.9	202	−2.330	58.0	252	0.662	56.4
153	0.161	54.1	203	−2.053	59.5	253	0.709	56.4
154	0.301	54.0	204	−1.739	60.0	254	0.605	56.0
155	0.517	53.6	205	−1.261	60.4	255	0.501	55.2
156	0.566	53.2	206	−0.569	60.5	256	0.603	54.0
157	0.560	53.0	207	−0.137	60.2	257	0.943	53.0
158	0.573	52.8	208	−0.024	59.7	258	1.223	52.0
159	0.592	52.3	209	−0.050	59.0	259	1.249	51.6
160	0.671	51.9	210	−0.135	57.6	260	0.824	51.6
161	0.933	51.6	211	−0.276	56.4	261	0.102	51.1
162	1.337	51.6	212	−0.534	55.2	262	0.025	50.4
163	1.460	51.4	213	−0.871	54.5	263	0.382	50.0
164	1.353	51.2	214	−1.243	54.1	264	0.922	50.0
165	0.772	50.7	215	−1.439	54.1	265	1.032	52.0
166	0.218	50.0	216	−1.422	54.4	266	0.866	54.0
167	−0.237	49.4	217	−1.175	55.5	267	0.527	55.1
168	−0.714	49.3	218	−0.813	56.2	268	0.093	54.5
169	−1.099	49.7	219	−0.634	57.0	269	−0.458	52.8
170	−1.269	50.6	220	−0.582	57.3	270	−0.748	51.4
171	−1.175	51.8	221	−0.625	57.4	271	−0.947	50.8
172	−0.676	53.0	222	−0.713	57.0	272	−1.029	51.2
173	0.033	54.0	223	−0.848	56.4	273	−0.928	52.0
174	0.556	55.3	224	−1.039	55.9	274	−0.645	52.8
175	0.643	55.9	225	−1.346	55.5	275	−0.424	53.8
176	0.484	55.9	226	−1.628	55.3	276	−0.276	54.5
177	0.109	54.6	227	−1.619	55.2	277	−0.158	54.9
178	−0.310	53.5	228	−1.149	55.4	278	−0.033	54.9
179	−0.697	52.4	229	−0.488	56.0	279	0.102	54.8
180	−1.047	52.1	230	−0.160	56.5	280	0.251	54.4
181	−1.218	52.3	231	−0.007	57.1	281	0.280	53.7
182	−1.183	53.0	232	−0.092	57.3	282	0.000	53.3
183	−0.873	53.8	233	−0.620	56.8	283	−0.493	52.8
184	−0.336	54.6	234	−1.086	55.6	284	−0.759	52.6
185	0.063	55.4	235	−1.525	55.0	285	−0.824	52.6
186	0.084	55.9	236	−1.858	54.1	286	−0.740	53.0
187	0.000	55.9	237	−2.029	54.3	287	−0.528	54.3
188	0.001	55.2	238	−2.024	55.3	288	−0.204	56.0
189	0.209	54.4	239	−1.961	56.4	289	0.034	57.0
190	0.556	53.7	240	−1.952	57.2	290	0.204	58.0
191	0.782	53.6	241	−1.794	57.8	291	0.253	58.6
192	0.858	53.6	242	−1.302	58.3	292	0.195	58.5
193	0.918	53.2	243	−1.030	58.6	293	0.131	58.3
194	0.862	52.5	244	−0.918	58.8	294	0.017	57.8
195	0.416	52.0	245	−0.798	58.8	295	−0.182	57.3
196	−0.336	51.4	246	−0.867	58.6	296	−0.262	57.0
197	−0.959	51.0	247	−1.047	58.0			
198	−1.813	50.9	248	−1.123	57.4			
199	−2.378	52.4	249	−0.876	57.0			
200	−2.499	53.5	250	−0.395	56.4			

Series K   SIMULATED DYNAMIC DATA WITH TWO INPUTS*

t	X_{1t}	X_{2t}	Y_t	t	X_{1t}	X_{2t}	Y_t
−2	0	0	58.3	30			65.8
−1			61.8	31			67.4
0			64.2	32	−1	−1	64.7
1			62.1	33			65.7
2	−1	1	55.1	34			67.5
3			50.6	35			58.2
4			47.8	36			57.0
5			49.7	37	−1	1	54.7
6			51.6	38			54.9
7	1	−1	58.5	39			48.4
8			61.5	40			49.7
9			63.3	41			53.1
10			65.9	42	1	−1	50.2
11			70.9	43			51.7
12	−1	−1	65.8	44			57.4
13			57.6	45			62.6
14			56.1	46			65.8
15			58.2	47	−1	−1	61.5
16			61.7	48			61.5
17	1	1	59.2	49			56.8
18			57.9	50			62.3
19			61.3	51			57.7
20			60.8	52	−1	1	54.0
21			63.6	53			45.2
22	1	−1	69.5	54			51.9
23			69.3	55			45.6
24			70.5	56			46.2
25			68.0	57	1	1	50.2
26			68.1	58			54.6
27	1	1	65.0	59			55.6
28			71.9	60	0	0	60.4
29			64.8	61			59.4

* 64 Observations

SERIES L   PILOT SCHEME DATA*

t	x_t	ε_t	t	x_t	ε_t	t	x_t	ε_t
1	30	−4	53	−60	6	105	55	−4
2	0	−2	54	50	−2	106	0	2
3	−10	0	55	−10	0	107	−90	8
4	0	0	56	40	−4	108	40	0
5	−40	4	57	40	−6	109	0	0
6	0	2	58	−30	0	110	80	−8
7	−10	2	59	20	−2	111	−20	−2
8	10	0	60	−30	2	112	−10	0
9	20	−2	61	10	0	113	−70	6
10	50	−6	62	−20	2	114	−30	6
11	−10	−2	63	30	−2	115	−10	4
12	−55	4	64	−50	4	116	30	−1
13	0	2	65	10	−2	117	−5	0
14	10	0	66	10	−2	118	−60	6
15	0	−2	67	10	−2	119	70	−4
16	10	−2	68	−30	0	120	40	−6
17	−70	6	69	0	0	121	10	−4
18	30	0	70	−10	2	122	20	−4
19	−20	2	71	−10	3	123	10	−3
20	10	0	72	15	0	124	0	−2
21	0	0	73	20	−2	125	−70	6
22	0	0	74	−50	4	126	50	−2
23	20	−2	75	20	0	127	30	−4
24	30	−4	76	0	0	128	0	−2
25	0	−2	77	0	0	129	−10	0
26	−10	0	78	0	0	130	0	0
27	−20	2	79	0	0	131	−40	4
28	−30	4	80	−40	4	132	0	2
29	0	2	81	−100	12	133	−10	2
30	10	0	82	0	8	134	10	0
31	20	−2	83	0	−12	135	0	0
32	−10	0	84	50	−15	136	80	−8
33	0	0	85	85	−15	137	−80	4
34	20	−2	86	5	−12	138	20	4
35	10	−2	87	40	−14	139	20	0
36	−10	0	88	10	−8	140	−10	2
37	0	0	89	−60	2	141	10	0
38	0	0	90	−50	6	142	0	0
39	0	0	91	−50	8	143	−20	2
40	0	0	92	40	0	144	20	−1
41	0	0	93	0	0	145	55	−6
42	0	0	94	0	0	146	0	−3
43	20	−2	95	−20	2	147	25	−4
44	−50	4	96	−30	4	148	20	−4
45	20	0	97	−60	8	149	−60	4
46	0	0	98	−20	6	150	−40	6
47	0	0	99	−30	6	151	10	4
48	40	−4	100	30	0	152	20	0
49	0	−2	101	−40	4	153	60	−6
50	50	−6	102	80	−6	154	−50	2
51	−40	0	103	−40	0	155	−10	2
52	−50	3	104	−20	2	156	−30	4

* 312 Observations

t	x_t	ε_t	t	x_t	ε_t	t	x_t	ε_t
157	20	0	209	−40	4	261	−25	4
158	0	0	210	40	−2	262	35	−2
159	20	−2	211	−90	8	263	70	8
160	10	−2	212	40	0	264	−10	−5
161	10	−2	213	0	0	265	100	−20
162	10	−22	214	0	0	266	−20	−8
163	50	−6	215	0	0	267	−40	0
164	−30	0	216	20	−2	268	−20	2
165	−30	6	217	90	−10	269	10	0
166	−90	12	218	30	−8	270	0	0
167	60	0	219	20	−6	271	0	0
168	−40	4	220	30	−6	272	−20	2
169	20	0	221	30	−6	273	−50	6
170	0	0	222	30	−6	274	50	−2
171	20	−2	223	30	−6	275	30	−4
172	10	−2	224	−90	6	276	60	−8
173	−30	2	225	10	2	277	−40	0
174	−30	4	226	10	2	278	−20	2
175	0	2	227	−30	4	279	−10	2
176	50	−4	228	−20	4	280	10	0
177	−60	4	229	40	−2	281	−110	13
178	20	0	230	10	−2	282	15	4
179	0	0	231	10	−2	283	30	−2
180	40	−8	232	10	−2	284	0	−1
181	80	−12	233	−100	12	285	25	−3
182	20	−8	234	10	6	286	−5	−1
183	−100	6	235	45	−2	287	−15	1
184	−30	6	236	30	−4	288	45	−4
185	30	0	237	30	−5	289	40	−6
186	−20	2	238	−15	−1	290	−50	2
187	−30	4	239	−5	0	291	−10	2
188	20	0	240	10	−1	292	−50	6
189	60	−6	241	−85	8	293	20	1
190	−10	−2	242	0	4	294	5	0
191	30	−4	243	0	0	295	−40	4
192	−40	2	244	60	−4	296	0	6
193	30	−2	245	40	−6	297	−60	8
194	−20	1	246	−30	0	298	40	0
195	5	0	247	−40	4	299	−20	2
196	−20	2	248	−40	6	300	130	−12
197	−30	4	249	50	−2	301	−20	−4
198	20	0	250	10	−2	302	0	−2
199	10	−1	251	30	−4	303	30	−4
200	−15	1	252	−40	2	304	−20	0
201	−75	8	253	10	0	305	60	6
202	−40	8	254	−40	4	306	10	−4
203	−40	6	255	40	−2	307	−10	1
204	90	−6	256	−30	2	308	−25	2
205	90	−12	257	−50	6	309	0	1
206	80	−14	258	0	3	310	15	−1
207	−45	−2	259	−45	6	311	−5	0
208	−10	0	260	−20	5	312	0	0

SERIES M   SALES DATA WITH LEADING INDICATOR*

t	Leading Indicator X_t	Sales Y_t	t	Leading Indicator X_t	Sales Y_t	t	Leading Indicator X_t	Sales Y_t
1	10.01	200.1	51	10.77	220.0	101	12.90	249.4
2	10.07	199.5	52	10.88	218.7	102	13.12	249.0
3	10.32	199.4	53	10.49	217.0	103	12.47	249.9
4	9.75	198.9	54	10.50	215.9	104	12.47	250.5
5	10.33	199.0	55	11.00	215.8	105	12.94	251.5
6	10.13	200.2	56	10.98	214.1	106	13.10	249.0
7	10.36	198.6	57	10.61	212.3	107	12.91	247.6
8	10.32	200.0	58	10.48	213.9	108	13.39	248.8
9	10.13	200.3	59	10.53	214.6	109	13.13	250.4
10	10.16	201.2	60	11.07	213.6	110	13.34	250.7
11	10.58	201.6	61	10.61	212.1	111	13.34	253.0
12	10.62	201.5	62	10.86	211.4	112	13.14	253.7
13	10.86	201.5	63	10.34	213.1	113	13.49	255.0
14	11.20	203.5	64	10.78	212.9	114	13.87	256.2
15	10.74	204.9	65	10.80	213.3	115	13.39	256.0
16	10.56	207.1	66	10.33	211.5	116	13.59	257.4
17	10.48	210.5	67	10.44	212.3	117	13.27	260.4
18	10.77	210.5	68	10.50	213.0	118	13.70	260.0
19	11.33	209.8	69	10.75	211.0	119	13.20	261.3
20	10.96	208.8	70	10.40	210.7	120	13.32	260.4
21	11.16	209.5	71	10.40	210.1	121	13.15	261.6
22	11.70	213.2	72	10.34	211.4	122	13.30	260.8
23	11.39	213.7	73	10.55	210.0	123	12.94	259.8
24	11.42	215.1	74	10.46	209.7	124	13.29	259.0
25	11.94	218.7	75	10.82	208.8	125	13.26	258.9
26	11.24	219.8	76	10.91	208.8	126	13.08	257.4
27	11.59	220.5	77	10.87	208.8	127	13.24	257.7
28	10.96	223.8	78	10.67	210.6	128	13.31	257.9
29	11.40	222.8	79	11.11	211.9	129	13.52	257.4
30	11.02	223.8	80	10.88	212.8	130	13.02	257.3
31	11.01	221.7	81	11.28	212.5	131	13.25	257.6
32	11.23	222.3	82	11.27	214.8	132	13.12	258.9
33	11.33	220.8	83	11.44	215.3	133	13.26	257.8
34	10.83	219.4	84	11.52	217.5	134	13.11	257.7
35	10.84	220.1	85	12.10	218.8	135	13.30	257.2
36	11.14	220.6	86	11.83	220.7	136	13.06	257.5
37	10.38	218.9	87	12.62	222.2	137	13.32	256.8
38	10.90	217.8	88	12.41	226.7	138	13.10	257.5
39	11.05	217.7	89	12.43	228.4	139	13.27	257.0
40	11.11	215.0	90	12.73	233.2	140	13.64	257.6
41	11.01	215.3	91	13.01	235.7	141	13.58	257.3
42	11.22	215.9	92	12.74	237.1	142	13.87	257.5
43	11.21	216.7	93	12.73	240.6	143	13.53	259.6
44	11.91	216.7	94	12.76	243.8	144	13.41	261.1
45	11.69	217.7	95	12.92	245.3	145	13.25	262.9
46	10.93	218.7	96	12.64	246.0	146	13.50	263.3
47	10.99	222.9	97	12.79	246.3	147	13.58	262.8
48	11.01	224.9	98	13.05	247.7	148	13.51	261.8
49	10.84	222.2	99	12.69	247.6	149	13.77	262.2
50	10.76	220.7	100	13.01	247.8	150	13.40	262.7

* 150 Observations

References

[1] C. C. Holt, F. Modigliani, J. F. Muth and H. A. Simon, *Planning Production, Inventories and Work Force*, Prentice–Hall, New Jersey, 1963.

[2] R. G. Brown, *Smoothing, Forecasting and Prediction of Discrete Time Series*, Prentice–Hall, New Jersey, 1962.

[3] *Short Term Forecasting*, I.C.I. Monograph No. 2, Oliver and Boyd, Edinburgh, 1964.

[4] P. J. Harrison, "Short-term sales forecasting," *Applied Stat.*, **14**, 102, 1965.

[5] K. J. Aström, "Numerical identification of linear dynamic systems from normal operating records," *Theory of Self-adaptive Control Systems*, 96, Plenum Press, 1966.

[6] A. W. Hutchinson and R. J. Shelton, "Measurement of dynamic characteristics of full-scale plant using random perturbing signals: an application to a refinery distillation column," *Trans. Inst. Chem. Eng.*, **45**, 334, 1967.

[7] P. A. N. Briggs, P. H. Hammond, M. T. G. Hughes and G. O. Plumb, "Correlation analysis of process dynamics using pseudo-random binary test perturbations," *Inst. Mech. Eng.*, *Advances in Automatic Control*, Paper 7, Nottingham, U.K., April 1965.

[8] W. A. Shewhart, *The Economic control of the Quality of Manufactured Product*, Macmillan, New York, 1931.

[9] B. P. Dudding and W. J. Jennet, "Quality control charts," *British Standard 600R*, 1942.

[10] E. S. Page, "On problems in which a change in a parameter occurs at an unknown point," *Biometrika*, **44**, 249, 1957.

[11] E. S. Page, "Cumulative sum charts," *Technometrics*, **3**, 1, 1961.

[12] G. A. Barnard, "Control charts and stochastic processes," *Jour. Royal Stat. Soc.*, **B21**, 239, 1959.

[13] S. W. Roberts, "Control chart tests based on geometric moving averages," *Technometrics*, **1**, 239, 1959.

[14] G. E. P. Box and G. M. Jenkins, "Some statistical aspects of adaptive optimization and control," *Jour. Royal Stat. Soc.*, **B24**, 297, 1962.

[15] G. E. P. Box and G. M. Jenkins, "Further contributions to adaptive quality control: simultaneous estimation of dynamics: non-zero costs," *Bull. Intl. Stat. Inst.*, *34th Session*, 943, Ottawa, Canada, 1963.

[16] G. E. P. Box and G. M. Jenkins, "Mathematical models for adaptive control and optimization," *A.I.Ch.E.–I.Chem.E. Symp. Series*, **4**, 61, 1965.

538

[17] G. E. P. Box, G. M. Jenkins and D. W. Bacon, "Models for forecasting seasonal and non-seasonal time series," *Advanced Seminar on Spectral Analysis of Time Series*, ed. B. Harris, 271, John Wiley, New York, 1967.

[18] G. E. P. Box and G. M. Jenkins, "Discrete models for feedback and feedforward control," *The Future of Statistics*, ed. D. G. Watts, 201, Academic Press, New York, 1968.

[19] G. E. P. Box and G. M. Jenkins, "Some recent advances in forecasting and control, I," *Applied Stat.*, **17**, 91, 1968.

[20] G. E. P. Box and G. M. Jenkins, "Discrete models for forecasting and control," *Encyclopedia of Linguistics, Information and Control*, 162, Pergamon Press, 1969.

[21] K. D. Oughton, "Digital computer controls paper machine," *Ind. Electron.*, **5**, 358, 1965.

[22] C. C. Holt, "Forecasting trends and seasonals by exponentially weighted moving averages," *O.N.R. Memorandum, No. 52*, Carnegie Institute of Technology, 1957.

[23] P. R. Winters, "Forecasting Sales by exponentially weighted moving averages," *Management Sci.*, **6**, 324, 1960.

[24] G. U. Yule, "On a method of investigating periodicities in disturbed series, with special reference to Wölfer's sunspot numbers," *Phil. Trans.*, **A226**, 267, 1927.

[25] J. W. Tukey, "Discussion emphasizing the connection between analysis of variance and spectrum analysis," *Technometrics*, **3**, 191, 1961.

[26] G. E. P. Box and W. G. Hunter, "The experimental study of physical mechanisms," *Technometrics*, **7**, 23, 1965.

[27] G. M. Jenkins and D. G. Watts, *Spectral Analysis and Its Applications*, Holden–Day, San Francisco, 1968.

[28] M. S. Bartlett, "On the theoretical specification of sampling properties of autocorrelated time series," *Jour. Royal Stat. Soc.*, **B8**, 27, 1946.

[29] M. G. Kendall, "On the analysis of oscillatory time series," *Jour. Royal Stat. Soc.*, **108**, 93, 1945.

[30] A. Schuster, "On the investigation of hidden periodicities," *Terr. Mag.*, **3**, 13, 1898.

[31] G. C. Stokes, "Note on searching for periodicities," *Proc. Royal Soc.*, **29**, 122, 1879.

[32] G. Walker, "On periodicity in series of related terms," *Proc. Royal Soc.*, **A131**, 518, 1931.

[33] C. M. Stralkowski, "Lower order autoregressive-moving average stochastic models and their use for the characterization of abrasive cutting tools," Ph.D. Thesis, University of Wisconsin, 1968.

[34] J. Durbin, "The fitting of time series models." *Rev. Int. Inst. Stat.*, **28**, 233, 1960.

[35] M. H. Quenouille, "Approximate tests of correlation in time series," *Jour. Royal Stat. Soc.*, **B11**, 68, 1949.

[36] G. M. Jenkins, "Tests of hypotheses in the linear autoregressive model," I. *Biometrika*, **41**, 405, 1954; II. *Biometrika*, **43**, 186, 1956.

[37] H. E. Daniels, "The approximate distribution of serial correlation coefficients," *Biometrika*, **43**, 169, 1956.

[38] A. M. Yaglom, "The correlation theory of processes whose nth difference constitute a stationary process," *Matem. Sb.*, **37** (79), 141, 1955.

[39] L. A. Zadeh and J. R. Ragazzini, "An extension of Wiener's theory of prediction," *Jour. of App. Phys.*, **21**, 645, 1950.

[40] R. E. Kalman, "A new approach to linear filtering and prediction problems," *Jour. of Basic Eng.*, Series **D82**, 35, 1960.

[41] R. E. Kalman and R. S. Bucy, "New results in linear filtering and prediction theory," *Jour. of Basic Eng.*, Series **D83**, 5, 1961.

[42] G. E. P. Box and D. R. Cox, "An analysis of transformations," *Jour. Royal Stat. Soc.*, **B26**, 211, 1964.

[43] J. F. Muth, "Optimal properties of exponentially weighted forecasts of time series with permanent and transitory components," *Jour. Amer. Stat. Assoc.*, **55**, 299, 1960.

[44] H. O. Wold, *A Study in The Analysis of Stationary Time Series*, Almquist and Wicksell, Uppsala, 1938 (2nd. ed. 1954).

[45] A. Kolmogoroff, "Sur l'interpolation et l'extrapolation des suites stationnaires," *C. R. Acad. Sci. Paris*, **208**, 2043, 1939.

[46] A. Kolmogoroff, "Stationary sequences in Hilbert space," *Bull. Math. Univ. Moscow 2*, No. 6, 1941.

[47] A. Kolmogoroff, "Interpolation und Extrapolation von stationären zufälligen folgen," *Bull. Acad. Sci. (Nauk) U.S.S.R.*, *Ser. Math.*, **5**, 3, 1941.

[48] N. Wiener, *Extrapolation, Interpolation and Smoothing of Stationary Time Series*, John Wiley, New York, 1949.

[49] P. Whittle, *Prediction and Regulation by Linear Least-Squares Methods*, English Universities Press, London, 1963.

[50] R. G. Brown and R. F. Meyer, "The fundamental theorem of exponential smoothing," *Operations Res.*, **9**, 673, 1961.

[51] L. Bachelier, "Theorie de la speculation," *Ann. Sci. Éc. norm. sup.*, *Paris*, Series 3, **17**, 21, 1900.

[52] R. L. Anderson, "Distribution of the serial correlation coefficient," *Ann. Math. Stat.*, **13**, 1, 1942.

[53] A. Schuster, "On the periodicities of sunspots," *Phil. Trans. Royal Soc.*, **A206**, 69, 1906.

[54] P. A. P. Moran, "Some experiments in the prediction of sunspot numbers," *Jour. Royal Stat. Soc.*, **B16**, 112, 1954.

[55] G. T. Wilson, "Factorization of the generating function of a pure moving average process," *SIAM Jour. Num. Analysis*, **6**, 1, 1969.

[56] R. A. Fisher, *Statistical Methods and Scientific Inference*, Oliver and Boyd, Edinburgh, 1956.

[57] G. A. Barnard, "Statistical inference," *Jour. Royal Stat. Soc.*, **B11**, 116, 1949.

[58] A. Birnbaum, "On the foundations of statistical inference," *Jour. Amer. Stat. Assoc.*, **57**, 269, 1962.

[59] C. R. Rao, *Linear Statistical Inference and Its Applications*, John Wiley, New York, 1965.

[60] G. E. P. Box and N. R. Draper, "The Bayesian estimation of common parameters from several responses," *Biometrika*, **52**, 355, 1965.

[61] S. S. Wilks, *Mathematical Statistics*, John Wiley, New York, 1962.

[62] G. W. Booth and T. I. Peterson, "Non-linear estimation," *IBM Share Program. Pa. No. 687 WL NLI*, 1958.

[63] D. W. Marquardt, "An algorithm for least squares estimation of non-linear parameters," *Jour. Soc. Ind. Appl. Math.*, **11**, 431, 1963.

[64] L. J. Savage, *The Foundations of Statistical Inference*, Methuen, London, 1962.

[65] H. Jeffreys, *Theory of Probability*, 3rd. ed., Clarendon Press, Oxford, 1961.

[66] G. E. P. Box and G. C. Tiao, *Bayesian Inference*, Addison-Wesley, Reading, 1973.

[67] G. M. Jenkins, contribution to the discussion of the paper "Relationships between Bayesian and confidence limits for predictors" by A. R. Thatcher, *Jour. Royal Statist. Soc.*, **B26**, 176, 1964.

[68] E. A. Cornish, "The multivariate *t*-distribution associated with a set of normal sample deviates," *Aust. Jour. Phys.*, **7**, 531, 1954.

[69] C. W. Dunnett and M. Sobel, "A bivariate generalization of Student's *t*-distribution, with tables for special cases," *Biometrika*, **31**, 153, 1954.

[70] G. A. Barnard, "The logic of least squares," *Jour. Royal Stat. Soc.*, **B25**, 124, 1963.

[71] R. L. Plackett, *Principles of Regression Analysis*, Clarendon Press, Oxford, 1960.

[72] F. J. Anscombe, "Examination of residuals," *Proc. 4th Berkeley Symp.*, **1**, 1, 1961.

[73] F. J. Anscombe and J. W. Tukey, "The examination and analysis of residuals," *Technometrics*, **5**, 141, 1963.

[74] C. Daniel, "Use of half normal plots in interpreting factorial experiments," *Technometrics*, **1**, 311, 1959.

[75] J. Durbin, "Testing for serial correlation in least-squares regression when some of the regressors are lagged dependent variables," *Econometrica*, **38**, 410, 1970.

[76] J. Durbin, "An alternative to the bounds test for testing serial correlation in least squares regression," *Econometrica*, **38**, 422, 1970.

[77] G. E. P. Box and D. A. Pierce, "Distribution of residual autocorrelations in autoregressive-integrated moving average time series models," *Jour. Amer. Stat. Assoc.*, **64**, 1509, 1970.

[78] M. S. Bartlett, *Stochastic Processes*, Cambridge University Press, Cambridge, 1955.

[79] A. Hald, *Statistical Theory with Engineering Applications*, John Wiley, New York, 1952.

[80] D. W. Bacon, "Seasonal time series," Ph.D. Thesis, University of Wisconsin, Madison, 1965.

[81] A. J. Young, *An Introduction to Process Control Systems Design*, Longman Green, New York, 1955.

[82] J. O. Hougen, *Experience and Experiments with Process Dynamics*, Chemical Engineering Progress Monograph Series, **60**, No. 4, 1964.

[83] K. D. Kotnour, G. E. P. Box and R. J. Altpeter, "A discrete predictor-controller applied to sinusoidal perturbation adaptive optimization," *Instr. Soc. Amer. Trans.*, **5**, 225, 1966.

[84] D. A. Pierce, "Distribution of residual correlations in dynamic/stochastic time series models," *University of Wisconsin Tech. Rep. 173*, August 1968.

[85]   G. E. P. Box, G. M. Jenkins and D. W. Wichern, "Least squares analysis with a dynamic model," *University of Wisconsin Technical Report 105*, 1967.

[86]   G. E. P. Box, "Use and abuse of regression," *Technometrics*, **8**, 625, 1966.

[87]   P. Whittle, "Estimation and Information in Stationary time series," *Arkiv für Mathematik*, **2**, 423, 1953.

[88]   G. A. Barnard, G. M. Jenkins and C. B. Winsten, "Likelihood inference and time series," *Jour. Royal Stat. Soc.*, **A125**, 321, 1962.

[89]   M. G. Kendall and A. Stuart, *The Advanced Theory of Statistics*, Vol. 3, Griffin, London, 1966.

[90]   G. Tintner, *The Variate Difference Method*, Principia Press, Bloomington, Indiana, 1940.

[91]   G. Tintner and J. N. K. Rao, "On the variate difference method," *Australian Journal of Statistics*, **5**, 106, 1963.

[92]   E. Slutsky, "The summation of random causes as the source of cyclic processes" (Russian), *Problems of Economic Conditions*, **3**, 1, 1927; English trans. in *Econometrica*, **5**, 105, 1937.

[93]   H. B. Mann and A. Wald, "On the statistical treatment of linear stochastic difference equations." *Econometrica*, **11**, 173, 1943.

[94]   T. C. Koopmans, "Serial correlation and quadratic forms in normal variables," *Ann. Math. Stat.*, **13**, 14, 1942.

[95]   T. C. Koopmans (ed.), *Statistical Inference in Dynamic Economic Models*, John Wiley, New York, 1950.

[96]   D. R. Cox, "Prediction by exponentially weighted moving averages and related methods." *Jour. Royal Stat. Soc.*, **B23**, 414, 1961.

[97]   E. J. Hannan, *Time Series Analysis*, Methuen, London, 1960.

[98]   U. Grenander and M. Rosenblatt, *Statistical Analysis of Stationary Time Series*, John Wiley, New York, 1957.

[99]   M. H. Quenouille, *Analysis of Multiple Time Series*, Hafner, New York, 1957.

[100] M. H. Quenouille, *Associated Measurements*, Butterworth, London, 1952.

[101] P. Whittle, *Hypothesis Testing in Time Series Analysis*, University of Uppsala publication, 1951.

[102] J. L. Doob, *Stochastic Processes*, John Wiley, New York, 1953.

[103] E. A. Robinson, *Multichannel Time Series Analysis*, Holden-Day, San Francisco, 1967.

[104] G. E. P. Box and G. C. Tiao, "Multiparameter problems from a Bayesian point of view," *Ann. Math. Statist.* **36**, 1468, 1968.

[105] P. M. Reilly, *Personal Communication*, 1967.

[106] B. Noble, *Methods Based on the Wiener–Hopf Technique for the Solution of Partial Differential Equations*, Pergamon Press, New York, 1958.

[107] G. T. Wilson, "Optimal Control—A General Method of Obtaining the Feedback Scheme which Minimizes the Output Variance, Subject to a Constraint on the Variability of the Control Variable," Technical Report No. 20, Dept. of Systems Engineering, University of Lancaster.

[108] D. W. Wichern, "The behaviour of the sample autocorrelation function for an integrated moving average process," Biometrika, **60**, 235, 1973.

[109] J. F. MacGregor, "Topics in the control of linear processes with stochastic disturbances," Ph.D. Thesis, University of Wisconsin, Madison, 1972.

[110] G. E. P. Box and J. F. MacGregor, "The analysis of closed-loop dynamic stochastic systems," Technometrics, **16**, 391, 1974.

[111] G. E. P. Box and J. F. MacGregor, "Parameter estimation for dynamic-stochastic models using closed-loop operating data," Proceedings of Sixth Triennial World Congress I.F.A.C. Boston, Sept., 1975, also Technometrics, **18**, 1976.

[112] G. C. Tiao, G. E. P. Box and W. J. Hamming, "Analysis of Los Angeles photochemical smog data: a statistical overview," Jour. Air Pollution Control Assoc., **25**, 260, 1975.

[113] H. E. Thompson and G. C. Tiao, "Analysis of telephone data: a case study of forecasting seasonal time series," Bell. Journ. of Econ. and Man. Sci., **2**, 515, 1971.

[114] G. E. P. Box and G. C. Tiao, "Intervention analysis with applications to economic and environmental problems," Jour. Amer. Stat. Assoc., **70**, 70, 1975.

[115] G. M. Jenkins, "The interaction between the muskrat and mink cycles in North Canada," Proceedings of the 8th International Biometric Conference, Editura Academiei Republicii Socialiste Romania, 1975, pp. 55–71.

[116] G. E. P. Box and G. C. Tiao, "Comparison of forecast and actuality," Applied Stat., **25**, 1976.

[117] G. E. P. Box, G. M. Jenkins and J. F. MacGregor, "Some recent advances in forecasting and control, II," Applied Stat., **23**, 158, 1974.

Indexes

Author Index

Subject Index

Part VI

This part of the book is a collection of Exercises and Problems for the separate chapters. We hope that these will further enhance the value of the book when used as a course text and also assist private study. A number of examples point to extensions of the ideas and act as a first introduction to methods such as Intervention Analysis.

Exercises and Problems

CHAPTER 2

2.1 The following are temperature measurements z made every minute on a chemical reactor:

200, 202, 208, 204, 204, 207, 207, 204, 202, 199, 201, 198, 200,
202, 203, 205, 207, 211, 204, 206, 203, 203, 201, 198, 200, 206,
207, 206, 200, 203, 203, 200, 200, 195, 202, 204.

(i) Plot the series,
(ii) plot z_{t+1} versus z_t,
(iii) plot z_{t+2} versus z_t.

After inspecting the graphs, do you think that the series is autocorrelated?

2.2 A stochastic process is such that its first two autocorrelations are non-zero and the remainder zero. State whether or not the process is stationary if
(i) $\rho_1 = 0.80$, $\rho_2 = 0.55$
(ii) $\rho_1 = 0.80$, $\rho_2 = 0.28$

2.3 Two stochastic processes z_{1t} and z_{2t} have the following autocovariance functions:

$$z_{1t}: \gamma_0 = 0.5, \ \gamma_1 = 0.2, \ \gamma_j = 0 \ (j \geqslant 2)$$
$$z_{2t}: \gamma_0 = 2.30, \ \gamma_1 = -1.43, \ \gamma_2 = 0.30, \ \gamma_j = 0 \ (j \geqslant 3)$$

Calculate the autocovariance function of the process $z_{3t} = z_{1t} + 2z_{2t}$ and verify that it is a valid stationary process.

2.4 Calculate c_0, c_1, c_2, r_1, r_2 for the series given in Exercise 2.1. Make a graph of $r_k, k = 0,1,2$.

2.5 On the supposition that $\rho_j = 0$ for $j > 2$,
(i) obtain approximate standard errors for r_1, r_2 and $r_j, j > 2$,
(ii) obtain the approximate correlation between r_4 and r_5.

2.6 Using the data of Exercise 2.1, calculate the periodogram for periods 36, 18, 12, 9, 36/5, 6 and draw up an analysis of variance table showing the mean squares associated with these periods and the residual mean square.

2.7 A *circular* stochastic process with period N is defined by $z_t = z_{t+N}$.
(i) Show that [27] when $N = 2n$, the latent roots of the autocorrelation matrix of z_t are

$$\lambda_k = 1 + 2 \sum_{i=1}^{n-1} \rho_i \cos \frac{\pi i k}{n} + \rho_n \cos \pi k$$

$k = 1, 2, \ldots, N$ and that the latent vectors corresponding to λ_k, λ_{N-k} are

$$l'_k = \left(\frac{\cos \pi k}{n}, \frac{\cos 2\pi k}{n}, \ldots, \cos 2\pi k \right)$$

$$l'_{N-k} = \left(\frac{\sin \pi k}{n}, \frac{\sin 2\pi k}{n}, \ldots, \sin 2\pi k \right)$$

(ii) Verify that as N tends to infinity, λ_k tends to $2g(k/n)$, where $g(f)$ is the spectral density function, showing that in the limit the latent roots of the autocorrelation matrix trace out the spectral curve.

CHAPTER 3

3.1 Write the following models in B notation:
  (a) $\tilde{z}_t - 0.5\tilde{z}_{t-1} = a_t$
  (b) $\tilde{z}_t = a_t - 1.3a_{t-1} + 0.4a_{t-2}$
  (c) $\tilde{z}_t - 0.5\tilde{z}_{t-1} = a_t - 1.3a_{t-1} + 0.4a_{t-2}$

3.2 For each of the models in Exercise 3.1, obtain
  (i) the first four ψ weights,
  (ii) the first four π weights,
  (iii) the covariance generating function,
  (iv) the first four autocorrelations,
  (v) the variance of $\tilde{z}_t$ assuming $\sigma_a^2 = 1.0$.

3.3 For each of the models of Exercise 3.1 and also for the following models:
  (d) $\tilde{z}_t - 1.5\tilde{z}_{t-1} + 0.5\tilde{z}_{t-2} = a_t$
  (e) $\tilde{z}_t - \tilde{z}_{t-1} = a_t - 0.5a_{t-1}$
  (f) $\tilde{z}_t - \tilde{z}_{t-1} = a_t - 1.3a_{t-1} + 0.3a_{t-2}$
  state whether it is (i) stationary, (ii) invertible.

3.4 Classify each of the four models (a)–(d) in Exercises 3.1 and 3.3 as a member of the class of ARMA (p,q) processes.

3.5 (i) Write down the Yule-Walker equations for models (a) and (d).
  (ii) Solve these equations to obtain ρ_1 and ρ_2 for the two models.
  (iii) Obtain the partial autocorrelation function for the two models.

3.6 For the AR(2) process $\tilde{z}_t - 1.0\tilde{z}_{t-1} + 0.5\tilde{z}_{t-2} = a_t$,
  (i) Calculate ρ_1,
  (ii) using ρ_0 and ρ_1 as starting values and the difference equation form for the autocorrelation function, calculate the values of ρ_k for $k = 2, \ldots, 15$,
  (iii) use the plotted function to estimate the period and damping factor of the autocorrelation function,
  (iv) check the values in (iii) by direct calculation using the values of ϕ_1 and ϕ_2.

3.7 (i) Plot the power spectrum $g(f)$ of the autoregressive process of Exercise 3.6

and show that it has a peak at a period which is close to the period in the autocorrelation function.

(ii) Graphically, or otherwise, estimate the proportion of the variance of the series in the frequency band between $f = 0.0$ and $f = 0.2$ cycles per data interval.

3.8 (i) Why is it important to factorize the autoregressive and moving average and autoregressive operators after fitting a model to an observed series?

(ii) It is shown in [115] that the number of mink skins z_t traded annually between 1848–1909 in North Canada is adequately represented by the AR(4) model

$$(1 - 0.82B + 0.22B^2 + 0.28B^4) \ln z_t = a_t$$

Factorize the autoregressive operator and explain what the factors reveal about the autocorrelation function and the underlying nature of the mink series.

CHAPTER 4

4.1 For each of the models
(a) $\qquad (1 - B)z_t = (1 - 0.5B)a_t$
(b) $\qquad (1 - B)z_t = (1 - 0.2B)a_t$
(c) $(1 - B)(1 - 0.5B)z_t = a_t$
(d) $(1 - B)(1 - 0.2B)z_t = a_t$
(e) $(1 - B)(1 - 0.2B)z_t = (1 - 0.5B)a_t$,
(i) obtain the first seven ψ weights,
(ii) obtain the first seven π weights,
(iii) classify as a member of the class of ARIMA (p,d,q) process.

4.2 For the five models of Exercise 4.1, and using where appropriate the results there obtained,
(i) write each model in random shock form,
(ii) write each model as a complementary function plus a particular integral in relation to an origin $k = t - 3$,
(iii) write each model in inverted form.

4.3

t	0	1	2	3	4	5	6	7	8	9	10	11	12	13	14
a_t	−0.3	0.6	0.9	0.2	0.1	−0.6	1.7	−0.9	−1.3	−0.6	−0.4	0.9	0.0	−1.4	−0.6

Given the series of random shocks a_t shown above and given that $z_0 = 20, z_{-1} = 19$,
(i) use the difference equation form of the model to obtain $z_1, z_2, \ldots, z_{14}$ for each of the five models in Exercise 4.1,
(ii) plot the derived series.

4.4 Using the inverted forms of each of the models in Exercise 4.1, obtain z_{12}, z_{13} and z_{14}, using only the values $z_1, z_2, \ldots, z_{11}$ derived in Exercise 4.3 and a_{12}, a_{13} and a_{14}. Confirm that the values agree with those obtained in Exercise 4.3.

4.5 If $\bar{z}_t = \sum\limits_{j=1}^{\infty} \pi_j z_{t-j+1}$, then for the models (a) and (b) of Exercise 4.1, which are of
the form $(1 - B)z_t = (1 - \theta B)a_t$, $\bar{z}_t$ is an exponentially weighted moving average.
For these two models, by actual calculation confirm that $\bar{z}_{11}$, $\bar{z}_{12}$ and $\bar{z}_{13}$ satisfy
the relations

$$z_t = \bar{z}_{t-1} + a_t \quad \text{(See Exercise 4.4)}$$
$$\bar{z}_t = \bar{z}_{t-1} + (1 - \theta)a_t$$
$$\bar{z}_t = (1 - \theta)z_t + \theta\bar{z}_{t-1}$$

4.6 If $w_{1t} = (1 - \theta_1 B)a_{1t}$ and $w_{2t} = (1 - \theta_2 B)a_{2t}$, show that $w_{3t} = w_{1t} + w_{2t}$ may be
written as $w_{3t} = (1 - \theta_3 B)a_{3t}$ and derive an expression for θ_3 and σ_{3a}^2 in terms
of the parameters of the other two processes. State your assumptions.

4.7 Suppose that $Z_t = z_t + b_t$, where z_t is a first-order autoregressive process $(1 - \phi B)z_t$
$= a_t$ and b_t is a white noise process with variance σ_b^2. What process does Z_t follow?
State your assumptions.

CHAPTER 5

5.1 For the models
   (a) $\bar{z}_t - 0.5\bar{z}_{t-1} = a_t$
   (b) $\nabla z_t = a_t - 0.5a_{t-1}$
   (c) $(1 - 0.6B)\nabla z_t = a_t$
generate forecasts for lead times $l = 1$ and $l = 2$
   (i)   from the difference equation,
   (ii) in integrated form (using the ψ-weights),
   (iii) as a weighted average of previous observations.

5.2 The following observations represent the values $z_{91}, z_{92}, \ldots, z_{100}$ from a series
fitted by the model $\nabla z_t = a_t - 1.1a_{t-1} + 0.28a_{t-2}$

        166, 172, 172, 169, 164, 168, 171, 167, 168, 172.

   (i) generate the forecasts $\hat{z}_{100}(l)$ for $l = 1, 2, \ldots, 12$ and draw a graph of the
        series values and the forecasts, (assume $a_{90} = 0$, $a_{91} = 0$),
   (ii) with $\hat{\sigma}_a^2 = 1.103$, calculate the estimated standard deviations $\hat{\sigma}(l)$ of the fore-
        cast errors and use them to calculate 80% probability limits for the forecasts.
        Insert these probability limits on the graph, on either side of the forecasts.

5.3 Suppose that the data of Exercise 5.2 represent monthly sales,
   (i) calculate the minimum mean square error forecasts for quarterly sales for
        1, 2, 3, 4 *quarters* ahead, using the data up to $t = 100$,
   (ii) calculate 80% probability limits for these forecasts.

5.4 Using the data and forecasts of Exercise 5.2, and given the further observation
$z_{101} = 174$,
   (i) calculate the forecasts $\hat{z}_{101}(l)$ for $l = 1, 2, \ldots, 11$ using the updating formula

$$\hat{z}_{t+1}(l) = \hat{z}_t(l + 1) + \psi_l a_{t+1}$$

   (ii) verify these forecasts using the difference equation directly.

5.5 For the model $\nabla z_t = a_t - 1.1a_{t-1} + 0.28a_{t-2}$ of Exercise 5.2,
   (i) write down expressions for the forecast errors $e_t(1)$, $e_t(2)$, ..., $e_t(6)$, from the same origin t,
   (ii) calculate and plot the autocorrelations of the series of forecast errors $e_t(3)$
   (iii) calculate and plot the correlations between the forecast errors $e_t(2)$ and $e_t(j)$ for $j = 1, 2, \ldots, 6$.

5.6 Let the vector $e' = (e_1, e_2, \ldots, e_L)$ have for its elements the forecast errors made 1, 2, ..., L steps ahead, all from the same origin t. Then if $a' = (a_{t+1}, a_{t+2}, \ldots, a_{t+L})$ are the corresponding uncorrelated shocks

$$e = Ma, \text{ where } M = \begin{bmatrix} 1 & 0 & 0 & . & . & . & 0 \\ \psi_1 & 1 & 0 & . & . & . & 0 \\ \psi_2 & \psi_1 & 1 & . & . & . & 0 \\ . & . & . & . & . & . & . \\ \psi_{L-1} & \psi_{L-2} & \psi_{L-3} & . & . & . & 1 \end{bmatrix}$$

Show [112, 116] Σ_e, the covariance matrix of the e's, is $\Sigma_e = MM'\sigma_a^2$ and hence that a test that a set of subsequently realized values $z_{t+1}, z_{t+2}, \ldots, z_{t+L}$ of the series do not jointly differ significantly from the forecasts made at origin t is obtained by referring

$$e'\Sigma_e^{-1}e = e'(MM')^{-1}e/\sigma_a^2 = a'a/\sigma_a^2 = \sum_{j=t+1}^{t+L} a_j^2/\sigma_a^2$$

to a chi-squared distribution with L degrees of freedom. (Note that a_{t+j} is the *one* step ahead forecast error calculated from $z_{t+j} - \hat{z}_{t+j-1}(1)$).

5.7 It was found that a quarterly economic time series was well represented by the model

$$\nabla z_t = 0.5 + (1 - 1.0B + 0.5B^2)a_t$$

with $\sigma_a^2 = 0.04$,
   (i) given $z_{48} = 130$, $a_{47} = -0.3$, $a_{48} = 0.2$, calculate and plot the forecasts $\hat{z}_{48}(l)$ for $l = 1, 2, \ldots, 12$,
   (ii) insert the 80% probability limits on the graph,
   (iii) express the series and forecasts in integrated form.

CHAPTER 6

6.1 Given the identified models and the values of the estimated autocorrelations of $w_t = \nabla^d z_t$ in the following table:

	Identified Model			Estimated Autocorrelations
	p	d	q	
(a)	1	1	0	$r_1 = 0.72$
(b)	0	1	1	$r_1 = -0.41$
(c)	1	0	1	$r_1 = 0.40$, $r_2 = 0.32$
(d)	0	2	2	$r_1 = 0.62$, $r_2 = 0.13$
(e)	2	1	0	$r_1 = 0.93$, $r_2 = 0.81$

(i) obtain preliminary estimates of the parameters analytically,
(ii) check these estimates using the charts and tables at the end of the book.
(iii) write down the identified models in backward shift operator notation with the preliminary estimates inserted.

6.2 For the $(2,1,0)$ process considered in (e) of Exercise 6.1, the mean and variance of $w_t = \nabla z_t$ are $\bar{w} = 0.23$ and $s_w^2 = 0.25$. If the series contains $N = 101$ observations
(i) show that a constant term needs to be included in the model,
(ii) express the model in the form $w_t - \phi_1 w_{t-1} - \phi_2 w_{t-2} = \theta_0 + a_t$ with numerical values inserted for the parameters.

6.3 The following table shows the first 16 values of the a.c.f. r_k and p.a.c.f. $\hat{\phi}_{kk}$ for a series of 60 observations of logged quarterly unemployment in the U.K.

k	1	2	3	4	5	6	7	8
r_k	0.93	0.80	0.65	0.49	0.32	0.16	0.03	−0.09
$\hat{\phi}_{kk}$	0.93	−0.41	−0.14	−0.11	−0.07	−0.10	0.05	−0.07

k	9	10	11	12	13	14	15	16
r_k	−0.16	−0.22	−0.25	−0.25	−0.21	−0.12	−0.01	0.10
$\hat{\phi}_{kk}$	0.12	−0.14	0.03	0.09	0.19	0.20	0.03	−0.11

(i) Draw graphs of the a.c.f. and p.a.c.f.
(ii) Identify a model for the series.
(iii) Obtain preliminary estimates for the parameters and for their standard errors.
(iv) Given $\bar{z} = 2.56$ and $s_z^2 = 0.01681$, obtain preliminary estimates for μ_z and σ_a^2.

6.4 The following table shows the first 10 values of the a.c.f. and p.a.c.f. of z_t and ∇z_t for a series of 56 observations consisting of quarterly measurements of Gross Domestic Product (G.D.P.) in the U.K.

	k	1	2	3	4	5	6	7	8	9	10
z	r_k	0.95	0.90	0.86	0.80	0.76	0.71	0.66	0.61	0.57	0.52
	$\hat{\phi}_{kk}$	0.95	0.02	−0.02	−0.06	0.00	−0.02	−0.05	−0.01	−0.02	−0.04

	k	1	2	3	4	5	6	7	8	9	10
∇z	r_k	0.01	0.09	0.17	0.02	−0.04	−0.01	−0.24	−0.18	−0.03	−0.08
	$\hat{\phi}_{kk}$	0.01	0.09	0.17	0.01	−0.08	−0.04	−0.25	−0.18	0.02	0.04

$\bar{z} = 97.0$; $s_z^2 = 128.6$; $\bar{w} = 0.66$; $s_w^2 = 0.7931$ $(w_t = \nabla z_t)$

(i) draw graphs of the a.c.f. and p.a.c.f. for z and ∇z,
(ii) identify a model for the series,
(iii) obtain preliminary estimates for the parameters.

6.5 The following table shows the first 10 values of the a.c.f. of z_t and ∇z_t for a series defined by $z_t = 1000 \log_{10} H_t$ where H_t is the price of hogs recorded annually by the U.S. Census of Agriculture on January 1 for each of the 82 years 1867–1948*

k	1	2	3	4	5	6	7	8	9	10
$r_k(z)$	0.85	0.67	0.56	0.52	0.51	0.46	0.42	0.38	0.36	0.32
$r_k(\nabla z)$	0.25	−0.25	−0.35	−0.12	0.03	0.13	0.18	0.02	−0.07	−0.10

Based on these autocorrelations, identify a model for the series and obtain preliminary estimates of the parameters.

CHAPTER 7

7.1

t	z_t	$w_t = \nabla z_t$	$a_t = w_t - 0.5 a_{t-1}$
0	40		a_0
1	42	2	$2 - 0.50 a_0$
2	47	5	$4 + 0.25 a_0$
3	47	0	$-2 - 0.13 a_0$
4	52	5	$6 + 0.06 a_0$
5	51	−1	$-4 - 0.03 a_0$
6	57	6	$8 + 0.02 a_0$
7	59	2	$-2 - 0.01 a_0$

The above table shows calculations for an (unrealistically short) series z_t for which the (0,1,1) model $w_t = \nabla z_t = (1 - \theta B)a_t$ with $\theta = -0.5$ is being entertained with an unknown starting value a_0.
(i) Confirm the entries in the table,
(ii) show that the conditional sum of squares

$$\sum_{t=1}^{7} (a_t | -0.5, a_0 = 0)^2 = S_*(-0.5|0) = 144.00$$

7.2 Using the data in Exercise 7.1,
(i) show (using least squares) that the value $\hat{a}_0$ of a_0 which minimizes $S_*(-0.5|a_0)$ is

$$\hat{a}_0 = \frac{(2)(0.50) + (4)(-0.25) + \cdots + (-2)(0.01)}{1^2 + 0.5^2 + \cdots + 0.01^2}$$

*Further details of these data are given in [99].

(ii) by first writing this model in the backward form $w_t = (1 - \theta F)e_t$ and recursively computing the e's, show that the value of a_0 obtained in (i) is the same as that obtained by the back forecasting method.

7.3 (i) Using the value of $\hat{a}_0$ calculated in Exercise 7.2, show that the unconditional sum of squares $S(-0.5)$ is 143.4,

(ii) show that for the (0,1,1) model, for large n,

$$S(\theta) = S_*(\theta|0) - \hat{a}_0^2/(1 - \theta^2)$$

7.4 For the process $w_t - \mu = (1 - \theta B)a_t$ show that for long series the variance-covariance matrix of the maximum likelihood estimates $\hat{\mu}, \hat{\theta}$ is approximately

$$n^{-1}\begin{bmatrix} (1 - \theta)^2\sigma_a^2 & 0 \\ 0 & 1 - \theta^2 \end{bmatrix}$$

7.5 (i) Problems were experienced in obtaining a satisfactory fit to a series, the last 16 values of which were recorded as follows:

129, 135, 130, 130, 127, 126, 131, 152, 123, 124, 131, 132, 129, 127, 126, 124.

Plot the series and suggest where the difficulty might lie.

(ii) In fitting a model of the form $(1 - \phi_1 B - \phi_2 B^2)z_t = (1 - \theta B)a_t$ to a set of data, convergence was slow and the coefficients in successive iterations oscillated wildly. Final estimates having large standard errors were obtained as follows: $\hat{\phi}_1 = 1.19$, $\hat{\phi}_2 = -0.34$, $\hat{\theta} = 0.52$. Can you suggest an explanation for the unstable behavior of the model? Why should preliminary identification have eliminated the problem?

(iii) In fitting a model $\nabla^2 z_t = (1 - \theta_1 B - \theta_2 B^2)a_t$ convergence was not obtained. The last iteration yielded the values $\hat{\theta}_1 = 1.81$, $\hat{\theta}_2 = 0.52$. Can you explain the difficulty?

7.6 For the ARIMA (1,1,1) model $(1 - \phi B)w_t = (1 - \theta B)a_t$, where $w_t = \nabla z_t$,
(i) write down the linearized form of the model,
(ii) set out how you would start off the calculation of the conditional non-linear least squares algorithm with start values $\phi = 0.5$ and $\theta = 0.4$ for a series whose first 9 values are shown below:

t	0	1	2	3	4	5	6	7	8
z_t	149	145	152	144	150	150	147	142	146

7.7 (i) Show that the second-order autoregressive model $\tilde{z}_t = \phi_1\tilde{z}_{t-1} + \phi_2\tilde{z}_{t-2} + a_t$ may be written in orthogonal form as $\tilde{z}_t = \dfrac{\phi_1}{1 - \phi_2}\tilde{z}_{t-1} + \phi_2\left(\tilde{z}_{t-2} - \dfrac{\phi_1}{1 - \phi_2}\tilde{z}_{t-1}\right)$
$+ a_t$, suggesting that the approximate estimates r_1 of $\dfrac{\phi_1}{1 - \phi_2}$ and $\hat{\phi}_2 = \dfrac{r_2 - r_1^2}{1 - r_1^2}$ of ϕ_2 are uncorrelated for long series.

(ii) Starting from the variance-covariance matrix of $\hat{\phi}_1$ and $\hat{\phi}_2$ or otherwise, show that the variance-covariance matrix of r_1 and $\hat{\phi}_2$ for long series is given approximately by

$$n^{-1}\begin{bmatrix} (1 - \phi_2^2)(1 - \rho_1^2) & 0 \\ 0 & 1 - \phi_2^2 \end{bmatrix}$$

CHAPTER 8

8.1 The following are the first 30 residuals obtained when a tentative model was fitted to a time series:

$t = 1-6$	0.78	0.91	0.45	−0.78	−1.90	−2.10
7–12	−0.54	−1.05	0.68	−3.77	−1.40	−1.77
13–18	1.18	0.02	1.29	−1.30	−6.20	−1.89
19–24	0.95	1.49	1.08	0.80	2.02	1.25
25–30	0.52	2.31	1.64	0.78	1.99	1.36

Plot the values and state any reservations you have concerning the adequacy of the model.

8.2 The residuals from a model $\nabla z_t = (1 - 0.6B)e_t$ fitted to a series of $N = 82$ observations yielded the following residual autocorrelations:

k	1	2	3	4	5	6	7	8	9	10
$r_k(\hat{e})$	0.39	0.20	0.09	0.04	0.09	−0.13	−0.05	0.06	0.11	0.02

(i) Plot the residual a.c.f. and determine whether there are any abnormal values.
(ii) Calculate the chi-squared statistic and check whether the residual autocorrelation function as a whole is indicative of model inadequacy.
(iii) What modified model would you now tentatively entertain, fit and check?

8.3 Suppose that a $(0,1,1)$ model $\nabla z_t = (1 - \theta B)e_t$, corresponding to the use of an exponentially weighted moving average forecast, with θ arbitrarily chosen to be equal to 0.5, was used to forecast a series which was in fact well fitted by the $(0,1,2)$ model $\nabla z_t = (1 - 0.9B + 0.2B^2)a_t$,
(i) calculate the autocorrelation function of the lead-1 forecast errors e_t obtained from the $(0,1,1)$ model,
(ii) show how this a.c.f. could be used to identify a model for the e_t series, leading to the identification of a $(0,1,2)$ model for the z_t series.

8.4 A long series containing $N = 326$ terms was split into two halves and a $(1,1,0)$ model $(1 - \phi B)\nabla z_t = a_t$ identified, fitted and checked for each half. If the estimates of the parameter for the two halves are $\hat{\phi}^{(1)} = 0.5$ and $\hat{\phi}^{(2)} = 0.7$, is there any evidence that the parameter ϕ has changed?

8.5 (i) Show that the variance of the mean of n observations from a stationary AR(1) process $(1 - \phi B)\tilde{z}_t = a_t$ is given by

$$\mathrm{var}[\bar{z}] \simeq \frac{\sigma_a^2}{n(1 - \phi)^2}$$

(ii) The yields from consecutive batches of a chemical process obtained under fairly uniform conditions of process control were shown to follow a stationary AR(1) process $(1 + 0.5B)\tilde{z}_t = a_t$. A technical innovation is made at a given point in time leading to 85 data points with mean $\bar{z}_1 = 41.0$ and residual variance $s_{1a}^2 = 0.1012$ before the innovation is made and 60 data points with $\bar{z}_2 = 43.5$, $s_{2a}^2 = 0.0895$ after the innovation. Is there any evidence that the innovation has improved the yield?

CHAPTER 9

9.1 Show that the seasonal difference operator $1 - B^{12}$, often useful in the analysis of monthly data, may be factorized as follows:

$$(1 - B^{12}) = (1 + B)(1 - \sqrt{3}B + B^2)(1 - B + B^2)(1 + B^2)(1 + B + B^2)$$
$$\times (1 + \sqrt{3}B + B^2)(1 - B)$$

Plot the zeros of this expression in the unit circle and show by actual numerical calculation and plotting of the results that the factors in the order given above correspond to sinusoids with frequencies (in cycles per year) of 6, 5, 4, 3, 2, 1, together with a constant term. (For example, the difference equation $(1 - B + B^2)x_t = 0$ with arbitrary starting values $x_1 = 0$, $x_2 = 1$ yields $x_3 = 1$, $x_4 = 0$, $x_5 = -1$ etc., generating a sine wave of frequency 2 cycles per year).

9.2 A method which has sometimes been used for "deseasonalizing" monthly time series employs an equally weighted 12 month moving average

$$\bar{z}_t = \frac{1}{12}\{z_t + z_{t-1} + \cdots + z_{t-11}\}.$$

(i) Using the decomposition $(1 - B^{12})/(1 - B) = 1 + B + B^2 + \cdots + B^{11}$, show that $12(\bar{z}_t - \bar{z}_{t-1}) = (1 - B^{12})z_t$.

(ii) The exceedance for a given month over the previous moving average may be computed as $z_t - \bar{z}_{t-1}$. A quantity u_t may then be calculated which compares the current exceedance with the average of similar monthly exceedances experienced over the last k years. Show that u_t may be written as

$$u_t = \left\{1 - \frac{B}{12}\frac{(1 - B^{12})}{(1 - B)}\right\}\left\{1 - \frac{B^{12}}{k}\frac{(1 - B^{12k})}{(1 - B^{12})}\right\}z_t$$

9.3 It has been shown [112] that monthly averages for the (smog producing) oxidant level in Azusa, California, may be represented by a model

$$(1 - B^{12})z_t = (1 + 0.2B)(1 - 0.9B^{12})a_t, \quad \sigma_a^2 = 1.0$$

(i) Compute and plot the ψ weights.
(ii) Compute and plot the π weights.

(iii) Calculate the standard deviation of the forecast three months and 12 months ahead.

(iv) Obtain the eventual forecast function.

9.4 The monthly oxidant averages in parts per hundred million in Azusa from January 1969–December 1972 were as follows:

	J	F	M	A	M	J	J	A	S	O	N	D
1969	2.1	2.6	4.1	3.9	6.7	5.1	7.8	9.3	7.5	4.1	2.9	2.6
1970	2.0	3.2	3.7	4.5	6.1	6.5	8.7	9.1	8.1	4.9	3.6	2.0
1971	2.4	3.3	3.3	4.0	3.6	6.2	7.7	6.8	5.8	4.1	3.0	1.6
1972	1.9	3.0	4.5	4.2	4.8	5.7	7.1	4.8	4.2	2.3	2.1	1.6

Using the model of Exercise 9.3, compute the forecasts for the next 24 months. (Approximate unknown a's by zeros.)

9.5 Thompson and Tiao [113] have shown that the outward station movements of telephones (logged data) in Wisconsin are well represented by the model

$$(1 - 0.5B^3)(1 - B^{12})z_t = (1 - 0.2B^9 - 0.3B^{12} - 0.2B^{13})a_t$$

Obtain and plot the autocorrelation function for $w_t = (1 - B^{12})z_t$ for lags 1, 2, ..., 24.

9.6 The following table shows the first 12 autocorrelations of various differences of a series of $N = 41$ observations consisting of quarterly deposits in a bank.

k	1	2	3	4	5	6	7	8	9	10	11	12	$\bar{w}$	s_w^2
$r_k(z)$	0.88	0.83	0.72	0.67	0.56	0.53	0.42	0.41	0.32	0.29	0.21	0.18	152.0	876.16
$r_k(\nabla z)$	−0.83	0.68	−0.70	0.70	−0.65	0.62	−0.70	0.69	−0.60	0.55	−0.60	0.62	2.46	97.12
$r_k(\nabla_4 z)$	0.21	0.06	0.15	−0.34	−0.10	0.03	−0.18	−0.13	0.02	−0.04	0.10	0.22	10.38	25.76
$r_k(\nabla\nabla_4 z)$	−0.33	−0.06	0.36	−0.45	0.08	0.22	−0.10	−0.05	0.14	−0.12	−0.05	0.28	0.42	38.19

Plot these autocorrelation functions and identify a model (or models) for the series. Calculate preliminary estimates for the parameters and for σ_a^2.

9.7 In the analysis of a series consisting of the logarithms of the sales of a seasonal product, it was found that differencing of the form $w_t = \nabla\nabla_{12} \ln z_t$ was needed to induce stationarity with respect to both monthly and seasonal variation. The following table shows the first 48 autocorrelations of the w_t series which contained $n = 102$ monthly observations:

	$\bar{w} = 0.241$					$s_w^2 = 106.38$						
Lags 1–12	−0.39	−0.24	0.17	0.21	−0.27	−0.03	0.26	−0.10	−0.20	0.07	0.44	−0.58
Lags 13–24	0.09	0.17	0.01	−0.24	0.16	0.04	−0.12	−0.01	0.11	0.08	−0.33	0.28
Lags 25–36	0.01	−0.14	−0.02	0.18	−0.13	0.04	−0.01	0.10	−0.13	−0.09	0.27	−0.22
Lags 37–48	0.00	0.09	0.02	−0.18	0.17	−0.05	0.00	−0.06	0.06	0.06	−0.13	0.11

Identify a suitable model (or models) for the series and obtain preliminary estimates of the parameters.

CHAPTER 10

10.1   In the following transfer-function models, X_t is the methane gas feedrate to a
       gas furnace, measured in cubic feet per minute and Y_t the % carbon dioxide
       in the outlet gas:

       (a) $Y_t = 10 + \dfrac{25}{1 - 0.7B} X_{t-1}$

       (b) $Y_t = 10 + \dfrac{22 - 12.5B}{1 - 0.85B} X_{t-2}$

       (c) $Y_t = 10 + \dfrac{20 - 8.5B}{1 - 1.2B + 0.4B^2} X_{t-3}$

       (i) Verify that the models are stable.
       (ii) Calculate the steady state gain g, expressing it in the appropriate units.

10.2   For each of the models (a), (b), and (c) of Exercise 10.1, calculate from the differ-
       ence equation and plot the responses to:
       (i)   a unit impulse $(0,1,0,0,0,0,\ldots)$ applied at time $t = 0$
       (ii) a unit step $(0,1,1,1,1,1,\ldots)$ applied at time $t = 0$
       (iii) a ramp input $(0,1,2,3,4,5,\ldots)$ applied at time $t = 0$
       (iv) a periodic input $(0,1,0,-1,0,1,\ldots)$ applied at time $t = 0$
       Estimate the period and damping factor of the step response to model (c).

10.3   Use equations (10.2.8) to obtain the impulse weights v_j of models (a), (b), and (c)
       of Exercise 10.1, and check that they are the same as the impulse response
       obtained in (i) of Exercise 10.2.

10.4   Express models (a), (b), and (c) of Exercise 10.1 in ∇ form.

10.5   (i) Calculate and plot the response of the two-input system

$$Y_t = 10 + \frac{6}{1 - 0.7B} X_{1,t-1} + \frac{8}{1 - 0.5B} X_{2,t-2}$$

       to the orthogonal and randomized input sequences shown below:

t	0	1	2	3	4	5	6	7	8
X_{1t}	0	-1	1	-1	1	1	1	-1	-1
X_{2t}	0	1	-1	-1	1	-1	1	-1	1

       (ii) Calculate the gains g_1 and g_2 of Y with respect to X_1 and X_2 respectively
       and express the model in ∇ form.

CHAPTER 11

11.1 If two series may be represented in ψ-weight form as

$$y_t = \psi_y(B)a_t \qquad x_t = \psi_x(B)a_t$$

(i) Show that the cross-covariance generating function

$$\gamma^{xy}(B) = \sum_{k=-\infty}^{\infty} \gamma_{xy}(k)B^k$$

is given by $\psi_y(B)\psi_x(F)\sigma_a^2$

(ii) Use the above result to obtain the cross covariance function between y_t and x_t when

$$y_t = (1 - \theta B)a_t \qquad x_t = (1 - \theta_1' B - \theta_2' B^2)a_t$$

11.2 After estimating a prewhitening transformation $\phi_x(B)\theta_x^{-1}(B)x_t = \alpha_t$ for an input series x_t and then computing the transformed output $\beta_t = \phi_x(B)\theta_x^{-1}(B)y_t$, cross correlations $r_{\alpha\beta}(k)$ were obtained as follows:

k	0	1	2	3	4	5	6	7	8	9
$r_{\alpha\beta}(k)$	0.05	0.31	0.52	0.43	0.29	0.24	0.07	-0.03	0.10	0.07

with $\hat{\sigma}_\alpha = 1.26$, $\hat{\sigma}_\beta = 2.73$, $n = 187$.
(i) Obtain approximate standard errors for the cross correlations,
(ii) calculate rough estimates for the impulse response weights v_j,
(iii) suggest a model form for the transfer function with rough estimates of its parameters.

11.3 It is frequently the case that the user of an estimated transfer function model $y_t = \omega(B)\delta^{-1}(B)B^b x_t$ will want to establish whether the steady state gain $g = \omega(1)\delta^{-1}(1)$ makes sense.
(i) For the first-order system

$$y_t = \frac{\omega_0}{1 - \delta B} x_{t-1}$$

show that an approximate standard error $\hat{\sigma}(\hat{g})$ of the estimate $\hat{g} = \hat{\omega}_0/1 - \hat{\delta}$ is given by

$$\frac{\hat{\sigma}^2(\hat{g})}{\hat{g}^2} \simeq \frac{\operatorname{var}[\hat{\omega}_0]}{\hat{\omega}_0^2} + \frac{\operatorname{var}[\hat{\delta}]}{(1 - \hat{\delta})^2} + \frac{2\operatorname{cov}[\hat{\omega}_0, \hat{\delta}]}{\hat{\omega}_0(1 - \hat{\delta})}$$

(ii) Calculate $\hat{g}$ and an approximate value for $\hat{\sigma}(\hat{g})$ when $\hat{\omega}_0 = 5.2$, $\hat{\delta} = 0.65$, $\hat{\sigma}(\hat{\omega}_0) = 0.5$, $\hat{\sigma}(\hat{\delta}) = 0.1$, $\operatorname{cov}[\hat{\omega}_0, \hat{\delta}] = 0.025$.

11.4 Consider the model

$$Y_t = \beta_1 X_{1,t} + \beta_2 X_{2,t} + N_t$$

where N_t is a non-stationary error term and $\nabla N_t = a_t - \theta a_{t-1}$. Show that the model may be rewritten in the form

$$Y_t - \overline{Y}_{t-1} = \beta_1(X_{1,t} - \overline{X}_{1,t-1}) + \beta_2(X_{2,t} - \overline{X}_{2,t-1}) + a_t$$

where $\overline{Y}_{t-1}$, $\overline{X}_{1,t-1}$, $\overline{X}_{2,t-1}$ are exponentially weighted averages so that for example:

$$\overline{Y}_{t-1} = (1 - \theta)(Y_{t-1} + \theta Y_{t-2} + \theta^2 Y_{t-3} + \cdots)$$

It will be seen that the fitting of this non-stationary model by maximum likelihood is equivalent to fitting the *deviations* of the independent and dependent variables from *local updated exponentially weighted moving averages* by ordinary least squares.

11.5 In an analysis [114] of monthly data y_t on smog-producing oxidant, allowance was made for two possible "interventions" I_1 and I_2 as follows:

I_1: In early 1960, diversion of traffic from the opening of the Golden State Freeway and the coming into effect of a law reducing reactive hydrocarbons in locally sold gasoline.

I_2: In 1966, the coming into effect of a law requiring all new cars to have modified engine design. In the case of this intervention, allowance was made for the well known fact that the smog phenomenon is different in summer and winter months.

In a pilot analysis of the data the following intervention model was used:

$$y_t = \omega_1 \xi_{1t} + \frac{\omega_2}{1 - B^{12}} \xi_{2t}^2 + \frac{\omega_3}{1 - B^{12}} \xi_{3t} + \frac{(1 - \theta B)(1 - \Theta B^{12})}{1 - B^{12}} a_t$$

where

$$\xi_{1t} = \begin{cases} 0, t < \text{Jan. 1960} \\ 1, t \geqslant \text{Jan. 1960} \end{cases} \quad \xi_{2t} = \begin{cases} 0, t < \text{Jan. 1966} \\ 1, t \geqslant \text{Jan. 1966} \end{cases} \quad \xi_{3t} = \begin{cases} 0, t < \text{Jan. 1966} \\ 1, t \geqslant \text{Jan. 1966} \end{cases}$$
$$\text{(summer months)} \qquad \text{(winter months)}$$

(i) Show that the model allows for
   (a) a possible step change in Jan. 1960 of size ω_1, possibly produced by I_1,
   (b) a "staircase function" of annual step size ω_2 to allow for possible summer effect of cumulative influx of cars with new engine design,
   (c) a "staircase function" of annual size ω_3 to allow for possible winter effect of cumulative influx of cars with new engine design.

(ii) Describe what steps you would take to check the representational adequacy of the model.

(iii) Assuming you were satisfied after (ii), what conclusions would you draw from the following results (estimates are shown with their standard errors below in parentheses)

$$\hat{\omega}_1 = -1.09, \quad \hat{\omega}_2 = -0.25, \quad \hat{\omega}_3 = -0.07, \quad \hat{\theta} = -0.24, \quad \hat{\Theta} = 0.55$$
$$(\pm 0.13) \qquad (\pm 0.07) \qquad (\pm 0.06) \qquad (\pm 0.03) \qquad (\pm 0.04)$$

11.6 A general transfer function model of the form

$$y_t = \sum_{j=1}^{k} \omega_j(B)\delta_j^{-1}(B)\xi_{jt} + \theta(B)\phi^{-1}(B)a_t$$

can include input variables ξ_j which are themselves time series and other inputs ξ_i which are indicator variables. The latter can estimate (and eliminate) the effects of interventions of the kind described in Exercise 11.5 and, in particular, are often useful in the analysis of sales data.

Let $\xi_t^{(T)}$ be an indicator variable which takes the form of a unit pulse at time T, that is

$$\xi_t^{(T)} = \begin{cases} 0, t \neq T \\ 1, t = T \end{cases}$$

For illustration, consider the models

(a) $y_t = \dfrac{\omega_1 B}{1 - \delta B} \xi_t^{(T)}$ (with $\omega_1 = 1.00$, $\delta = 0.50$)

(b) $y_t = \left\{ \dfrac{\omega_1 B}{1 - \delta B} + \dfrac{\omega_2 B}{1 - B} \right\} \xi_t^{(T)}$ (with $\omega_1 = 1.00$, $\delta = 0.50$, $\omega_2 = 0.30$)

(c) $y_t = \left\{ \omega_0 + \dfrac{\omega_1 B}{1 - \delta B} + \dfrac{\omega_2 B}{1 - B} \right\} \xi_t^{(T)}$

                   (with $\omega_0 = 1.50$, $\omega_1 = -1.00$, $\delta = 0.50$, $\omega_2 = -0.50$)

Compute recursively the response y_t for each of these models at times $t = T$, $T + 1$, $T + 2$, ... and comment on their possible usefulness in the estimation and/or elimination of effects due to such phenomena as advertising campaigns, promotions and price changes.

CHAPTER 12

12.1 In the chemical process described in Example 2 on page 440, 30 successive values of viscosity N_t which occurred during a period when the control variable (gas rate) X_t was *held fixed* at its standard reference origin, were recorded as follows:

1–10	92	92	96	96	96	98	98	100	100	94
11–20	98	88	88	88	96	96	92	92	90	90
21–30	90	94	90	90	94	94	96	96	96	96

Reconstruct and plot the error sequence ε_t and adjustments x_t which would have occurred if the optimal feedback control scheme

$$x_t = -10\varepsilon_t + 5\varepsilon_{t-1} \tag{1}$$

had been applied during this period. It is given that the dynamic model is

$$\mathscr{Y}_t = 0.5\mathscr{Y}_{t-1} + 0.10x_{t-1} \tag{2}$$

and that the error signal may be obtained from

$$\varepsilon_t = \varepsilon_{t-1} + \nabla N_t + \mathscr{Y}_t \tag{3}$$

Your calculation sequence should proceed in the order (2), (3), and (1) and initially you should assume that $\varepsilon_1 = 0$, $\mathcal{Y}_1 = 0$, $x_1 = 0$. Can you devise a more direct way to compute ε_t from N_t?

12.2 Given the following combinations of disturbance and transfer-function models

(a)
$$\nabla N_t = (1 - 0.7B)a_t$$
$$(1 - 0.4B)Y_t = 5.0X_{t-1+}$$

(b)
$$\nabla N_t = (1 - 0.5B)a_t$$
$$(1 - 1.2B + 0.4B^2)Y_t = (20 - 8.5B)X_{t-1+}$$

(c)
$$\nabla^2 N_t = (1 - 0.9B + 0.5B^2)a_t$$
$$(1 - 0.7B)Y_t = 3X_{t-1+}$$

(d)
$$\nabla N_t = (1 - 0.7B)a_t$$
$$(1 - 0.4B)Y_t = 5.0X_{t-2+}$$

(i) Design the minimum mean square error feedback control schemes associated with each combination of disturbance and transfer-function model.

(ii) For case (d), derive an expression for the error ε_t and for its variance in terms of σ_a^2.

(iii) For case (d), design a nomogram suitable for carrying out the control action manually by a process operator.

12.3 In a treatment plant for industrial waste, the strength z_t of the influent is measured every 30 minutes and can be represented by the model $\nabla z. = (1 - 0.5B)a_t$. In the absence of control the strength of the effluent Y_t is related to that of the influent z_t by

$$\tilde{Y}_t = \frac{0.3B}{1 - 0.2B}\tilde{z}_t$$

An increase in strength in the waste may be compensated by an increase in the flow u_t of a chemical to the plant according to the model

$$\tilde{Y}_t = \frac{21.6B^2}{1 - 0.7B}\tilde{u}_t$$

Show that minimum mean square error control is obtained with the control equation

$$\tilde{u}_t = -\frac{0.3}{21.6}\left\{\frac{(0.7 - 0.2B)(1 - 0.7B)}{(1 - 0.2B)(1 - 0.5B)}\right\}\tilde{z}_t$$

that is $\tilde{u}_t = 0.7\tilde{u}_{t-1} - 0.1\tilde{u}_{t-2} - 0.0139(0.7\tilde{z}_t - 0.69\tilde{z}_{t-1} + 0.14\tilde{z}_{t-2})$.

12.4 A pilot feedback control scheme, based on the following disturbance and transfer-function models:

$$\nabla N_t = a_t$$
$$(1 - \delta B)\mathcal{Y}_t = \omega_0 X_{t-1+} - \omega_1 X_{t-2+}$$

was operated, leading to a series of adjustments x_t and errors ε_t. It was believed that the noise model was reasonably accurate but that the parameters of the transfer function model were of questionable accuracy.

(i) Given the first 10 values of the x_t, ε_t series shown below

t	1	2	3	4	5	6	7	8	9	10
x_t	25	42	3	20	5	-30	-25	-25	20	40
ε_t	-7	-7	-6	-7	-4	1	3	4	0	-3

set out the calculation of the residuals $a_t(t = 2,3, \ldots, 10)$ for $\delta = 0.5$, $\omega_0 = 0.3$, $\omega_1 = 0.2$ and for arbitrary starting values $\mathscr{y}_1^0$ and x_0^0.

(ii) Calculate the values $\hat{\mathscr{y}}_1$, $\hat{x}_0$ of $\mathscr{y}_1^0$ and x_0^0 which minimize the sum of squares $\sum_{t=2}^{10} (a_t | \delta = 0.5$, $\omega_0 = 0.3$, $\omega_1 = 0.2$, $\mathscr{y}_1^0$, $x_0^0)^2$ and the value of this minimum sum of squares.

12.5 Consider [111] a system like that illustrated in Figure 12.6 for which the process transfer function is gB and the noise model is $(1 - B)N_t = (1 - \theta B)a_t$ so that the error ε_t at the output is given by

$$(1 - B)\varepsilon_t = g(1 - B)X_{t-1+} + (1 - \theta B)a_t$$

Suppose the system is controlled by a known discrete 'integral' controller

$$(1 - B)X_{t+} = -c\varepsilon_t$$

(i) Show that the errors ε_t at the output will follow the ARMA(1,1) process

$$(1 - \phi B)\varepsilon_t = (1 - \theta B)a_t, \qquad \phi = 1 - gc$$

and hence that the problem of estimating g and θ using data from a pilot control scheme is equivalent to that of estimating the parameters in this model.

(ii) Show also that the optimal scheme is such that $c = c_0 = (1 - \theta)/g$ and hence that if the pilot scheme used in collecting the data happens to be optimal already, then $1 - \theta$ and g cannot be separately estimated.